ELECTRONIC CONCEPTS

Electronic Concepts is a clear, self-contained introduction to modern microelectronics. Analog and digital circuits are stressed equally from the outset, and the applications of particular devices and circuits are described within the context of actual electronic systems. A combination of bottom–up and top–down approaches is used to integrate this treatment of devices, circuits, and systems.

The author begins with an overview of several important electronic systems, discussing in detail the types of signals that circuits are used to process. In the following chapters he deals with individual devices such as the bipolar junction transistor and the metal-oxide semiconductor field-effect transistor. For each device he presents a brief physical description and demonstrates the use of different models in describing the device's behavior in a particular circuit application. Throughout the book, he uses SPICE computer simulations extensively to supplement analytic descriptions.

The book contains over 500 circuit diagrams and figures, over 400 homework problems, and over 100 simulation and design exercises. It includes many worked examples and is an ideal textbook for introductory courses in electronics. It can also be used for self-study. Laboratory experiments related closely to the material covered in the book are available via the World Wide Web.

Jerrold Krenz received his Ph.D. from Stanford University and is Associate Professor of Electrical and Computer Engineering at the University of Colorado, Boulder. He is the author of several books, including *Microelectronic Circuits: A Laboratory Approach* and *An Introduction to Electrical Circuits and Electronic Devices: A Laboratory Approach*.

ELECTRONIC CONCEPTS

AN INTRODUCTION

JERROLD H. KRENZ
University of Colorado, Boulder

CAMBRIDGE
UNIVERSITY PRESS

PUBLISHED BY THE PRESS SYNDICATE OF THE UNIVERSITY OF CAMBRIDGE
The Pitt Building, Trumpington Street, Cambridge, United Kingdom

CAMBRIDGE UNIVERSITY PRESS
The Edinburgh Building, Cambridge CB2 2RU, UK   http://www.cup.cam.ac.uk
40 West 20th Street, New York, NY 10011-4211, USA   http://www.cup.org
10 Stamford Road, Oakleigh, Melbourne 3166, Australia
Ruiz de Alarcón 13, 28014 Madrid, Spain

First published 2000

Printed in the United States of America

Typeface Sabon 10/13 pt. and Futura   *System* LaTeX 2_ε [TB]

A catalog record for this book is available from the British Library.

Library of Congress Cataloging in Publication Data
Krenz, Jerrold H., 1934-
    Electronic concepts : an introduction / Jerrold H. Krenz.
        p.   cm.
    Includes bibliographical references (p.
    ISBN 0-521-66282-6 (hb.)
    1. Electronics.   2. Electronic circuits – Computer simulation.
I. Title.
TK7816.K73   1999
621.381 – dc21   99-30407
                                                      CIP

ISBN 0 521 66282 6 hardback

CONTENTS

.

PREFACE

The field of electronics or microelectronics today encompasses a vast quantity of knowledge and practice. The topics that can be covered in a basic course must, by necessity, be limited to avoid a mere encyclopedic cataloging of various electronic circuits and systems. There are, however, a set of underlying concepts that one needs to grasp to understand electronics. It is the goal of the author to provide students and instructors with an accessible treatment of those modern electronic concepts along with appropriate applications. Applications are considered essential to grasp the utility of general concepts as well as to appreciate their limitations. The approach used in the text is to cover a limited number of topics well, as opposed to a cursory coverage of a very wide range of topics that may do little more than leave one with an extensive vocabulary.

The text provides more than adequate material for a one-semester, junior-level electronics course. A good working knowledge of linear circuits along with a reasonable understanding of calculus and physics is required. Although there is a progression in the complexity of the material covered, the text provides a flexibility in selecting the material to cover. The author has attempted to provide sufficient descriptive material to indicate not only what is being done but also to show how a particular circuit is used. Examples with detailed solutions utilizing analytic solutions and computer simulations conclude most sections. In addition, numerous references are cited to allow the interested student to learn more about a particular topic.

An unique feature of the book is its introductory chapter, which provides an overview of several electronic systems. This chapter makes the learning task more interesting for the students and serves to introduce them to the signals that electronic circuits are used to process. Electronic circuits were developed to fulfill particular needs. At the same time, the evolution of electronic systems depended on what could be done with the electronic devices then available. The text provides a combination of a top–down and a bottom–up approach. Systems are considered as well as basic physical concepts that are necessary for understanding devices.

The text provides a transition from the coverage of introductory circuit theory courses that tend to deal only with two-terminal linear devices for which simple circuit models are valid ($v = iR$, etc.). Although, for circuit theory, it is seldom necessary to distinguish between the model and the device it describes, this is not the case for electronic devices. In *Electronic Concepts*, a student is gradually introduced to the use of different models used for a single device. The particular model employed depends on the nature of the circuit in which the device is used and the signals involved. The treatment of circuits with nonlinear devices and with three (and at times four) terminals is recognized as a significant conceptual leap for students.

The second chapter, The Semiconductor Junction Diode, includes a brief qualitative discussion of semiconductor physics. Although it is recognized that an in-depth knowledge of semiconductor physics is very important, the author feels that this can best be accomplished in a concurrent or subsequent theory course. *Electronic Concepts* discusses electrons and holes moving as the result of potential differences, thereby providing a basis for an intuitive understanding of semiconductor devices. Load lines, the diode equation, and various diode models used to approximate the behavior of diodes are introduced. The basic principles of photovoltaic cells and light-emitting diodes are discussed as well as important applications.

The bipolar junction transistor is introduced in Chapter 3 before the coverage of field-effect devices in Chapter 4. Treating field-effect devices first does have a certain appeal because, for some applications, the field-effect models are simpler than those of bipolar junction transistors. However, the bipolar junction transistor is a direct extension of the junction diode, and this type of transistor is considerably more convenient for doing laboratory experiments. An understanding of individual semiconductor devices, as viewed from their terminals, as well as the concepts related to using devices for amplification and switching is stressed.

Chapter 4 provides a brief qualitative physical discussion of MOSFET devices and introduces approximate analytical expressions for their terminal behavior. Although the behavior of analog circuits based on the small-signal behavior of devices is covered, the main thrust of this chapter is digital circuits. Both the static and dynamic behaviors of logic gates using device configurations suitable for integrated circuits are determined. Following a treatment of bistable circuits, semiconductor memories are discussed.

Negative feedback, along with operational amplifiers, is the subject of Chapter 5. The feedback nature of operational amplifier circuits is stressed because the frequently used "ideal op amp" treatment of basic circuit texts generally glosses over the feedback nature of op amp circuits. Negative feedback, although introducing a higher order of complexity, is shown to offer many improvements over circuits without feedback. It is also emphasized that if feedback is not used properly, undesirable behavior can occur. Analog design techniques using op amps are highlighted in this chapter.

A concluding chapter on electronic power supplies treats rectifiers, filters, electronic regulators, and batteries. A knowledge of this material, all too frequently omitted in basic electronics courses, is necessary for the design of nearly all electronic systems. This chapter may be covered immediately after Chapter 2 if the electronic regulator section is omitted.

Appendix A on the fabrication of integrated circuits carries one beyond the electronics circuits emphasis of the text. It provides a glimpse of the physical and chemical techniques used in the fabrication process and a perspective on the actual physical structures and sizes of devices. Appendix B, The Design Process, carries through the design of a few sample electronic circuits. Explanations are provided for each step so that the student may appreciate the rationale for the design decisions.

Computer simulations are used throughout the text. It is assumed that students are familiar with SPICE, that is, that they have used it in a linear circuits course (if not, numerous basic reference texts are available). Circuit files, common for all versions of SPICE, are included for all simulation examples. Although the text uses Probe (MicroSim) graphs, similar presentations can be obtained with other programs.

Problems requiring analytical solutions as well as computer simulations are also included. There are considerably more problems and simulations than can be used for a one-semester course, and thus instructors can vary assignments from semester to semester and reduce the use of solutions from previous classes. Computer simulations are limited to circuits that can be run on personal computers with the student version of the PSPICE program. Open-ended, design-type exercises are also included.

Laboratory experiments that relate directly to the theory of each chapter are also available on the World Wide Web:
http://www.cup.org/titles/66/0521662826.html

A portable document format (.pdf extension) is used for the experiments so that they may readily be downloaded and printed. Detailed experimental steps are employed to guide a student through a set of measurements and observations, and minimum effort is required on the part of the instructor. Alternatively, portions of experiments may be used for classroom demonstrations.

Boulder, CO

CHAPTER ONE

ELECTRONIC SYSTEMS: A CENTURY OF PROGRESS

Our daily lives are shaped by electronic systems. In the home we have a myriad of electronic accessories: radios, TVs, VCRs, hi-fis, camcorders, cassette and CD players, telephone answering machines, microwave ovens, and personal computers. Not so obvious but just as much a part of our lives are sophisticated electronic controls such as the microprocessor engine control of our car. We utilize a telephone system that functions with electronic devices to amplify and transfer telephone signals. Our conversations are carried around the world using a combination of microwave or fiber-optic links and satellites. Electronic radar systems are relied on for a safe flight from one airport to the next, and electronic sensors and computers "fly" a modern jet airplane. Modern medical practice depends on extremely complex diagnostic and monitoring electronic systems. Moreover, the commercial and industrial sectors could no longer function without electronic communications and information processing systems. The video monitor is a pervasive reminder of the new electronic world.

For better and at times for worse, electronics has changed our lives. Although we are in constant touch with what is happening around the world, we are also at the peril of weapons of unimaginable destructive power that rely on electronic developments. An understanding of electronics is imperative not only for designing and using electronic systems but for directing the evolution of electronic systems so that they serve to improve the human condition.

It has been stated that to move forward we must know where we have been. The 20th century is the era of electronics – it was only after 1900 that the devices we now describe as electronic appeared. The use of the term *electronics* in the current sense did not occur until 1930 (Süsskind 1966). This introductory chapter starts with a very brief overview of electronic devices and is followed by a discussion of wireless systems: radio. The first application of electronic devices, the vacuum tube diode invented in 1904 and the triode invented in 1906, was for radio receivers. Radio communications was not only nearly a decade old at the time the tube was invented, but most of the systems of the first decade of the 1900s did not use tubes. The vacuum tube, without exaggeration, can be described as having revolutionized radio communications, resulting in the generation of coherent transmitting signals and highly sensitive and selective receivers. The vacuum tube, following its first telephone use in 1913, also became an important component of telephone systems. With vacuum tube amplifiers and multiplexing circuits, long-distance telephone service greatly expanded. With the development of digital systems made possible by the transistor and integrated circuits in the latter half of the 20th century, telephone switching and transmission systems were again significantly improved.

The development of electronic devices, on the one hand, depended on a knowledge of basic physical principles: the behavior of electrons in a vacuum and the interaction of electrons with matter. On the other hand, electron devices were frequently developed to fulfill perceived needs. The characteristics of electronic devices dictated those applications that could be realized. Television, discussed in Section 1.4, illustrates the interrelatedness of the development of electronic devices and circuits with a particular application. An analog television system was developed in the 1930s and was commercially introduced in the late 1940s. Over the rest of the 20th century, television was based on this analog system, and the only enhancement was the introduction of a subcarrier for color information. At the close of the 20th century, a digital system, totally different, and therefore incompatible with the analog system, was developed. Although this digital system, from a transmission perspective, is considerably more efficient, the signal processing required is very complex. Without the development of very-large-scale integrated (VLSI) circuits during the 1980s that could do the encoding and decoding, digital TV would not have been possible.

The electromagnetic spectrum (Section 1.5) is used for a variety of radio, TV, and other communications services. Although early radar systems can be traced back to the 1930s, it was the impetus of World War II that resulted in a rapid development of this technology. New electronic devices capable of transmitting and detecting extremely high-frequency signals ($f > 1000$ MHz) were invented. Communications satellites, first launched in the 1960s, also relied on these extremely high-frequency (microwave) devices.

Digital electronic circuits have revolutionized computing. Early computers, until about the mid-1960s, relied on vacuum tube circuits. These computers, from today's perspective, not only had minuscule processing capabilities, but, owing to the limited reliability of vacuum tubes, were frequently down. Solid-state devices resulted not only in a tremendous improvement in reliability but made possible machines with much greater computing capabilities. With ultra-large-scale integrated circuits, desktop computers emerged with a computing capability that a decade earlier was available only in large mainframe machines.

Needless to say, electronic devices and circuits have become common for many applications in addition to those discussed. Power electronics is dependent on electronic switching devices and circuits. Frequency and voltage transformations, as well as alternating-to-direct-current and direct-to-alternating-current conversions can often be efficiently achieved using electronic systems. In medical electronics, a variety of electronic sensing circuits have been developed along with computer systems to process and display the data. Furthermore, electronic systems, such as heart pacemakers, have been perfected to augment body functions. Electronic sensing and control systems dependent on simple microprocessors are now used in applications ranging from programmable thermostats to automobile ignition and fuel systems. More complex sensing and control systems involving large computing capabilities are used for automated manufacturing systems. Although it is beyond this introductory chapter to discuss these and other applications, it should be recognized that similar electronic devices and circuits are often used by these different systems. A knowledge of basic concepts, the subject of this text, is a prerequisite for understanding both the simplest and the most esoteric of electronic systems.

1.1 ELECTRONIC DEVICES: AN OVERVIEW

The thermionic valve or vacuum tube was developed in Great Britain by Sir John Ambrose Fleming (Pierce 1950; Shiers 1969). This tube relied on what is known as the *Edison effect*, a current being produced by the hot filament of a light bulb. Fleming, through a series of experiments with bulbs having an electrode near the hot filament, deduced that this current was due to negative electric charges. We now understand the current to be due to electrons emitted by the hot filament that are collected by the electrode. To the extent that only electrons are responsible for this current, the current to the electrode is only in one direction; in a high-vacuum tube, a current corresponding to the movement of positive charges does not occur.

THE DIODE

Fleming's valve consisted of a hot filament (corresponding to the incandescent filament of a light bulb) heated by a current produced by an external battery. The emitted electrons were then collected by a plate surrounding the filament (Figure 1.1). Even though the physical current is that due to electrons traveling from the filament to the plate, the plate current i_P, is, by convention, a positive quantity because a current is defined in terms of the movement of hypothetical positive charges. A positive plate voltage v_P attracts electrons, thus increasing the current, whereas a negative plate voltage repels electrons, yielding either

Figure 1.1: Vacuum tube diode and typical characteristic.

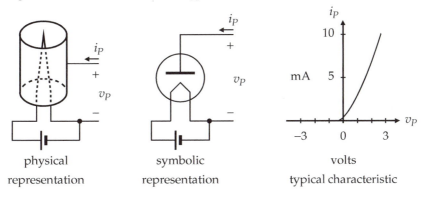

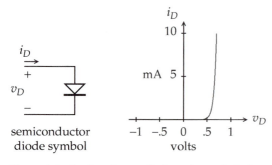

semiconductor
diode symbol

Figure 1.2: Semiconductor diode and a typical characteristic.

a very small or zero current. This nonlinear effect results in a current in only one direction ($i_P \geq 0$). For significant negative voltages, the current of a well-evacuated tube is essentially zero.

At about the same time that Fleming introduced his vacuum tube, Greenleaf W. Pickard was experimenting with a point-contact semiconductor detector (Douglas 1981). This device may be considered the precursor of modern solid-state devices. In addition to the detector using silicon developed by Pickard, a similar detector using Carborundum was developed by Henry H. C. Dunwoody in 1906. Point-contact diodes were extensively used until the junction semiconductor diode was introduced in the 1950s.

A semiconductor diode has a nonlinear characteristic, as does the vacuum tube (Figure 1.2). The current of the diode i_D increases very rapidly with diode voltage v_D (for an ideal semiconductor diode it may be shown that the current has an exponential dependence on the diode voltage). The rectification property of a diode, which allows a current in only one direction, was first used for the detection of radio signals. The detection problem provided the impetus for the development of vacuum tube and semiconductor diodes. Represented in Figure 1.3 is a basic radio receiver with a typical amplitude-modulated carrier signal. Although carrier frequencies of 50 to 100 kHz were common for early communications systems, the present radio broadcast band consists of signals with carrier frequencies of 540 to 1600 kHz. For an *on–off* system (continuous wave or CW), the carrier is simply keyed on and off to form a pattern of dots and dashes. However, for amplitude modulation (AM), the amplitude of the carrier signal is varied in accordance with the modulating signal; for example, that of a voice signal produced with a microphone.

It should be noted that the period corresponding to the carrier frequency is generally much smaller than that associated with the time scale over which appreciable variations in the modulating signal occur. In a radio receiver, the

Figure 1.3: An elementary diode radio detector.

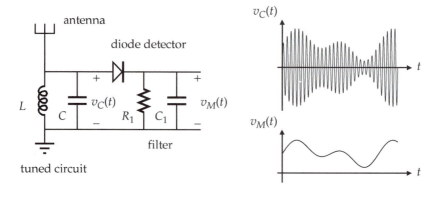

energy received by the antenna is coupled to the tuned circuit which, ideally, excludes all other signals with different carrier frequencies. A diode rectifier is then used to convert the carrier signal $v_C(t)$ to a signal with a single polarity. For the circuit shown, the capacitor C_1 tends to smooth the detected signal. Without the capacitor, a signal similar to the top half of $v_C(t)$ would result.

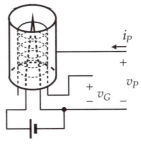 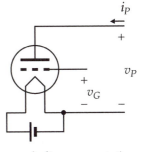

physical representation   symbolic representation

Figure 1.4: A triode vacuum tube.

THE VACUUM TRIODE

The next significant development that, in effect, ushered in the electronics age, was Lee De Forest's addition of a control electrode or grid to Fleming's vacuum diode. This resulted in the triode vacuum tube. A sketch of the physical device, which is referred to as a triode because it has three elements, is presented in Figure 1.4. The third element, the grid, is a cagelike wire structure surrounding the filament of the tube. An externally applied grid potential regulates the plate current of the tube.

For normal operation, the grid is at a negative potential (relative to that of the filament), which tends to repel electrons emitted by the hot filament. The more negative the grid potential v_G, the smaller the plate current i_P for a given plate voltage v_P (Figure 1.5). Because electrons are repelled by a negative grid potential, the grid current is essentially zero. (The exceedingly small grid current that does occur is due to positive ions produced by ionizing electron collisions with the air molecules of the imperfect vacuum. Although the grid current of De Forest's early tube may have been significant, those of later tubes with good vacuums were truly negligible.) As a result of this essentially zero grid current, the power utilized by the grid circuit is extremely close to zero. Herein lies the worth of the triode vacuum tube. Its plate current and voltage are not only controlled by the grid voltage, but essentially zero power is required to do the controlling. It is not a perpetual-motion device (a power source is required for the plate circuit) but, for many applications, it is the next best thing!

To illustrate the utility of a vacuum tube triode, consider the typical characteristic of Figure 1.5 and suppose that a constant current source of 10 mA is connected between the filament and plate of the tube ($i_P = 10$ mA). For a particular value of grid voltage, the resultant plate voltage corresponds to the intersection of the curve corresponding to that grid voltage with the 10-mA coordinate (shown as a

Figure 1.5: The plate characteristic of a typical triode vacuum tube.

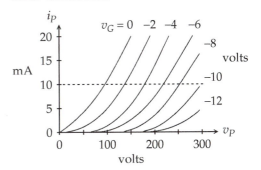

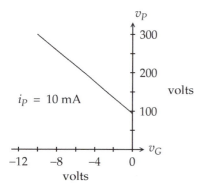

Figure 1.6: The transfer characteristic of the triode of Figure 1.5.

dashed line in Figure 1.5). A grid voltage of −4 V, for example, results in a plate voltage of 180 V; a grid voltage of −6 V in a plate voltage of 220 V, and so forth. The transfer characteristic of Figure 1.6 is thus obtained. Of particular importance is that a relatively small change in grid voltage results in a fairly large change in plate voltage. The slope of the characteristic of Figure 1.6 is approximately −20. This implies that a 1-V change in v_G results in a change of −20 V in v_P. The minus sign signifies that an increase in v_G results in a decrease in v_P. This circuit therefore has a voltage gain with a magnitude of approximately 20.

The first triode vacuum tube of De Forest was used to detect radio signals (in place of the diode of Figure 1.3); it was initially described as an oscillation valve. However, because vacuum tube triodes have the ability to amplify as well as to detect radio signals, tubes were soon used for a multitude of applications, including the generation of high-frequency radio signals.

THE TRANSISTOR AND INTEGRATED CIRCUITS

Solid-state devices, transistors, have replaced vacuum tubes for most, but not all, electronic applications. The symbolic representation and typical characteristic of a modern metal-oxide semiconductor field-effect transistor (MOSFET) are given in Figure 1.7. For the device shown, free electrons from the source of the MOSFET semiconductor device flow to its drain. In a manner analogous to that of the grid of the vacuum tube, the free-electron current is controlled by the gate potential of the MOSFET device. The gate current, like the grid current of a triode, is essentially zero. The free electrons, however, are produced by a doped semiconductor rather than by a hot filament, thus resulting in a much more efficient device. Furthermore, the voltage levels required for a typical MOSFET application are considerably smaller than those of a typical triode vacuum tube circuit.

Figure 1.7: The metal-oxide semiconductor field-effect transistor (MOSFET).

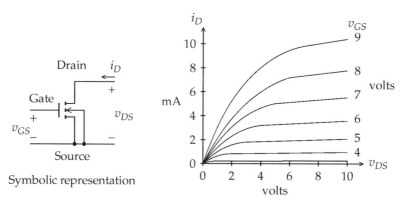

Symbolic representation

In addition to MOSFET devices, the bipolar junction transistor (BJT) is also extensively used in modern electronic circuits. Germanium bipolar junction transistors were developed shortly after the invention of the point-contact transistor in 1948. With the development of silicon processing techniques during the 1950s, germanium and silicon transistors tended to replace vacuum tubes for most applications by the 1960s. It was, however, the introduction of the integrated circuit, a single semiconductor wafer initially limited to a few tens of transistors, that has had the most profound effect on electronic systems. This effect has been characterized by some as revolutionary (Noyce 1977).

Vacuum tubes generally consisted of only one, two, or possibly three electronic devices enclosed by a single glass envelope. These tube circuits were generally mounted on a metal chassis that had sockets relying on spring contacts to hold the vacuum tubes. This permitted vacuum tubes to be readily replaced – an all-too-frequent need. Connections between the sockets and other components were achieved through hand-soldered wires. Small components, such as resistors and capacitors, were often supported directly by their leads while forming connections between components.

Even the earliest commercially produced transistors, introduced during the 1950s, were considerably more reliable than the vacuum tubes they replaced. Hence, transistors could be wired directly into a circuit, thereby eliminating the need for sockets. This led to the printed circuit board utilizing copper foil conductors bonded to a phenolic base. Transistors, as well as other components, were mounted directly on the printed circuit board, and a dip-type soldering process was used for electrical connections to the copper foil. Because transistors are much smaller than vacuum tubes and tend to dissipate considerably less power, a much higher density of components was possible.

A batch process was soon developed in which several transistors were simultaneously fabricated on a single semiconductor wafer. The wafer was then cut to obtain individual transistors, leads were attached, and the transistors were encapsulated in a package suitable for their application. During the assembly process, individual transistors were tested, and faulty ones were discarded. With the improvement of processing techniques, the yield of well-functioning devices greatly increased.

In retrospect, it now seems obvious to question why the individual transistors of a semiconductor wafer were separated. Why not develop a process for electrically isolating the devices from each other to replace the isolation that had been achieved by cutting them apart? The devices could then be interconnected on the semiconductor wafer to form what we now refer to as an integrated circuit. At the end of the 1950s, this idea was realized (Meindl 1977). As is often the case, several individuals working independently were involved in developing the earliest integrated circuits. However, Jack Kilby is frequently credited with having "invented" the integrated circuit (Kilby 1976). In 1958, he demonstrated a hand-fabricated phase-shift oscillator and a flip-flop using germanium transistors. Resistors consisted of appropriately doped semiconductors, whereas capacitors utilized reverse-biased semiconductor junctions. These demonstration circuits established the feasibility of a concept that was rapidly exploited.

1.2 WIRELESS COMMUNICATION: A NEW ERA

The first use of the triode vacuum tube was for wireless communication. Lee De Forest, its inventor, described the tube as an oscillation valve – that is a device for detecting wireless or radio signals. (As an aside, it should be noted that Lee De Forest's autobiography has the subtitle of *Father of Radio*. This parentage is not widely accepted.) A close relationship of electronic devices to radio characterized the first half of the 20th century. The related professional organization in the United States was the Institute of Radio Engineers founded in 1912. It was not until 1963 that the designation "radio" was dropped when this organization merged with the Institute of Electrical Engineers to form the Institute of Electrical and Electronic Engineers (IEEE).

Maxwell's equations, the kernel of electromagnetic theory, provide the basis on which wireless communication, that is radio, is based. James Clerk Maxwell built on the work of Coulomb, Oersted, Ampère, Henry, Faraday, and Gauss in formulating these now well-known equations. Through a series of experimental observations and theoretical deductions, Heinrich Rudolf Hertz demonstrated the validity of Maxwell's equations. Hertz published the first text on electrodynamics in 1892 *Untersuchungen über die Ausbreitung der elektrischen Kraft* (*Electric Waves*, the title of an English translation by D. E. Jones). Following the death of Hertz in 1894, the lectures on the studies of Hertz by Oliver Joseph Lodge laid the groundwork for a much wider understanding of electromagnetic principles. Lodge and Ferdinand Braun were responsible for developing the concept of resonant tuning and demonstrating the importance of having the transmitter and receiver of a system tuned to the same frequency (Aitken 1976, Jolly 1975, Kurylo and Süsskind 1981, McNicol 1946). Concurrently, Oliver Heaviside is credited with putting Maxwell's equations into their presently utilized form (Nahin 1988, 1990).

A difficulty encountered in performing early electromagnetic experiments was that of obtaining a suitable detector of high-frequency signals. An early detector was the coherer, basically a small glass tube filled with loosely packed metal filings. The operation of this device relied on the nonlinear nature of the resistance of the filings. For small currents the filings had a high resistance, whereas for larger currents the filings tended to cohere, resulting in a small resistance. A mechanical tapping of the coherer was necessary to restore the high resistance after the termination of a large current. For a receiver, the alternating current produced by an electromagnetic signal caused the filings to cohere. This effect was detected by a low-voltage direct-current circuit connected to the coherer. Edouard Branly developed several different coherers and appears to have been the first to use the term radio (in this context) by proposing the name *radioconductor* for the coherer.

It was Guglielmo Marconi who in 1895 refined and assembled the appropriate apparatus and demonstrated that it could be used for signaling (Jolly 1972, Masini 1995). Not being successful in interesting his Italian government in this new means of communication, he traveled to England, where the British post office was receptive. Recognizing the commercial importance of wireless

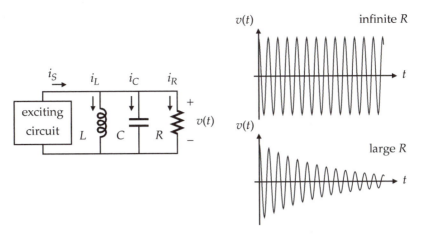

Figure 1.8: A resonant circuit.

telegraphy, he took out patents and formed the Marconi Wireless Signal Company. Progress was rapid: in 1901 he succeeded in sending a wireless signal across the Atlantic.

ELECTRICAL TUNING

An important aspect of radio communication is that of tuning; that is, to utilize a circuit that has an optimal response at a particular signal frequency. This is generally achieved with an inductor–capacitor circuit such as that of the parallel circuit of Figure 1.8. The resistance R is included to account for circuit losses (the resistance of the inductor) and energy that might be radiated as a result of an antenna connected to the circuit. Early wireless transmitters used a current impulse i_S produced by a spark gap to initiate the voltage oscillations of the circuit. Consider the case for which the circuit has previously been excited and the current i_S is zero. This implies that the sum of the currents of the individual elements must be zero, as given by the following:

$$i_S(t) = i_L + i_C + i_R = 0$$

$$i_L = \frac{1}{L} \int v\, dt, \qquad i_C = C\frac{dv}{dt}, \qquad i_R = \frac{v}{R} \qquad (1.1)$$

These two equations may be combined and then differentiated to produce a single second-order differential equation:

$$\frac{1}{L} \int v\, dt + C\frac{dv}{dt} + \frac{v}{R} = 0$$

$$\frac{d^2v}{dt^2} + \frac{1}{RC}\frac{dv}{dt} + \frac{v}{LC} = 0 \qquad (1.2)$$

For an ideal circuit with no loss ($R \to \infty$), a constant-amplitude oscillating voltage is a valid solution of the differential equation as follows:

$$v(t) = V_m \cos \omega_0 t, \qquad \omega_0 = \frac{1}{\sqrt{LC}} \qquad (1.3)$$

Hence, once this lossless circuit is excited by an external current, its voltage will continue to oscillate indefinitely.

For a circuit with loss (finite R), damped sinusoidal oscillations occur given by

$$v(t) = V_m \, e^{-\alpha t} \cos \omega_0 t$$

$$\alpha = \frac{1}{2RC}, \qquad \omega_0 = \sqrt{\frac{1}{LC} + \left(\frac{1}{2RC}\right)^2} \tag{1.4}$$

The current impulse of a spark was used for the earliest wireless transmitters. Modern transmitters (radio and TV stations, citizens band transceivers, cellular telephones, etc.) rely on essentially the same principle except that an electronic exciting circuit is utilized that generally provides a current impulse for each oscillating cycle.

How does this circuit manage to continue to oscillate when the exciting current no longer exists? To answer this question, we must recall that inductors and capacitors store electrical energy. Let e_C and e_L be the instantaneous stored energies of the capacitor and inductor, respectively.

$$e_C = \frac{1}{2}Cv^2, \qquad e_L = \frac{1}{2}Li^2 \tag{1.5}$$

Consider the idealized case ($R \to \infty$) for which the amplitude of the voltage is constant (Eq. (1.3)).

$$i_L = \frac{1}{L}\int v \, dt = \frac{V_m}{\omega_0 L}\sin \omega_0 t$$

$$e_L = \frac{1}{2}\frac{V_m^2}{\omega_0^2 L}\sin^2 \omega_0 t = \frac{1}{2}C V_m^2 \sin^2 \omega_0 t \quad \text{because} \quad \frac{1}{\omega_0^2} = LC \tag{1.6}$$

The total energy $e_C + e_L$ is constant for this circuit:

$$e_C + e_L = \frac{1}{2}C V_m^2 = \frac{1}{2}\frac{V_m^2}{\omega_0^2 L} \tag{1.7}$$

It will be noted that when the stored energy of the capacitor is a maximum, that of the inductor is zero and vice versa (Figure 1.9).

In effect, there is an interchange of energy between the capacitor and the inductor of the circuit. For a circuit with a finite resistance, the electrical energy is gradually dissipated by the resistor; that is, the electrical energy is converted to thermal energy (or radiated if the resistor represents the effect of an antenna).

VACUUM TUBE CIRCUITS

Following its invention, the vacuum tube triode was extensively improved, and numerous electronic circuits were developed that greatly increased the tube's utility. Armstrong's invention of regeneration in 1912, the use of positive feedback to increase the gain of a circuit, increased the sensitivity of receivers. For example, using Armstrong's regeneration principle, it is possible to build a shortwave receiver with but a single vacuum tube (or transistor) that is capable of receiving signals from all over the world. A modification of this circuit was also used

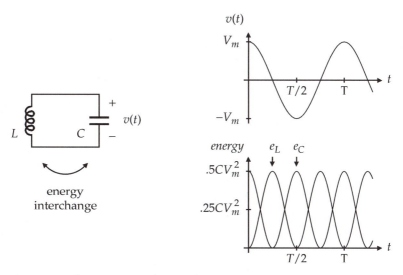

Figure 1.9: The energy interchange of a resonant circuit.

to produce radio-frequency oscillations that allowed the replacement of radio transmitters relying on either spark-gap or mechanical alternator generators.

Although Armstrong was the first to submit a patent application for regeneration, the application was immediately challenged by De Forest. It was claimed by De Forest that this effect was discovered a year earlier in his laboratory, albeit the technician's curt notebook entry for the circuit was "no good." A lifelong animosity surrounding legal challenges over patents ensued between these two radio pioneers. Armstrong was initially granted a patent for regeneration and successfully resisted the early challenges of De Forest. Eventually, however, De Forest won through challenges that were carried all the way to the U.S. Supreme Court. Nevertheless, Armstrong is generally accepted as the circuit's inventor, and the historical record indicates that Armstrong had a better understanding of the circuit than De Forest. The Institute of Radio Engineers (IRE) honored Armstrong for the regeneration invention with its medal of honor in 1918. When Armstrong attempted to return the medal in 1934 after losing De Forest's patent challenge, the IRE board of directors not only refused to accept the return of the medal (a unanimous decision) but reaffirmed its initial citation (Lewis 1991).

THE SUPERHETERODYNE RECEIVER

Among Armstrong's numerous inventions is the superheterodyne radio receiver. His earlier regenerative receiver, although sensitive, was prone to behave erratically (it frequently burst into oscillation). Tuned circuits, such as the parallel resonant circuit of Figure 1.8, are required to select a desired radio signal and to reject other signals. In addition, tuned circuits are used to enhance the gain of radio-frequency amplifiers. To tune a given circuit, its capacitance, inductance, or both must be changed. Several amplifiers, each with a tuned circuit, are often needed. The tuning of a radio thus required the simultaneous adjustment of several circuits – a tuning knob was needed for each circuit.

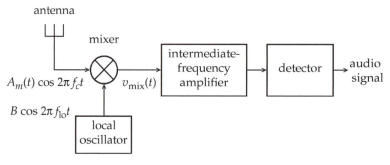

Figure 1.10: A superheterodyne receiver.

Armstrong recognized that the carrier frequency of a signal could be changed through a nonlinear mixing process (Figure 1.10). Consider the case for an amplitude-modulated signal $A_m(t) \cos 2\pi f_c t$ derived from an antenna system. A second high-frequency signal generated by the local oscillator of the receiver, $B \cos 2\pi f_{lo} t$, is also required. Suppose, initially, that the mixer results in an output voltage $v_{mix}(t)$ that is the product of its two inputs (a standard multiplier symbol is shown in Figure 1.10) as expressed by

$$
\begin{aligned}
v_{mix}(t) &= A_m(t) \cos 2\pi f_c t \cdot B \cos 2\pi f_{lo} t \\
&= \frac{1}{2} A_m(t) B[\cos 2\pi (f_{lo} + f_c)t + \cos 2\pi (f_{lo} - f_c)t]
\end{aligned}
\tag{1.8}
$$

The preceding result was obtained using the trigonometric identities for the cosine of the sum and difference of two angles as follows:

$$
\begin{aligned}
\cos(\alpha + \beta) &= \cos \alpha \cos \beta - \sin \alpha \sin \beta \\
\cos(\alpha - \beta) &= \cos \alpha \cos \beta + \sin \alpha \sin \beta \\
\cos \alpha \cos \beta &= \frac{1}{2}[\cos(\alpha + \beta) + \cos(\alpha - \beta)]
\end{aligned}
\tag{1.9}
$$

The output voltage of the mixer consists of two signals, one multiplied by $\cos 2\pi (f_{lo} + f_c)t$ and the other multiplied by $\cos 2\pi (f_{lo} - f_c)t$. These signals are two distinct amplitude-modulated signals, one having a carrier frequency of $f_{lo} + f_c$ and the other of $f_{lo} - f_c$. The amplitude of each is proportional to the amplitude of the original signal, that is, $A_m(t)$.

Consider the case for a typical AM broadcast receiver that might be tuned to receive an amplitude-modulated signal with a carrier frequency of 1350 kHz. Suppose that its local oscillator is generating an 1800-kHz signal. The output of the mixer would consist of two amplitude-modulated signals, one with a carrier frequency of 450 kHz and the other with a carrier frequency of 2250 kHz. If the intermediate-frequency amplifier is tuned to a frequency of 450 kHz, the component with a carrier frequency of 450 kHz would be amplified, whereas the 2250-kHz carrier signal would be lost. The 450-kHz signal would be detected after being amplified, thus yielding an audio output signal corresponding to the amplitude modulation $A_m(t)$ of the received signal. (A level shifting, generally achieved with a coupling capacitor, is also necessary to recover the audio signal.)

What is the advantage of a superheterodyne receiver? Again, consider the broadcast receiver with an intermediate-frequency amplifier tuned to a fixed frequency of 450 kHz. The carrier frequency of the signal to which the receiver responds depends on the receiver's local oscillator frequency. To receive a signal of 550 kHz (the lower end of the broadcast band), a local oscillator frequency of 1000 kHz is required. This results in signals with carrier frequencies of 450 kHz and 1550 kHz being produced by the mixer. The 450-kHz signal is amplified, and the 1550-kHz signal is rejected. To receive a 1600-kHz signal (the upper end of the broadcast band), a local oscillator frequency of 2050 kHz is required, which, in turn, produces mixer output signals with frequencies of 450 kHz and 2500 kHz. The advantage of this receiver is that tuning is achieved by changing the local oscillator frequency (a range of 1000 to 2050 kHz is required). Although this necessitates that the inductance, capacitance, or both of the circuit be changed, the resonant frequency of only a single circuit needs to be changed. Even for an improved receiver, in which a tuned circuit is employed for the input of the mixer, a mechanical tracking system is used to tune the two circuits simultaneously with a single tuning knob.

1.3 THE TELEGRAPH AND TELEPHONE: WIDE-SCALE INTERCONNECTIONS

The telephone, invented by Alexander Graham Bell in 1876, predated electronic devices by over a quarter of a century (Bruce 1973, Sharlin 1963, Pupin 1926). By the time of the invention of the vacuum tube triode in 1906, the telephone was widely used throughout urban areas. The operation of the telephone was predicated on an earlier electrical communication system, the telegraph. Bell was attempting to develop a multiplexing system to transmit several telegraph signals simultaneously on a single telegraph line when he went "astray" and invented the telephone. It is not, however, inappropriate that Bell should be associated with the telephone because he, his father, and his grandfather were highly respected speech specialists (elocution experts).

THE TELEGRAPH

The telegraph system of Morse, invented in 1837, depended on the earlier work of Volta, Oersted, and Ampère, among others. Volta (after whom the voltage unit is named) devised the first battery, an "electrochemical pile" consisting of zinc and silver discs separated by brine-soaked cloth or paper. Oersted observed that an electric current produces a magnetic field, and Ampère established the mathematical theory relating magnetic fields to electric currents. This led to the development of the electromagnetic responder, the basis of the telegraph receiver. An elementary telegraph system is shown in Figure 1.11. Morse's success depended on the then available electrical devices, which he assembled into a telegraph system. His prime contribution, that for which he is generally remembered, was a binary coding system for signaling. An *on* state corresponded to a current (the key depressed), and an *off* state corresponded to no current. Furthermore,

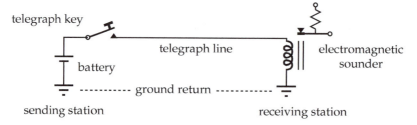

Figure 1.11: An elementary telegraph system.

on periods were broken into short and long intervals, that is, the dots and dashes that we now refer to as Morse code.

Morse's original telegraph system utilized a receiver consisting of a pencil activated with an electromagnet, which made a trace on a moving strip of paper. A telegraph operator would then decode the marks, thus recovering the original message. Operators, however, soon discovered that they could directly decode the message by listening to the clicks of the printer – the marked tape was unnecessary. The electromagnetic sounder was developed to optimize the decoding, one type click being associated with the electromagnet's being activated and the other with its being deactivated, thus distinguishing the *off*-to-*on* from the *on*-to-*off* transition of the current.

A revised code, the American Morse code, was used for wired telegraph systems in the United States, whereas a second version of this code, the International Morse code, was adopted for wireless communication and is still extensively used for shortwave radio communications by radio amateurs and others. The same *on–off* signaling principle utilized in the early telegraph forms the basis of today's modern fiber-optic systems. The light from a light-emitting laser, which is turned on and off, is transmitted through an optical fiber to a receiver, a light-detecting diode. Not only is the speed of the fiber-optic system much greater, but the electrical signals, such as those produced by a telephone, are directly encoded into *on–off* signals.

BASIC TELEPHONE SYSTEM

The telephone of Figure 1.12 works on a similar principle to that of the telegraph. In place of the key, a microphone is used to modulate the current of the circuit. Bell initially utilized a microphone that consisted of a diaphragm attached to a needle immersed in an acidic solution. The motion of the needle, the result of sound waves striking the diaphragm, caused the resistance of the circuit to

Figure 1.12: A basic telephone system.

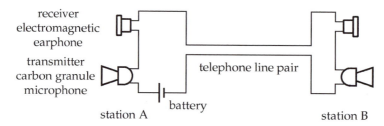

fluctuate. The fluctuating resistance, in turn, resulted in circuit current fluctuations. As for the telegraph system, an electromagnet was used for the receiver. In place of metallic contacts to produce the clicks of a sounder, an iron diaphragm acted upon by a magnetic field was used. A permanent magnet and an electromagnet energized by the fluctuating current of the telephone circuit jointly produced the magnetic field.

An improved telephone microphone in which carbon granules were used in place of the acidic solution was soon introduced. The motion of the microphone's diaphragm produced pressure fluctuations on the granules, which, in turn resulted in a fluctuation in resistance. Although modified versions of the carbon granule microphone have been introduced, this type of microphone is still used for phones over 100 years after being first introduced.

The circuit of Figure 1.12 is bidirectional, with two transmitters and receivers, so that either party can transmit while the other listens. In place of the earth ground return of the telegraph system, a second wire is used to complete the telephone circuit. A ground return, although reducing the amount of wire required, resulted in erratic and unpredictable effects. Today, a pair of wires is universally used for all local telephone connections (for example, to connect one's home phone to the telephone office). With a two-wire circuit, interference (cross talk) caused by electric and magnetic coupling between different telephone circuits is greatly reduced.

In addition to radio applications, early triode vacuum tubes were also used for long-distance repeater telephone amplifiers. As a result of insulation and wire losses, telephone signals are attenuated, that is, they become weaker as the length of the telephone line is increased. Before the advent of the vacuum tube amplifier, a mechanical-type amplifier was developed that consisted of a tightly coupled telephone receiver and a carbon granule microphone. Unfortunately, it badly distorted the telephone signal. The triode vacuum tube amplifier, first used as a telephone repeater in 1913, was the ideal solution for extending long-distance telephone service. Triode vacuum tube amplifiers were used in 1915 for the first U.S. transcontinental telephone line, although a set of mechanical-type amplifiers were held in reserve (Fagen 1975).

ANALOG TELEPHONE SIGNALS

The basic telephone system of Figure 1.12 is an analog system because the voltage differences of the circuit may take on any value within a set of prescribed limiting values. Circuit voltages fluctuate in accord with the audio signal produced by the speaker's microphone. This differs from the telegraph system in which signaling depends only on an *on–off* voltage condition.

An important characteristic of analog signals is their frequency spectrum (Figure 1.13). A sinusoidal signal has only a single frequency component, its periodic frequency, whereas a nonsinusoidal periodic signal has a fundamental frequency component as well as a set of harmonic components. The amplitude and phase of each frequency component depend on the periodic waveform of the signal: the more rapid the variations of the signal, the larger the harmonic amplitudes. From an information perspective, a periodic signal is not very interesting because, if

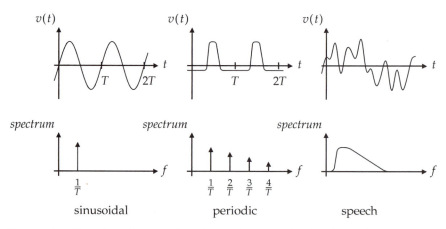

Figure 1.13: Time-dependent signals with the relative amplitudes of their frequency spectrum.

one has "seen" one period of the signal, one has "seen" them all. The nonperiodic speech signal has a frequency spectrum that tends to be continuous. As a result of electrical limitations of a telephone system, those that are unavoidable as well as those intentionally introduced, the frequency spectrum of telephone signals is generally limited. For the U.S. system, a spectrum of approximately 300 to 3400 Hz is utilized, and frequency components outside this range are filtered out.

An electronic system, such as an amplifier, must be capable of responding to all desired frequency components of a signal. Because a signal may be considered to be composed of a multitude of sinusoidal signals (based on a Fourier series representation of a periodic signal or a Fourier transform of a nonperiodic signal), an electronic system may be designed to reproduce sinusoidal signals faithfully. Furthermore, testing is usually done with sinusoidal signals, and operating specifications are given in terms of sinusoidal signals. For the speech signal of Figure 1.13, an electronic system must have a uniform frequency response over the frequency spectrum of the signal and be capable of responding to the amplitude range of the signal without significant distortion.

Generally, a dedicated pair of wires is used to connect a subscriber's telephone to a local telephone switching office. If the subscriber is calling a second subscriber connected to the same office, a direct connection is established through the switching equipment of the telephone office. Until the 1970s, switching was primarily through mechanically positioned contacts. For a subscriber calling a different office, for example an individual in another city, an interoffice connection was necessary. Depending on the circumstances, this may have required intermediate switching connections to reach the final destination.

In addition to amplifying signals, electronic systems are used for multiplexing telephone signals. A single pair of wires, or any other type of transmission system, is generally used to carry several telephone signals simultaneously. This, before the advent of digital telephone circuits, was achieved by translating the frequency spectrum of individual telephone signals (Figure 1.14). This process is similar to that employed for radio systems in which a high-frequency carrier is used to "carry" a lower-frequency modulating signal. The modulating signals

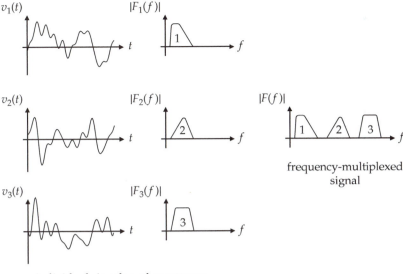

individual signals and spectrums

Figure 1.14: Frequency multiplexing of telephone signals.

are the individual telephone signals. In the United States, Bell Telephone Laboratories was responsible for much of the early improvements in vacuum tubes and vacuum tube circuits (Fagen 1975). Hence, a parallel, simultaneous development of electronic systems occurred for both early telephone and wireless systems.

DIGITAL TELEPHONE SYSTEMS

Although an analog connection is still generally used for connecting a subscriber and a telephone office, digital signals are used within switching offices and for transmitting telephone signals between offices. Analog-to-digital and digital-to-analog converters are used at each subscriber's office connection. With digital signals, switching by entirely electronic means is readily achieved, thereby eliminating erratic connections that may occur with mechanically activated contacts. With the development of specialized integrated circuits, the wide-scale usage of all-electronic digital telephone systems is now common.

As indicated in Figure 1.15, a sampling process is used to convert an analog signal to a digital signal. The U.S. telephone system uses 8000 samples per second; that is, the telephone signal is sampled every 125 μs. The resultant samples are

Figure 1.15: Sampling of a telephone signal.

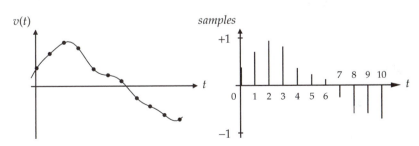

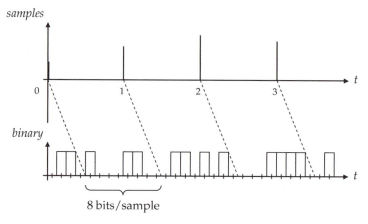

Figure 1.16: The conversion of sample amplitudes to a binary coded signal. An expanded time scale has been utilized for clarity.

a set of amplitudes that represent the analog signal. It may be shown that if the sampling rate is at least twice the highest frequency component of the analog signal, the samples contain all the information of the original analog signal; that is, the samples may be used to regenerate the original analog signal without error.

For the U.S. telephone system, the sample amplitudes are converted to a set of eight binary quantities. The normalized range of ± 1 for the samples of $v(t)$ of Figure 1.15 is divided into 256 (2^8) discrete levels, and each sample is assigned the level to which it is the closest. A nonuniform set of levels is utilized so that high-level and low-level signals tend to be reasonably well preserved. The digital representation of the signal of Figure 1.15 is indicated in Figure 1.16. Because 8 bits are used for each sample, the bits have to be squeezed into the 125-μs (1/8000 s) sampling interval. This results in a time interval of 15.625 μs being available for each bit, which corresponds to 64,000 bits per second. Although the digital signal requires a much higher transmission rate than the analog signal it represents, only an *on–off* condition needs to be transmitted.

Two digital signal paths, one for each direction of the telephone connection, are required. Electronic logic circuits are utilized for "switching" these signals. A modern digital telephone office is, in essence, a specialized computer with a very large number of input and output connections. Efficient modern telephone transmission systems utilize time-multiplexed digital signals. The time of each group of 8 bits representing a sample of a signal is reduced. For a 50-percent reduction in time, the bits of a second signal, similarly modified, could be inserted between the bits of the first signal. At the end of the transmission system, the signals that occur at different time intervals can be separated.

A typical first-level time-multiplexing system combines 24 digital telephone signals. One frame of the resultant digital signal consists of the 8 bits corresponding to a single sample of each signal (a total of 192 bits) plus one framing bit used to identify the frame. If an *on–off* sequence (*on* one frame, *off* the next, etc.) is used for the framing bit, the beginning of a frame may be readily identified at the receiving end of a time-multiplexed transmission system. The overall bit rate of the time-multiplexed 24 telephone signals is therefore 1,544,000 bits per second ($193 \times 8000/\text{s}$). For higher capacity systems, such as fiber-optic transmission lines, first-level groups are combined to produce higher-level groups, and these groups in turn are combined with similar higher-level groups to produce a super-level group. Transmission rates as high as one billion bits per second can be carried by modern high-capacity systems. Over 15,000 simultaneous telephone conversations can be carried by a system having a capacity of one billion bits per second. Alternatively, other digitally encoded signals, such as produced by television systems, can be time-multiplexed with the telephone signals.

1.4 TELEVISION: TIME-DEPENDENT VISUAL IMAGES

Although it was not until the 1950s that commercial television broadcasting became common, the research on which television is based was initiated in the 1920s (Abramson 1987; Fink 1952; Fisher and Fisher 1996; Zworykin and Morton 1940). Two approaches were tried: a mechanical system using a set of rotating scanning disks and a totally electronic system. Only the electronic system proved successful. Vladimir K. Zworykin, the inventor of the first television pickup tube, the iconoscope, is responsible for many of the early electronic television developments. Although several television systems were demonstrated in the late 1930s, commercial broadcasting in the United States began only after World War II.

The conversion of an optical image to an electronic signal is considerably more complex than that for sound waves. A microphone, such as that of the telephone system of Figure 1.12, produces a varying voltage and current. The varying component of the voltage (or current) is proportional to the fluctuations in atmospheric pressure caused by the sound waves that impinge on the diaphragm of the microphone. Hence, if $p(t)$ is the fluctuation in pressure, the microphone voltage $v(t)$ is proportional to $p(t)$, that is

$$\text{audio signal:} \quad v(t) \propto p(t) \tag{1.10}$$

A telephone or radio system needs only to transmit a replica of this time-dependent voltage.

An image has a spatial as well as a time dependence. Hence, its brightness is a function of three quantities:

$$\text{image signal:} \quad b(x, y, t) \tag{1.11}$$

One approach to dealing with the spatial dependence would be to divide the image into a set of discrete elements, as indicated in Figure 1.17. Each element could be designated by its row m and column n coordinate, and its time-dependent brightness $b_{mn}(t)$. For each element, a voltage might be developed as follows:

$$v_{11}(t) \propto b_{11}(t)$$
$$\vdots$$
$$v_{mn}(t) \propto b_{mn}(t) \tag{1.12}$$
$$\vdots$$
$$v_{MN}(t) \propto b_{MN}(t)$$

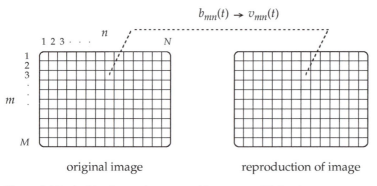

$$b_{mn}(t) \rightarrow v_{mn}(t)$$

| original image | reproduction of image |

Figure 1.17: A video image decomposed into a set of finite elements.

A total of $M \times N$ separate voltages would thus be needed. Because a standard-resolution television image corresponds to approximately 350,000 elements (pixels), this many time-dependent voltage signals would be required. This approach is not only impractical but is totally unnecessary.

The instantaneous time dependence of each picture element, as implied by Eq. (1.12), is not needed. Motion pictures (introduced commercially at the beginning of the 20th century) rely on a mechanical projection process. Following the projection of a single image, the screen is darkened while the film advances to the next image. Therefore, one views a succession of fixed images, 24 images/s being the commonly used rate. The brightness of a particular element changes in discrete steps from one frame to the next rather than continuously. Hence, for this system, a set of only 24 brightness values per second is required for each image element. Although this process does not provide an exact replica of the original image, it is, from a viewer's perspective, adequate. A higher image projection rate is not generally perceived as a significant improvement.

ANALOG TELEVISION

Because discrete time elements are acceptable for reproducing elements of an image, a sequential transmission of the brightness of individual elements achieved with a scanning process may be used. Imagine that a brightness detector is moved across an image that is to be transmitted. For the solid scanning line, a-to-b of Figure 1.18, corresponding to a left-to-right movement, the brightness remains uniformly constant (white). When the detector reaches the end of the trace (right side of the image), it is rapidly moved back to the left side of the image. For this segment, the retrace, the brightness signal $b(t)$ is set to a black value (a zero signal value of Figure 1.18). The brightness varies for the next scanning trace, c-to-d, and the center portion of the trace corresponds to the light gray of the outer box of the image pattern. In general, a different time-dependent brightness pattern is obtained for each scanning trace.

Although, for the sake of clarity, only 10 scanning traces are indicated in Figure 1.18, a much larger number of traces are obviously necessary to form a high-quality television image. The North American system utilizes a field consisting of 262.5 scanning traces. Only about 240 of these traces form part of the visible picture; the other traces occur during a vertical retrace. A second set

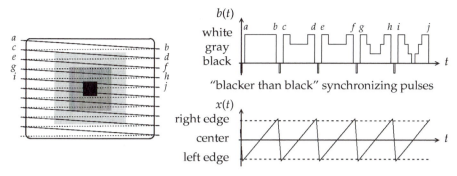

Figure 1.18: A video scanning process and corresponding video signal.

of 262.5 scanning traces falls between the tracings of the first field (a scanning process known as interlacing). A complete picture frame consists of 525 lines and is accomplished in 1/30 s. This results in a horizontal scanning rate of 15,750 lines per second (30 frames/s × 525 lines/frame). The scanning of each line and its retrace is thus accomplished in approximately 63.5 μs. Because the brightness of an image could vary many times from bright to dark over a single scanning line, the scanning signal $b(t)$ could vary extremely rapidly.

CATHODE-RAY TUBE DISPLAY

For an electronic video system, the brightness is converted to a varying voltage. In addition to the brightness signal, synchronizing pulses ("blacker than black") are added to the signal. These pulses control the sweep generators of a receiver used to reproduce the image. In 1929, Zworykin demonstrated an all-electronic receiver television picture tube. Except for the color feature of modern picture tubes, this 1929 invention incorporated essentially all basic features of modern television picture tubes and computer monitors. The television display is the descendant of the cathode-ray tube invented by Ferdinand Braun in 1897 (Kurylo and Süsskind 1981). Although Braun shared the 1909 Nobel Prize in physics with Marconi for his contribution to the invention of the radio, it is the former invention for which he is most frequently remembered (in German, a cathode-ray tube is known as a *Braunsche Röhre*). These tubes rely on a screen that gives off visible light when struck by an electron beam (Figure 1.19). An electron beam is used to write on the screen, that is, to reproduce the scanning traces used to generate the video signal.

A cathode, heated by a hot filament, is the electron source. Through a set of electrodes constituting an electron gun, the electrons are accelerated and then focused into a very small-diameter beam by means of an axial magnetic field produced by an external focusing coil. After leaving this region of the electron gun, the electron beam is further accelerated by an anode potential of 10,000 to 30,000 V connected between the cathode and the metallized internal surface of the tube. Through variations in the potential of a control electrode next to the cathode of the electron gun, the current of the beam and hence the intensity of the spot it produces can be varied. In order to write on the screen, that is, to generate the scanning traces used to produce the video signal, the electron beam needs to

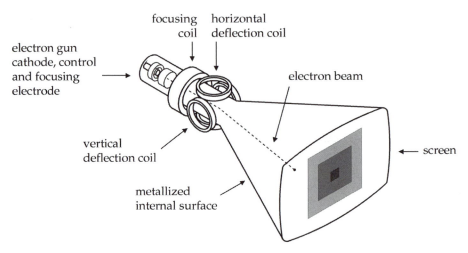

Figure 1.19: A monochrome cathode-ray tube with magnetic focusing and deflection.

be suitably deflected. Although either electric or magnetic deflection fields may be used for this purpose, television tubes utilize a set of orthogonal magnetic fields produced by an external set of coils (analog oscilloscope tubes use a set of electric fields produced by a set of internal deflection plates). The two coils with a vertical axis, one above and the other below the tube, produce a vertical magnetic field. Because the electron beam has an axial velocity, the electrons will experience a horizontal force ($\mathbf{F} = -e\mathbf{v} \times \mathbf{B}$, a cross-product relationship). As a result of this force, a horizontal deflection of the beam occurs that is proportional to the current of the coil. The set of coils with a horizontal axis produces a horizontal magnetic field, which, in turn, produces a vertical deflection of the electron beam. Hence, the instantaneous spot position is dependent on the instantaneous values of the currents of the coils, and the spot's intensity is dependent on the control electrode voltage.

As the beam moves across the screen, the beam current, and thus the brightness of the phosphorescence, is varied in accordance with the video brightness signal. The synchronizing pulses are used to control oscillators that produce the appropriate deflection currents. At a given instant, the electron beam produces but a single spot. However, as a result of the persistence of the screen phosphor, the screen continues to give off light after the beam has moved on. Furthermore, a viewer's persistence of vision gives the impression of a complete image with a continuous time dependence.

A color television display tube uses three separate electron beams that strike different screen phosphors to produce the three primary colors of red, green, and blue. Two systems are common. In the delta-type tube, the electron guns are aligned so that their individual axes form a triangle about the center line of the tube, whereas in the in-line tube the electron guns are in a horizontal line with the center gun ("green") coinciding with the axis of the tube. Both systems utilize a picture tube with an internal metal mask about 1 cm from the screen. The mask restricts an individual electron beam to striking only a single color phosphor.

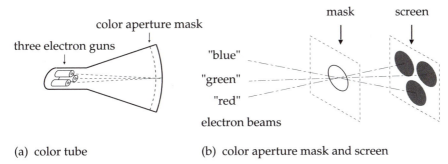

three electron guns

color aperture mask

(a) color tube

mask screen

"blue"
"green"
"red"
electron beams

(b) color aperture mask and screen

Figure 1.20: A delta-type color television tube with a detail showing its color aperture mask and screen phosphors.

A delta-type color television tube is depicted in Figure 1.20. Included is a detailed drawing of its color aperture mask and its three color phosphorus dots. The three electron beams of the tube, the intensity of each associated with a primary color, are simultaneously accelerated, focused, and deflected by electrodes and coils, which are not shown. The beams are so aligned that they converge at the plane of the aperture mask. On passing through the mask, an individual beam is defined so that it will strike only a single color phosphorus dot on the inside of the screen. Hence, the three dots (a triad) will have emission intensities corresponding to the primary colors. The dots, because they are extremely close to each other (within about 0.1 mm), are perceived as a single colored emission. Three video signals, one for each primary color, are required to control the currents of the individual electron beams. An in-line tube uses three side-by-side electron guns in a horizontal plane and a mask with a set of vertical slots adjacent to a screen having vertical phosphor stripes.

VIDEO CAMERA DEVICES

Following the invention of the picture tube by Zworykin in 1923, numerous camera pickup devices were developed (Hashimoto, Yamamoto, and Asaida 1995; Weimer 1976). The iconoscope relied on the photoemission of electrons by alkali metals. Visible light consists of photons with wavelengths λ of 0.38 to 0.78 μm. The energy of a photon is given by hf, where h is Planck's constant (6.625×10^{-34} J$\cdot$s) and f is the frequency of the light. Because the product of frequency and wavelength is equal to the velocity of light ($f\lambda = c$), the following is obtained for the energy of a photon:

$$\lambda = 0.55 \ \mu\text{m}$$
$$E_{\text{photon}} = hf = hc/\lambda = 3.61 \times 10^{-19} \ \text{J} \tag{1.13}$$
$$E_{\text{photon}}/e = 2.26 \ \text{V}$$

A photon with a wavelength of 0.55 μm, the wavelength for which the sensitivity of the eye is the greatest, has an energy of 2.26 eV. For the cesium and potassium alkali metals used in iconoscope tubes, this photon energy is sufficient to liberate an electron (electron photoemission).

An iconoscope utilizes a mosaic of photosensitive elements deposited on an extremely thin mica insulating sheet (Figure 1.21). Each element of the mosaic is

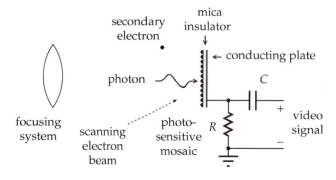

Figure 1.21: A video iconoscope.

insulated from its neighbors as well as from the conducting plate deposited on the backside of the mica insulation. The light image is focused on the photosensitive mosaic by a set of lenses. When photons forming the image strike a photosensitive element, electrons are emitted, leaving the element positively charged. An electron beam is used to scan the mosaic. The positively charged elements (those that have been illuminated) will absorb electrons from the beam thus being discharged. Because the element and the conducting plate form a capacitor, a capacitive current to the plate occurs at this instant. As a consequence, a video voltage is developed across a resistor connected between the plate and ground.

An early improvement on the original iconoscope was the image orthicon. This tube uses an electrode to accelerate the photoemission electrons. When these accelerated electrons strike their target, several secondary electrons are emitted. Hence, an "electron gain" is achieved, increasing the illumination sensitivity of the tube. Magnetic coils are used to focus the photoemission electrons and to deflect the beam used to scan the target.

Modern camera pickup devices no longer rely on vacuum-type devices. Instead, large-scale integrated circuits with photosensitive devices are used to generate video signals. A photon of visible light (Eq. (1.13)) has a very high likelihood of generating an electron-hole pair of charges when absorbed by a semiconductor such as silicon. These charges, if generated in the vicinity of the transition region of a junction diode, will be physically separated, resulting in charges that can produce a voltage or current in an electronic circuit that depends on the intensity of the incident light. In essence, a light-sensitive integrated circuit consists of a mosaic of photodiodes, one for each picture element of the video image. A typical rectangular camera detector designed to produce a conventional TV signal has a diagonal dimension of approximately 1 cm and consists of about 350,000 photodiodes.

Although one photodiode is used for each picture pixel, a direct wire connection to each diode is not practical. The "reading" of the diodes is accomplished with an array of charge-coupled devices functioning on the same principle as the MOSFET device of Figure 1.7. This circuit detects the light-produced charge of each diode through a sequential process. A simplified schematic of a commonly used camera pickup device, an interline-transfer charge-coupled device, is indicated in Figure 1.22. For the standard-resolution North American TV system (this system will be assumed for the numerical quantities of the discussion that follows), an array of approximately 500 horizontal rows and 700 vertical columns of diodes along with MOSFET charge-transfer switches and charge-coupled devices (CCDs) are required. In addition, a set of horizontal charge-transfer devices

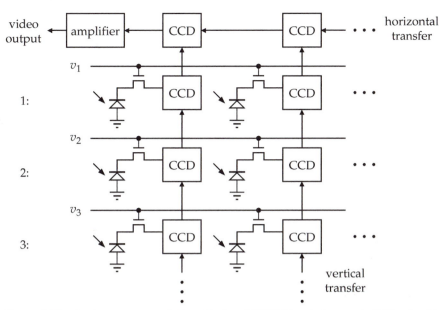

Figure 1.22: An image array of photodiodes, MOSFET switches, and CCDs. In this interline-type device, only the photodiodes are exposed to the incident light because optically produced electron-hole pairs would interfere with the operation of the other devices.

is needed to combine the outputs of the vertical CCDs. Precise timing of the gate voltages of the MOSFET devices and the control signals of the charge-coupled devices (not shown in Figure 1.22) is required. The output of this circuit is a single time-dependent video voltage.

The photodiodes are charge-integrating devices, that is, their charge accumulation depends on the intensity of the light and time interval over which they are allowed to charge. Charging of a photodiode occurs during the interval when its MOSFET device is inactive, which corresponds to a low-level gate voltage. When the MOSFET device is activated, the charge generated by the photodiode is transferred to its adjacent vertical CCD. After this transfer, the diode will again generate charge at a rate dependent on its incident light intensity. Approximately one field interval, 1/30 s, is available for this charge-generation process.

To explain the operation of the image device of Figure 1.22, a time corresponding to the beginning of a frame will be assumed. The MOSFET devices of the odd-numbered rows having voltages of v_1, v_3, and so forth, are activated, transferring the charge generated by each of the photodiodes of these rows to the adjacent vertical column of CCDs. During the next 1/60 s, the discrete charges are moved up the vertical column of CCDs in a step-by-step fashion. These devices, when operated in this fashion, are frequently referred to as a "bucket brigade" circuit. At the top of the integrated circuit, the charges of each vertical column of CCDs are transferred to the horizontal line of devices. Again, the individual charges are transferred in a step-by-step fashion by the horizontal row of CCDs to produce the video output signal. To obtain the video signal of the second field corresponding to the interlaced lines of a standard frame, the MOSFET devices of the even-numbered rows are activated, and the process is repeated.

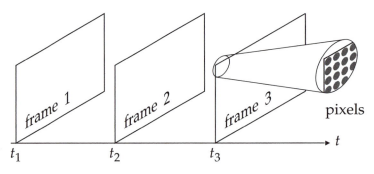

Figure 1.23: A video image considered as a sequence of frames composed of individual picture elements or pixels.

Many modifications of the basic image device of Figure 1.22 are used. For a color system, an array of three photodiodes, each with an on-chip lens filter for the three primary colors, is used for each pixel. Three video signals, each corresponding to a primary color, are thereby produced. The array size depends on the application. For example, approximately 20,000 pixels are adequate for a video telephone, and as many as 2 million pixels are needed for a high-definition TV camera device.

DIGITAL TELEVISION

The development of very-large-scale digital integrated circuits has not only made digital encoding of TV images possible, but it has made possible very complex encoding techniques inconceivable with analog systems. As a result, digital-encoded TV signals can be transmitted much more efficiently than analog-encoded signals. Not only do high-definition TV (HDTV) systems rely on digital technologies but also the transmission of limited-quality images over conventional telephone networks (videophones).

As indicated in Figure 1.23, a video image may be treated as a sequence of individual picture frames occurring at discrete, equally spaced time intervals. Each frame is composed of an array of pixels. Only an intensity quantity needs to be specified for each pixel of a monochrome image, whereas an intensity and a set of color quantities needs to be specified for a color image. For a basic digital encoding system, two quantized numbers to represent the intensity and color might be assigned to each pixel. Suppose, for example, that two 8-bit numbers were used; each pixel would require 16 bits of data. For a system with 350,000 pixels per frame and 30 frames per second, a data bit rate of 168×10^6 bits/s (168 Mb/s) would be required. This is not only an inordinately high bit rate – only a single or at most a few TV signals could be transmitted by a single optical fiber – but is totally unnecessary. With digital processing, the bit rate can be reduced to a much more acceptable value.

From an information perspective, the preceding 168 Mb/s digital system is similar to an analog system. Each pixel is treated as having an intensity and color that is not only independent of the quantities of adjacent pixels of the same frame but also independent of the quantities of a similarly located pixel of the frames that occur before and after its frame. This, of course, is not the case for a typical video signal. An image generally consists of small-size regions that have either identical pixels or have pixels that differ only slightly from each other. Only for an abrupt transition between two visual objects do pixels differ significantly from their neighbors. Regions of each frame (typically 8×8 pixels) can be encoded as a group, and thus very little information is required if a region is uniform. Often one entire frame may be the same as the previous frame or differ only slightly from it. If, for example, a frame is identical to the previous frame, from an information perspective it is only necessary to transmit "ditto," or if it differs only slightly, "ditto" except for new values of those pixels that are different. To achieve this encoding, which significantly reduces the bit rate required for transmitting a TV image, extensive digital memory and processing capabilities are required.

Associated with each pixel frame is a set of arrays of quantized data representing the intensity and color (generally one luminance and two chrominance components). These arrays may be obtained through a sampling of a conventional video signal or from a charge-coupled image sensor such as that of Figure 1.22 followed by an analog-to-digital output converter. A digital-encoding system converts the data of the logic arrays to a single unambiguous bit stream that is adequate to regenerate the arrays at the receiving end. These regenerated arrays, when appropriately scanned, produce the output video display. Because multiple logic data arrays are used, it is not necessary to transmit the data at a uniform rate as is the case for an analog system.

The digital encoding and decoding techniques utilized are extremely complex; they reflect the complexity of the logic functions that can be achieved with very-large-scale integrated circuits (Anastassiou 1994; Netravali and Haskell 1995; Netravali and Lippman 1995; Schäfer and Sikora 1995). Variable-length coding is used for the color values of each pixel because not all color combinations are equally likely. The resultant chrominance array and luminance array are then divided into regions corresponding to blocks of 8×8 pixels. Transformed values using a discrete-cosine transform are then obtained for each region. If the region should happen to be uniform, only one of the 64 transform values will exist – all others will be zero – and thus only a single quantity needs to be encoded and transmitted. Even for a more general case, many of the transform values will either be zero or sufficiently small that they may be ignored without a significant degradation of the video image. A variable-length coding scheme is then employed to encode these coefficients into a minimum-length set of binary bits (entropy coding). At the receiving end of the transmission system, the digital signal must be decoded to obtain the transformed values followed by an inverse transform operation to recover the original luminance and chrominance quantities associated with each pixel.

A temporal correlation between sequential video frames generally exists. As a consequence, the sequential frames are not treated identically, as is the case for an analog system. A complete set of encoded values is transmitted only for certain frames – frames designated as being intracoded (I frames). In between, there are predictive-coded frames (P frames). The P frames have regions that are determined by a previous intracoded frame or a previous predictive-coded frame. A difference of the transformed values of the two frames is used to specify each region. In addition, an encoding scheme that accounts for a motion of the scene is also utilized. Finally, sandwiched between the I and P frames are bidirectionally predictive, interpolated frames (B frames). The values of the data for these frames are determined from either the closest earlier or later I or P frame or from both an earlier and a later frame. The precise manner in which the sequential frames are encoded is determined during the encoding process. Sufficient control data are transmitted to allow the decoder at the receiving end to interpret the data stream properly.

With digital encoding, a standard television picture and audio signal can be transmitted with a bit rate as low as 1.5 Mb/s (Motion Picture Experts Group–1, MPEG–1, standard). The transmission of high-definition TV using a conventional "over-the-air" channel is dependent on a very efficient digital encoding of the video signal. The Grand Alliance System for U.S. broadcasting will provide a TV image with twice the resolution of a conventional image and a 16 × 9 format (16 units wide by 9 units high), which corresponds to motion picture inputs. This system achieves a basic bit rate of slightly less than 20 Mb/s.

1.5 THE ELECTROMAGNETIC SPECTRUM: A MULTITUDE OF USES

An electromagnetic wave propagating through the atmosphere or space is characterized by an oscillation frequency and a wavelength. At a given point in space, the amplitude of its electric and magnetic field oscillates in time with a frequency that is that of the transmitting oscillator. Wavelength relates to the spatial dependence of the electromagnetic wave. At a given instant (as if one were to take a snapshot of the wave), the electric and magnetic fields have a periodic spatial dependence. The wavelength of the radiation is the distance over which the wave tends to repeat itself.

For a plane wave (electric and magnetic fields in a common plane perpendicular to the direction of propagation), the product of frequency f and wavelength λ is equal to the velocity of light c:

$$f\lambda = c = 3 \times 10^8 \text{ m/s}$$
$$f = c/\lambda \quad \text{or} \quad \lambda = c/f \tag{1.14}$$

A signal with a frequency of 1.0 MHz, for example, has a wavelength of 300 m (100 MHz, 3 m; 10,000 MHz or 10 GHz, 3 cm). The higher the frequency, the shorter the wavelength.

The early electromagnetic experiments and demonstrations by Hertz and Lodge used radiation with wavelengths of a few meters or less so as to conform to the confines of a laboratory or lecture room. However, it was discovered by Marconi and other early pioneers that longer-wavelength radiation gave better results for long-distance communication ($\lambda > 1000$ m or $f < 300$ kHz, judging from descriptions of the apparatus used; techniques for measuring frequency or wavelength were not available). The early "wisdom" at the beginning of the 20th century was that only radiation having wavelengths of greater than 200 m is of value for long-distance communication. However, as equipment was improved, it was discovered that shorter-wavelength radiation ($\lambda < 200$ m) is also of value.

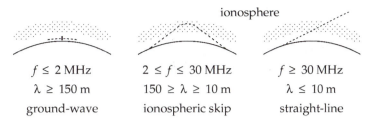

Figure 1.24: Radio propagation modes.

At present, with the exception of the AM broadcast band ($540 \leq f \leq 1600$ kHz, $556 \geq \lambda \geq 187.5$ m), almost all communications systems use these shorter wavelengths.

FREQUENCY SPECTRUM

The frequency of a radiated signal tends to influence how it is propagated. Low-frequency, that is, long-wavelength signals such as used for AM broadcasting, tend to follow the curvature of the earth (Figure 1.24). Although daytime distances of up to 100 km may result, much longer nighttime distances can occur as a result of ionospheric refraction. Higher-frequency, shorter-wavelength signals tend to be reflected from the earth's ionosphere, frequently returning to the earth several thousand kilometers from the transmitter. Although this makes long-distance communication possible, the transmission mode (ionospheric skip) depends on highly variable ionospheric conditions. The properties of the ionosphere that influence the reflection of signals not only vary between night and daytime, but they depend on sunspot activity. Higher-frequency, shorter-wavelength signals, such as used for television broadcasting and cellular phones, tend to travel in straight lines, providing line-of-sight communication.

The electromagnetic spectrum utilized for communications systems is indicated in Figure 1.25. Audio signals may contain frequency components up to

Figure 1.25: The electromagnetic spectrum. Logarithmic scales have been used for frequency and wavelength.

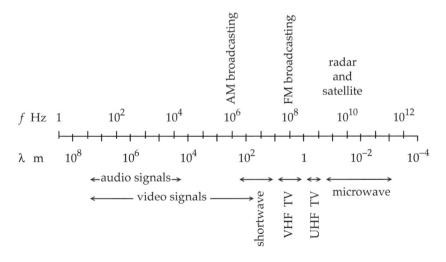

approximately 20 kHz (the upper frequency response of the human ear). On the other hand, analog video signals have frequency components up to approximately 4.5 MHz.

A carrier frequency is required for a communications system, depending on the propagation of electromagnetic waves. In addition, a spectrum width is required for the transmission of information such as an audio or video signal. The width of an AM broadcast signal is limited to 10 kHz; that is, ± 5 kHz about the carrier frequency of a signal. As a result, 106 stations can fit into this band. Because of the limited range of these stations (especially during daylight hours), a large number of broadcast stations is possible. Within the United States, stations are assigned frequencies by the Federal Communications Commission. The electromagnetic spectrum is treated as a public good to be used by commercial interests on a shared basis. Stations are separated by 10 kHz and have carrier frequencies evenly divisible by 10 kHz.

Shortwave ($10 \text{ m} \leq \lambda \leq 150 \text{ m}$) broadcasting occupies the frequency spectrum above the AM broadcast band up to approximately 30 MHz. Although very long distances can be achieved under favorable conditions, transmission conditions can be erratic. Furthermore, only fairly narrow bandwidths, such as those used for AM broadcasting, can be utilized.

Frequency-modulated (FM) broadcasting utilizes much higher carrier frequencies (88–108 MHz in the United States) and requires a much wider spectrum width than AM broadcasting. A much higher-fidelity audio signal having frequency components of up to 20 kHz is transmitted. In addition, a wide-bandwidth modulation system is used to reduce the effects of noise. The result is that a spectrum width of 200 kHz is utilized. Although only 5 such signals could fit in the AM broadcast band, 100 signals fit in the 20-MHz-wide FM band (carrier frequencies are separated by 200 kHz).

A 6-MHz spectrum width is utilized for U.S. commercial TV broadcasting. Channels 2 through 6 are between 54 and 88 MHz, and channels 7 to 13 are between 174 and 216 MHz. These are the VHF (very-high-frequency) channels. The UHF (ultrahigh-frequency) channels of 14 through 69 occupy a frequency range of 470 through 806 MHz. Interspersed with the TV channels and FM broadcast bands are a myriad of communications services (aircraft, police, fire, etc.). Cellular phones use carrier frequencies above 806 MHz.

RADAR

An important application of high-frequency, short-wavelength radiation is radar (radio detection and ranging) (James 1989). Early radar systems were developed in the 1930s using transmitting frequencies of 30 MHz or less (Page 1962; Swords 1986). A basic system is indicated in Figure 1.26. Radiation from a high-power pulsed transmitter is directed (beamed) toward an object to be detected. A small amount of the radiation incident on the object will tend to be reflected back to the receiver (an echo). A cathode-ray tube is frequently used for the receiver display. A linear horizontal sweep is used for a time axis, and the vertical deflection is the instantaneous output of the receiver. The first pulse of

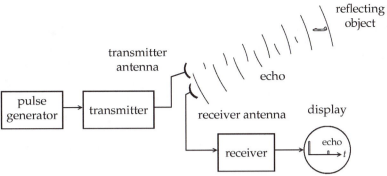

Figure 1.26: An elementary radar system.

the display indicated in Figure 1.26 is the transmitted signal, whereas the second is the received echo, which occurs at a later time.

The process of ranging depends on the finite propagation velocity of the electromagnetic radiation. In 1 s, radiation travels 3×10^8 m; in one μs, 300 m. Therefore, if the reflecting object is at a distance of 300 m, 1 μs will be required for the radiation to reach the object, whereas another 1 μs will be required for the echo to return to the receiver. Hence, the echo on the display will occur 2 μs after the transmitted pulse (therefore, each 1-μs delay corresponds to a range of 150 m).

The direction of an object can be ascertained if a well-focused, narrow beam is used. This also increases the range of the system because less energy is "wasted." Optimal antenna sizes depend on the wavelength of the radiation. For an AM broadcast station, a quarter-wavelength vertical antenna is common ($\lambda/4 = 75$ m for a 1-MHz carrier frequency). An antenna that has a dimension comparable to the wavelength of the signal it is radiating tends to have little directivity; its radiation tends to be uniformly distributed. To concentrate the radiation pattern, an antenna that is large compared with the wavelength of the radiation is necessary. Hence, practical-sized radar antennas that have narrow beams and that can be mechanically or electrically oriented require very short microwave radiation ($\lambda < 0.3$ m or $f > 1000$ MHz).

Conventional electronic devices, both vacuum tubes and transistors, tend to perform very poorly (if at all) at microwave frequencies. For a frequency of 1000 MHz, the period is only 1 ns. This requires that the response time of the electronic device be considerably less than 1 ns. The finite transit time of the electrons of these devices imposes an upper limit to the frequency at which these devices may be used.

The klystron amplifier and oscillator invented by Russell and Sigurd Varian provides a source of extremely high-frequency radiation (Ginzton 1976). Its operation depends on the very effect that limits the performance of conventional devices, that is, the finite transit time of electrons. Through the use of a cavity resonator (in place of a conventional LC circuit), electrons of a high-velocity electron stream are velocity modulated – the velocity of some electrons is increased, whereas that of others is decreased. The electron stream is allowed to drift, thus allowing the faster electrons to catch up with slower electrons (bunching). What was initially a stream with electrons uniformly distributed becomes a stream consisting of bunches of electrons. The bunched electrons are then used to interact with the electric fields of a second cavity, producing an amplified version of the signal used to excite the first cavity. An alternative type of klystron uses a static reflecting electric field to return the bunched electrons to the original cavity. The electron stream, for proper conditions, results in positive feedback (regeneration) and sustained oscillations.

Shortly after the invention of the klystron in the United States, John Randall and Henry Boot invented the cavity magnetron in Great Britain (Boot and Randall 1976). This tube, which also depends on electron transit times, utilizes a circumferential arrangement of resonant cavities around a cylindrical cathode. An axial magnetic field is used to force electrons emitted by the cathode into circumferential trajectories. Even the earliest tubes proved to be ideally suited for providing pulses of microwave radiation of very high power (as much as 500 kW at a frequency of 3000 MHz and 100 kW at 10,000 MHz).

For a radar system such as that of Figure 1.26, the same antenna can be used for transmitting and receiving because, ideally, the received echo occurs after the transmitter is switched off. A direct connection of the transmitter and receiver to a common antenna is obviously not possible, for the high-power signal from the transmitter would destroy the sensitive receiver circuits (literally burn them out). An extremely rapidly switching transmitter–receiver circuit, faster than can be achieved with a mechanical switching system, is needed. Gaseous plasma discharges initiated by the high-power pulses of the transmitter, however, can have a sufficiently rapid response. Plasma discharges are generally used in conjunction with a set of resonant circuits to "connect" the transmitter to the antenna and to reflect the transmitted signal simultaneously from the receiver. When the transmitter is switched off (at the end of its pulse), the discharge is dissipated. Instead of the near short circuits caused by the discharges, open circuits now occur. For an appropriately constructed circuit, the receiver is directly connected to the antenna.

An alternative echo indicator, the position-plan indicator, is also common. This indicator utilizes a timing line that originates at the center of the screen of the cathode-ray tube and sweeps outward. The line is simultaneously rotated as the orientation of the antenna is changed. The received echo is used to increase the intensity of the line at a distance from the center corresponding to the delay in the echo. With a long-persistence screen, a two-dimensional picture of the radar echoes is formed. A familiar example of this type of system is the radar image formed by reflections of storm clouds that is presented on television weather reports.

COMMUNICATIONS SATELLITES

One of the first serious proposals for using artificial earth-circling satellites for relaying electromagnetic messages was by the well-known science fiction writer Arthur C. Clarke (Clarke 1945; Pierce 1968). Clarke proposed placing satellites in an equatorial, geostationary orbit, an orbit that is at a distance from the surface of the earth of approximately 36,000 km, thereby resulting in the satellite's remaining over the same spot on the earth (the satellite and the earth are thus rotating at the same angular velocity). Such an orbit is now frequently referred to as a Clarke orbit (see Figure 1.27). A detailed technical proposal for communications satellites was put forth in 1955 by John R. Pierce of Bell Telephone Laboratories (Pierce 1955). (Pierce was also a science fiction writer under the pseudonym J.J. Coupling.)

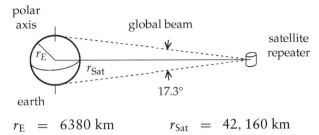

$$r_E = 6380 \text{ km} \qquad r_{Sat} = 42,160 \text{ km}$$

Figure 1.27: A satellite with a geostationary equatorial orbit.

It was not until 1957 that the first artificial satellite, Sputnik, was launched into a low-earth orbit, causing it to circle the earth approximately every 90 min. Its 20-MHz beacon transmitter fascinated listeners around the world. The first successful experimental communications satellite, Echo – a joint venture of Bell Telephone Laboratories and the National Aeronautics and Space Administration (NASA) – was launched in 1960. This satellite consisted of a passive 100-ft-diameter reflecting balloon that was inflated when the satellite reached its orbit. As with all early satellites, it was in a low-earth orbit.

The Telstar I satellite with an active electronic repeater, launched in 1962, may be considered the predecessor of modern communications satellites. Although this satellite was not in a geostationary orbit (it had a 158-min orbit), Telstar I, followed by Telstar II, demonstrated the feasibility of using satellites for long-distance communication. A signal with a carrier frequency of 6390 MHz was beamed from the earth to the satellite. At the satellite, the signal from the earth was amplified, translated to a new carrier frequency of 4170 MHz, and radiated back to the earth. Electronic circuits using transistors and a single microwave traveling-wave-tube amplifier were utilized. In addition, semiconductor photovoltaic cells were used for the electrical power source. The wide-bandwidth repeater system of Telstar provided the first experimental trans-Atlantic live television transmission (Bell System Technical Journal 1963; O'Neill 1985; Solomon 1962).

Before the advent of communications satellites, coaxial cable and microwave relay systems were used for the long-distance transmission of wide-bandwidth communications signals such as television. Because a set of earth-based microwave relay stations was limited to line-of-sight distances, over 100 relay stations were necessary to span the continental United States. No such system was available for crossing oceans. Ocean telephone cables with built-in repeater amplifiers (the first set of Atlantic cables, designed for 36 telephone conversations, was laid in 1956) had a bandwidth inadequate for television transmission (O'Neill 1985).

Geostationary communications satellites in an equatorial orbit, a Clarke orbit, are now widely used to relay television transmissions as well as numerous other communications services. An electronic repeater amplifier similar to that of the Telstar satellites is common (Figure 1.28). It will be noted that a mixer is used to

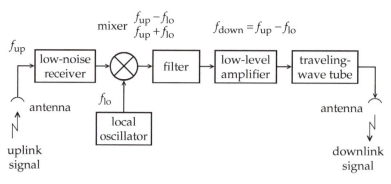

Figure 1.28: A satellite repeater amplifier.

translate the carrier frequency of the uplink signal. This frequency translation is imperative for a satellite repeater amplifier because an extremely large amplification of the signal is required. If the uplink signal frequency were not translated, even a minuscule amount of leakage of the transmitted signal back to the receiver antenna could result in positive feedback and oscillations. Translating the carrier frequency using a mixer, as in a superheterodyne receiver, along with an appropriate set of frequency-selective filters, eliminates this problem.

As an illustration of a repeater amplifier, consider the communications satellites used for relaying U.S. television transmissions, which are frequently received using 2- to 3-m-diameter dishes (Baylin and Gale 1986; Easton and Easton 1988; Fthenakis 1984). For channel 1 of this system, $f_{up} = 5,945$ MHz and $f_{down} = 3,720$ MHz. This requires that the satellite repeater have a local oscillator frequency of 2225 MHz. However, because $f_{up} - f_{down}$ has been chosen the same for all the channels of this band, which are relayed by a single satellite, only a single local oscillator and wide-bandwidth receiver are required (a backup receiver is available should a failure occur). Separate filters and amplifiers are used for each channel – an output power from the traveling-wave-tube amplifier of 10 W is typical. Although each channel corresponds to a different frequency, an overlapping of channels is achieved by using different circular electric-field polarizations (clockwise or counterclockwise), thus conserving spectrum space. On the earth, highly directional antennas (which, as a consequence, have high gains) are used to receive these satellite signals. Although there are many satellites in an equatorial orbit, they are located at different longitudes and can thus be selected by pointing the earth-based receiving antenna in the appropriate direction. Highly directional antennas, along with high-power transmitters, are used to beam the uplink signal to the appropriate satellite.

Another satellite repeater system uses uplink frequencies in the 14-GHz band and downlink frequencies of 11–12 GHz. These satellites are frequently used for relaying European television transmissions (Stephenson 1991). A combination of a moderate satellite transmitter power ($\approx$50–100 W) and a rather narrow transmitting beam providing a "country size" pattern results in signals that can be received using relatively small receiving dishes (60-cm diameter or less). Such antennas mounted on the south side of buildings are now common in Europe (as opposed to the 2- to 3-m diameter dishes in the United States).

A more recent satellite system, introduced in the mid-1990s, uses multiplexed digitally encoded TV signals (Elbert 1997). These satellites have downlink frequencies in the 12-GHz range and can be received using a small-diameter dish. As a result of very efficient digital encoding and multiplexing, a single satellite is capable of relaying hundreds of TV signals. In addition, data channels are available for controlling subscriber access to the channels.

1.6 COMPUTERS: TRANSISTORS BY THE MILLIONS

A desire to replace energy consumptive and fragile vacuum tubes provided the impetus for the applied research that resulted in the transistor. Not only has the transistor replaced vacuum tubes in most applications, but, as is frequently the case for a new invention, it gave rise to a new and unforeseen application: large-scale digital computers. Although the earliest electronic computer (ENIAC) and its immediate successors relied on vacuum tubes, present computers using integrated circuits (including the personal computer, without which most of us would now be lost) would be inconceivable without the transistor.

Although large-scale electronic digital computers are a relatively recent development, the concepts on which they are based are much older (Goldstine 1972; Shurkin 1984; Slater 1987). The 17th-century mechanical adding machine of Blaise Pascal, an adaptation of the ancient abacus, gave rise to the mechanical multiplier and other calculating machines that were widely used until the introduction of electronic calculators in the 1970s. It was in the 19th century that Charles Babbage set forth the basic concepts on which modern computers are based. His initial Difference Engine was soon followed by a more advanced Analytical Engine. He was able to visualize very clearly a set of separate computing units working together in a synchronized manner to carry out a sequential set of operations. Unfortunately, his designs could not be realized physically at the time. Babbage's conception included an arithmetic unit, an input reader for punched cards, an output device, an internal memory, and external memory storage (an attendant who would insert a set of punched cards was to be summoned by a bell). It was not until 1943 that Howard Aiken achieved an electromechanical realization of Babbage's concepts in the Mark I computer at Harvard.

A second type of computer, also based on concepts from the 17th century, utilized the principle of logarithms that was originally introduced by John Napier. An understanding of logarithms led to the slide rule, which relies on an analog process to add or subtract logarithms, thus multiplying or dividing numbers. This approach to computing evolved into the Differential Analyzer developed by Vannevar Bush in the 1930s and to the analog computer using electronic amplifying and integrating circuits. These computers have now been superseded by analog simulation programs that can be run with considerably less effort on digital computers.

LOGIC CIRCUITS

Although one would not presently consider using electromechanical relays for performing logic functions in a computer (as in Aiken's Mark I computer), relays

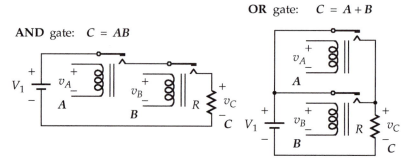

Figure 1.29: Electromechanical logic gates using relays.

are extensively used for numerous control systems that are based on a limited set of logical operations. Furthermore, an understanding of relay circuits that have either open or closed contacts one can clearly visualize is invaluable for understanding electronic logic circuits. The electromagnet of a relay is either energized or not; for a logic application it is only momentarily in an "in-between" state. Its switch contacts are therefore either closed (conducting) or open (nonconducting). This corresponds to the logic system expounded by George Boole in 1854 in which variables took on but one of two states. Nearly a century later, Claude E. Shannon applied this logic system, now generally described as Boolean algebra, to relay and switching circuits (Shannon 1938).

Consider the relay circuits of Figure 1.29. A relay will be assumed to be energized if its coil voltage (v_A or v_B) is equal to V_1 and not energized for a zero input voltage. Using positive logic, $A = 1$ if $v_A = V_1$ and $A = 0$ if $v_A = 0$. Voltage values other than zero and V_1 are excluded from consideration. For the **AND** gate, the output voltage v_C is equal to V_1 if both relays are energized and zero otherwise. If one associates the Boolean variable C with the output, $C = AB$. A parallel relay connection of the second circuit of Figure 1.29 results in an **OR** operation.

Semiconductor diodes, each with a characteristic such as that of Figure 1.2, may also be used to construct logic gates. When the diode is conducting ($i_D > 0$), its voltage, v_D, is relatively small, whereas when it is not conducting ($v_D < 0$), its current is essentially zero. The diode may therefore be treated as either conducting or nonconducting, reverse biased. For many applications, the diode behaves as a switch that is either closed (conducting) or open (reverse biased).

The diode circuits of Figure 1.30 perform the Boolean **OR** and **AND** operations. If both inputs of the **OR** gate, v_A and v_B, are zero, the output voltage v_C will also be zero. If, however, one input is equal to V_1 (for example, $v_A = V_1$, corresponding to $A = 1$), the output voltage v_C will be equal to $V_1 - v_D$. To the extent that v_D can be ignored, v_C is nearly equal to V_1, corresponding to $C = 1$. If both inputs are equal to V_1, the output will remain equal to approximately V_1. Hence, a logic **OR** operation results, and $C = A + B$. A battery with a potential of V_1 corresponding to a logic **1** signal is required for the diode **AND** gate of Figure 1.30. If either input is zero (a logic **0**), the output voltage is equal to the small diode voltage, a value close to zero corresponding to a logic **0**. Only if both

OR gate: $C = A + B$ AND gate: $C = AB$

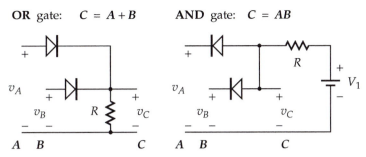

Figure 1.30: Logic gates using diodes.

inputs are equal to V_1 ($A = B = 1$) is the output voltage v_C also equal to V_1 ($C = 1$). This corresponds to the **AND** operation $C = AB$.

A relay with a set of normally closed contacts may be used to form a **NOT** gate (Figure 1.31). When $v_A = 0$ ($A = 0$), the relay is not energized and $v_B = V_1$ ($B = 1$). Energizing the relay ($v_A = V_1$, corresponding to $A = 1$) opens the relay contacts, and $v_B = 0$ ($B = 0$). Hence, $B = \bar{A}$. Also shown in Figure 1.31 is a circuit using a MOSFET device such as that of Figure 1.7. When $v_A = 0$ ($A = 0$), the drain current i_D of the MOSFET device is zero, resulting in $v_B = V_1$ ($B = 1$). On the other hand, when $v_A = V_1$ ($A = 1$), the drain current i_D will be large. For a well-designed circuit, a large voltage will be developed across R, namely $i_D R$, and only a small voltage will be developed across the MOSFET device. For v_B small, the output will be interpreted as a logic 0. Hence $B = \bar{A}$ for both circuits.

Although combinations of **AND**, **OR**, and **NOT** gates can perform many logic functions, memory gates are also needed for a digital computer. An electromechanical relay memory gate might be used for a simple control application. A memory gate using MOSFET devices is shown schematically in Figure 1.32. Consider, initially, the condition for which the input signals v_R and v_S are zero. Because the drain currents of the devices with these inputs are zero, these devices will not affect the behavior of the remaining circuit. The middle transistors with gate-to-source voltages of v_A and v_B form a set of **NOT** gates; the input gate-to-source voltage of one device is equal to the output drain-to-source voltage of the other. If the gate-to-source voltage v_B of the left-hand device is sufficiently small, its drain-to-source voltage v_A will be equal to V_1. For a properly designed circuit, this will result in a sufficiently large current of the right-hand device to cause its drain-to-source voltage v_B to be small. Hence, $A = 1$ and $B = 0$. The other

Figure 1.31: Electromechanical and transistor **NOT** gates.

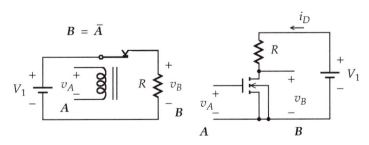

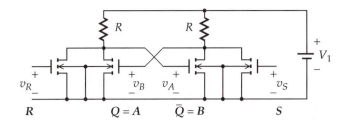

Figure 1.32: A transistor flip-flop memory.

memory state corresponds to v_A being small and $v_B = V_1$ ($A = 0$ and $B = 1$). Hence, the memory has two stable states. Now consider the effect of the R and S inputs. If v_A is small ($A = 0$), an input of $v_R = V_1$ will have only the negligible effect of reducing v_A to an even smaller voltage. The voltage v_B will remain equal to V_1. Only if v_A is initially equal to V_1 will an input of $v_R = V_1$ have an effect, namely, changing the state of the memory so that v_A is small ($A = 0$). An input $v_S = V_1$ will change the state of the memory, if necessary, so that $v_A = V_1$ ($A = 1$).

A BASIC COMPUTER

Logic gates using MOSFET or other electronic devices may be combined to produce a set of elementary logic functions (Figure 1.33). Both the **NOR** and **NAND** functions, having outputs that are the complements of **OR** and **AND** gates, respectively, can be obtained from a circuit modification of the basic MOSFET **NOT** gate of Figure 1.31. The **OR** and **AND** functions are obtained with an additional **NOT** gate. A buffer function may be achieved using two **NOT** gates. Although this realization does not result in a change of logic levels, it is often used to restore the voltage levels of a logic signal that may have been degraded by a logic system. The buffer shown in Figure 1.33 has an enable input. When this input is at a high level, the logic output of the buffer is that of its input, whereas for a low-level enable input, the output is an open circuit. For an open-circuit condition, the logic level at the output will depend on the other devices sharing its output. When MOSFET devices are used, a simple buffer implementation consists of a single MOSFET device in series with the input and output terminals (this realization will not restore degraded logic levels). The series MOSFET device behaves as an open or closed switch in accordance with its controlling gate-source voltage.

An **RS** memory element, again often realized using MOSFET devices, is also indicated in Figure 1.33. This memory, having a clock input, is an enhancement of the basic flip-flop circuit of Figure 1.32. Synchronous logic systems are generally

Figure 1.33: Symbols used for basic logic functions.

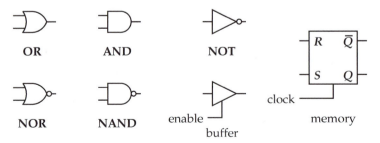

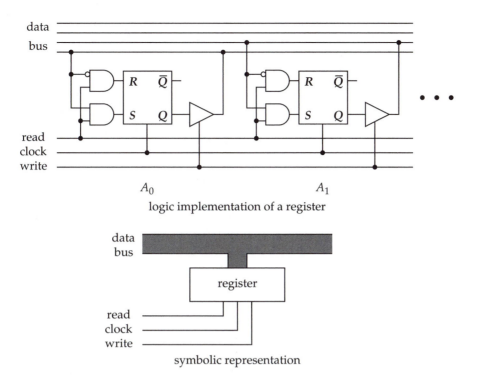

logic implementation of a register

symbolic representation

Figure 1.34: A data register and bus. As a result of the input and output gates of the memories, a memory may either read the logic level of the line to which it is connected or write the level of its output to the line. Although only two memory elements are shown, the number of individual memories is equal to the number of lines of the data bus, that is, its width.

used for computers. A clocked memory has an internal response to its inputs during one condition of the clock signal but does not change its output state, if required by the inputs, until a later condition of the clock signal. Often, the leading or falling edge of the clock signal is used to separate in time the output response of the memory from that required by its inputs.

The operation of a computer relies on data registers sharing a common data bus (Figure 1.34). The data bus, as its name implies, is used to transfer data between different logic units of a computer. For this type of system to function properly, it is imperative that only one logic circuit attempt to write to the data bus at any instant. A register consists of a group of memories functioning as a unit that may either read data from or write data to the bus. It is the logic buffer at the output of each memory that will cause a register to write to the bus when its enable input is at a high level. On the other hand, the input **AND** gates, when their read inputs are high, produce a set of inputs for each memory. An inversion bubble is indicated for the input of the **AND** gate connected to the R input of the memory – an indication that a **NOT** gate is in series with the input connection. One memory element is required for each data line of the bus, and bus widths from 4 to 64 lines are common.

An accumulator register, a modified data register, is depicted in Figure 1.35. The operation of this register is determined by the content of the instruction

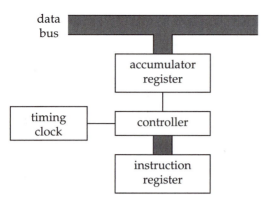

Figure 1.35: An accumulator register controlled by an instruction register. The single "line" between the accumulator and the controller represents a set of logic control signals.

register, the content being interpreted by the controller that provides the appropriate logic control signals to the accumulator. Consider, initially, a very limited accumulator, the data register of Figure 1.34, with its controlling inputs provided by the controller. The three functions of the data register (doing nothing, reading, and writing) are determined by the content of the accumulator register. For example, if a 4-bit instruction is utilized, **0000** might correspond to doing nothing, **0001** to reading, and **0010** to writing. The controller would translate these logic words into appropriate read and write signals for the accumulator register. The read operation followed at some later time by a write operation might be used to save a data word on the bus at the read time while the bus is used for other data transfers.

More complex accumulator functions can be obtained through modifications of a basic data register – enhancements achieved with additional logic elements (Taub 1982). It is desirable to be able to "clear" the output of all memories, that is, to set each to a low level, $Q = 0$. A high-level signal at each memory's reset input R will achieve this. If these inputs are through an **OR** gate, the functioning of the memory for other inputs will be unchanged. For the example suggested, an instruction register word of **0011** might be used to implement a clear operation. Another useful operation is that of complementing the content of each memory of the register. A set of logic gates that transfers the outputs of each memory back to its inputs could be used, the Q output to the R input and the $\overline{Q}$ to the S input. This could correspond to an instruction register of **0100**. If it is desired to set the register, that is, change all memory outputs to a high level, $Q = 1$, a clear operation could be followed by a complement operation. Another important function is to shift in a forward direction the contents of memories, that is, move that of A_0 to A_1, that which was originally in A_1 to A_2, and so forth. Alternatively, a shift in the reverse direction might be desired. Furthermore, when shifting the contents of the memories, the content of the last memory might either be discarded or moved to the first register. Again it is the word of the instruction register that determines if a shifting will occur, its direction, and the possible destination of the last memory.

Only one data word is stored and operated on by an accumulator. As such, it is not possible to compare two data quantities (for example to test if they are the same) or to combine two data quantities (for example, add them together if they represent numbers). An arithmetic–logic unit (ALU) indicated for the elementary computer of Figure 1.35 is required for these type of operations. The operation of the ALU, which shares a common data bus with an accumulator as well as other registers, is determined by the content of the instruction register through the controller. The ALU contains additional registers that it can use, for example,

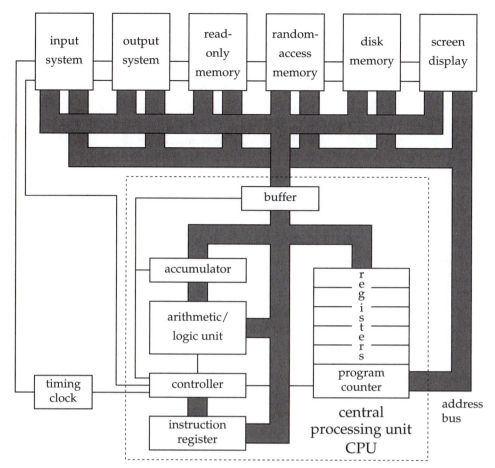

Figure 1.36: An elementary computer. For small computers, the central processing unit is on a single integrated circuit, a microprocessor. The data bus is that connected to the buffer of the central processing unit, the CPU.

when adding quantities as well as registers that it can set according to certain results called flags. Furthermore, as a result of the data bus, the ALU may access systems external to the processing unit as well as registers associated with other memory systems of the computer. This unit represents a significant step in logic complexity over that of an accumulator register.

The information processing capability of the computer is achieved by its central processing unit (CPU), albeit additional processing is generally performed by the individual systems shown at the top of Figure 1.36. In addition to the data bus, the computer of Figure 1.36 has an address bus. Each register that may write to the data bus has a unique address, an address represented by a set of logic levels of the individual lines making up the address bus. It is in this fashion that the CPU communicates with the other logic systems of a computer – for example, a keyboard or a screen display.

Of importance is the programmable nature of a computer. When the elementary computer of Figure 1.36 is running a program, instructions are transferred, one-by-one, to the instruction register from a memory of the computer. At each step, the program counter is incremented so that it will address the next instruction. An instruction could call for reading a data word that might be at the next memory location, or, alternatively, might be at some other location. For the latter case, a new address, specified by the instruction, is transferred to the address bus. If it is desired, for example, to add two quantities, these quantities would first need to be obtained, one transferred to the accumulator and the other to the ALU. Following the instructions that accomplish these moves, an instruction to perform the add operation would be required. At the completion of the addition, the sum that might be in the accumulator register would need to be transferred to an external memory location unless it would be used for the next operation.

An extremely elementary set of instructions, that is, a machine language program, is required by the CPU. These instructions might be generated from an assembly language program providing a set of detailed symbolic instructions, whereas memory locations and the actual machine code are generated by an assembler program. Alternatively, a higher-level program might be used which, in turn, would generate the machine code. Even the simplest operations, it will be noted, tend to require a lengthy sequence of elementary machine instructions. It is only because of the rapid response of modern processing units that complex computational procedures can be realized.

MEMORIES

Flip-flop circuits using vacuum tubes, similar to the circuit of Figure 1.32 using MOSFET devices, were used for the internal memory of the CPUs of early electronic computers. Owing to the cost and complexity of vacuum tube circuits, another type of memory circuit was required for the memory arrays of these computers. These circuits included crystal-excited mercury delay tubes and special cathode-ray tube storage units (costs as high as $1/ bit were common). While working on the Whirlwind computer at MIT, Jay W. Forrester put forth the idea of using a three-dimensional system of ferrite core elements, which became the standard for computers produced during the 1960s (Forrester 1951).

A ferrite core memory element with two windings is depicted in Figure 1.37. Its operation depends upon the magnetic material of the core having a magnetization curve (magnetic flux density B versus magnetizing force H) with a hysteresis loop. Because the magnetizing force H is proportional to the current i, the flux density B versus the current i displays a similar hysteresis characteristic. If, for the characteristic shown, the current is reduced to zero after having had a value of I_0 or $-I_0$, the resultant magnetic flux density will be B_0 or $-B_0$, respectively. This magnetic retention, the same principle on which a permanent magnet is based, gives the core its memory capability. One direction of

Figure 1.37: A magnetic ferrite core memory element.

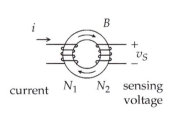
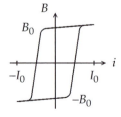

the magnetic flux density, say B_0, may be associated with a Boolean logic value of **1** and the other direction, $-B_0$, with **0**.

The state of the core depends on the most recent value of current, either I_0 or $-I_0$ (a **1** or **0** if a Boolean designation is used for the current). In contrast to most other memory elements, including commonly used semiconductor memories, a magnetic core memory will retain its data when the computer is shut off. A voltage-sensing winding is used to determine the magnetization of the core, that is, its logic state. The winding's induced voltage v_S depends on the time rate of change of the winding's flux linkages. For a nonchanging current, including i remaining zero, the voltage is zero regardless of the value of the magnetic flux density. For a voltage-sensing winding with N_2 turns, the flux linkage is $N_2\lambda_2$, where λ_2 is the total flux.

$$v_S = N_2 \frac{d\lambda_2}{dt} \tag{1.15}$$

The flux linkage λ_2 is the surface integral of the flux density B over the cross-sectional area of the core:

$$\lambda_2 = N_2 \int B \, dA \approx B A_{\text{cross section}} \tag{1.16}$$

If the flux density can be approximated as being uniform over the cross section of the toroid, the flux is simply B multiplied by the area $A_{\text{cross section}}$. Because the voltage depends on a time derivative, it is necessary to interrogate the memory with a current pulse to ascertain the memory state. Consider the situation for which $B = B_0$ for $i = 0$. A current pulse of I_0 will have very little effect on the magnetic field, and hence v_S will be very small. If, on the other hand, the initial value of magnetic flux density was $-B_0$, a current pulse of I_0 will change the flux density to B_0. This results in a large change in flux density (from $-B_0$ to B_0) and hence a much larger voltage v_S. Thus the voltage, when the current is pulsed, provides the indication of the original state of the memory.

A complication arising from using a current to ascertain the original state of the memory is that the state of the memory might be changed in the process. For the positive current pulse considered, the magnetic flux is changed to B_0 if it was originally $-B_0$. Hence, if the voltage indicates that the memory was changed, it is necessary, if it is assumed to be desirable to preserve the original data, to reset the memory, that is, to apply a pulse of $-I_0$ to the current winding.

Although magnetic core memories are no longer used, magnetic storage using a moving magnetic media is still extensively used for external data storage. Magnetic tapes and disks (both flexible diskettes and hard drives) rely on the magnetization of a ferrite material (Figure 1.38). For magnetic tape systems, either a single read–write head or separate read and write heads are used. Furthermore, several parallel tracks, each having its own head or set of heads, are utilized to increase the data-recording density. A tape head consists of a soft-iron circular magnetic circuit with a small gap that is either in direct contact with the tape or separated from the tape by a very thin moving air layer. A current produces a fringing magnetic field at the gap of the magnetic circuit that magnetizes the tape. Unlike the magnetic core memory, for magnetic tape systems an interrogating

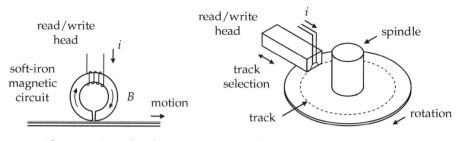

Figure 1.38: Magnetic tape and disk systems.

current is not needed to read previously recorded data. As the tape passes the tape head, a changing magnetic flux produced by the motion of the tape results in an induced voltage. This voltage can, in turn, be correlated with the stored data. Reading the data, it should be noted, does not change the recorded data.

A disk memory storage system works in a similar fashion to that of a tape system, except that data are recorded on a set of circular tracks. A floppy disk consists of a plastic disk with a thin ferrite surface layer. The spinning motion of the disk produces a changing magnetic flux, which, in turn, produces a voltage in the winding of the head. A hard-disk memory uses a set of aluminum disks mounted on a common spindle shaft. The disks have ferrite surfaces on both sides. In disk systems, data are recorded in a serial fashion, that is, bit by bit, filling a complete sector of a track. It is therefore necessary, when reading data, to read the complete data string of a sector.

Very-large-scale integrated circuits are now used for the CPU of smaller computers as well as most of a computer's peripheral tasks. The difficulty of memory storage, which limited the capacity of early computers, has been largely mitigated by integrated circuit memories. One type of memory uses an array of individual flip-flops such as that of Figure 1.32, a memory system described as a static memory. It is not only possible to read the memory without altering its state, but the memory element will remain in a particular state as long as it is adequately powered.

A much simpler memory element is possible, namely, an elementary capacitor. A capacitor, if charged to a given potential, will tend to retain its stored charge and hence retain its voltage. This principle was first used by John Atanasoff in a special-purpose early electronic computer; a memory prototype was demonstrated in 1939 (Mackintosh 1988). Atanasoff used a rotating disk memory consisting of individual capacitors connected between a common conductor at the center of the disk and individual outer contacts. The electronic memory circuit was successively connected to each capacitor as the disk rotated. As for modern integrated circuit memories using capacitor storage, a regenerating circuit was needed because, as a result of dielectric losses, a charge leakage occurred. It is necessary to read a capacitor memory element periodically and restore the capacitor's voltage to its original value.

An integrated circuit capacitor type of memory (generally designated as a dynamic memory) consists of a two-dimensional array of capacitors connected by means of a transistor to a read–write bus. For this circuit, only a single transistor is required for each memory element (for each bit of storage) – an internal transistor capacitance serves as the memory capacitor. For a flip-flop memory element, at least three transistors are required for each memory element (generally four are used). Hence, for a large memory array, a dynamic memory system requires the smallest number of transistors even though a regenerating (refresh) system is needed. As a result, memory costs per bit of storage capacity have fallen dramatically from the early costs of $1/bit (by a factor on the order of one million in inflation-adjusted dollars).

1.7 INTEGRATED CIRCUITS: SHRINKING DEVICE SIZES AND INCREASED COMPLEXITY

Since the invention of semiconductor devices and integrated circuits, the evolution of electronics has been toward ever smaller devices (Riordan and Hoddeson 1997). As a result, the number of devices formed on a single integrated circuit has been steadily increasing, whereas the overall average cost of a typical integrated circuit has not changed greatly. Hence, the cost of a single transistor of an integrated circuit has been steadily declining – its present cost can be expressed in millionths of a cent.

The first integrated circuits, consisting of a few semiconductor devices and resistors, performed basic functions such as that of a logic gate or an elementary amplifier. The next step in the complexity of integrated circuits resulted in circuits that accomplished basic small-scale system functions such as a logic adder or an operational amplifier. Now ultra-large-scale integrated circuits with device counts in the millions, which include dynamic random-access memories and microprocessors, are common. Besides the possibility being able to put more devices on a single integrated circuit, greatly improved electrical characteristics have been obtained as a result of the smaller device sizes. With smaller devices and shorter interconnecting leads, circuit capacitances are reduced. Therefore, smaller charge quantities need to be moved to produce a desired voltage change, thereby reducing device currents. As a result, not only is the power dissipated by individual devices reduced, but logic circuits can be operated much more rapidly.

When commercially produced integrated circuits were being introduced in 1965, Gordon Moore, then of Fairchild Semiconductor, predicted a very rapid increase in the number of devices that could be crammed onto a single integrated circuit (Moore 1965). At that time, the largest integrated circuit had only 64 components (a combination of diodes, transistors, and resistors) with each transistor, on the average, occupying an area of approximately $50 \times 50 \ \mu$m. By today's standards, these integrated circuit transistors would be described as monstrous. On the basis of the scant record of the previous 6 years, from a single planar transistor in 1959 to integrated circuits with 64 components in 1965, Moore predicted that the number of components per integrated circuit would double each year for at least the next 10 years. This prediction implied that integrated circuits with 2^{10} (1024) times as many devices would be possible in 1975.

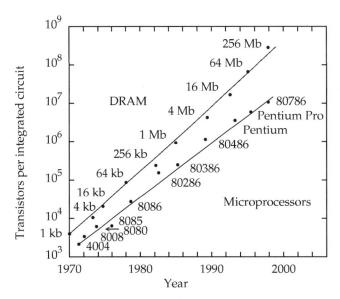

Figure 1.39: Transistors per integrated circuit of dynamic random access memories (DRAM) and Intel microprocessors.

Indeed, an integrated circuit with approximately 65,000 transistors was produced in 1975, proving Moore correct.

The planar process of producing transistors made it possible to form interconnections by evaporating metal-film conductors onto the semiconductor wafer. Another major improvement was the use of photographic techniques to place intricate masking patterns on the semiconductor so that after a mask is formed, dopants can be introduced with a diffusion process. A succession of masks and appropriate diffusion steps are used to form semiconductor devices and other components. These processes have come to be characterized by extremely high levels of accuracy and reliability. Concurrently, as the result of automated techniques, the cost per individual integrated circuit remained small even with the increased levels of complexity.

In 1975, Moore predicted that the number of devices per integrated circuit would continue to increase but at a somewhat slower rate with a doubling time of approximately 18 months (Moore 1976; Noyce 1977).* As can be seen in Figure 1.39, this has been nearly the case, and the doubling times since 1975 for dynamic random-access memories as well as for microprocessors have been just under 2 years (Schaller 1997). In 1998, memory chips with approximately 269 million transistors, one for each stored bit, and microprocessors with 10 million transistors, were available.

How long will this trend continue? When will the doubling every 1-to-2 years of the number of transistors (modern digital integrated circuits tend to use only transistors) on an integrated circuit end? It seems to be generally accepted that the present fabrication technologies will be extended so that this trend will continue to about 2010. However, experts do not agree on the extent that this trend can be pushed beyond 2010.

For a particular technology, physical barriers limit the size to which devices can be reduced (Service 1996). Optical lithography is presently being used for ultra-large-scale integrated circuits. This process is utilized to form a pattern or mask on a sheet of glass with a thin coating of light-absorbing metal. Ultraviolet light is then shone through the mask and focused with a series of lenses onto the semiconductor. The surface of the semiconductor is coated with a photoresistant material, and a change in its chemical structure occurs for those portions

---

* By this time, Moore, one of the founders of Intel, was its president.

struck by light photons. By means of a chemical process, either the chemically altered or original photoresistant surface material is removed so that what remains forms a mask for doping or other fabrication steps used to form devices and interconnections. The size of the semiconductor features that can be obtained in this fashion is comparable to the wavelength of the light source used. With short-wavelength ultraviolet light, features as small as 0.13 μm are expected to be possible. Unfortunately, this may be the limit for optical lithography because, for the wavelengths needed to produce smaller features, conventional quartz lenses can no longer be used. Quartz lenses absorb, rather than transmit photons with shorter wavelengths. Although X-ray lithography utilizing photons with considerably shorter wavelengths is a possibility, producing the patterning masks presents a formidable challenge. One problem associated with producing ultra-large-scale integrated circuits that have millions of extremely small transistors is the capital cost of the fabrication facilities. Because the cost of these facilities increases with the complexity of the integrated circuits, this is likely to be another limiting factor.

Even if new technologies are developed that make it possible to fabricate devices with smaller dimensions, other fundamental limitations can be foreseen (Keyes 1992). The behavior of transistors is premised on electrons behaving in a well-ordered fashion. When devices are large compared with the quantum-mechanical wavelength of the electrons involved, classical concepts of charge, current, and voltage apply. As dimensions are reduced, a quantum "fuzziness" occurs because electrons are able to tunnel through barriers that are intended to confine their movement. These effects tend to occur when device dimensions shrink to approximately 10 nm (an order of magnitude smaller then the limit imposed by optical lithography). This limitation for conventional devices, however, opens new device possibilities that depend on only a single or possibly a few electrons. The quantum-mechanical properties of an electron can be exploited with a conducting island that is sufficiently small to permit only a single electron to occupy an island (Ferry and Goodnick 1997; Glanz 1997; McEuen 1997; Service 1997). The repulsive force of an electron inhibits a second electron from sharing the island. Transistor-like devices have been proposed in which a semiconductor island, a *quantum dot*, is used to control the current of a device. Single-electron devices, should they prove practical, will offer a fabricating challenge in producing devices with nanometer dimensions. If this fabrication hurdle is overcome, integrated circuits with possibly trillions of devices on a single chip will be possible.

REFERENCES

Abramson, A. (1987). *The History of Television, 1880 to 1941*. Jefferson, NC: McFarland & Co.
Aitken, H. G. J. (1976). *Syntony and Spark – The Origins of Radio*. New York: John Wiley & Sons.
Anastassiou, D. (1994). Digital television. *Proceedings of the IEEE*, **82**, 4, 510–19.
Baylin, F. and Gale, B. (1986). *Satellites Today – The Guide to Satellite Television* (2nd ed.). Columbus, OH: Howard W. Sams & Co. and Universal Electronics.

Bell System, The (1963). The Telstar experiment. *The Bell System Technical Journal*, **42**, 4, parts 1–3, entire journal.

Boot, H. A. H. and Randall, J. T. (1976). Historical notes on the cavity magnetron. *IEEE Transactions on Electron Devices*, **ED-23**, 7, 724–9.

Bruce, R. V. (1973). *Bell: Alexander Graham Bell and the Conquest of Solitude*. Ithaca, NY: Cornell University Press.

Clarke, A. C. (1945). Extra-terrestrial relays. *Wireless World*, **51**, 305–8.

Douglas, A. (1981). The crystal detector. *IEEE Spectrum*, **18**, 4, 64–7.

Easton, A. T. and Easton, S. (1988). *The Complete Sourcebook of Home Satellite TV*. New York: Putnam Publishing.

Elbert, B. R. (1997). *The Satellite Communication Applications Handbook*. Boston: Artech House.

Fagen, M. D. (Ed.) (1975). *A History of Engineering and Science in the Bell System*. Murray Hill, NJ: Bell Telephone Laboratories.

Ferry, D. K. and Goodnick, S. M. (1997). *Transport in Nanostructures*. New York: Cambridge University Press.

Fink, D. G. (1952). *Television Engineering* (2d ed.). New York: McGraw–Hill Book Co.

Fisher, D. E. and Fisher, M. J. (1996). *Tube: The Invention of Television*. Washington: Counterpoint.

Forrester, J. W. (1951). Digital information storage in three dimensions using magnetic cores. *Journal of Applied Physics*, **22**, 1, 44–8.

Fthenakis, E. (1984). *Manual of Satellite Communications*. New York: McGraw–Hill Book Co.

Ginzton, E. L. (1976). The $100 idea. *IEEE Transactions on Electron Devices*, **ED-23**, 7, 714–23.

Glanz, J. (1997). Quantum cells make a bid to outshrink transistors. *Science*, **277**, 5328, 898–9.

Goldstine, H. H. (1972). *The Computer: From Pascal To Von Neumann*. Princeton, N J: Princeton University Press.

Hashimoto, Y., Yamamoto, M., and Asaida, T. (1995). Cameras and display systems. *Proceedings of the IEEE*, **83**, 7, 1032–43.

James, R. J. (1989). A history of radar. *IEE Review*, **35**, 9, 343–9.

Jolly, W. P. (1972). *Marconi*. New York: Stein and Day.

Jolly, W. P. (1975). *Sir Oliver Lodge*. Rutherford, NJ: Fairleigh Dickinson University Press.

Keyes, R. W. (1992). The future of solid-state electronics. *Physics Today*, **45**, 8, 42–8.

Kilby, J. S. (1976). Invention of the Integrated Circuit. *IEEE Transactions on Electron Devices*, **ED-23**, 7, 648–54.

Kurylo, F. and Süsskind, C. (1981). *Ferdinand Braun: A Life of the Nobel Prize Winner and Inventor of the Cathode-Ray Oscilloscope*. Cambridge, MA: MIT Press.

Lewis, T. (1991). *Empire of the Air: The Men Who Made Radio*. New York: Edward Burlingame Books (HarperCollins Publishers).

Mackintosh, A. R. (1988). Dr. Atanasoff's computer. *Scientific American*, **259**, 2, 90–6.

Masini, G. (1995). *Marconi*. New York: Marsilo Publishers.

McEuen, P. L. (1997). Artificial atoms: New boxes for electrons. *Science*, **278**, 5344, 1729–30.

McNicol, D. (1946). *Radio's Conquest of Space: The Experimental Rise in Radio Communication*. New York: Murray Hill Books.

Meindl, J. D. (1977). Microelectronic circuit elements. *Scientific American*, **237**, 3, 70–81.

Moore, G. E. (1965). Cramming more components onto integrated circuits. *Electronics*, **38**, 18, 114–17.

Moore, G. E. (1976). Microprocessors and integrated electronic technology. *Proceedings of the IEEE*, **64**, 6, 837–41.

Nahin, P. J. (1988). *Oliver Heaviside: Sage in Solitude*. New York: IEEE Press.

Nahin, P. J. (1990). Oliver Heaviside. *Scientific American*, **262**, 6, 122–29.

Netravali, A. N. and Haskell, B. G. (1995). *Digital Pictures: Representation, Compression, and Standards*. New York: Plenum Press.

Netravali, A. and Lippman, A. (1995). Digital television: A perspective. *Proceedings of the IEEE*, **83**, 6, 834–42.

Noyce, R. N. (1977). Microelectronics. *Scientific American*, **237**, 3, 63–9.

O'Neill, E. F. (Ed.) (1985). *A History of Engineering and Science in the Bell System: Transmission Technology (1925–1975)*. Murray Hill, NJ: AT&T Bell Laboratories.

Page, R. M. (1962). *The Origin of Radar*. Garden City, NY: Anchor Books.

Pierce, J. R. (1950). Electronics. *Scientific American*, **183**, 4, 30–9.

Pierce, J. R. (1955). Orbital radio relays. *Jet Propulsion*, **25**, 4, 153–7.

Pierce, J. R. (1968). *The Beginnings of Satellite Communications*. San Francisco, CA: San Francisco Press.

Pupin, M. I. (1926). Fifty years' progress in electrical communications. *Science* **64**, 1670, 631–8.

Riordan, M. and Hoddeson, L. (1997). *Crystal Fire: The Birth of the Information Age*. New York: W. W. Norton & Company.

Schäfer, R. and Sikora, T. (1995). Digital video coding standards and their role in video communications. *Proceedings of the IEEE*, **83**, 6, 907–24.

Schaller, R. R. (1997). Moore's law: Past present, and future. *IEEE Spectrum*, **36**, 6, 52–9.

Service, R. F. (1996). Can chip devices keep shrinking? *Science*, **274**, 5294, 1834–6.

Service, R. F. (1997). Making single electrons compute. *Science*, **275**, 5298, 303–4.

Shannon, C. E. (1938). A symbolic analysis of relay and switching circuits. *Transactions of the American Institute of Electrical Engineers*, **57**, 713–23.

Sharlin, H. I. (1963). *The Making of the Electrical Age*. London: Abelard–Schuman.

Shiers, G. (1969). The first electron tube. *Scientific American*, **220**, 3, 104–12.

Shurkin, J. (1984). *Engines of the Mind: A History of the Computer*. New York: W. W. Norton & Co.

Slater, R. (1987). *Portraits in Silicon*. Cambridge, MA: MIT Press.

Solomon, L. (1962). *Telstar: Communication Breakthrough by Satellite*. New York: McGraw–Hill Book Co.

Stephenson, D. J. (1991). *Newsnes Guide to Satellite TV* (2d ed.). Oxford: Newsnes (Butterworth–Heinemann, Ltd.).

Süsskind, C. (1966). The origin of the term "electronics." *IEEE Spectrum*, **3**, 5, 72–9.

Swords, S. S. (1986). *Technical History of the Beginnings of Radar*. London: Peter Peregrinus, Ltd.

Weimer, P. K. (1976). A historical review of the development of television pickup devices (1930–1976). *IEEE Transactions on Electron Devices*, **ED-23**, 7, 739–52.

Taub, H. (1982). *Digital Circuits and Microprocessors*. New York: McGraw-Hill Book Co.

Zworykin, V. K. and Morton, G. A. (1940). *Television – The Electronics of Image Transmission*. New York: John Wiley & Sons.

PROBLEMS

1.1 Consider the plate characteristic of Figure 1.5. Obtain the transfer characteristic of Figure 1.6 for plate currents of 5 and 15 mA.

1.2 Assume that the plate voltage of the triode of Figure 1.5 is supplied by a constant voltage battery. Determine the current transfer characteristic (a graph similar to that of Figure 1.6 except that i_P is the dependent variable) for plate potentials of 100, 150, and 200 V.

1.3 The drain characteristic of a typical MOSFET device is given in Figure 1.7. Determine the voltage transfer characteristic (similar to that of the triode vacuum tube) for drain currents of 2, 4, and 6 mA.

1.4 Assume that the drain voltage of the MOSFET device of Figure 1.7 is held constant. Determine the current transfer characteristic for drain potentials of 2, 6, and 10 V.

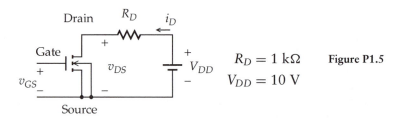

$R_D = 1 \text{ k}\Omega$ **Figure P1.5**

$V_{DD} = 10 \text{ V}$

1.5 Consider the case for which the drain circuit of the MOSFET device of Figure 1.7 consists of a resistor R_D and a battery V_{DD} (Figure P1.5). As a result of the load circuit (R_D and V_{DD}), $v_{DS} = V_{DD} - i_D R_D$. Alternatively, $i_D = (V_{DD} - v_{DS})/R_D$. If the straight line corresponding to i_D is drawn on the drain characteristic of Figure 1.7, the drain current i_D and drain-to-source voltage v_{DS} may be obtained for any particular value of gate-to-source voltage, v_{GS}. Obtain curves of i_D and v_{DS} versus v_{GS} for the circuit.

1.6 Repeat Problem 1.5 for $V_{DD} = 8$ V and $R_D = 2$ kΩ.

1.7 A parallel LC circuit, similar to that of Figure 1.8, is frequently used to tune an AM radio's input circuit. Consider the case in which a fixed inductor is used along with a variable capacitance that has a minimum capacitance of 50 pF (this includes the unavoidable capacitance of the circuit to which it is connected).

a) What is the inductance required if the circuit is to be resonant at the upper end of the broadcast band (1600 kHz) with the minimum capacitance?

b) What is the maximum value of capacitance required if the circuit is to be able to tune to the lowest frequency of the broadcast band (550 kHz)?

1.8 As in Problem 1.7, a parallel LC circuit is to be used for an AM radio. A circuit with a maximum capacitance of 150 pF is to be used. Determine the value of L and the minimum value of capacitance required to tune the broadcast band of 550 to 1600 kHz.

1.9 The input circuit of an FM receiver is tuned to a frequency of 100 MHz. A resonant circuit similar to that of Figure 1.8 is used that has a capacitance of 30 pF.

a) What is the inductance required for the circuit?
b) What are the minimum and maximum values of capacitance required for tuning an FM band of 88 to 108 MHz?
c) An approximate formula for a simple air-core inductor is $L(\mu H) = 0.04n^2r$, where r is the radius expressed in centimeters. What is the approximate number of turns of wire n required for a coil with a radius of 0.25 cm?

1.10 Verify that Eq. (1.4) is indeed a valid solution of Eq. (1.2).

1.11 Obtain a solution of Eq. (1.2) that has an initial voltage ($t = 0$) of V_m and an initial derivative that is zero.

1.12 Assume that Eq. (1.4) is the desired solution of Eq. (1.2). Obtain expressions for the capacitive and inductive stored energies e_C and e_L.

1.13 Consider the telegraph system of Figure 1.11 and assume that a current of 2 mA (I_{on}) is required to activate the sounder. However, once activated, the sounder will remain on until its curent falls below 1 mA (I_{off}). Assume a sounder resistance of 1 kΩ, a wire resistance of 20 Ω/km, a capacitance of 5 nF/km, and a line length of 200 km. Ignore the ground resistance.

a) Determine the potential of the battery V_B required to achieve a solenoid current of $2I_{on}$ after the key has been closed for a long time interval.
b) An estimate of the dynamic behavior of the circuit may be obtained by assuming that a single capacitor with a capacitance equal to that of the entire length of line is located at its center and that half of the total line resistance is located on each side of the center. Use the battery potential determined in part (a), to estimate the time after the key is closed that the sounder is activated (when its current reaches I_{on}). Assume the line was initially uncharged.
c) Estimate the time required for the solenoid current to reach 90 percent of its final value, that is, $1.8I_{on}$.
d) Assume the key has been closed for a very long time and that it is suddenly opened. Estimate the time required for the sounder to be deactivated, that is, for its current to fall to I_{off}. (Note: These times set a limit to how fast code, that is, information can be sent.)

1.14 Repeat Problem 1.13 but assume a ground resistance equal to that of the telegraph line.

1.15 Assume a telegraph line pair is used for a telegraph system (as for the telephone line pair of Figure 1.12). Other than for a larger capacitance, a wire-to-wire capacitance of 20 nF/km, assume that the parameters of Problem 1.13 apply. Repeat Problem 1.13 for this condition.

1.16 Repeat Problem 1.15 for lengths of 100 and 400 km. Are the times obtained linearly related to the length of the telegraph line?

1.17 Submarine cables using coaxial conductors are characterized by a fairly large capacitance. Repeat Problem 1.13 for a center-wire resistance of 2 Ω/km and a capacitance of 0.1 μF/km. The resistance of the outer conductor may be ignored.

1.18 Repeat Problem 1.17 for the 2,000-mile trans-Atlantic cable.

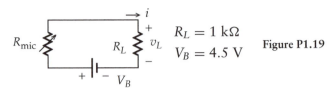

Figure P1.19

1.19 Consider the carbon granule microphone circuit of Figure P1.19. The resistor R_L represents the other parts of a telephone system, the earphones, the telephone lines, and the other microphone. Assume that as a result of an external sound wave, the following occurs for the resistance of the microphone:

$$R_{mic} = R_0 + R_1 \sin \omega t$$
$$R_0 = 100 \ \Omega, \qquad R_1 = 10 \ \Omega$$

a) Determine expressions for $i(t)$ and $v_L(t)$.
b) Show that $v_L(t) \approx V_0 + V_1 \sin \omega t$.
c) Determine numerical values of V_0 and V_1.
d) The approximate average signal power delivered to the load resistor R_L is $V_1^2/(2 R_L)$. Determine this quantity.

1.20 Repeat Problem 1.19 for $R_L = 100 \ \Omega$.

1.21 Using the circuit and parameters of Problem 1.19, determine the value of R_L for which the approximate average signal power delivered to the load resistor is a maximum. What is the value of the power for this condition?

1.22 A particular earphone has been found to have a nominal resistance of 50 Ω. The following subjective data were obtained for its response for a sinusoidal voltage with a frequency of 800 Hz:

Audio signal	Voltage (peak-to-peak)
perceptible	1 mV
soft	20 mV
medium	200 mV
loud	1.5 V

What are the electrical powers associated with each of the responses?

1.23 A small loudspeaker with a nominal resistance of 8 Ω, when excited by a sinusoidal voltage with a frequency of 500 Hz, was found to produce the following subjective response data (at a distance of 1 m):

Audio signal	Voltage (peak-to-peak)
perceptible	2 mV
soft	50 mV
medium	400 mV
loud	6 V

What are the electrical powers associated with each of the responses?

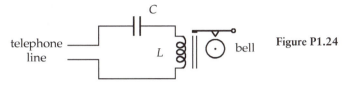

Figure P1.24

1.24 A large-amplitude sinusoidal signal is generally used for "ringing" telephones. Consider a ringing frequency of 20 Hz. A series LC circuit (preelectronic telephones) is used in which the inductance of the electromagnet of the bell is the inductor of a resonant circuit, as shown in Figure P1.24.

a) Consider the case for which $L = 5$ H. What is the value of C that results in the maximum magnitude of ringer current?

b) When the receiver is removed, this circuit is disconnected from the line. However, the ringer circuit of other telephones (extension telephones or telephones sharing a party line) are still connected to the line. What is the complex impedance of the ringer circuit for an audio signal with a frequency of 1 kHz? Assume an inductor resistance of 250 Ω.

1.25 Consider the video scanning system of Figure 1.18. Sketches of the intensity signal $b(t)$ are desired for the following test patterns:

a) A centered black vertical bar with a width of approximately that of the black square of Figure 1.18. The remainder of the screen is white.

b) A similar centered horizontal black bar.

c) A pattern with both bars.

1.26 Repeat Problem 1.25 with the white and black regions interchanged.

1.27 Suppose that the display of Figure 1.18 consists of two diagonal black bars (corner-to-corner) with widths of approximately that of the center black square of the original figure. Sketch the intensity signal $b(t)$ for this condition.

1.28 The North American TV system uses a horizontal scanning rate of 15,750 lines/s. Assume that 90 percent of the horizontal scan period is associated with the active display and the remaining 10 percent corresponds to the off-screen portion and the retrace. Consider a test pattern of 10 equal-width vertical black bars separated by equal-width white spaces.

a) Sketch the video signal for a single horizontal sweep.

b) What is the time interval of the active display?

c) What are the period and frequency of the square-wave video signal that produces the vertical bars?

d) Suppose a sinusoidal signal were to be used with the same peak-to-peak amplitude instead of the square-wave signal to produce the pattern. How would the pattern change?

1.29 Repeat Problem 1.28 for 100 vertical black bars.

1.30 The vertical resolution of a television display is limited by the number of scanning lines used. For the North American system of 525 lines, this limits the display to approximately 250 horizontal black lines (separated by white lines). For an aspect ratio of 4:3 (33 percent wider than high display), (4/3)(250) vertical lines would correspond to a comparable horizontal resolution. If it is assumed that 90 percent of the horizontal scan period is associated with an active display, determine the period and frequency of the corresponding portion of the video signal. Why would the display appear nearly unchanged if a sinusoidal rather than a square-wave voltage were to be used for the video signal? Because a horizontal resolution greater than the vertical resolution would not appreciably contribute to the quality of the picture, the frequency determined for the display is the upper frequency response required by a TV system.

1.31 The European TV system uses a vertical scanning frequency of 50 Hz (the same frequency as that of the European electric power system). For 625 interlaced lines per frame, this results in a horizontal scanning frequency of 15,625 lines/s. With appropriate assumptions, repeat Problem 1.30 for this system.

1.32 A half-wavelength dipole is commonly used for a transmitting antenna. Determine the length of the dipole for each of the following:

AM broadcast	1.0 MHz
Shortwave	10 MHz
Citizens band	27 MHz
FM broadcast	100 MHz
VHF Television – channel 2	57 MHz
– channel 13	213 MHz
UHF Television – channel 14	473 MHz
– channel 69	803 MHz
Satellite TV – channel 1	3720 MHz
X-band radar	10,000 MHz

For an AM broadcast transmitter, a vertical quarter-wavelength antenna is common (a ground "image" provides the other quarter-wavelength section). A parabolic reflector is often used to concentrate the radiation of microwave antennas.

1.33 To generate high-frequency signals, it is generally necessary that the response time of an electronic device be small compared with the period of the signal. Assume a response time that is 10 percent of the period that is required. What are the required response times for the signals of Problem 1.32?

1.34 Capacitance is unavoidable in all circuits. For an electronic circuit using discrete devices, a stray capacitance of 10 pF from a node to the common node of the circuit is typical. Determine the reactance introduced by a 10-pF capacitance for each of the following signals:

Audio signal – middle C	262 Hz
– upper limit	20 kHz
AM broadcast carrier	1.0 MHz
Video signal	4.5 MHz
Citzens band carrier	27 MHz
FM broadcast carrier	100 MHz

What is the rms value of the current for an rms voltage of 1.0 V?

1.35 Repeat Problem 1.34 for an integrated circuit with a stray capacitance of only 0.1 pF.

1.36 Consider the satellite repeater of Figure 1.27 with a "global beam." Assume that the satellite uses a transmitter with an output power of 10 W, that the power is radiated uniformly within the beam, and that it is zero outside the beam.

a) What is the incident power density at the equator of the earth?
b) What is the incident power density at 45° north or south latitude?

1.37 A relay circuit (such as that of Figure 1.29) has two inputs, A and B, and an output of C. The output is to be V_1 for either input equal to V_1 and zero if both inputs are zero or V_1 (an exclusive **OR** operation). Design a relay system to accomplish this. Assume that relays with normally *on* as well as normally *off* contacts are available.

1.38 The exclusive **OR** function of the relay circuit of Problem 1.37 is that used for controlling a single light (or set of lights) with two switches (generally referred to as a three-way circuit). Design a circuit using single-pole, double-throw switches to accomplish this. No relays are required.

1.39 Repeat Problem 1.37 for a comparator function; that is, the output is V_1 if both inputs are the same.

1.40 A relay logic system has three inputs, A, B, and C and a single output D. The output is to be high, V_1, if, and only if, a single input is high. Design a circuit to achieve this. Relays with multiple contacts are available.

1.41 Repeat Problem 1.40 for an output that is to be V_1 if, and only if, two inputs are equal to V_1.

1.42 A relay depends on a magnetic field to move its armature, thus moving its contacts. Assume that a current of I_{on} is required to achieve this. Once the armature is activated, an iron magnetic path exists and a much smaller curent will continue to keep the armature activated. Consider a relay

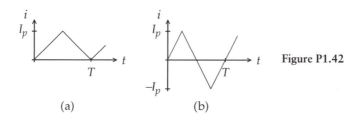

(a)   (b)   Figure P1.42

that is turned on for a current of I_{on} (or greater) but remains on for a current greater than I_{off}. For this relay, $I_{off} = 0.5 I_{on}$. Suppose a periodic triangular wave is used for an exciting current and that $I_p = 2 I_{on}$. Assume that the relay responds essentially instantly to its current i.

a) Determine the fraction of a period for which the relay is activated for the current of Figure P1.42(a).

b) Determine the response of the relay for the current of Figure P1.42(b). Assume the relay has a response that depends on the magnitude of its current.

1.43 Repeat Problem 1.42 for sinusoidal signals that have the same peak amplitudes I_p.

COMPUTER SIMULATIONS

C1.1 Consider the telegraph system of Problem 1.13 in which R is the total resistance of the wire and C is the total capacitance of the line. Instead of approximating the system with two resistors of $R/2$ and one capacitor of C at the center of the line, use five T-type circuits to approximate the line (Figure C1.1). Obtain a transient solution for the transmission line and sounder. Assume an input step function voltage with a peak value of 20 V. Obtain on a single graph the voltages across each capacitor and the sounder. What are the times necessary for these voltages to reach 50 percent of their final values? (Note: A transient time increment of 10 μs and an overall duration of 5 ms should be adequate.)

Figure C1.1

C1.2 The model of the transmission line used for Simulation C1.1 ignored the effect of the inductance of the line. As a result of magnetic field lines that loop the wire, an actual line has a series inductance that affects its behavior. Assume the transmission line of Simulation C1.1 has a series inductance of 10 mH/km. The behavior of this line can be modeled by adding one-tenth of the total inductance of the line to the $R/10$ resistance of each T section of Figure C1.1. Repeat Simulation C1.1 for a series inductance.

C1.3   Consider the resonant circuit of Figure 1.8 with $L = 100\ \mu H$, $C = 253$ pF, and $R = 20\ k\Omega$. A circuit with these component values has a resonant frequency f_0 of approximately 1 MHz ($1/2\pi\sqrt{LC}$). The behavior of this circuit for a single exciting current pulse $i_S(t)$ that has an amplitude of 1.0 mA and a duration of 0.5 μs is to be determined. A transient time increment of 10 ns is appropriate for a solution. An overall duration of 10 periods, that is, 10 μs, will show the decay of the oscillation amplitude. Obtain a graph of $v(t)$. Also, on a second graph, obtain plots of the energy stored by the inductor, the energy stored by the capacitor, and the sum of these two quantities.

C1.4   For a particular application, it is found that an audio signal can be approximated by a voltage source consisting of the following four sinusoidal terms:

$$v_S(t) = V_1 \sin 2\pi f_1 t + V_2 \sin 2\pi f_2 t + V_3 \sin 2\pi f_3 t + V_4 \sin 2\pi f_4 t$$

$$V_1 = 1\ V, \qquad V_2 = 1\ V, \qquad V_3 = 0.5\ V, \qquad V_4 = 0.2\ V$$

$$f_1 = 400\ Hz, \quad f_2 = 1\ kHz, \quad f_3 = 2.5\ kHz, \quad f_4 = 3.5\ kHz$$

This voltage source has a Thévenin equivalent resistance R_T of 100 kΩ and is connected to a load capacitance C_L of 2 nF. As a result of the resistance and capacitance, the voltage across the capacitance $v_C(t)$ is a distorted and delayed version of $v_S(t)$.

a) Obtain $v_C(t)$ for a time duration of 10 ms. To ensure sufficient resolution of the output voltage, a transient time step of 5 μs or less is recommended.

b) Estimate the time by which $v_C(t)$ is delayed from $v_S(t)$. From a plot of both quantities, the delay time can be estimated by averaging the time difference of the zero crossings.

c) Using a trial and error method, determine the maximum capacitance C_L yielding a voltage $v_C(t)$ that looks reasonably like $v_S(t)$; a subjective decision will be necessary.

C1.5   A logic **OR** gate using semiconductor diodes is indicated in Figure 1.30. For a SPICE simulation, diodes with default parameters may be used. This requires two SPICE circuit lines:

```
D1   5   6   MYDIODE
.MODEL   MYDIODE   D
```

The first line specifies a diode, a D-type element. The diode indicated points from the first node number, 5, to the second node number, 6, and has been assigned the arbitrary model name MYDIODE. The .MODEL statement references this model name and specifies a diode model D. All diodes of the circuit can use the same model name; only a single .MODEL statement is then necessary. Determine the dependence of v_C on the one input voltage v_A ($0 \geq v_A \leq 5$ V) for values of 0, 2.5, and 5 V for v_B. Assume $R = 1$ kΩ. The response for all three values of v_B may be obtained simultaneously by using three independent circuits, each with a different constant voltage source for v_B.

C1.6 Repeat Simulation C1.5 for the diode **AND** gate of Figure 1.30. The voltage V_1 is a constant voltage of 5 V.

CHAPTER TWO

THE SEMICONDUCTOR JUNCTION DIODE: THE BASIS OF MODERN ELECTRONICS

Although the invention of the transistor at Bell Telephone Laboratories in 1947 was destined to revolutionize the field of electronics, the initial response to the public announcement on June 30, 1948, was anything but overwhelming. The *New York Times* carried the news in its daily column "The News of Radio," which dealt with new radio shows. Near the end of the column appeared the following (*New York Times* 1948):

> A device called a transistor which has several applications in radio where a vacuum tube ordinarily is employed, was demonstrated yesterday at Bell Telephone Laboratories, 463 West Street, where it was invented.

Three additional brief paragraphs completed the announcement. A more fitting debut of the transistor was provided by three letters to *Physical Review* directed toward the scientific and technical community (Bardeen and Brattain 1948; Brattain and Bardeen 1948; Shockley and Pearson 1948). "The Transistor – A Crystal Triode" appeared as the cover article of the September 1948 issue of *Electronics* (Fink and Rockett 1948). Although the terms *crystal triode* and *semiconductor triode* were used to describe the transistor, by way of analogy with the vacuum tube triode (Figure 1.4), the name *transistor* has prevailed. The term *transistor*, that is <u>trans</u>fer re<u>sistor</u>, is attributed to John R. Pierce (already mentioned in relation to satellite communication) (Shockley 1976). Thus commenced the era of solid-state electronics in which transistors not only play an important role but within which entirely new electronic devices have been developed.

The roots of the solid-state electronics era reach back to the 1920s with the development of quantum theory (Weiner 1973). A. H. Wilson developed a theoretical model for semiconductors in which he introduced the concept of a "hole," that is, an absence of a valence electron (Wilson 1931). William Shockley, who, along with John Bardeen and W. H. Brattain shared the 1956 Nobel Prize in physics for the invention of the transistor, published the seminal treatise on semiconductors, *Electrons and Holes* (Shockley 1950), which provides an overview of the theory. The development of the transistor also depended on fabrication techniques developed in the 1930s and 1940s. An extremely difficult task was that of obtaining highly pure semiconductor samples – a purity much beyond that required for essentially any other need.

To describe the internal functioning of a transistor, an understanding of a semiconductor junction diode is necessary. A bipolar junction transistor (BJT), for example, consists of two interacting diodes, whereas a metal-oxide field-effect transistor (MOSFET) is fabricated from two noninteracting diodes. Only a limited physical description will be presented, for a rigorous theoretical development would require a book-length treatment. This physical description, however, should prove adequate for gaining an appreciation of the basic mechanisms of semiconductor devices. A more thorough subsequent study of semiconductor physics is encouraged.

2.1 ELECTRONS AND CONDUCTION: A LOOK AT ELEMENTARY PROCESSES

An understanding of electron devices requires a knowledge of the behavior of electrons, which are influenced by electric and at times magnetic fields. The movement of electrons in a vacuum determines the characteristics of vacuum tubes, whereas the movement of electrons in semiconductors, a process considerably more complex than in a vacuum, determines the characteristics of semiconductor devices. In addition, an understanding of the conduction process for metals is necessary.

The charge of an electron, even though very minute, has been directly measured, the earliest measurement being performed by Millikan in his classic oil drop experiment of 1906. For electrons to contribute to the current of a material, they must have sufficient energy to be free from the electron valence shell of an atom (Figure 2.1). In metals, electrons have sufficient energy for this to occur at normal temperatures. On the other hand, this is not the case for insulators, which have nearly complete valence shells. Because the charge of an electron is very small, the quantity of electrons necessary to produce a significant current is extremely large. Consider, for example, a current of 1 A, that is, a charge flow of 1 C/s. If individual charges of 1.6×10^{-19} C (the magnitude of the electron's charge) are to produce a current of 1 A, the following relation, where N is the number of charges per second, must be satisfied:

$$1.6 \times 10^{-19} N = 1 \quad \text{(A)}$$

$$N = \frac{1}{1.6 \times 10^{-19}} = 6.25 \times 10^{18} \quad \text{(electrons/s)} \tag{2.1}$$

Figure 2.1: The ionization of an atom.

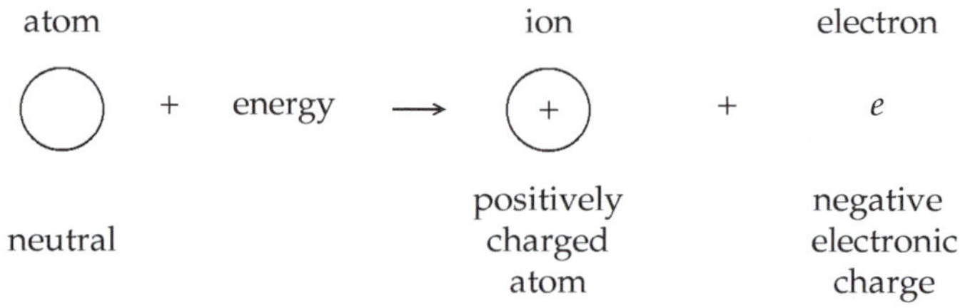

| atom | | | ion | | electron |

neutral → positively charged atom + negative electronic charge

Over one billion-billion electrons per second are thus required. Even for a current of 1 nA (10^{-9} A), a current that is about as small as may be readily measured, 6.25×10^9 electrons per second are required. Although the behavior of individual electrons of a device will be discussed, it should be kept in mind that millions and more likely billions of electrons will be involved.

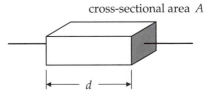

cross-sectional area A

Figure 2.2: A sample of material used for calculating conductance.

A macroscopic parameter that distinguishes a metal or insulator from a semiconductor is its conductivity σ (or its reciprocal, resistivity, ρ). Consider the sample of homogeneous material of Figure 2.2 that has a cross-sectional area of A and a length of d. The conductance of the sample G (or its resistance R), may be expressed in terms of its physical dimensions and its conductivity σ as follows:

$$G = \sigma A/d \ \text{(S)}, \qquad R = 1/G \ (\Omega) \tag{2.2}$$

Conductance (or resistance) depends on the physical size and shape of a material, whereas conductivity is an innate property of the material. It should be noted that the longer the sample, the smaller its conductance (the larger its resistance), whereas the larger its area, the larger its conductance (the smaller its resistance). Using the mks system of units in which dimensions are expressed in meters, conductivity σ has the dimensions of S/m.

Materials used for electronic devices may be classified into three broad, general categories of metals, semiconductors, and insulators according to the conductivity of the material (Table 2.1). Metals such as silver, copper, aluminum, or alloys of these and other elements, are characterized by high values of conductivity. These elements have valence electrons that, for most chemical reactions, are "lent" to other elements. In a metal, these electrons, as a result of their thermal energy, become free of the atom to which they were originally bound. As a result of the large number of free electrons, metals are very good conductors – sufficiently good conductors that for many circuit applications their conductivity is treated as being infinite (for example, the zero-resistance wire).

Although electrons tend to be free to move about within a metal, their movement is characterized by numerous collisions with the atoms of the metal. As a

TABLE 2.1 CONDUCTIVITY OF MATERIALS

	Material	Conductivity, σ (S/m)
Conductors	Silver	6.3×10^7
	Copper	5.8×10^7
	Aluminum	3.5×10^7
	Other	$> 10^6$
Semiconductors	Doped silicon or germanium	10^{-3} to 10^5
	Intrinsic silicon	4.7×10^{-4}
	Intrinsic germanium	2.2
Insulators		10^{-16} to 10^{-9}

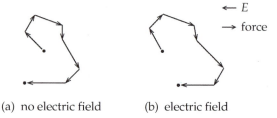

(a) no electric field   (b) electric field

Figure 2.3: The motion of an electron in a metal.

result of their thermal energy, electrons tend to move with relatively large velocities. However, at each collision, the direction of an electron's velocity tends to change in a random manner (Figure 2.3(a)). Owing to this random process, a particular electron tends to move away from its original position. If, however, all electrons of a small finite volume are considered, the number that move to the right, for example, will be about the same as the number that move to the left. Hence, as a result of thermal motion and collisions, the average drift velocity of an ensemble of electrons is zero. (This random motion, however, results in very small current fluctuations that limit the sensitivity of electronic devices.)

An electric field produced by an external potential (such as a potential applied between the ends of the sample of Figure 2.2) modifies the motion of the individual electrons (Figure 2.3(b)). During each movement between collisions, the electron is slightly accelerated in the direction of the force produced by the electric field ($F = -eE$). For the electric fields that are common for electronic devices, the effect of the acceleration is small, the motion of the electrons remains primarily due to thermal processes as in the case without an electric field. But, because, in the presence of an electric field, each electron will tend to drift slightly in the direction of the force, when all electrons are considered, an average drift velocity will be observed in the direction of the force produced by the electric field. For electric fields that are not abnormally large, the average drift velocity v_d may be shown to be proportional to the electric field E. Because the drift velocity for electrons is in the opposite direction of the electric field, a minus sign is required:

$$v_d = -\mu_n E \ \text{(m/s)} \tag{2.3}$$

The proportionality constant μ_n is known as the mobility. Because velocity has the dimensions of meter per second and the electric field of volt per meter, mobility has the dimensions of meter squared per volt second.

The conductivity of a material such as a metal depends on the mobility and the number of electrons that take part in the conduction process, that is, the number of free electrons. An increase in either of these quantities increases the conductivity of the material. Consider the sample of Figure 2.4 in which the number density of free electrons is n_e (number/m^3). Assume that an electric field is applied to the sample that results in a free electron drift velocity of v_d. The special case for which the center region has a length of v_d (1 s) is of interest. If only the drift mechanism is considered, an electron starting at the left edge of the center

Figure 2.4: A sample used for determining conductivity.

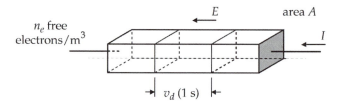

region will in 1 s reach the right edge of the center region. Not only will this electron reach this edge, but all free electrons within the center region will cross into the right-hand region. For a free electron density of n_e, this is $n_e v_d A$ (density × volume) electrons. The charge will be this quantity times the charge of an electron, namely $-e$. Because a time interval of 1 s was assumed, the current is numerically equal to the charge.

$$I = -e n_e v_d A$$
$$j = I/A = -e n_e v_d \ (\text{A/m}^2)$$

(2.4)

A macroscopic quantity, a current density j, has been introduced. Using the expression for mobility, an equation for the current density in terms of the electric field is obtained as follows:

$$j = -e n_e(-\mu_n E) = e n_e \mu_n E = \sigma E$$
$$\sigma = e n_e \mu_n$$

(2.5)

Because current density and electric field are related by the conductivity of a material (the basic definition of conductivity), an expression for conductivity in terms of electron density and mobility results.

As a result of having a large number of free electrons, metals have a large conductivity. On the other hand, insulators do not, for normal conditions, have many free electrons. Insulators are characterized by atoms that have nearly complete (or in the case of inert elements, complete) electron valence shells. In chemical reactions, these elements tend to accept valence electrons from other atoms. For normal temperatures, thermal energies are insufficient to produce free electrons in these materials. However, by the application of extremely large electric fields, valence electrons may gain sufficient energy to escape from valence bonds and contribute to a current. Although this breakdown condition results in a current, it is usually destructive and is not (intentionally) utilized for electron devices.

Semiconductors consist of atoms whose valence shells have neither as many free electrons as metals nor are as nearly complete as insulators. Compounds of atoms in the center of the periodic table have valence shells that are only half filled (or half empty) and are therefore used to produce semiconductors. It is from these materials that the array of solid-state electronics devices, including integrated circuits, are formed.

EXAMPLE 2.1

Consider a length of AWG 22 (American Wire Gauge) copper wire that might be used in the laboratory for a connection to an experimenter board (AWG 22 wire has a diameter of 0.64 mm). The wire is conducting a current of 1 A. What is the conductance and resistance of the 1-m length of wire? What is the drift velocity and the mobility of the free electrons?

SOLUTION Equation (2.2) may be used to determine the conductance and resistance of the wire.

$$A = \frac{\pi}{4}d^2 = \frac{\pi}{4}(0.64 \times 10^{-3} \text{ m})^2 = 3.22 \times 10^{-7} \text{ m}^2$$

$$G = \sigma A/d = (5.8 \times 10^7 \text{ S/m})(3.22 \times 10^{-7} \text{ m}^2)/(1 \text{ m}) = 18.7 \text{ S}$$

$$R = 1/G = 0.054 \ \Omega$$

For a current of 1 A, the potential difference is 0.054 V and, for a wire length of 1 m, the magnitude of the electric field E is 0.054 V/m. Copper has an atomic mass of 63.5 g and a density of 8.96 g/cm^3. Because one atomic mass contains Avogadro's number of atoms (6.02×10^{23}), the number of atoms of 1 cm^3, n', is readily determined as follows:

$$n' = \frac{(6.02 \times 10^{23} \text{ atoms})(8.96 \text{ g/cm}^3)}{63.5 \text{ g}} = 8.49 \times 10^{22} \text{ atoms/cm}^3$$

$$n = 8.49 \times 10^{28} \text{ atoms/m}^3$$

If one free electron for each copper atom is assumed, $n_e = 8.49 \times 10^{28}/\text{m}^3$. Equation (2.4) yields the drift velocity:

$$v_d = \frac{-I}{e n_e A} = \frac{-1 \text{ C/s}}{(1.6 \times 10^{-19} \text{ C})(8.49 \times 10^{28}/\text{m}^3)(3.22 \times 10^{-7} \text{ m}^2)}$$

$$= -2.29 \times 10^{-4} \text{ m/s} \quad \text{(opposite direction to } E)$$

Equation 2.3 may be used to determine the mobility of the free electrons through the previously determined value of E as follows:

$$\mu = -v_d/E = (-2.29 \times 10^{-4} \text{ m/s})/(-0.054 \text{ V/m})$$

$$= 4.24 \times 10^{-3} \text{ m}^2/\text{V} \cdot \text{s}$$

EXAMPLE 2.2

Compare the drift velocity of Example 2.1 with that associated with the thermal motion of the free electrons. Assume a temperature T of 300 K (27°C).

SOLUTION The average thermal energy of an electron is $(3/2)kT$ (where k is the Boltzmann constant, 1.38×10^{-23} J/K).

$$\frac{1}{2}m\overline{v^2} = \frac{3}{2}kT, \qquad \overline{v^2} = \frac{3kT}{m}$$

An electron with this energy will have a speed v_{typical} of $\sqrt{\overline{v^2}}$.

$$v_{\text{typical}} = \sqrt{\frac{3kT}{m}} = \sqrt{\frac{(3)(1.38 \times 10^{-23} \text{ J/K})(300 \text{ K})}{9.107 \times 10^{-31} \text{ kg}}}$$

$$= 1.16 \times 10^5 \text{ m/s}$$

For this case $|v_d/v_{\text{typical}}| = 1.97 \times 10^{-9}$, that is, the drift velocity is only about two billionths of the thermal speed of an electron with an average energy of $(3/2)kT$.

TABLE 2.2 SEMICONDUCTOR ELEMENTS

III	Z Atomic Number Element Atomic Weight Electron Configuration IV	V
5 B Boron 10.8 2–3	6 C Carbon 12.0 2–4	7 N Nitrogen 14.0 2–5
13 Al Aluminum 27.0 2–8–3	14 Si Silicon 28.1 2–8–4	15 P Phosphorus 31.0 2–8–5
31 Ga Gallium 69.7 2–8–18–3	32 Ge Germanium 72.6 2–8–18–4	33 As Arsenic 74.9 2–8–18–5
49 In Indium 115.8 2–8–18–18–3	50 Sn Tin 118.7 2–8–18–18–4	51 Sb Antimony 121.8 2–8–18–18–5

2.2 SEMICONDUCTORS: THE ROLE OF ELECTRONS AND HOLES

It is elements of columns III, IV, and V of the periodic table, those elements with 3, 4, or 5 valence electrons, respectively, that are utilized to produce semiconductors (Table 2.2). Most semiconductors are fabricated from the crystalline form of these elements, and amorphous and polycrystalline forms used for limited applications. The crystalline form of carbon, the diamond, is a structure with extremely tightly bound electrons that has no value as a semiconductor. However, silicon and germanium, each with four valence electrons, are valuable semiconductors in their crystalline forms. Germanium is primarily of interest from a historical perspective – the earliest commercial semiconductor devices were fabricated from germanium. It was not until improved technologies were developed that silicon devices and integrated circuits replaced those of germanium. The electrical characteristics of silicon semiconductors, including their ability to operate at higher temperatures, are generally preferable to those of germanium.

Rather than lending valence electrons (as do metals) or borrowing valence electrons (as do insulators), the crystalline forms of silicon or germanium tend to form covalent bonds similar to organic compounds of carbon. These elements, which have four valence electrons and a nucleus with the remaining nonvalence

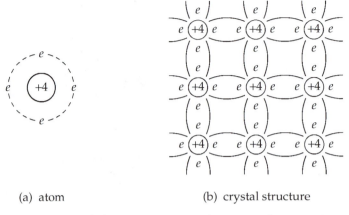

(a) atom   (b) crystal structure

Figure 2.5: Symbolic representation of a tetravalence semiconductor. The symbol e represents the negative electronic charge of an electron.

electrons, may be represented symbolically, as in Figure 2.5(a). The nucleus of protons and neutrons along with the nonvalence electrons will have a net positive charge of $+4$ unit electronic charges. This charge, combined with the four negative unit electronic charges of the four valence electrons, results in a neutral atom.

If silicon (melting point of 1415 °C) or germanium (937 °C) is slowly cooled from its liquid state, a crystalline form of the element can be obtained. Each atom forms four covalent bonds with its four equally distant neighbors, resulting in a structure that is repeated uniformly throughout the crystal. A two-dimensional representation is given in Figure 2.5(b).

It should be emphasized that a three-dimensional model in which each atom is equally distant from the four atoms with which it forms a covalent bond is necessary to represent the actual crystal. In addition, the number of the atoms involved in any device of reasonable size is immense. A silicon crystal, for example, contains 5.0×10^{28} atoms/m^3. A cube 1 mm on a side, such as might be used for a discrete device, contains 5.0×10^{19} atoms.

Conduction occurs when valence electrons gain sufficient energy to break their covalent bonds, thus becoming free electrons. At room temperature (approximately 300 K or 27 °C) the average thermal energy of a valence electron is very small compared with that needed to break a bond. Thermal energy, however, is not distributed uniformly among the electrons – some electrons have more and some less than the average quantity. As a consequence, a few electrons have sufficient energy to break their covalent bonds and become free electrons. This quantity of electrons is very small compared with the total number of valence electrons. For silicon, at room temperature, the free-electron density is about 1.5×10^{16} per cubic meter, that is, one free electron for each 1.33×10^{13} valence electrons. This quantity, which depends on thermal energies, increases rapidly with increasing temperature. The symbol n_i is used to designate the number of free electrons, and the subscript i stands for intrinsic. Intrinisic silicon (or germanium) is used to designate an extremely pure semiconductor.

AN INTRINSIC SEMICONDUCTOR

As for metals, the free electrons contribute to the conduction process when an electric field is present. For a mobility of μ_n (the subscript n is used to denote electrons, negative charges), the current density due to the free electrons may readily be calculated by

$$v_d = -\mu_n E \quad \text{free electrons}$$
$$j_n = -en_i v_d = en_i \mu_n E \tag{2.6}$$

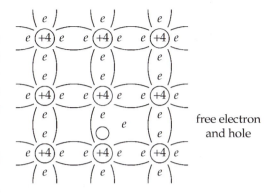

free electron and hole

Figure 2.6: Symbolic representation of a semiconductor with a thermally generated electron-hole pair.

This current density is not, as for the case of metals, the total current density. For each free electron there is an incomplete valence bond (Figure 2.6). The deficiency of an electron that would otherwise form a covalent bond is known as a *hole*. A valence electron that is part of an adjacent bond may readily move into the vacancy or hole, leaving an incomplete valence bond, a hole, behind. Once a vacancy or hole exists, valence electrons may, as a result of their thermal energy, continually reorder themselves, giving rise to what appears to be a motion of the vacancy or hole. With an electric field, valence electrons will drift in a direction opposite to that of the electric field, resulting in a drift of the hole in the direction of the field, that is, the hole behaves as if it were a particle with a positive charge. As for electrons, the drift velocity of the hole is found to be proportional to the electric field. The symbol μ_p (the subscript p stands for positive owing to the positive-charge-like behavior of a hole) is used for its mobility:

$$v_d = \mu_p E \quad \text{holes}$$
$$j_p = en_i v_d = en_i \mu_p E \tag{2.7}$$

Because holes drift in the direction of the electric field, no minus sign is required – unlike for the free electrons. Furthermore, for pure or intrinsic material, one hole exists for each free electron.

The total current density, that is, the current density that gives rise to an external current, is the sum of that due to the free electrons and that due to the holes and is given by

$$j = j_n + j_p = en_i(\mu_n + \mu_p) \tag{2.8}$$

The current density j is related to E by the conductivity of the semiconductor as follows:

$$j = \sigma E$$
$$\sigma = en_i(\mu_n + \mu_p) \tag{2.9}$$

As a result of thermal processes, free electron-hole pairs are continually being generated. The generation rate of free electron-hole pairs depends on the average thermal energy of the valence electrons and hence on the temperature of the crystal. Their number density, however, does not continually increase because, concurrent with the generation process, free electrons and holes are "lost"

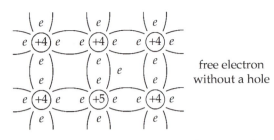

free electron
without a hole

Figure 2.7: Symbolic representation of a semiconductor with a donor atom impurity.

through recombination. Recombination depends primarily on the density of free electrons and holes – the greater their density, the more likely it is that a free electron and hole will approach each other and recombine. As a result of the temperature dependence of the generation rate, the intrinisic density n_i is also extremely temperature sensitive. In silicon, for example, n_i tends to double for each 11 C° temperature increase. Consequently, the operation of semiconductor devices may be very sensitive to temperature changes that occur not only as a result of ambient temperature changes but also as a result of electrical power being dissipated by the devices (the conversion of electrical energy to thermal energy).

Conduction in intrinisic semiconductors is the result of equal numbers of two types of carriers: free electrons and holes. Electronic devices, however, are usually constructed from semiconductors in which conduction for a particular region of the device is due primarily to a single type of carrier that is either free electrons or holes but not both. Such materials are obtained through a process of doping in which selected impurities are introduced into the crystal lattice of intrinisic material. These materials are designated as extrinisic semiconductors.

AN *n*-TYPE SEMICONDUCTOR

Consider the case in which a silicon atom of a crystal is replaced by an atom with five valence electrons such as antimony (Figure 2.7). Four valence electrons will be used to complete the covalent bonds with their neighbor atoms. The fifth electron is not only not needed for a bond, but very little thermal energy is necessary to free this electron from its pentavalent atom. Once free, this electron may migrate throughout the crystal and become indistinguishable from the other free electrons (in fact, all electrons are "indistinguishable" from each other). Although this electron contributes to the conduction process, it leaves behind a positively charged, albeit immobile, nucleus. Because a free electron is contributed by the pentavalent atom, this atom is known as a donor atom (N_d is used to designate the density of donor atoms). In addition to antimony, both arsenic and phosphorus (each with five valence electrons) are used as donor atoms.

In addition to free electrons donated by donor atoms, there are the electrons and holes that can be associated with the intrinisic material. The number of free electrons increases as donor atoms are added to a semiconductor, which, in turn increases the probability that free electrons will combine with holes. As the doping density is increased, the equilibrium density of the holes is reduced as a result of recombination. If n and p are used to designate the number density of free electrons and holes, it may be shown that the following is valid:

$$np = n_i^2 \tag{2.10}$$

The product np is constant and is, as would be expected, equal to n_i^2, the value of np for an intrinsic semiconductor.

For a density of donor atoms much larger than n_i ($N_d \gg n_i$), the density of free electrons will be approximately equal to the density of donor atoms ($n = N_d$). This implies that the density of holes will be very small:

$$p = n_i^2/n \ll n_i \qquad \text{for } n \gg n_i \qquad (2.11)$$

Hence, for this condition, conduction will primarily be the result of free electrons, that is,

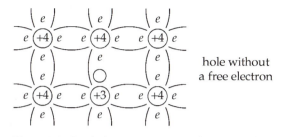

hole without a free electron

Figure 2.8: Symbolic representation of a semiconductor with an acceptor atom impurity.

$$j = j_n + j_p = j_n \qquad \text{for } p \ll n_i$$
$$= e\mu_n N_d E \qquad (2.12)$$

Because free electrons are primarily responsible for conduction, this type material is known as an *n*-type (negative) semiconductor.

A *p*-TYPE SEMICONDUCTOR

In a similar fashion, a semiconductor in which conduction is primarily due to holes may be formed. Consider the case in which a silicon atom of an intrinsic silicon crystal is replaced by an atom with three valence electrons such as boron or aluminum (Figure 2.8). Only three valence bonds of the original silicon atom will be completed – an incomplete bond occurs as a result of the trivalence atom. For normal conditions, adjacent valence electrons will have sufficient energy to move into the incomplete bond. Hence, a hole that is free to migrate throughout the crystal will be produced by each trivalent atom introduced into the crystal. Because these atoms tend to accept valence electrons, they are known as acceptor atoms (N_a is used to specify their density).

As the doping density of acceptor atoms is increased, the additional holes will tend to combine with free electrons, thus reducing the free electron density. For the case in which the acceptor atom density is much larger than n_i ($N_a \gg n_i$), the density of holes will be approximately equal to the density of donor atoms ($p = N_a$). Hence, the free electron density n will be very small:

$$n = n_i^2/p \ll n_i \qquad \text{for } p \gg n_i \qquad (2.13)$$

For this condition, conduction will be primarily the result of holes.

$$j = j_n + j_p = j_p \qquad \text{for } n \ll n_i$$
$$= e\mu_p N_a E \qquad (2.14)$$

Because it is primarily holes that are responsible for conduction, this type of material is known as *p*-type (positive) semiconductor.

With appropriate doping, one can produce either *n*- or *p*-type semiconductors in which it is either free electrons (negative charges) or holes (the equivalent of positive charges) that are primarily responsible for conduction. It is the transition from one type of semiconductor to the other type within a single crystal that is of interest for fabricating useful semiconductor devices.

EXAMPLE 2.3

An n-type silicon semiconductor at a temperature of 300 K has a donor atom density of 10^{13} atoms/cm^3.

a. What is the ratio of donor to silicon atoms?
b. What is the conductivity of the semiconductor?
c. What is the resistance of a sample with a cross-sectional area of 1 mm^2 and a length of 0.1 mm?
d. Compare the results of parts (b) and (c) with that for intrinsic silicon semiconductor ($\mu_n = 1500$ cm^2/V $\cdot$ s, $\mu_p = 450$ cm^2/V $\cdot$ s).

SOLUTION A silicon crystal has 5.0×10^{28} atoms/m^3 (n_0).

a. $N_d/n_0 = (10^{19}/\text{m}^3)/(5 \times 10^{28}/\text{m}^3) = 2 \times 10^{-10}$. There is only one donor atom for each 5×10^9 (5 billion) silicon atoms (n_0/N_d).

b. Assume a free-electron density n equal to the donor density N_d of $10^{19}/\text{m}^3$ ($N_d \gg n_i$). Equation (2.12) yields the following:

$$j = e\mu_n N_d E$$
$$\sigma = e\mu_n N_d = (1.6 \times 10^{-19} \text{ C})(0.15 \text{ m}^2/\text{V} \cdot \text{s})(10^{19}/\text{m}^3) = 0.24 \text{ S/m}$$

c. If the conductivity of the semiconductor is known, the conductance and resistance of the sample may be determined as follows:

$$
\begin{aligned}
G &= \sigma A/d \\
&= (0.24 \text{ S/m})(1.0 \times 10^{-6} \text{ m}^2)/(0.1 \times 10^{-3} \text{ m}) \\
&= 2.4 \times 10^{-3} \text{ S} \\
R &= 1/G = 417 \ \Omega
\end{aligned}
$$

d. Both holes and free electrons contribute to the conductivity of an intrinsic semiconductor (Eq. (2.9)):

$$
\begin{aligned}
\sigma &= en_i(\mu_n + \mu_p) \\
&= (1.6 \times 10^{-19} \text{ C})(1.5 \times 10^{16}/\text{m}^3)[(0.15 + 0.045) \text{ m}^2/\text{V} \cdot \text{s}] \\
&= 4.68 \times 10^{-4} \text{ S/m}
\end{aligned}
$$

$$\frac{\sigma_{\text{doped}}}{\sigma_{\text{intrinsic}}} = \frac{0.24 \text{ mho/m}}{4.68 \times 10^{-4} \text{ mho/m}} = 513$$

The conductance of the intrinsic sample is 4.68×10^{-6} S, and its resistance is 214 kΩ. Only one donor atom for each 5 billion silicon atoms has a significant effect!

EXAMPLE 2.4

Consider a p-type silicon semiconductor at a temperature of 300 K.

a. What is the acceptor atom density necessary for a sample to have the same conductivity as the n-type semiconductor of Example 2.3?
b. What is the ratio of the acceptor to silicon atoms?

SOLUTION The mobility of holes is considerably less than that of free electrons.

a. Using Eqs. (2.12) and (2.14), the following is valid for semiconductors having equal conductivities:

$$\mu_n N_d \; (n\text{-type}) = \mu_p N_a (p\text{-type})$$
$$N_a/N_d = \mu_n/\mu_p = (0.15 \mathrm{m}^2/\mathrm{V} \cdot \mathrm{s})/(0.045 \mathrm{m}^2/\mathrm{V} \cdot \mathrm{s}) = 3.33$$
$$N_a = 3.33 \times 10^{19}/\mathrm{m}^3$$

b. This corresponds to one acceptor atom for each 1.5×10^9 silicon atoms (n_0/N_a).

EXAMPLE 2.5

Assume that n_i of silicon doubles for each 11 C° temperature change. What is n_i for temperatures of 0, 50, 100, and 150 °C?

SOLUTION Let T_D equal the temperature change for which n_i doubles (11 C°). If n_{i0} is the value of n_i for $T = T_0$, the following applies for any other temperature:

$$n_i = n_{i0}(2)^{(T-T_0)/T_D}$$

For an 1 C° change, n_i doubles; for a 22 C° change, n_i quadruples; for a 33 C° change, n_i is eight times its original value, etc. Alternatively, an exponential relationship can be written for n_i as

$$n_i = n_{i0}e^{a(T-T_0)}$$

If $n_i = n_{i1}$ for $T = T_1$ and $n_i = n_{i2}$ for $T = T_2$, the following is obtained:

$$n_{i1} = n_{i0}e^{a(T_1-T_0)}$$
$$n_{i2} = n_{i0}e^{a(T_2-T_0)}$$
$$\frac{n_{i2}}{n_{i1}} = e^{a(T_2-T_1)}$$

Let $T_2 - T_1 = T_D$, the temperature increment for doubling.

$$\frac{n_{i2}}{n_{i1}} = 2 = e^{aT_D}$$
$$a = \frac{\ln 2}{T_D} = \frac{0.693}{11\,\mathrm{C}°} = 0.063/\mathrm{C}°$$

If a is known, the density for any temperature may be calculated. Because only a temperature difference is involved, either Celsius or absolute temperatures may be used.

$$T = 0°\mathrm{C}, \quad n_i = (1.5 \times 10^{16})e^{(0.063/\mathrm{C}°)[(0-27)\mathrm{C}°]}$$
$$= 2.74 \times 10^{15}/\mathrm{m}^3$$
$$T = 50°\mathrm{C}, \quad n_i = 6.39 \times 10^{16}/\mathrm{m}^3$$
$$T = 100°\mathrm{C}, \quad n_i = 1.49 \times 10^{18}/\mathrm{m}^3$$
$$T = 150°\mathrm{C}, \quad n_i = 3.49 \times 10^{19}/\mathrm{m}^3$$

EXAMPLE 2.6

Consider a semiconductor with donor doping. If N_d is much larger than n_i, it is reasonable to assume that $n = N_d$. This, however, is not valid if the inequality $N_d \gg n_i$ is not valid. Use the assumption that charge neutrality prevails and obtain an expression for n and p for any value of N_d.

SOLUTION Charge neutrality implies the following:

$$N_d + p - n = 0$$

Each pentavalent donor atom is responsible for one excess positive electronic charge. Through Eq. (2.10), which also applies, the following is obtained:

$$np = n_i^2, \qquad p = n_i^2/n$$

$$N_d + n_i^2/n - n = 0$$

$$n^2 - nN_d - n_i^2 = 0$$

$$n = N_d/2 \pm \sqrt{(N_d/2)^2 + n_i^2}$$

Only the plus sign of the square root applies because n must be positive. If $N_d \gg n_i$, then $n \approx N_d$ and, for $N_d = 0$, $n = n_i$, as expected. An expression for holes may also be obtained as follows:

$$n = n_i^2/p$$

$$N_d + p - n_i^2/p = 0$$

$$p^2 + pN_d - n_i^2 = 0$$

$$p = -N_d/2 \pm \sqrt{(N_d/2)^2 + n_i^2}$$

Again, because p is greater than zero, the positive sign for the square root applies.

2.3 THE JUNCTION DIODE: A QUINTESSENTIAL SEMICONDUCTOR DEVICE

An understanding of the theory and operation of a junction diode is basic to the understanding of essentially all semiconductor devices and integrated circuits. As a result of their nonlinear characteristic, junction diodes are used extensively in modern electronic circuits. A semiconductor diode has a current-versus-voltage characteristic that results in a large current for a very small voltage of one polarity but essentially zero current for a potential of the opposite polarity (Figure 1.2). This property is of value for rectification and detection in analog circuits and for performing logic functions in digital circuits. Furthermore, a semiconductor junction diode is a basic element used in fabricating active semiconductor devices, bipolar junction transistors, and field-effect transistors as well as integrated circuits.

A junction diode with an abrupt transistion from p-type to n-type semiconductor material will be considered (Figure 2.9). The p-type material is the result

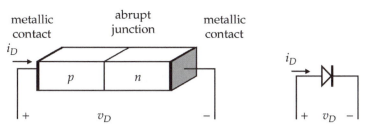

Figure 2.9: A semiconductor junction diode.

of a doping with acceptor atoms ($N_a \gg n_i$), whereas the n-type material is doped with donor atoms ($N_d \gg n_i$). For this diode, a positive value of external potential v_D results in a movement of the holes of the left-hand region to the right and free electrons of the right-hand region to the left. Because electrons have a negative charge, the crossing of the junction by both types of carriers results in a positive diode current i_D. Only a relatively small external potential (1 V or less) is needed to produce an appreciable current. On the other hand, there are few carriers that contribute to a current in the opposite direction (free electrons in the p-type material and holes in the n-type material). As a result, the diode current for a negative external potential tends to be very small.

The doping, along with a convenient coordinate system, is indicated in Figure 2.10. It should be stressed that the entire diode is a single crystal – it is only the doping that differs on each side of the junction ($x = 0$). Metallic contacts are included at each end of the diode to provide a connection to an external circuit. At a sufficient distance from the junction (in a typical diode, junction effects extend at most only a few microns from the junction) the density of carriers, holes, and free electrons depends on the doping densities. In the p-type material, $p = N_a$, and in the n-type material, $n = N_d$ (Figure 2.10(c)). Mi-

nority carriers are also present in drastically reduced quantities: $n = n_i^2/N_a$ in the p-type material and $p = n_i^2/N_d$ in the n-type material. Logarithmic scales have been used in Figure 2.10 to show this. In silicon, for example, if $N_d = 10^{21}/m^3$, $p = 10^{21}/m^3$ and $n = 2.25 \times 10^{11}/m^3$ because $n_i = 1.5 \times 10^{16}/m^3$. Hence, $n/p = 2.25 \times 10^{-10}$. It will initially be assumed that there is no external connection to the diode, that is, the diode current i_D is zero.

Figure 2.10: Doping and carrier densities (logarithmic scales) of a semiconductor junction diode.

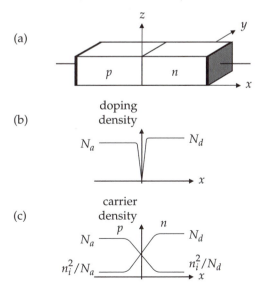

To gain an understanding of a distribution of charge carriers that occurs in the region of the junction, it is instructive to imagine a distribution that does not actually exist. This imagined distribution is of value because it allows us to obtain the distribution that actually does exist for an equilibrium condition; furthermore, it serves to demonstrate the physical mechanism that gives rise to this distribution. Suppose that the density of holes is N_a throughout the p-type material to the left of the junction and that the density of free electrons is N_d throughout the region to the right of the junction. For this condition, charge neutrality exists throughout the diode, each acceptor atom has a hole in its immediate vicinity (a charge-neutral pair), and each donor atom has a free electron in its immediate vicinity. As a consequence of charge neutrality, there would not be an electric field within the crystal. A drift of carriers, which depends on an electric field as discussed in the previous section, would therefore not occur.

Diffusion is another mechanism resulting in the ordered motion of holes and free electrons in a semiconductor. Consider the case of holes just to the left of the diode junction. As a result of their thermal energies, some holes will move to the left away from the junction. But, holes that move from the left-hand p-type region toward the junction will tend to compensate for this movement. Other holes near the junction will tend to move to the right and cross the junction. Two important events occur as a result. To the right of the junction, the density of free electrons (n-type material) is very large. Hence, the holes that entered this region will have a high likelihood of recombining with free electrons and thus being "lost" as carriers. Secondly, as a result of the minute density of holes on the right of the junction, few holes will diffuse across the junction in the opposite direction. The net effect is that holes diffuse from the p-type region on the left of the junction to the right of the junction, where they tend to recombine.

A similar situation occurs for the free electrons on the right side of the junction. They diffuse across the junction, where they tend to recombine with holes. The diffusion of the free electrons across the junction results in a transfer of negative charges to the left of the junction. Concurrently, the diffusion of holes across the junction results in a transfer of positive charges to the right. As a result of recombination, there are donor atoms on the right of the junction lacking free electrons to balance their charge, and acceptor atoms on the left of the junction lacking holes to balance their charge. The result is a net charge density on each side of the junction (Figure 2.11(a)).

The charge density in the vicinity of the junction gives rise to an electric field. Suppose that a positive test charge is placed at the junction of the diode ($x = 0$). The test charge will be repelled by the positive charges on its right and attracted by the negative charges on its left. The test charge experiences a force in the negative x direction corresponding to an electric field E_x with a negative value (Figure 2.11(b)). Because the crystal was initially neutral, the total positive charge on the right side of the junction must be equal to the magnitude of the negative charge on the left side

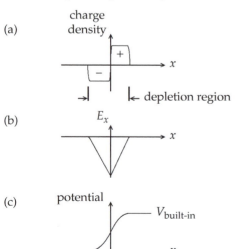

Figure 2.11: Charge density (linear scale), electric field, and potential profile of a junction diode.

(a)

(b)

(c)

of the junction. Hence, sufficiently removed from either side of the junction, the electric field will vanish because the net charge "seen" by a test charge will be zero.

Let us return to a consideration of the diffusion of holes and electrons across the junction. As a result of the electric field, a hole crossing the junction from the left to the right experiences a force that retards its motion. Similarly, a free electron that crosses the junction in the opposite direction also experiences a retarding force because the force on an electron is in the opposite direction. The electric field, therefore, retards the movement of holes and electrons across the junction until an equilibrium condition is established. The condition of a uniform density of holes on the left and free electrons on the right that was originally postulated does not occur.

THE BUILT-IN POTENTIAL

The electric field in the vicinity of the junction gives rise to a potential difference defined by

$$V = -\int E_x \, dx \tag{2.15}$$

For convenience, the arbitrary constant of integration has been assumed to be such that the potential is zero at the left of the junction where E_x vanishes (Figure 2.11(c)). The built-in potential across the junction, the potential on the right-hand side where E_x vanishes, is designated as $V_{\text{built-in}}$. The region near the junction in which a charge density and electric field exists is depleted of mobile charges (holes or free electrons) that would otherwise neutralize the charges of doping atoms. This region is known as the *depletion region* of the diode.

One who has carefully followed the development up to this point might be tempted to suggest that the potential difference of the junction $V_{\text{built-in}}$ be used to power an external circuit (Figure 2.12). The result would be a semiconductor battery. The terminal potential v_D, however, is not $V_{\text{built-in}}$ – it is zero. This may be established by drawing on elementary thermodynamics principles. Suppose that a potential did exist ($v_D \neq 0$; the assumed polarity is not important). If this were the case, a current would exist, and electric power would be delivered to the load resistor, which might represent a small electric motor. Useful work could therefore be obtained from the junction diode. The diffusion process of the

Figure 2.12: Junction diode with a resistor load.

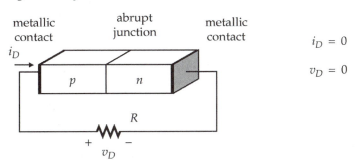

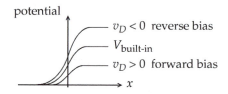

$v_D < 0$ reverse bias

$V_{\text{built-in}}$

$v_D > 0$ forward bias

x

Figure 2.13: Potential profile of a junction diode with an external potential.

junction diode that gives rise to the potential difference of $V_{\text{built-in}}$ is a thermal process – it is "powered" by the thermal energy of the diode's surroundings. Hence, a diode producing work does not violate the first law of thermodynamics; energy is conserved. It is the thermal energy of the diode's surroundings that would be converted to work. Unfortunately, this assumed effect violates the second law of thermodynamics: the impossibility of producing work through the use of a single thermal source. A conventional heat engine rejects heat to a lower temperature sink. Because this does not occur for the diode being considered, one must conclude that the external potential for a resistor load is zero ($v_D = 0$, $i_D = 0$).

A useful junction diode has external metallic contacts (Figures 2.9 and 2.12). As a result of the junction of dissimilar materials (a metal and doped semiconductor), each of these junctions not only gives rise to a potential difference, but the two potential differences precisely cancel the potential of the junction diode. Hence, for a zero diode current, $i_D = 0$, the diode voltage is zero ($v_D = 0$).

AN EXTERNAL POTENTIAL

An externally applied potential v_D has a direct effect on the potential difference across the junction of the diode (Figure 2.13). For small currents, the potential differences across the metal-to-semiconductor contacts will tend to remain constant, and if the potential difference across the n- and p-type regions is negligible (large conductivities), the external potential will, depending on its polarity, either directly add or subtract from the built-in potential of the diode.

Consider the case for which v_D is positive, that is, the situation for which the potential difference of the junction is reduced. It was the original potential difference that inhibited the movement of majority carriers (holes on the left and free electrons on the right) from crossing the junction. A reduced potential difference will allow majority carriers to cross the junction more readily, thus resulting in a positive external diode current ($i_D > 0$). As v_D is increased, thus reducing the potential difference across the junction, the external current tends to increase rapidly (Figure 2.14).

On the other hand, a negative value of v_D tends to increase the potential difference across the junction of the diode, thus reducing the already extremely small current due to majority carriers. The resultant diode current i_D, although very small in magnitude, is not zero. For negative values of v_D, there is a small negative current due to the crossing of the junction by the minority carriers. The minority carriers, the free electrons of the p-type material and the holes of the n-type material, regardless of the potential difference, readily cross the junction. For significant negative values of v_D, the diode current i_D tends to reach a negative value that is independent of v_D. It should be noted that

Figure 2.14: Current versus voltage characteristic of a junction diode.

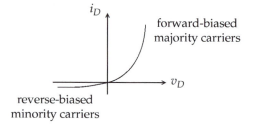

i_D

forward-biased majority carriers

v_D

reverse-biased minority carriers

the magnitude of the current for this condition is extremely small for most diodes.

To summarize, a positive value of v_D can result in a large diode current. Furthermore, this current will tend to increase very rapidly with only a very small increase in v_D. This is the forward-biased condition in which the diode current is primarily due to majority carriers. A negative value of v_D ($v_D < 0$) results in an extremely small magnitude of current, a current that is primarily due to minority carriers. For many applications, this current, corresponding to the reverse-biased condition, is extremely small.

Although the qualitative description of a junction diode presented here provides a physical feel for the operation of a diode, numerous materials are available for those wishing to gain a better quantitative perspective of diodes (Adler, Smith, and Longini 1964; Gray et al. 1964; Kittel 1996; Milnes 1980; Muller and Kamins 1986; Neudeck 1989; Pierret 1988; Streetman 1990; Sze 1981; Weaver 1986; Wolfe, Holonyak, and Stillman 1989).

2.4 THE JUNCTION DIODE: ITS TERMINAL CHARACTERISTICS

An idealized physical shape, essentially a rectangular block (Figures 2.9, 2.10, 2.12), has been referred to when discussing junction diodes. Structures typical of a discrete diode and an integrated circuit diode are presented in Figure 2.15. For the diodes shown, the transition in doping was achieved by the process of diffusing doping atoms into a semiconductor crystal (other processes are also used). The p-type region of the discrete diode could be, for example, the result of a high-temperature diffusion of boron atoms into a silicon crystal that had been previously grown from a silicon melt with donor atoms (an n-type semiconductor).

Although the p-type region has both donor and acceptor atoms, it behaves as p-type material for an acceptor density that is much larger than the donor density. How can acceptor atoms cancel the effect of donor atoms? Suppose that a crystal has equal densities of donor and acceptor atoms. For this condition, the free electrons (the result of donor atoms) will tend to combine with the holes (the result of acceptor atoms). If it is assumed that the doping densities are not excessively high, the net result will be a crystal that behaves as intrinsic material, that is, its density of holes and free electrons will be n_i. Additional acceptor atoms ($N_a > N_d$) will result in a density of holes that is in excess of the available free electrons. The net result is a p-type semiconductor. This result may be quantitatively shown (Example 2.7).

Figure 2.15: Discrete and integrated circuit junction diodes.

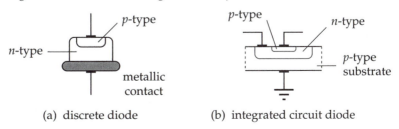

(a) discrete diode   (b) integrated circuit diode

A similar molecular diffusion process may be used to fabricate the integrated circuit diode of Figure 2.15(b). For the integrated circuit shown, all devices are formed within the *p*-type substrate. The lower *n*-type region is the result of a diffusion of donor atoms into the substrate. A second diffusion process in which acceptor atoms are diffused into the *n*-type region is used to form a diode. Through a series of masking operations and oxidizing steps, the geometry of the diode can be very precisely controlled. It will be noted that two junction diodes have been formed, an upper diode (that which will be used as a circuit element) and a lower *n*-to-*p*-type substrate diode. For normal operation of the integrated circuit, all potentials are greater than (or possibly equal to) the substrate potential. Therefore, the current of the lower diode, which is reverse biased, will be extremely small. The lower diode is used, in essence, to "insulate" the circuit element from the substrate. An alternative type of integrated circuit can be fabricated using an *n*-type substrate; it is then necessary that all circuit potentials be less than that of the substrate.

CURRENT OF A DIODE

The terminal current of a junction diode is the sum of two current components: one due to minority carriers and the other due to majority carriers (Figure 2.16). A quantitative theoretical treatment of an idealized junction diode results in the following theoretical dependence of current on voltage:

$$i_D = I_s e^{ev_D/nkT} - I_s$$

where

I_s = reverse saturation current, A

e = electronic charge, 1.6×10^{-19} C

k = Boltzmann constant, 1.38×10^{-23} J/K (2.16)

T = absolute temperature, K

n = dimensionless ideality factor (size of 1 to 2)

Figure 2.16: Current components of a junction diode.

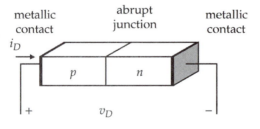

invariant current   ←   holes from the *n*-type semiconductor
due to minority carriers   free electrons from the *p*-type semiconductor

voltage dependent current   →   holes from the *p*-type semiconductor
due to majority carriers   free electrons from the *n*-type semiconductor

The $-I_s$ component of current is due to minority carriers; it does not depend on v_D. On the other hand, the exponential component of current that depends on majority carriers is strongly influenced by the external diode potential. Because the ability of majority carriers to cross the diode junction depends on their thermal energies, a current relationship that depends on ev_D/kT is not surprising. The exponential dependence arises as a result of the carrier's thermal energy distribution. Finally, the dimensionless ideality factor depends on fabrication details of the diode. Values of 1 to 2 are typical for discrete devices, and a value fairly close to 1 is common for modern silicon integrated circuits. It will be noted that $i_D = 0$ for $v_D = 0$ as expected on the basis of the thermodynamic arguments of the previous section.

It is convenient to introduce a thermal potential for kT/e:

$$V_T = kT/e \tag{2.17}$$

This quantity does indeed have the dimension of volts, that is, kT, an energy expressed in joules, divided by e, a charge expressed in coulombs (potential is joules per coulomb). For $T = 293$ K (20 °C), a value of approximately 25 mV is obtained for V_T as follows:

$$V_T = 0.0253 \text{ V} \approx 25 \text{ mV} \quad \text{for } T = 293 \text{ K} \tag{2.18}$$

For convenience, it will be assumed in this text that $V_T = 25$ mV unless an alternative temperature has been explicitly specified.

The diode current may be written in terms of V_T as

$$i_D = I_s(e^{v_D/nV_T} - 1) \tag{2.19}$$

For a typical diode, the terminal potential v_D must be fairly large compared with nV_T for an appreciable current. A typical discrete low-power silicon junction diode might require a potential of 0.7 V to result in a current of 1 mA. If $n = 1$, then $v_D/v_T = 28$, and $e^{28} = 1.45 \times 10^{12}$. This implies an extremely small value for I_s. By ignoring -1 compared with e^{28}, the following is obtained:

$$I_s = i_D/e^{28} = i_D e^{-28} = 6.9 \times 10^{-16} \text{ A} \tag{2.20}$$

Hence, the value of I_s is microscopic compared with the forward current of the diode. The following approximations are therefore appropriate for the current of the diode:

$$
\begin{aligned}
i_D &= -I_s & v_D &\ll 0, & \text{reverse-biased} \\
&= I_s e^{v_D/nV_T} & v_D &\gg nV_T, & \text{forward-biased}
\end{aligned} \tag{2.21}
$$

The reverse-biased current for the diode considered, $-I_s$, is for most applications negligible – it is generally too small to be measured by conventional techniques.

In Figure 2.17 the current-versus-voltage characteristic of a typical discrete silicon junction diode is given ($I_s = 10^{-15}$ A, $n = 1$). The logarithmic current scale of Figure 2.17(b) results in a straight-line relationship for a forward-biased

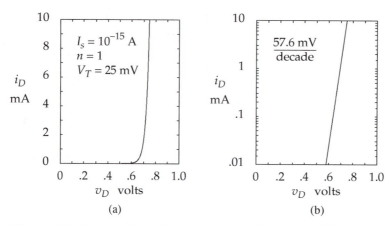

Figure 2.17: Linear and logarithmic current scales for a forward-biased junction diode.

diode as follows:

$$i_D = I_s e^{v_D/nV_T}$$

$$\log i_D = \log I_s + \frac{v_D}{nV_T} \log e \tag{2.22}$$

$$= \log I_s + 0.434 v_D / n V_T$$

The slope of the characteristic for a logarithmic current scale is thus $0.434/nV_T$ if it is assumed that base 10 logarithms are used. For the diode considered, its current increases by a factor of 10 for each 57.6 mV increase in v_D ($nV_T/0.434 = 0.025/0.434 = 0.0576$ V). This occurs regardless of the initial value of current as long as the idealized current expression is valid. A deviation from this behavior occurs for excessively large currents that result in a significant voltage being developed across the semiconductor material on each side of the junction.

A convenient quantity to know for a junction diode is the increase in terminal voltage v_D required to double its current when forward biased. Let i_{D1} and v_{D1} represent one current–voltage set and i_{D2} and v_{D2}, a second current–voltage set as follows:

$$i_{D1} = I_s e^{v_{D1}/nV_T}, \qquad i_{D2} = I_s e^{v_{D2}/nV_T}$$

$$i_{D2}/i_{D1} = e^{(v_{D2}-v_{D1})/nV_T} \tag{2.23}$$

For $i_{D2}/i_{D1} = 2$, a doubling, the following is obtained for $v_{D2} - v_{D1}$:

$$e^{(v_{D2}-v_{D1})/nV_T} = 2$$

$$v_{D2} - v_{D1} = nV_T \ln 2 = 0.693 n V_T \tag{2.24}$$

If $n = 1$, $v_{D2} - v_{D1} = 17.3$ mV ($V_T = 25$ mV). An increase of 17.3 mV doubles the current, an increase of 2×17.3 mV $= 34.6$ mV quadruples the current, an increase of 3×17.3 mV $= 51.9$ mV results in a current eight times the original value, and so forth. Very small changes in voltage result in rather large changes in current. Concurrently, even a modest change in current has only a small effect on the diode voltage.

The theoretical expression for diode current (Eq. (2.16)) has an explicit temperature dependence. In addition, the quantity I_s also depends on temperature because it depends on the availability of minority carriers. The density of minority carriers, it will be recalled, depends on n_i^2 (Figure 2.10(c)). If n_i changes owing to change in temperature, so too will I_s change. For silicon, n_i tends to double for each 11 C° temperature increase. This implies that the concentration of minority carriers quadruples for each 11 C° temperature increase. As a consequence, a quadrupling of I_s occurs for an increase of approximately 11 C° in temperature. A doubling of I_s occurs for each approximately 5.5 C° temperature increase.

For the reverse-biased condition, the ideal diode's current is $-I_s$. However, if the magnitude of the reverse-bias voltage is excessive, a breakdown condition will occur, which, depending on the circuit in which the diode is used, can result in an excessive current. Often, a breakdown current will destroy the diode, resulting in a near short circuit for the diode (a very small resistance).

SPICE MODEL

A SPICE (Simulation Program with Integrated Circuit Emphasis) computer simulation provides a convenient means of illustrating the temperature dependence of a junction diode (Figure 2.18). It should be noted that the value of I_s specified in the .MODEL statement (10^{-15} A) is interpreted by the program to be the value of I_s for a diode temperature of 27 °C (unless the nominal temperature of the program has been changed). The .TEMP statement causes I_s to be evaluated for each temperature, taking into account the change in the concentration of minority carriers of the diode. The .TEMP statement results in a set of simulations. First, T is set equal to 0 °C, and the .DC statement is processed. This is followed by setting T equal to 50 °C, and the .DC statement is again processed. A set of solutions, one for each temperature specified, is thus obtained.

The MicroSim PSPICE program for personal computers will be used for the simulation examples of the text (Tuinenga 1995). With a .PROBE statement in the circuit file, a set of data is produced from which the Probe graphics postprocessor will generate graphical displays of data. When the current of the diode is requested, the plot of Figure 2.19 is obtained. It should be noted that the circuit of Figure 2.18, although suitable for a numerical simulation, is impractical for a laboratory determination because excessively large diode currents occur. For example, a current of nearly 10 A is obtained for $T = 150$ °C and $v_D = 0.85$ V. This current would most likely destroy a diode that had a value of 10^{-15} A for I_s. A vertical scale modification was used to limit the current to 10 mA.

Figure 2.18: A SPICE circuit file for a junction diode.

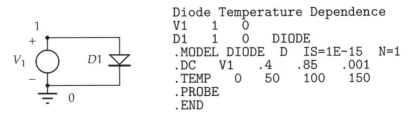

```
Diode Temperature Dependence
V1    1    0
D1    1    0      DIODE
.MODEL DIODE   D   IS=1E-15   N=1
.DC    V1    .4    .85    .001
.TEMP    0    50    100    150
.PROBE
.END
```

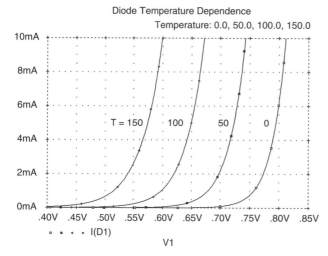

Figure 2.19: SPICE solution.

As the temperature of the diode is increased, so too is its current for a given value of v_D. If the current of the diode is held constant, the diode's terminal voltage decreases as the junction temperature is increased. On the basis of the data of Figure 2.19, the diode displays a voltage sensitivity of about $-1.5\,\mathrm{mV/C°}$ for a fixed current. Although a junction temperature of $150\,°\mathrm{C}$ may seem much beyond that which might be encountered, a junction temperature considerably higher than the ambient temperature often occurs as a result of the diode's dissipating electrical power. To limit the temperature of the junction of a diode to a safe level, a maximum power dissipation limit that depends on the physical construction of the diode is generally given.

EXAMPLE 2.7

Consider a semiconductor that has both acceptor and donor doping. Assume $N_a \gg N_d$ and that each dopant atom results in one charge carrier.
  a. Obtain a solution for the density of holes using the method of Example 2.6, which is based on charge neutrality.
  b. Show that this expression yields $p = n = n_i$ for $N_d = N_a$.
  c. Show that $p \approx N_a - N_d$ for $N_a - N_d \gg n_i$.

SOLUTION
  a. Charge neutrality implies the following:

$$N_d - N_a + p - n = 0$$

From Eq. (2.10), the following is obtained:

$$np = n_i^2, \qquad n = n_i^2/p$$
$$N_d - N_a + p - n_i^2/p = 0$$
$$p^2 + (N_d - N_a)p - n_i^2 = 0$$
$$p = (N_a - N_d)/2 \pm \sqrt{[(N_a - N_d)/2]^2 + n_i^2}$$

Because p must be positive, only the plus sign of the square root applies.
  b. If $N_a = N_d$, then $p = n_i$ and $n = n_i^2/p = n_i$.
  c. If $N_a - N_d \gg n_i$, then n_i^2 may be ignored in the square-root term.

$$p \approx (N_a - N_d)/2 + (N_a - N_d)/2 = (N_a - N_d)$$

Figure 2.20: Logarithmic plot of data of Example 2.8.

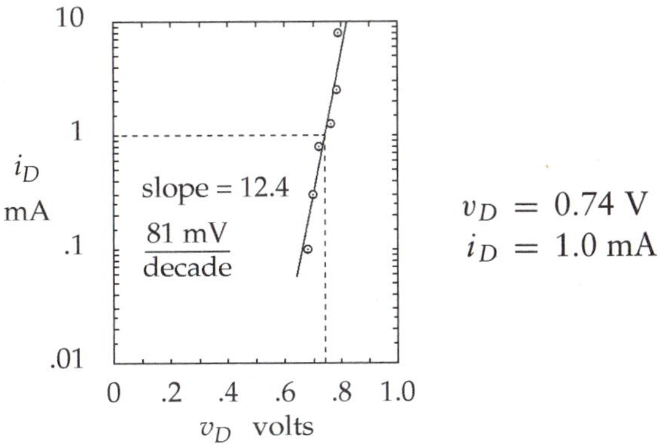

EXAMPLE 2.8

The following experimental data were obtained for a discrete silicon junction diode:

i_D	v_D	i_D	v_D
8.0 mA	0.79 V	0.80 mA	0.72 V
2.5 mA	0.78 V	0.40 mA	0.70 V
1.25 mA	0.76 V	0.10 mA	0.68 V

It is desired to do a SPICE simulation for a circuit using this diode. Determine values of I_s and n suitable for .MODEL parameters.

SOLUTION Figure 2.20 is a plot of the logarithm of the current versus voltage. A best fit straight line is indicated that has a slope of 12.4 per V (81 mV per decade). From Eq. (2.22), the following is obtained:

$$\text{slope} = 0.434/nV_T$$
$$nV_T = 0.434/\text{slope} = 35 \text{ mV}$$

If it is assumed that $V_T = 25$ mV, a value of 1.4 is obtained for n. Using a point on the best fit straight line (not an actual data point), the following is obtained:

$$nV_T = 35 \text{ mV}, \quad i_D = 1.0 \text{ mA}, \quad v_D = 0.74 \text{ V}$$
$$i_D = I_s e^{v_D/nV_T}$$
$$I_s = i_D e^{-v_D/nV_T} = (10^{-3}\text{A})e^{(-0.74\text{V})/(0.035\text{V})} = 6.6 \times 10^{-13} \text{ A}$$

2.5 A CIRCUIT WITH A DIODE: DEALING WITH A NONLINEAR ELEMENT

Basic circuit theories, namely Kirchhoff's current and voltage laws, apply for circuits with linear as well as nonlinear elements. Other circuit concepts, including superposition and Thévenin and Norton equivalent circuits, are not applicable when nonlinear elements are present. This does not imply that these powerful

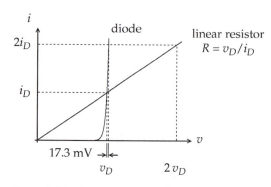

Figure 2.21: Current versus voltage for a junction diode ($n = 1$, $V_T = 25$ mV) and a linear resistor.

linear circuit concepts are of no value for analyzing electronic circuits (i.e., that they may now be set aside). Linear circuit concepts will be used for those parts of circuits with linear elements. In addition, it is frequently possible to approximate the behavior of a nonlinear element over some limited range of voltage and current as a combination of linear elements; techniques to treat a nonlinear element as a piecewise linear element will be developed. Thus, one can fully utilize linear circuit concepts for determining the behavior of an electronic circuit over a restricted range of voltages (or currents).

Only a very small increase in the terminal voltage of a forward-biased junction diode ($0.693\,nV_T$) will double its current (Figure 2.21). Suppose that a particular terminal voltage of v_D (approximately 0.7 V for a silicon diode) results in a current of i_D. An increase of only 17.3 mV ($n = 1$, $V_T = 25$ mV) doubles its current to $2i_D$. Compare this to the behavior of the linear resistor of Figure 2.21, which for a voltage of v_D also has a current of i_D ($R = v_D/i_D$). To double its current, a voltage of $2v_D$ (for example, approximately 1.4 V) is required.

An elementary series circuit consisting of a battery, resistor, and junction diode will be used to illustrate a set of approaches for solving a diode circuit. On the basis of Kirchhoff's voltage law, the circuit of Figure 2.22 yields the following:

$$V_A = i_D R + v_D \tag{2.25}$$

If $V_A > 0$, the diode is forward biased, and the following is obtained, if it is assumed that $v_D \gg nV_T$:

$$
\begin{aligned}
i_D &= I_s e^{v_D/nV_T} \quad \text{forward-biased diode} \\
V_A &= I_s R e^{v_D/nV_T} + v_D
\end{aligned}
\tag{2.26}
$$

This is a transcendental equation in terms of v_D, which does not yield an analytical solution for v_D. Alternatively, an equation in terms of i_D can also be obtained as follows:

$$
\begin{aligned}
v_D &= nV_T \ln(i_D/I_s) \\
V_A &= i_D R + nV_T \ln(i_D/I_s)
\end{aligned}
\tag{2.27}
$$

This too is a transcendental equation that does not yield an analytic solution for i_D. Only if explicit values for V_A, R, and the diode parameters are specified can a numerical solution of Eqs. (2.26) or (2.27) be obtained.

Figure 2.22: A series circuit with a diode.

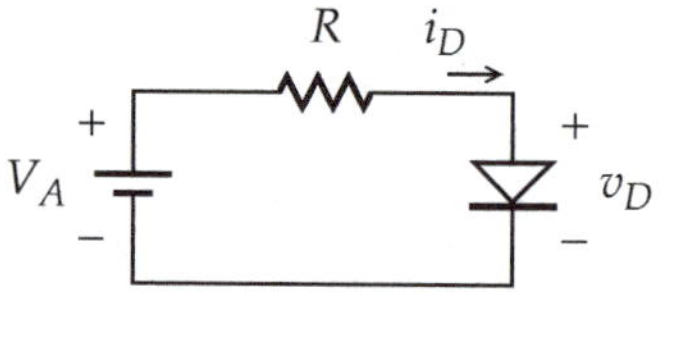

LOAD LINE

Before proceeding with a numerical solution, it is instructive to pursue an alternative approach that can not only simplify the numerical process but will suggest a

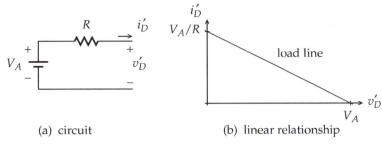

(a) circuit   (b) linear relationship

Figure 2.23: The linear circuit external to the diode.

set of simplifying approximations that will eventually be utilized. As a first step, consider the linear circuit of the battery and resistor (Figure 2.23). Primed quantities for the diode terminal voltage and current have been introduced because only when the diode is in the circuit is the solution v_D and i_D. The following can be written for the circuit:

$$V_A = i'_D R + v'_D$$
$$i'_D = (V_A - v'_D)/R$$

(2.28)

Because the circuit is linear, a solution for i'_D is linear with respect to v'_D. This linear (straight-line) relationship is shown in Figure 2.23(b). It will be noted that $i'_D = 0$ for $v'_D = V_A$ and $i'_D = V_A/R$ for $v'_D = 0$. This type of circuit response will be repeatedly encountered when working with electronic circuits; the straight-line relationship of Figure 2.23(b) is generally referred to as a "load line" (nonlinear load lines will also be encountered). The battery and resistor circuit external to the diode requires that the solution for the circuit with the diode fall on the load line.

A current-versus-voltage relationship is also known for the diode. This nonlinear relationship is plotted on the graph with the load line (Figure 2.24). The diode curve gives the locus of points for which the current and voltage satisfy the diode characteristic. Hence, at the intersection of the diode curve and the load line, the current and voltage simultaneously satisfy the diode characteristic and the battery–resistor circuit. The intersection is therefore the desired solution, the actual voltage and current of the series circuit, namely, v_D and i_D.

AN ITERATIVE APPROACH

Although a graphical approach may not necessarily be convenient (it requires an accurate plotting of the junction diode characteristic), the conceptual process, in which one thinks in terms of a graphical solution, is extremely helpful in obtaining a numerical solution. This can be illustrated with a numerical example. Suppose $V_A = 3$ V, $R = 300$ Ω, and the diode has parameters of $I_s = 10^{-15}$ A and $n = 1$ (the diode of Figure 2.17).

Figure 2.24: Diode characteristic and load line.

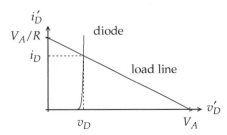

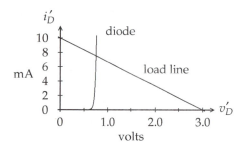

Figure 2.25: Diode characteristic and load line ($V_A = 3$ V, $R = 300$ Ω).

This results in the load line and diode characteristic of Figure 2.25. If one looks at the graph (which has been accurately drawn), it may be seen that $v_D \approx 0.8$ V and $i_D \approx 7$ mA. An iterative numerical solution may now be used to obtain a more accurate diode voltage and current. The voltage of a forward-biased diode is not very sensitive to rather large changes in diode current (this is obvious looking at Figure 2.25). Hence, let us assume a value of 7 mA for the diode current of the first iteration. On the basis of this estimated value of current i_{D1}, a corresponding value of diode voltage v_{D1} may be calculated:

$$v_{D1} = nV_T \ln(i_{D1}/I_s) = 0.743 \text{ V} \qquad (2.29)$$

This value of diode voltage may now be used to calculate a corrected value of current i_{D2} by utilizing the external battery–resistor circuit as follows:

$$i_{D2} = (V_A - v_{D1})/R = 7.52 \text{ mA} \qquad (2.30)$$

Continuing, one can obtain a new value of diode voltage v_{D2}:

$$v_{D2} = nV_T \ln(i_{D2}/I_s) = 0.741 \text{ V} \qquad (2.31)$$

With this value of voltage, the next iteration may be carried out as follows:

$$i_{D3} = (V_A - v_{D2})/R = 7.53 \text{ mA}$$
$$v_{D3} = nV_T \ln(i_{D3}/I_s) = 0.741 \text{ V} \qquad (2.32)$$

This voltage is the same voltage of the previous iteration (to the nearest millivolt) and is thus the desired numerical solution:

$$v_D = 0.741 \text{ V} \qquad i_D = 7.53 \text{ mA} \qquad (2.33)$$

A rapid numerical convergence occurred; only three iterations were needed to obtain a voltage with an error of less than 1 mV.

Although a more accurate solution (for example, to the nearest microvolt) is possible, such a solution would be, from a practical perspective, meaningless. In the first place, V_A may not be known this precisely. But, more importantly, the diode's terminal voltage is temperature sensitive. On the basis of the Figure 2.19 data, a temperature change of only 1 C° results in a voltage change of approximately −1.5 mV for a constant diode current. If typical temperature fluctuations are taken into account, for example ±10 C°, a variation of ±15 mV would be expected for the diode voltage, yielding a comparable variation in the solution for v_D. Hence, a solution accurate to even 1 mV is rarely justified.

As is well known, iterative numerical solutions will not necessarily converge rapidly or may not converge at all. To illustrate this, consider an alternative attempt to obtain a numerical solution for the diode circuit. Let us start by assuming a value of 0.8 V for the diode voltage (v_{D1}) and use this to calculate a corresponding diode current, i_D as follows:

$$i_{D1} = I_s e^{v_{D1}/nV_T} = 79.0 \text{ mA} \tag{2.34}$$

From the graph of Figure 2.25, it can be seen that this estimate of current is totally unreasonable. Nevertheless, let us continue (as would be the case of a "nonthinking" numerical algorithm routine that might be used). The battery–resistor circuit yields a new estimate for the diode voltage:

$$v_{D2} = V_A - i_{D1} R = -20.7 \text{ V} \tag{2.35}$$

For this voltage, the diode is reverse biased, and its current $(-I_s)$ is essentially zero. This approach does not yield a solution, for the numerical values are diverging.

SPICE SOLUTION

In the preceding numerical example, the potential source V_A was assumed to be constant (a battery). For many electronic applications it is often the case that the functional dependence of circuit voltages and currents on an independent voltage (or current) source is desired. This is the case, for example, if the voltage source has a time dependence. A set of solutions is thus required, one solution for each value that the voltage source might have (in practice, a series of closely spaced voltages). A SPICE simulation is ideally suited to obtain this type of solution. The circuit file of Figure 2.26 will produce a solution for v_A with a range of ± 5 V (0.05-V increments).

The SPICE solution using .PROBE is indicated in Figure 2.27. It will be noted that the solution for the diode voltage v_D tends to consist of two straight lines:

$$v_D \approx \begin{cases} v_A & \text{for } v_A < 0.7 \text{ V} \\ 0.7 \text{ V} & \text{for } v_A > 0.7 \text{ V} \end{cases} \tag{2.36}$$

A similar approximation applies for the voltage across the resistor v_R as follows:

$$v_R \approx \begin{cases} v_A - 0.7 \text{ V} & \text{for } v_A > 0.7 \text{ V} \\ 0 & \text{for } v_A < 0.7 \text{ V} \end{cases} \tag{2.37}$$

These results are not surprising if one recognizes that the diode characteristic, for the forward-biased condition, could be approximated by a vertical straight line (Figure 2.28).

$$\begin{aligned} v_D &= v_{D(\text{on})} & \text{if} \quad i_D > 0 \\ i_D &= 0 & \text{if} \quad v_D > v_{D(\text{on})} \end{aligned} \tag{2.38}$$

Figure 2.26: SPICE circuit for a junction diode circuit.

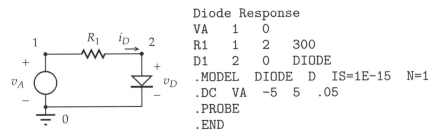

```
Diode Response
VA    1    0
R1    1    2    300
D1    2    0    DIODE
.MODEL  DIODE  D  IS=1E-15  N=1
.DC   VA   -5   5   .05
.PROBE
.END
```

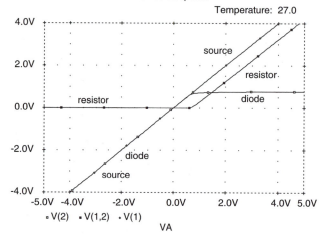

Figure 2.27: SPICE solution of a diode circuit.

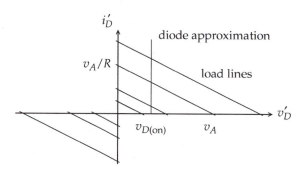

Figure 2.28: Piecewise linear diode approximation and load lines.

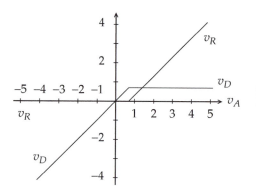

Figure 2.29: Solution for the diode circuit using a piecewise linear approximation for the diode.

This is a piecewise linear approximation for the diode – a type of approximation that is extensively used in obtaining solutions for electronic circuits with diodes as well as other nonlinear elements.

Several load lines, each corresponding to a different value of v_A, are indicated on the graph with the piecewise linear diode approximation. As may be seen from this graph, if $v_A > v_{D(\text{on})}$, then $v_D = v_{D(\text{on})}$, whereas if $v_A < v_{D(\text{on})}$, then $v_D = v_A$. This yields the approximate solution for v_D and v_R ($= v_A - v_D$) of Figure 2.29.

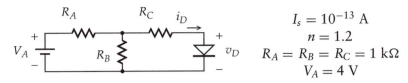

Figure 2.30: Diode circuit of Example 2.9.

$$I_s = 10^{-13} \text{ A}$$
$$n = 1.2$$
$$R_A = R_B = R_C = 1 \text{ k}\Omega$$
$$V_A = 4 \text{ V}$$

To summarize, the current expression with an exponential voltage dependence is generally used for computer simulations of diode circuits. Hence, the diode parameters I_s and n need to be known. However, for an analytic circuit solution, a piecewise linear approximation is often used to approximate the diode characteristic. Depending on the nature of the circuit, either the approximation used in this section to produce the result of Figure 2.29 or a more complex approximation, which will be discussed in the next chapter, may be used.

EXAMPLE 2.9

A junction diode with the parameters indicated is used in the circuit of Figure 2.30. Using an iterative approach, determine the diode voltage and current.

SOLUTION The linear part of the circuit, V_A, R_A, R_B, and R_C, may be replaced by a Thévenin equivalent circuit (Figure 2.31) as follows:

$$V_{Th} = \frac{R_B V_A}{R_A + R_B} = 2 \text{ V}$$
$$R_{Th} = R_A \| R_B + R_C = 1.5 \text{ k}\Omega$$

A value of 0.7 V will be assumed for the initial estimate of the diode voltage. This yields the following for i_{D1} and the corresponding voltage v_{D1}:

$$i_{D1} = (V_{Th} - 0.7 \text{ V})/R_{Th} = 0.867 \text{ mA}$$
$$v_{D1} = nV_T \ln(i_{D1}/I_s) = 0.686 \text{ V}$$

A new value of diode current i_{D2} and its corresponding voltage may now be obtained as follows:

$$i_{D2} = (V_{Th} - v_{D1})/R_{Th} = 0.876 \text{ mA}$$
$$v_{D2} = nV_T \ln(i_{D2}/I_s) = 0.687 \text{ V}$$

A corrected diode current of 0.875 mA and a diode voltage of 0.687 V are the desired solution (to the nearest millivolt for v_D).

Figure 2.31: Thévenin equivalent circuit of Example 2.9.

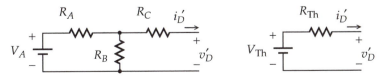

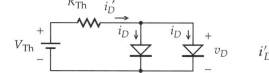

Figure 2.32: Diode circuit of Example 2.10(a).

$$i'_D = 2i_D$$

EXAMPLE 2.10

a. Repeat Example 2.9 for two diodes in parallel.
b. Repeat Example 2.9 for two diodes in series.

SOLUTION

a. The Thévenin equivalent circuit of Figure 2.32 will be used.

$$i'_D = 2I_s e^{v_D/nV_T}, \quad v_D = nV_T \ln(i'_D/2I_s)$$

Again, an initial value of 0.7 V will be assumed for the diode voltage as follows:

$$i'_{D1} = (V_{Th} - 0.7 \text{ V})/R_{Th} = 0.867 \text{ mA}$$
$$v_{D1} = nV_T \ln(i'_{D1}/2I_s) = 0.707 \text{ V}$$

This yields the following for a new current i'_{D2} and voltage v_{D2}:

$$i'_{D2} = (V_{Th} - v_{D1})/R_{Th} = 0.862 \text{ mA}$$
$$v_{D2} = nV_T \ln(i'_{D2}/2I_s) = 0.666 \text{ V}$$

The next iteration yields the following:

$$i'_{D3} = (V_{Th} - v_{D2})/R_{Th} = 0.889 \text{ mA}$$
$$v_{D3} = nV_T \ln(i'_{D3}/2I_s) = 0.666 \text{ V}$$

Hence, $i_D = i'_{D3}/2 = 0.445$ m A and $v_D = 0.666$ V.

b. The circuit of Figure 2.33 applies.

$$i_D = I_s e^{v'_D/2nV_T}, \quad v'_D = 2nV_T \ln(i_D/I_s)$$

Because the circuit has two diodes in series, a voltage of 1.4 V will be assumed for the initial value of v'_D.

$$i_{D1} = (V_{Th} - 1.4 \text{ V})/R_{Th} = 0.400 \text{ mA}$$
$$v'_{D1} = 2nV_T \ln(i_{D1}/I_s) = 1.327 \text{ V}$$

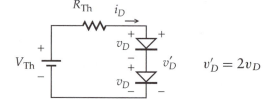

Figure 2.33: Diode circuit of Example 2.10(b).

$$v'_D = 2v_D$$

These values will be used for the next iteration as follows:

$$i_{D2} = (V_{Th} - v'_{D1})/R_{Th} = 0.449 \text{ mA}$$
$$v'_{D2} = 2nV_T \ln(i_{D2}/I_s) = 1.334 \text{ V}$$

The next iteration yields the following:

$$i_{D3} = (V_{Th} - v'_{D2})/R_{Th} = 0.444 \text{ mA}$$
$$v'_{D3} = 2nV_T \ln(i_{D3}/I_s) = 1.333 \text{ V}$$

A final iteration yields $i_D = 0.445$ mA and $v'_D = 1.333$ V. Hence, the diode voltage is 0.667 V.

EXAMPLE 2.11

Determine and sketch the dependence of v_{OUT} on v_{IN} for the circuit of Figure 2.34 with two diodes. Assume that the diodes may be approximated as having a constant potential difference of $v_{D(on)}$ when forward biased.

SOLUTION An input voltage of at least $v_{D(on)}$ is necessary to forward bias D_1. Hence, for $v_{IN} < v_{D(on)}$, the current of D_1 is zero and $v_{OUT} = 0$. Consider the case for which $v_{IN} > v_{D(on)}$ but sufficiently small that $v_{OUT} < v_{D(on)}$. The current of D_2 is zero, and diode D_1 can be replaced by a battery with a potential of $v_{D(on)}$ for this case (Figure 2.35). Superposition may be used to determine v_{OUT} by

$$v_{OUT} = \frac{R_2 v_{IN}}{R_1 + R_2} - \frac{R_2 v_{D(on)}}{R_1 + R_2} = (v_{IN} - v_{D(on)})/2$$

Figure 2.34: Diode circuit of Example 2.11.

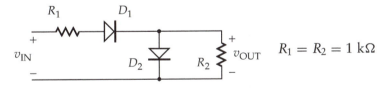

$$R_1 = R_2 = 1 \text{ k}\Omega$$

Figure 2.35: Equivalent diode circuit for $v_{IN} > v_{D(on)}$ and $v_{OUT} < v_{D(on)}$.

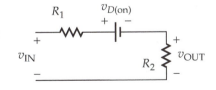

Figure 2.36: Solution of Example 2.11.

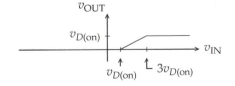

When v_{OUT} reaches $v_{D(on)}$, diode D_2 conducts and v_{OUT} is limited to $v_{D(on)}$. As may be seen from the preceding expression, this occurs for $v_{IN} > 3v_{D(on)}$. The following is therefore obtained for v_{OUT}:

$$v_{OUT} = \begin{cases} 0 & \text{for } v_{IN} < v_{D(on)} \\ (v_{IN} - v_{D(on)})/2 & \text{for } v_{D(on)} < v_{IN} < 3v_{D(on)} \\ v_{D(on)} & \text{for } v_{IN} > 3v_{D(on)} \end{cases}$$

2.6 MODELING THE JUNCTION DIODE: THE ROLE OF APPROXIMATIONS

In the previous section an exponential current-versus-voltage relationship was used for semiconductor junction diodes to solve circuits with diodes as given by

$$i_D = I_s(e^{v_D/nV_T} - 1) \tag{2.39}$$

This relationship, which may be derived from a theoretical analysis of an idealized semiconductor junction, does fairly well in predicting the behavior of actual diodes. An exponential relationship, however, tends to require numerical, iterative-type circuit solutions. Fortunately, sufficiently accurate solutions can often be obtained by approximating the behavior of the diode with a relatively simple circuit model. The exponential relationship tends to be used only for computer simulations; approximate circuit models are nearly always used for "pencil-and-paper"-type analyses. Approximate circuit models not only simplify calculations, but, more importantly, they provide a much better conceptual understanding of the behavior of diode circuits. In addition, the design of diode circuits relies on one's ability to conceptualize the behavior of diodes.

THE IDEAL DIODE SWITCH MODEL

An approximate diode model, suitable for some applications, is that of a circuit element generally referred to as an ideal diode. It has the idealized characteristic of Figure 2.37. The voltage across an ideal diode, when it is forward biased $(i_D > 0)$, is zero, and its current, when reverse biased $(v_D < 0)$, is zero. This model can be used to replace an actual semiconductor diode in a circuit when the voltages of the circuit are large compared with the small forward voltage of an actual diode.

Because either the voltage or current of an ideal diode is zero, the diode may be treated as the ideal switch of Figure 2.38. The switch is either open,

Figure 2.37: Actual and idealized characteristics of a diode.

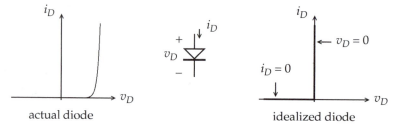

actual diode   idealized diode

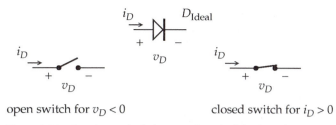

open switch for $v_D < 0$ closed switch for $i_D > 0$

Figure 2.38: Simulating the behavior of an ideal diode with a switch.

corresponding to a reverse-biased condition, or closed, corresponding to a forward-biased condition. The condition of the switch depends on its voltage or current which, in turn, depends on the other elements of the circuit. It is necessary to make a binary logic decision as to the condition of the switch (decide if it is open or closed). The difficulty is that of ascertaining the correct condition. For some circuits, the choice may be obvious, whereas for others, especially if a circuit has several diodes, the choice may not be so obvious. If the decision is not obvious, it is necessary to guess and proceed using a trial-and-error solution. One solves the circuit for the initial decision; a simple analytic solution is generally possible if there are no other nonlinear elements. If it was assumed that the diode was reverse biased (an open switch), its resultant voltage must be negative. Therefore, if the circuit yields a positive voltage, the decision was in error. Similarly, if the diode was assumed to be conducting (closed switch), a positive diode current is necessary; a negative current implies the decision was in error. If the circuit has more than a single diode, an open or closed decision is required for each diode. However, only one set of decisions will tend to yield a circuit solution that is valid for each diode. If the resultant diode current and voltage are both zero, then either diode condition is valid and they both yield the correct circuit solution.

To illustrate the process of obtaining a solution for a circuit in which ideal diodes are used to simulate the behavior of actual diodes, consider the elementary logic circuit of Figure 2.39. Although the actual logic levels might be 0 and 5 V for this circuit, suppose a solution is desired for $v_B = 3$ V, an intermediate voltage level. Assume $v_A = 5$ V. To start, an initial guess that both diodes are reverse biased, that is, that they behave as open switches, will be tried. It is obvious that $v_C = 0$ for this condition. Furthermore, because the diode currents are zero, $v_{D1} = v_A = 5$ V and $v_{D2} = v_B = 3$ V. This is not a valid solution because the diode voltages must be negative for the reverse-biased condition.

To continue, let us make the guess that both diodes are forward biased, that is, they behave as closed switches (Figure 2.40). Summing the currents at the center mode results in the following solution for v_C:

Figure 2.39: Diode logic circuit.

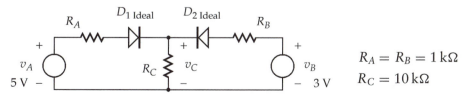

$R_A = R_B = 1\,\mathrm{k}\Omega$

$R_C = 10\,\mathrm{k}\Omega$

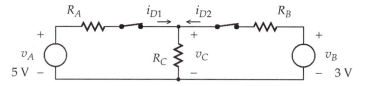

Figure 2.40: The equivalent circuit for Figure 2.39 on the assumption that both diodes are forward biased.

$$\frac{v_A - v_C}{R_A} + \frac{v_B - v_C}{R_B} = \frac{v_C}{R_C}$$

$$v_C \left(\frac{1}{R_A} + \frac{1}{R_B} + \frac{1}{R_C} \right) = \frac{v_A}{R_A} + \frac{v_B}{R_B} \tag{2.40}$$

$$v_C = 3.81 \text{V}$$

This implies the following for the diode currents:

$$i_{D1} = (v_A - v_C)/R_A = 1.19 \text{ mA}$$
$$i_{D2} = (v_B - v_C)/R_B = -0.81 \text{ mA} \tag{2.41}$$

Because the current of D_2 is negative, the initial assumption of both diodes being forward biased (closed switches) is obviously not valid. One diode must be forward biased and the other reverse biased. Suppose it is assumed that D_1 is conducting (a closed switch) and that D_2 is reverse biased (an open switch):

$$i_{D2} = 0$$
$$i_{D1} = v_A/(R_A + R_C) = 0.455 \text{ mA}, \quad v_C = 4.55 \text{ V} \tag{2.42}$$
$$v_{D2} = v_B - v_C = -1.55 \text{ V}$$

This is a valid condition, and it may readily be shown that the alternative assumption for the diodes is not valid. Solving the circuit for this last condition, however, is not necessary because only one set of assumptions for the conditions of the diodes will yield a valid solution for this circuit (unless both the current and voltage are zero).

CONSTANT FORWARD-BIASED VOLTAGE DIODE MODEL

The reader might be tempted to argue that, from a practical perspective, the ideal diode model is of limited value. For the potentials of the circuit just considered (Figure 2.39), the forward voltage of a typical semiconductor junction diode is not negligible. A better approximation for the diode, introduced in the previous section, was the assumption of a constant forward-biased voltage of $v_{D(\text{on})}$ (≈ 0.7 V for silicon diodes). A circuit model consisting of an ideal diode in series with a battery having a potential of $v_{D(\text{on})}$ results in this characteristic (Figure 2.41). Again, after this model is substituted in the circuit for the actual diode, the circuit can readily be analyzed by assuming either an open or closed condition for the switch associated with the ideal diode of the model.

Let us again consider the diode logic circuit of Figure 2.39, but this time a diode model that results in a forward-biased voltage of $v_{D(\text{on})} = 0.7$ V will be used (Figure 2.42). The diode conditions (open or closed switch) for the circuit in which ideal behavior of the diodes was assumed is a reasonable starting point. There is a high likelihood (but not a certainty) of the solution with ideal diodes being the correct set of conditions for the modified diode model. Therefore, it will be assumed that $D_{1\,\text{Ideal}}$ is forward biased (closed switch) and that $D_{2\,\text{Ideal}}$ is reverse biased (open switch).

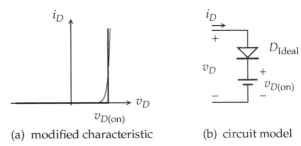

(a) modified characteristic   (b) circuit model

Figure 2.41: The constant forward-biased voltage diode model.

$$i_{D2} = 0$$
$$i_{D1} = (v_A - v_{D(\text{on})})/(R_A + R_C) = 0.391 \text{ mA}, \quad v_C = 3.91 \text{ V} \tag{2.43}$$
$$v_{D2} = v_B - v_C = -0.91 \text{ V}$$

For this solution the current of D_1 is positive, and the voltage of D_2 is less than $v_{D(\text{on})}$, as required. Although it is this diode model that is extensively used for analyzing electronic circuits, a further improvement is necessary for some applications.

DIODE MODEL WITH A SERIES RESISTOR

The diode model just considered (Figure 2.41) yields a forward-biased diode voltage that is independent of the current of the diode ($i_D > 0$). Although the change in voltage of a semiconductor junction diode is small for modest changes in current, a knowledge of this small change is at times important. This suggests a diode model that incorporates an equivalent diode resistance. Consider the diode characteristic of Figure 2.43 with the approximation indicated. Instead of a vertical straight line for the diode current, a straight line with a slope α that approximately matches that of the exponential is introduced:

$$i_D = \begin{cases} \alpha(v_D - V_\gamma) & \text{for } v_D > V_\gamma \\ 0 & \text{for } v_D < V_\gamma \end{cases} \tag{2.44}$$

The condition $v_D > V_\gamma$ results in the following:

$$v_D = V_\gamma + i_D/\alpha = V_\gamma + i_D r_d \tag{2.45}$$

The equivalent diode resistance r_d is thus the reciprocal of the slope of the straight line used to approximate the exponential behavior of the diode. When the ideal

Figure 2.42: A circuit using the constant forward-biased voltage diode model.

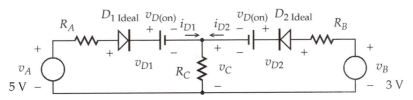

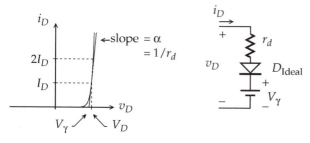

Figure 2.43: A diode model with an equivalent resistance.

diode of Figure 2.43 is forward biased (a closed switch), the diode voltage is that given by Eq. (2.45). However, when $v_D < V_\gamma$, the ideal diode of the model is reverse biased and it behaves as an open circuit. This model, although providing a more accurate solution than the previous models, will not result in the same solution as the exponential diode relationship unless the solution happens to fall at the point at which the curves coincide (either cross or are tangent).

The problem that must be addressed when using this model is that of determining values for the parameters of V_γ and r_d. Unfortunately the appropriate values of the parameters depend on the circuit in which the diode is to be used, that is, on the resultant current. Consider the exponential current-versus-voltage relationship for a forward-biased junction diode in which $-I_s$ may be ignored:

$$i_D = I_s e^{v_D/nV_T}$$
$$\frac{di_D}{dv_D} = \frac{1}{nV_T} I_s e^{v_D/nV_T} = \frac{i_D}{nV_T} \tag{2.46}$$

As may be seen from the expression for the derivative of i_D, it is not possible to associate a unique slope with a diode characteristic.

Consider the case for a particular diode voltage and current V_D and I_D (note the capital letters). At this point, the slope of the diode characteristic is I_D/nV_T. A diode model consisting of a straight line that is tangent to this point will be, for values of i_D and v_D that are close to I_D and V_D, appropriate. The equivalent resistance of the model r_d is thus nV_T/I_D. For a current of 1 mA, $r_d = 25 \ \Omega$ ($V_T = 25$ mV and $n = 1$). The equivalent voltage V_γ, is readily obtained using the slope of the tangent line as follows:

$$\text{slope} = I_D/nV_T = I_D/(V_D - V_\gamma)$$
$$V_D - V_\gamma = nV_T, \qquad V_\gamma = V_D - nV_T \tag{2.47}$$

Hence, the voltage V_γ is very near to V_D (only 25 mV removed for $V_T = 25$ mV and $n = 1$). It should be noted that for $i_D = 2I_D$, the diode model predicts a diode voltage of $V_D + nV_T$, whereas the exponential expression yields a slightly smaller voltage of $V_D + 0.693nV_T$.

EXAMPLE 2.12

Assume that the behavior of the diodes of the circuit of Figure 2.39 can be approximated with ideal diodes. The voltage v_A varies over the range of 0 to 5 V with the extremes of 0 and 5 V corresponding to valid logic levels.

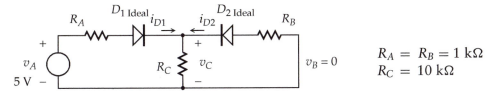

Figure 2.44: Circuit for Example 2.12(a).

$$R_A = R_B = 1\ k\Omega$$
$$R_C = 10\ k\Omega$$

Determine the condition of the diodes (reverse or forward biased) as a function of v_A for the following values of v_B:

a. $v_B = 0$ V.
b. $v_B = 2.5$ V.
c. $v_B = 5.0$ V.

SOLUTION

a. The circuit of Figure 2.44 applies. For $v_C > 0$, $D_{2\,Ideal}$ is reverse biased (an open switch) and $D_{1\,Ideal}$ is forward biased (a closed switch). This occurs for $v_A > 0$.

b. The circuit of Figure 2.45 applies for $v_B = 2.5$ V. If $D_{1\,Ideal}$ is reverse biased (an open switch), v_C is determined by v_B and is independent of v_A.

$$v_C = \frac{R_C v_B}{R_B + R_C} = 2.273\ V$$

Hence, $D_{1\,Ideal}$ is reverse biased for $v_A < 2.273$ V. For both diodes forward biased (closed switches), the following is obtained:

$$\frac{v_A - v_C}{R_A} + \frac{v_B - v_C}{R_B} = \frac{v_C}{R_C}$$
$$v_C = v_A/2.1 + v_B/2.1 = 0.476v_A + 1.190\ V$$

This condition is valid only if $i_{D2} > 0$.

$$i_{D2} = (v_B - v_C)/R_B$$
$$= 1.310 - 0.476v_A \quad mA$$

This implies $v_A < 2.752$ V. To summarize,

$v_A < 2.273$ V; $D_{1\,Ideal}$reverse-biased, $D_{2\,Ideal}$forward-biased
2.273 V $< v_A < 2.752$ V; both diodes forward biased
$v_A > 2.752$ V; $D_{1\,Ideal}$forward-biased, $D_{2\,Ideal}$reverse-biased

c. $v_B = 5.0$ V. If $D_{1\,Ideal}$ is reverse biased, the following is obtained for v_C:

Figure 2.45: Circuit for Example 2.12(b).

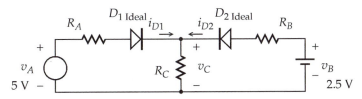

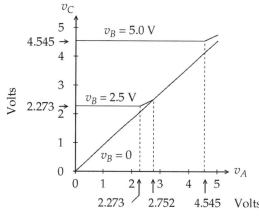

Figure 2.46: Voltage transfer characteristic of Example 2.13.

$$v_C = \frac{R_C v_B}{R_B + R_C} = 4.545 \text{ V}$$

Therefore, this condition applies for $v_A < 4.545$ V. For $v_A > 4.545$ V, both diodes are forward biased.

EXAMPLE 2.13

Determine the voltage transfer characteristic v_C as a function of v_A for Example 2.12 ($v_B = 0$, 2.5, and 5.0 V).

SOLUTION

a. For $v_B = 0$,

$$v_C = \frac{R_C v_A}{R_A + R_C} = 0.909 v_A$$

b. For $v_B = 2.5$ V,

$$v_A < 2.273 \text{ V}, \qquad v_C = 2.273 \text{ V}$$
$$2.273 \text{ V} < v_A < 2.752 \text{ V}, \qquad v_C = 0.476 v_A + 1.190 \text{ V}$$
$$v_A > 2.752 \text{ V}, \qquad v_C = \frac{R_C v_A}{R_A + R_C} = 0.909 v_A$$

c. For $v_B = 5.0$ V,

$$v_A < 4.454 \text{ V}, \qquad v_C = 4.545 \text{ V}$$
$$v_A > 4.545 \text{ V}, \qquad \frac{v_A - v_C}{R_A} + \frac{v_B - v_C}{R_B} = \frac{v_C}{R_C}$$
$$v_C = 0.476 v_A + 2.381 \text{ V}$$

EXAMPLE 2.14

Junction diodes are often used in integrated circuits to produce a desired voltage level that is nearly independent of the supply voltage of the integrated circuit. The circuit of Figure 2.47 results in a voltage of approximately 1.4 V.

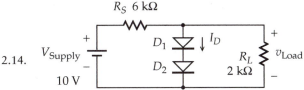

Figure 2.47: Diode circuit of Example 2.14.

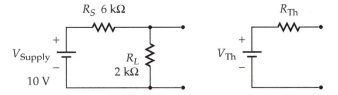

Figure 2.48: Thévenin equivalent circuit of Example 2.14.

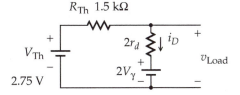

Figure 2.49: Equivalent circuit using a diode model with a resistance.

Determine the variation in load voltage v_{Load} for a ± 1 V variation in supply voltage, V_{Supply}. Assume $v_{D(\text{on})} = 0.7$ V, $n = 1$, and $V_T = 25$ mV for the diodes.

SOLUTION With the exception of the diodes, the circuit of Figure 2.47 may be replaced by a Thévenin-equivalent circuit (Figure 2.48) as follows:

$$V_{\text{Th}} = \frac{R_L V_{\text{Supply}}}{R_S + R_L} = 2.50 \text{ V}, \qquad R_{\text{Th}} = R_L \| R_S = 1.50 \text{ k}\Omega$$

Using the constant forward-biased voltage model for the diodes, a diode current I_D can be obtained for $V_{\text{Supply}} = 10$ V.

$$V_{\text{Th}} = I_D R_{\text{Th}} + 2 v_{D(\text{on})}$$
$$I_D = \left(V_{\text{Th}} - 2 v_{D(\text{on})}\right) / R_{\text{Th}} = 0.733 \text{ mA}$$

The load voltage is 1.40 V for this condition. To determine the effect of a change in supply voltage, a diode model with an equivalent resistance is required. It will be assumed that the diode current remains close to 0.733 mA (I_D).

$$r_d = n V_T / I_D = 34.1 \ \Omega$$

If the diode voltage V_D is assumed to be equal to $v_{D(\text{on})}$ for a current of I_D, then $V_\gamma = v_{D(\text{on})} - 0.025 \text{ V} = 0.675$ V. For $V_{\text{Supply}} = 11$ V, a $+1$-V change, $V_{\text{Th}} = 2.75$ V. The ideal diodes of Figure 2.49 have been assumed to be forward biased (closed switches).

$$i_D = (V_{\text{Th}} - 2 V_\gamma)/(R_{\text{Th}} + 2 r_d) = 0.893 \text{ mA}$$
$$v_{\text{Load}} = 2 i_D r_d + 2 V_\gamma = 1.411 \text{ V}$$

The load voltage increased by 11 mV (less than 1 percent). For $V_{Supply} = 9$ V, -1-V change, $V_{Th} = 2.25$ V.

$$i_D = 0.574 \text{ mA}, \qquad v_{Load} = 1.389 \text{ V}$$

This results in a load voltage decrease of 11 mV.

2.7   THE PHOTOVOLTAIC CELL: PHOTON–SEMICONDUCTOR INTERACTIONS

A photovoltaic cell converts the energy of photons from an external source (such as the sun) to an electrical current. Willoughby Smith is credited with the discovery of the selenium photoconductor in 1873 (Wolf 1981). Selenium photovoltaic cells, even though less than 1 percent efficient in converting visible light to electrical energy, have been used for photographic light meters throughout the 20th century. It was not until 1953, a few years after the invention of the transistor, that the junction diode photovoltaic cell was developed (Chapin, Fuller, and Pearson 1954; Smits 1976). Even the earliest of these silicon photovoltaic cells achieved efficiencies as high as 6 percent, a considerable improvement over selenium photovoltaic cells.

PHOTONS

A photovoltaic cell relies on the interaction of photons of visible radiation, or of radiation of adjacent spectral bands, with the valence electrons of the semiconductor from which it has been fabricated (Figure 2.50). The relative spectral intensity of the sun is a maximum for a wavelength of approximately 0.5 μm (the same wavelength for which the response of the human eye is at a maximum). The visible spectrum, it will be noted, is relatively narrow (about 0.38 to 0.78 μm). For semiconductor interactions, it is the energy of individual photons that is of particular importance.

$$\begin{aligned} E_{photon} &= hf = hc/\lambda \text{ J} \\ &= hc/q\lambda = 1.242/\lambda_{\mu m} \text{ eV} \end{aligned}$$

Figure 2.50: Relative spectral intensity of sun and photon energy.

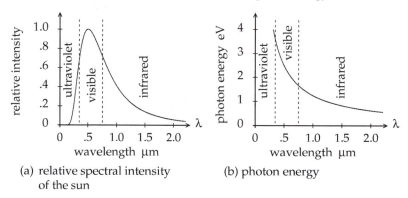

(a) relative spectral intensity of the sun

(b) photon energy

where

h = Planck's constant, 6.625×10^{-34} J·s
f = frequency, Hz
c = velocity of light, 3×10^8 m/s   (2.48)
λ = wavelength, m
$\lambda_{\mu\text{m}}$ = wavelength, μm

The photon's energy expressed in electron volts is equivalent to its energy expressed in joules divided by the electronic charge q (1.6×10^{-19} C).

The band-gap energy of a semiconductor E_g is the energy required for a valence electron of an intrinsic semiconductor to break its bond and form an electron–hole pair. For silicon, this energy is 1.1 eV. Hence, a photon with an energy of 1.1 eV or more has the potential of producing a free electron–hole pair through an interaction with a valence electron of a silicon semiconductor. As may be seen from Figure 2.50(b), radiation with wavelengths of less than 1.13 μm (this includes the entire visible spectrum of the sun) consists of sufficiently energetic photons.

During the development of the transistor it was recognized that energetic photons could produce free electron–hole pairs near the surface of the semiconductor. For a photovoltaic cell, this effect is optimized by fabricating a junction diode with a very large surface area. An n- on p-type photovoltaic cell is illustrated in Figure 2.51. For this device, an extremely thin n-type region is diffused into a heavily doped p-type substrate. Very narrow metallic contacts are used to provide the electrical connection to the n-type region and an antireflection coating is used to reduce optical losses.

The operation of a photovoltaic cell is dependent upon photons generating electron–hole pairs in the vicinity of the semiconductor junction. Consider the case for which photons traverse the thin n-type region (typically only a fraction of a micron thick) and interact with valence electrons of the p-type region to produce free electrons and holes. For the heavily doped p-type semiconductor, the effect of the additional holes will not be significant. However, the generation of additional free electrons, minority carriers in the p-type region, will have a significant effect if it occurs near the junction. As a result of the potential difference across the semiconductor junction, minority carriers readily cross the junction and give rise to the reverse-biased current $-I_s$ of a conventional junction diode. The photon-generated free-electron minority carriers of the p-type region will also tend to

Figure 2.51: Silicon photovoltaic cell.

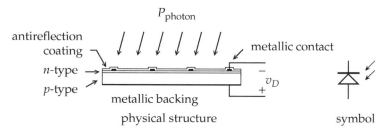

physical structure

symbol

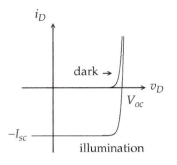

Figure 2.52: Current versus voltage of a photovoltaic cell.

cross the diode junction, thus contributing to the reverse-biased diode current. A similar situation occurs when photons interact with valence electrons of the thin n-type region, except it is now holes that are minority carriers. These holes will also contribute to the magnitude of the reverse-biased current of the junction.

The illumination of a photovoltaic cell can therefore be accounted for in the current expression of a junction diode with a current component $-I_{\mathrm{photon}}$.

$$i_D = I_s(e^{v_D/nV_T} - 1) - I_{\mathrm{photon}} \qquad (2.49)$$

The current I_{photon} depends on the intensity of the illumination and its spectral distribution. Furthermore, the current depends on the optical properties of the semiconductor. The current-versus-voltage characteristic of Eq. (2.49) is indicated in Figure 2.52. For a practical photovoltaic cell, the photon generation rate of minority carriers must be very much larger than the rate minority carriers are generated by thermal processes. This implies that I_{photon} is very much larger than I_s (the $-I_s$ term of Eq. (2.48) can therefore be ignored). The characteristic of Figure 2.52 is particularly interesting because the product of i_D and v_D is negative in the III quadrant of the graph. A negative value of $i_D v_D$ implies that electrical power is being supplied by the photovoltaic cell in much the same manner that a battery supplies electrical power (the product of the current into a battery's positive mode and its voltage is negative for a passive resistive load). It is not surprising, therefore, that "solar battery" has been used to describe this device (Raisbeck 1955).

Consider the situation in which a photovoltaic cell is used to supply electrical power to a resistor load R_L. The diode of Figure 2.53 is a junction diode with a characteristic corresponding to the dark response of the photovoltaic cell. The current source I_{photon} accounts for the effect of the photon-generated minority carriers. For illuminating radiation with a fixed spectral distribution, I_{photon} tends to be linearly proportional to the power density of the incident radiation. The load line for the resistor is a straight line through the origin of the diode characteristic.

$$i_D' = -v_D'/R_L \quad \text{load line of resistor} \qquad (2.50)$$

The minus sign accounts for the direction used for the current in Figure 2.53. As for the case of the diode circuits considered in the previous section, the intersection of the load line and the photovoltaic cell characteristic results in a solution for the circuit (Figure 2.54).

Because i_D is a negative quantity, the power delivered to the load resistor is positive. Energy of the illuminating radiation (generally that from the sun) is converted to electrical energy. Because not all photons generate minority carriers (long-wavelength photons have insufficient energy) and many photons

Figure 2.53: Equivalent circuit of a photovoltaic cell.

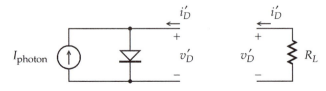

have excessive energy (greater than E_g), a concurrent thermal heating of the diode occurs. This thermal energy is removed through the ambient environment of the photovoltaic cell. An increase in temperature of the junction, it will be noted, has the undesirable effect of increasing the reverse saturation current of the diode, I_s, which tends, for a given circuit, to decrease the diode's terminal voltage.

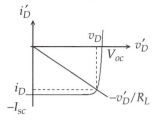

Figure 2.54: Photovoltaic cell characteristic and load line.

Two quantities, I_{sc} and V_{oc}, the short-circuit current and open-circuit voltage, respectively, are indicated in Figures 2.52 and 2.54.

$$I_{sc} = I_{photon} \quad \text{because } v_D = 0 \tag{2.51}$$

An expression for the open-circuit voltage V_{oc} is obtained from Eq. (2.48) by setting $i_D = 0$ and assuming $I_s \ll I_{photon}$.

$$0 = I_s e^{V_{oc}/nV_T} - I_{photon}$$
$$V_{oc} = V_T \ln(I_{photon}/I_s) \tag{2.52}$$

As may be seen in Figure 2.54, $|i_D| < I_{sc}$ and $v_D < V_{oc}$. Therefore, the electrical power supplied by the photovoltaic cell $|i_D|v_D$, is less than $I_{sc}V_{oc}$. The easy-to-calculate quantity $I_{sc}V_{oc}$ is useful for estimating the upper limit of the power that a photovoltaic cell might supply.

A numerical iterative type solution may readily be obtained for a photovoltaic cell with a load resistor (Figure 2.55). This is readily accomplished by transforming the current source I_{photon} and the load resistor R_L, to a Thévenin equivalent circuit (Figure 2.55(b)).

$$R_{Th} = R_L, \qquad V_{Th} = I_{photon} R_L \tag{2.53}$$

Using this circuit, the diode voltage v_D may readily be obtained by the procedure of the preceding section. Once v_D is known, the original circuit of Figure 2.55(a) may be used to calculate the electrical power supplied to the load (v_D^2/R_L) or any other quantity of interest. It should be noted that the Thévenin equivalent circuit of Figure 2.55(b) cannot be used to calculate the current of R_L or the power that it dissipates.

Photovoltaic cells have not only been extensively used to power communication satellites, but a major research and development effort has been directed toward utilizing these cells for terrestrial applications. Besides providing electrical power in remote locations (for example, to power communications repeaters),

Figure 2.55: The equivalent circuit for a photovoltaic cell with a load resistor.

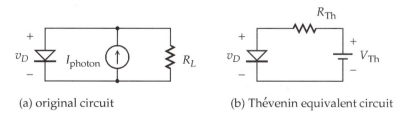

(a) original circuit

(b) Thévenin equivalent circuit

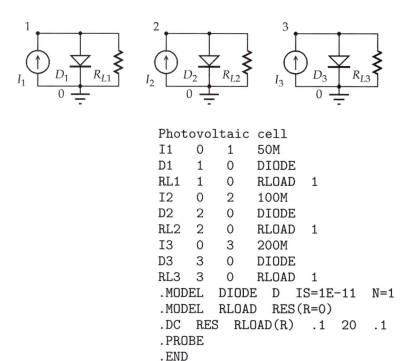

```
Photovoltaic cell
I1      0    1    50M
D1      1    0    DIODE
RL1     1    0    RLOAD   1
I2      0    2    100M
D2      2    0    DIODE
RL2     2    0    RLOAD   1
I3      0    3    200M
D3      3    0    DIODE
RL3     3    0    RLOAD   1
.MODEL   DIODE   D   IS=1E-11   N=1
.MODEL   RLOAD   RES(R=0)
.DC   RES   RLOAD(R)   .1   20   .1
.PROBE
.END
```

Figure 2.56: The circuit and SPICE file of Example 2.15.

large arrays of photovoltaic cells have been used on an experimental basis for producing electrical power that would otherwise be produced through the combustion of fossil fuels. Although solar radiation is essentially free (a zero fuel cost), the cost of collecting it remains high. Intricate fabricating processes and the need for highly purified semiconductor materials account for the high cost of photovolatic cells.

There has been a concerted effort to improve the efficiency of photovoltaic cells through the use of alternative semiconductor materials, multiple layer junctions, and solar concentrating systems. In addition, alternative fabricating techniques along with polycrystalline and amorphous semiconductor materials have been used to reduce costs. It is expected that photovoltaic cells will eventually play a significant role in the generation of electrical power (Hubbard 1989).

EXAMPLE 2.15

A small photovoltaic cell at different illumination levels has currents of 50, 100, and 200 mA for I_{photon}. The cell has parameters of $I_s = 10^{-11}$ A and $n = 1$.

a. Determine I_{sc}, V_{oc}, and I_{sc}, V_{oc} of the cell for each of the illumination levels.

b. Use SPICE to determine the load resistance R_L that results in a maximum electrical power output for each level of illumination. What is the diode voltage, as well as the load current and power for each of these conditions?

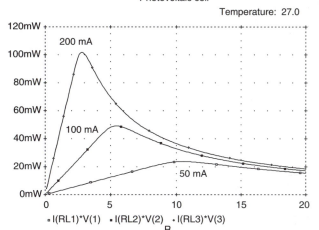

Figure 2.57: SPICE solution for Example 2.15.

SOLUTION

a. $I_{sc} = 50$, 100, and 200 mA for the three illumination levels. Equation (2.52) yields the open-circuit voltage V_{oc} ($V_T = 0.025$ V).

$$V_{oc} = 0.558 \text{ V}, \quad I_{\text{photon}} = 50 \text{ mA}$$
$$= 0.576 \text{ V}, \quad I_{\text{photon}} = 100 \text{ mA}$$
$$= 0.593 \text{ V}, \quad I_{\text{photon}} = 200 \text{ mA}$$

The products I_{sc}, V_{oc} are 27.9, 57.6, and 118.6 mW.

b. Three circuits, each with a different independent current source, will be used to determine the output power. A sweep of the load resistance, which has been specified through a .MODEL statement, results in the following .PROBE graph for power:

I_{photon}	$R_L(\text{max})$	$P_L(\text{max})$	v_D	i_L	V_{oc}/I_{sc}
50 mA	10.5 Ω	23.8 mW	0.513 V	45.8 mA	11.2 Ω
100 mA	5.4 Ω	49.2 mW	0.516 V	95.5 mA	5.76 Ω
200 mA	2.80 Ω	102 mW	0.534 V	191 mA	2.97 Ω

It will be noted that V_{oc}/I_{sc} yields a fairly accurate value of R_L for the maximum output power (within 7 percent). Furthermore, the values of $I_{sc}V_{oc}$ are within 15 percent of the maximum powers.

EXAMPLE 2.16

To increase the output power of a photovoltaic system, photovoltaic cells are generally assembled into an array. Consider the case in which cells with the parameters of Example 2.15 are used and the illumination results in $I_{\text{photon}} = 200$ mA.

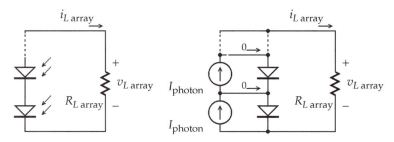

Figure 2.58: Equivalent circuit of 12 photovoltaic cells connected in series.

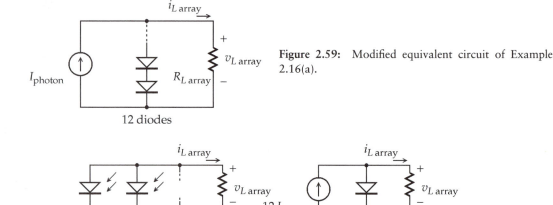

Figure 2.59: Modified equivalent circuit of Example 2.16(a).

Figure 2.60: Equivalent circuit of Example 2.16(b).

a. Suppose that 12 cells are connected in series to increase the terminal voltage of the array. Determine the open-circuit voltage and short-circuit current of the array. What is the value of the load resistance that results in a maximum power? What is the power and terminal voltage?

b. Suppose that the cells are connected in parallel to increase the output current. Repeat part (a) for this condition.

SOLUTION

a. A series circuit results in the equivalent circuit of Figure 2.58 where $R_{L\,array}$ is the load resistance. For identical photovoltaic cells, the currents of each of the diodes of the equivalent circuit will be equal. Hence, the currents of the horizontal connections between the current sources and the diodes will be zero. This results in the equivalent circuit of Figure 2.59.

$$I_{sc\,array} = I_{photon} = 200 \text{ mA}$$
$$V_{oc\,array} = 12V_T \ln(I_{photon}/I_s) = 12V_{oc} = 7.1 \text{ V}$$

Maximum power will be obtained for $R_{L\,array}$ being equal to 12 times the resistance that resulted in a maximum for a single cell.

$$R_{L\,array} = 12R_L(\text{max}) = 33.6\ \Omega, \quad P_{L\,array} = 12P_L(\text{max}) = 1.22 \text{ W}$$

b. Connecting the cells in parallel results in the following equivalent circuit:

$$I_{sc \, array} = 12 I_{photon} = 2.4 \text{ A}, \quad V_{oc \, array} = V_{oc} = 0.593 \text{ V}$$

To obtain maximum output power, $R_{L \, array} = R_L(\max)/12 = 0.233 \, \Omega$ and $P_{L \, array} = 12 P_L(\max) = 1.22 \text{ W}$.

2.8 LIGHT-EMITTING AND LASER DIODES: OPTICAL COMMUNICATION

In a photovoltaic semiconductor diode, electron–hole pairs are generated as a result of an interaction of incident photons with the semiconductor lattice. The complementary reaction can also occur, that is, the energy released when a free electron and hole recombines can be given off as a photon. Although this latter process does not occur in silicon junction diodes, it does occur for diodes fabricated from other semiconductors such as gallium arsenide and gallium phosphide.

In 1907, British scientist Henry J. Round observed the emission of light (luminescence) from a point-contact silicon carbide crystal diode of the type he was using for the detection of radio signals (Loebner 1976; Round 1907). Electroluminescence, was "rediscovered" in the Soviet Union in 1922 by Oleg V. Lasev. It was not, however, until 1951 that this low-voltage behavior of a forward-biased silicon carbide diode was explained by a U.S. scientist, Kurt Lehovec (Lehovec, Accardo, and Jamgochian 1951). Lehovec recognized that the electroluminescence of these diodes was due to the recombination of carriers that had crossed the semiconductor junction, that is, photons were being emitted when electrons and holes recombined. Practical light-emitting diodes (LEDs) were developed in the late 1960s and were first used for displays of calculators and other scientific instruments in the early 1970s.

LIGHT-EMITTING DIODES

A set of typical forward-biased current-versus-voltage characteristics of low-power LEDs having different wavelength emissions is given in Figure 2.61. It may be seen that considerably greater forward-bias voltages are required for these diodes than for the silicon junction diodes that have been previously considered.

Figure 2.61: Typical terminal characteristics of low-power LEDs.

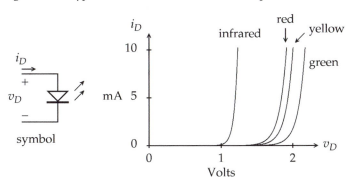

This should not be surprising. It will be noted, for example, that yellow light corresponds to a wavelength of approximately 0.59 μm. On the basis of Eq. (2.48), this implies a photon energy of approximately 2.1 eV. This energy corresponds to the band-gap energy of the semiconductor, the energy required to produce an electron–hole pair, and, consequently, to the energy that is given off when an electron and hole recombine. A larger band-gap energy (that of silicon is only 1.1 eV) translates into a larger forward-bias voltage. Furthermore, a direct recombination process, which does not occur in silicon, is required for the emission of a photon. The energy given off when an electron and hole recombine in silicon is transferred to the crystal lattice as thermal energy (a simple heating occurs).

To produce semiconductors with the band-gap energies required for visible and near-infrared emissions, crystals of III–V compounds are generally used (see Table 2.2). A gallium (III) arsenide (V) crystal, for example, consists of a lattice in which half of the atoms are gallium with three valence electrons and the other half are arsenic atoms with five valence electrons. The semiconductor crystal has a band-gap energy of 1.43 eV, whereas a crystal of gallium phosphide (phosphorous also has five valence electrons) has a band-gap energy of 2.26 eV. Semiconductors with band-gap energies between 1.43 and 2.26 eV (corresponding to photon wavelengths of 0.87 to 0.55 μm) can be produced by using a mixture of arsenic and phosphorus for the valence five atoms. For near-infrared emissions ($\lambda \approx 0.9$ μm), a smaller band-gap energy is required. This is obtained with other semiconductor alloys as well as semiconductors produced from II–VI compounds.

LIGHT-EMITTING DIODE APPLICATIONS

The earliest utilization of LEDs was the single "spot" indicator replacing incandescent pilot lamps. This was followed by the seven-segment LED display for the readout of numerical values that replaced much more expensive gaseous discharge tubes. Each segment in this display is a separate light-emitting diode that can be activated with a forward-biased current (Figure 2.62). Individual resistors are used for each diode to limit the diode current (alternatively a current-limited driver could be used). For the red diode of Figure 2.61, a current of 10 mA provides a reasonably intense emission. A terminal voltage of approximately 1.8 V is required. Thus, if the logic output voltage is 5 V, a resistor of 320 Ω (3.2 V/10 mA) is required. For other voltage levels, such as a transistor-transistor-logic gate with

Figure 2.62: Seven-segment LED display.

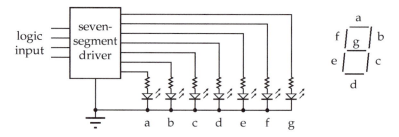

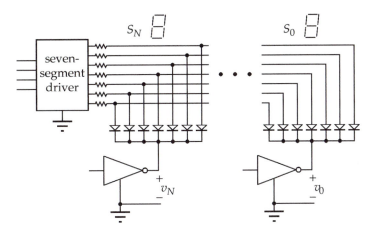

Figure 2.63: Multiplexed seven-segment numerical display.

a high output voltage of approximately 3.7 V, a different resistor value would be required.

A light-emitting display of several numerical digits such as used for early calculators posed an interesting engineering challenge (McWhorter 1976). A 10-digit display, for example, consists of 70 individual LEDs that would, in addition to a common connecting conductor, require 70 connections or wires (additional connections are required for diode decimal points). A multiplexing scheme (Figure 2.63) is frequently used to get around the wiring problem, albeit a more complex integrated circuit driver for the display is required. To understand its operation, assume that the output of the seven-segment decoder is such as to produce a given numerical digit (for example, if all inputs are high an "8" is obtained). If v_N that is associated with digit S_N is high, none of the diodes connected to it will conduct (they will either have a zero voltage or be reverse biased); thus, all segments will remain dark. If, however, $v_N \approx 0$, then the diodes connected to high-output lines will be forward biased and will be lit. The inputs of the inverters may be used to activate the numerical digits sequentially. In sequencing from one digit to the next, the output of the seven-segment driver must change appropriately. It should be noted that this multiplexing scheme is dependent on the diode property of the display segments; it would not work, for example, if the display elements behaved as simple resistors. For the circuit shown, the diodes of each seven-segment display have a common negative (cathode) terminal. Alternatively, the diodes could have a common positive (anode) terminal, which would require a different external circuit.

Another common application of LEDs is as a signal-level display such as is frequently used on tape recorders and hi-fi audio amplifiers. Assume the *on* voltage of the LEDs of Figure 2.64 is $v_{D(on)}$ (approximately 1.6 V if the red diodes of Figure 2.61 are used). When v_{IN} reaches $v_{D(on)}$, the first diode, D_0, will become forward

Figure 2.64: LED level indicator.

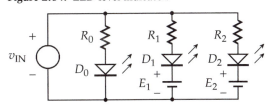

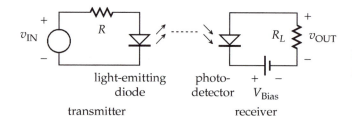

Figure 2.65: An LED transmitter and photodetector receiver.

light-emitting diode   photo-detector V_{Bias}

transmitter   receiver

biased, thus emitting light for $v_{\text{IN}} > v_{D(\text{on})}$. As a result of the biasing voltages of the other two diodes, their current will remain essentially zero (no emissions). However, when v_{IN} exceeds $v_{D(\text{on})} + E_1$, the current of D_1 will increase, thus turning on D_1. The circuit of D_2 will remain essentially zero for $v_{\text{IN}} < v_{D(\text{on})} + E_2$ ($E_2 > E_1$). It will turn on for $v_{\text{IN}} > v_{D(\text{on})} + E_2$; all three diodes will be on for this condition. For an audio-level indicator, the controlling voltage v_{IN} could be obtained using a rectifier and filter circuit, the output of this circuit being proportional to the amplitude of the audio signal. An improved audio-level indicator uses a set of electronic comparators for activating the LEDs. For this circuit, the diodes, if they are activated, will have a uniform brightness. Although only three diodes are shown in Figure 2.64, six or more diodes, each being activated at a different input level, are common.

Another application of an infrared LED is the ever-present remote control (TV, VCR, etc.). An infrared photodetector of the unit being controlled is used to detect the signal of the remote control. A photodiode operates on essentially the same principle as a photovoltaic cell except, to enhance its sensitivity, the photodiode is reverse biased with an external circuit. Generally, a photovoltaic diode has a much smaller area than a photovoltaic cell. Consider the transmitter–receiver circuit of Figure 2.65. When v_{IN} exceeds the forward bias required for a significant diode current (approximately 1.2 V for the infrared diode of Figure 2.61), an infrared emission occurs. If properly directed, a portion of the emission will be absorbed by the photodetector, and the generation of electron–hole pairs will result in an external current. This yields a small output voltage of the receiver v_{OUT}.

The operation of this system is complicated by background infrared radiation, primarily that which accompanies visible radiation (both natural and that produced by lamps). The photodetector responds to the signal of the infrared transmitter of the remote control as well as to that of the background radiation. To distinguish the transmitted signal from that produced by the background radiation, an *on–off* modulation of the current of the LED and hence its radiated signal is utilized (Figure 2.66). Even if the transmitted signal is small compared

Figure 2.66: Modulation of a remote control light-emitting diode.

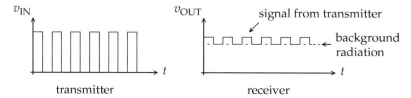

v_{IN}

transmitter

v_{OUT} signal from transmitter
background radiation

receiver

with the background radiation level, its presence may be readily detected. To the extent that the background signal is nonvarying (or only slowly varying), it may be eliminated with a simple high-pass filter (for example, a circuit with a series capacitor). In addition, a band-pass filter is also used to reduce the effect of the background radiation further.

Consider the case for a square wave with a periodic frequency of 40 kHz, which is a frequency used for numerous remote controls. The output voltage of the receiver v_{OUT} may be treated as consisting of a Fourier series of terms. The largest-amplitude sinusoidal term has a fundamental frequency of 40 kHz. Smaller-amplitude harmonics are also present (because a square wave has only odd harmonics, these terms will have frequencies of 120, 200 kHz, etc.). The output of a detector following the filter may be used to indicate the presence or absence of a signal.

To accomplish the various control functions (for a TV control, for example, power, mute, channel selection, etc.), a coding system is required. A commonly used system utilizes 0.5-ms bursts of the 40-kHz square-wave signal – each burst is thus 20 periods of the 40-kHz signal. Either a single or double burst (double bursts are separated by 0.5 ms) is used to transmit a sequence that may be interpreted by the receiver as a logic 0 or 1. A sequence of single or double bursts may be used to transmit a binary-coded word corresponding to a particular control function (Figure 2.67). The code word, after an extended off interval, is repeated as long as the control is activated.

Closely related to the development of the LED was the introduction of the semiconductor junction laser (an acronym for light amplifiction and stimulated emission of radiation). Whereas the radiation of an LED lacks spectral and spatial coherence, the radiation of a laser is highly coherent. The radiation of a laser consists of an extremely narrow spectrum of wavelengths with a highly directed spatial radiation pattern. The laser is an outgrowth of an earlier device, the maser (microwave amplification and stimulated emissions of radiation) invented by Charles H. Townes and his coworkers in 1954 (Gordon, Ziegler, and Townes 1954; Schawlow 1976). Townes, along with Schawlow, also developed the first laser in 1958 (Bromberg 1988; Schawlow and Townes 1958).

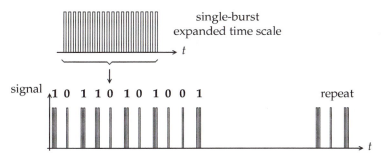

Figure 2.67: Example of coding used by a remote control (power control function for a particular TV set is shown).

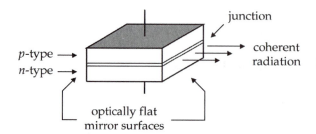

Figure 2.68: Basic structure of a junction diode laser.

A laser relies on an optical resonator, generally a Gabry–Perot mirror cavity, for its operation. A structure similar to that of an LED's is used for a semiconductor junction laser (it is also known as an injection laser). As is the case for many inventions, the idea for a semiconductor laser was not unique to a single research group. Three groups nearly simultaneously demonstrated the operation of a gallium–arsenide laser in 1962 (Hall 1976). There was Robert Hall and his coworkers and General Electric (Hall et al. 1962), Marshall Nathan and his colleagues at IBM (Nathan et al. 1962) and T.M. Quist at MIT's Lincoln Laboratory (Quist et al. 1962). Although the first laser diodes were cooled with liquid nitrogen and were limited to intermittent pulsed operation, efficient laser diodes that operated continuously at room temperature and higher were introduced in the 1970s.

The basic structure of a semiconductor laser is shown in Figure 2.68. A laser relies on an optical resonator, generally a Gabry–Perot mirror cavity, for its operation, and a semiconductor structure similar to that of an LED's is used for a junction laser. At low junction currents, the device tends to behave as an LED except that its radiation tends to be trapped within the diode structure. However, as the diode current is increased, a threshold current is reached at which radiation reflected from the mirrors is sufficient for laser operation to occur. For normal operation, a curent larger than the threshold current is required. As a result of the mirror at one end being slightly transparent, radiation is obtained from one end of the diode. To reduce the threshold current, heterojunction devices with very thin junction regions in which the laser interaction occurs have been developed.

The coherent radiation of semiconductor lasers is used for several applications in which the radiation of LEDs would not be suitable. The reading of a compact disc (audio or data) requires a laser source that can be focused to a sufficiently small beam necessary for detecting a binary bit pattern that has dimensions only slightly larger than the wavelength of visible light. Only a laser can provide sufficiently coherent radiation to accomplish this. Another application of semiconductor lasers is fiber-optic communication systems (Agrawal 1992; Hecht 1987; Palais 1992). Although LED transmitters could be used for short-distance, low-capacity systems, it is the semiconductor junction laser that is used for long-distance, high-capacity systems. Although numerous long-distance wire-type communications systems are still being utilized, all the new systems are fiber-optic systems.

fiber-optic
transmission line

laser transmitter   photodetector receiver

Figure 2.69: A fiber-optic communications system.

A semiconductor laser is basically an *on–off* type device that is not well suited for the direct transmission of analog signals but is ideally suited for digital signals. Except for the analog connection between an individual subscriber and the telephone office, all modern telephone systems utilize digital signals. It is a digital telephone signal that is routed through a telephone office, and, if necessary, is transmitted over short- and long-distance transmission systems. Hence, telephone signals are ideally suited for fiber-optic systems using semiconductor lasers.

A communications system using a fiber-optic transmission line is shown in Figure 2.69. The optical fiber is a very-small-diameter, low-loss glass filament that generally has a graded refraction index that confines the light rays to the center of the fiber. Because losses of 0.2 dB/km and less can be achieved, a 50-km-long optical fiber has a loss of only 10 dB, that is, 10 percent of the initial optical signal is available at the end of fiber. Regenerative repeaters in which the weak optical signal is detected, amplified, and used to drive a semiconductor laser coupled to the next section of fiber are used for long-distance systems.

It is the high rate at which semiconductor diode lasers can be pulsed that results in the high capacity of fiber-optic systems. Consider the case of a typical telephone system (Bigelow 1991). The analog signal produced by a subscriber's telephone is sampled at a rate of 8000 samples per second, and each sample is quantized to one of 256 discrete levels. Therefore, to transmit each sample with a binary code, 8 bits ($2^8 = 256$) are required. Hence, a single telephone signal requires a transmission of 64,000 bits/s (8000 samples/s × 8 bits/sample). If only a single telephone signal were to be transmitted (which would never be done with a fiber-optic system), a time interval of 15.625 μs (1/64,000 s) could be used for transmitting each bit. Commonly used electronic digital systems, however, work with much shorter bit times – times measured in nanoseconds rather than microseconds. Hence, each bit can be transmitted in a much shorter time interval than 15.625 μs. If this is done, the intervening time between the "compacted" groups of eight bits may be used for other telephone signals – a process known as time multiplexing. High-capacity fiber-optic systems with bit rates of 400 Mbits/s, that is, a time interval of 2.5 ns for each bit, are common. For this bit rate, 6000 simultaneous telephone conversations can be transmitted along with additional framing pulses that are required for separating the individual signals at the end terminal of the system. Because a fiber-optic system works only in one direction, two systems are used for a normal two-way telephone system. Fiber-optic "cables" generally consist of bundles of many fibers; hence, numerous two-way circuits are carried by each cable.

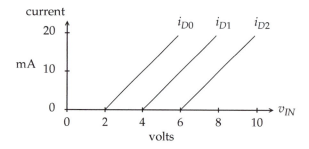

Figure 2.70: LED current of Example 2.17(a).

EXAMPLE 2.17

Consider the level indicator of Figure 2.64 that uses yellow light-emitting diodes with $v_{D(on)} = 2.0$ V.

$$R_0 = R_1 = R_2 = 200 \ \Omega, \quad E_1 = 2.0 \text{ V}, \quad E_2 = 4.0 \text{ V}$$

a. Determine and sketch the currents of the diodes ($0 \leq v_{IN} \leq 10$ V).
b. Design a diode-limiting circuit that will limit the currents of the LEDs to 5 mA. Assume that diodes with an *on* voltage of 0.7 V are available.

SOLUTION

a. Consider the diodes individually (Figure 2.70).

$$D_0 : v_D < 2.0 \text{ V}, \quad i_{D0} = 0$$
$$v_D \geq 2.0 \text{ V}, \quad i_{D0} = \left(v_D - v_{D(on)}\right)/R_0 = 5(v_D - 2) \text{ mA}$$
$$D_1 : v_D < 4.0 \text{ V}, \quad i_{D1} = 0$$
$$v_D \geq 4.0 \text{ V}, \quad i_{D1} = \left(v_D - v_{D(on)}\right)/R_1 = 5(v_D - 4) \text{ mA}$$
$$D_2 : v_D < 6.0 \text{ V}, \quad i_{D2} = 0$$
$$v_D \geq 6.0 \text{ V}, \quad i_{D2} = \left(v_D - v_{D(on)}\right)/R_2 = 5(v_D - 6) \text{ mA}$$

b. A shunt diode–battery circuit may be used to limit the current of the LEDs. Suppose two 100 Ω series resistors are used for R_0, R_1, and R_2. A current of 5 mA implies a voltage of 0.5 V across each of these series resistors. Therefore, a limiting voltage of 2.5 V (0.5 V + $v_{D(on)}$) is required for D_0. This implies a diode in series with a 1.8-V battery (Figure 2.71). Series voltages of 3.8 and 5.8 V are required to limit the currents of D_1 and D_2, respectively.

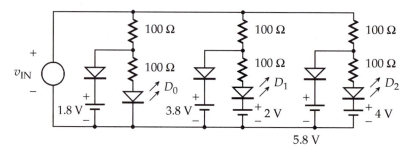

Figure 2.71: Zener diode limiting for level indicator of Example 2.17(b).

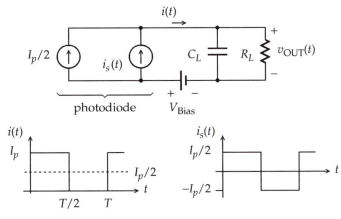

Figure 2.72: Equivalent circuit of diode for Example 2.18.

EXAMPLE 2.18

Consider the infrared transmitter and receiver of Figure 2.65. The response of the receiver circuit is limited by the capacitance of the diode and circuit. Suppose that the effect of the diode capacitance (a nonlinear quantity) and that of the circuit can be accounted for with a single capacitor C_L of 20 pF connected in parallel with R_L. Assume the peak photodetector current is 1 μA and that it has a frequency of 40 kHz (Figure 2.66). Assume no background radiation.

a. Determine and sketch $v_{OUT}(t)$ for $R_L = 1$ MΩ.

b. Repeat for $R_L = 100$ kΩ.

SOLUTION

a. The current pulses of the diode $i(t)$ may be treated as the sum of two terms, a constant of $I_p/2$ and a symmetrical square-wave signal $i_s(t)$. By superposition, the constant current term of $I_p/2$ results in a component of $I_p R_L/2$ for the output voltage (because V_{Bias} is in series with the current sources, it has no effect on v_{OUT}). To determine the effect of $i_s(t)$, the current source and R_L may be replaced by a Thévenin equivalent circuit (Figure 2.73). If C_L were zero, the component of output voltage due to $i_s(t)$ would be equal to $i_s(t) R_L$ (a voltage with a square-wave form). The capacitance results in

Figure 2.73: Thévenin equivalent circuit for $i_s(t)$ solution.

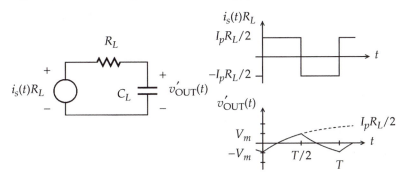

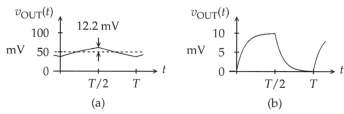

Figure 2.74: Solution of Example 2.18.

a distortion of the voltage (Figure 2.73). Because $i_s(t)R_L$ is symmetrical about zero volts, the same symmetry would be expected for $v'_{OUT}(t)$ for a steady-state response. The voltage $v'_{OUT}(t)$ must satisfy the differential equation of the circuit:

$$\frac{V_p}{2} = R_L C_L \frac{dv'_{OUT}}{dt} + v'_{OUT}, \qquad V_p = \frac{I_p R_L}{2}$$

$$\frac{dv'_{OUT}}{dt} + \frac{dv'_{OUT}}{R_L C_L} = \frac{V_p}{2 R_L C} \qquad 0 \le t \le T/2$$

The following may be shown to satisfy the differential equation and result in an initial voltage of V_m (a yet-to-be-determined quantity):

$$v'_{OUT}(t) = V_p/2 - (V_p/2 + V_m)e^{-t/R_L C_L}$$

This voltage must be equal to V_m at $t = T/2$.

$$V_m = V_p/2 - (V_p/2 + V_m)e^{-T/2 R_L C_L}$$
$$V_m(1 + e^{-T/2 R_L C_L}) = V_p/2(1 - e^{-T/2 R_L C_L})$$
$$V_m = \left(\frac{V_p}{2}\right) \frac{1 - e^{-T/2 R_L C_L}}{1 + e^{-T/2 R_L C_L}}$$

For this circuit, $R_L C_L = 20\ \mu s$ and $T/2 R_L C_L = 0.5$.

$$V_p/2 = 50\ \text{mV}, \qquad V_m = 12.2\ \text{mV}$$

This corresponds to a peak-to-peak value of 24.4 mV for $v'_{OUT}(t)$ as well as for $v_{OUT}(t)$. For no capacitance, the peak-to-peak value would be 100 mV. The output voltage, $v_{OUT}(t)$, is equal to $v'_{OUT}(t)$ plus 50 mV (Figure 2.74).

b. For $R_L = 100\ \text{k}\Omega$, $V_p/2 = 5\ \text{mV}$, $R_L C_L = 2.0\ \mu s$, $T/2 R_L C = 5$, and $V_m = 4.93\ \text{mV}$. The peak-to-peak value of $v_{OUT}(t)$ is 9.87 mV, nearly the 10 mV that is obtained for $C_L = 0$. The output voltage $v_{OUT}(t)$ is equal to $v'_{OUT}(t)$ plus 5 mV.

REFERENCES

Adler, R. B., Smith, A. C., and Longini, R. L. (1964). *Introduction to Semiconductor Physics*. New York: John Wiley & Sons.

Agrawal, G. P. (1992). *Fiber-Optic Communication Systems*. New York: John Wiley & Sons.

Bardeen J. and Brattain, W. H. (1948). The transistor, a semi-conductor triode (letter to the editor). *Physical Review*, **74**, 2 (July 15), 230–1.

Bigelow, S. J. (1991). *Understanding Telephone Electronics* (3d ed.). Carmel, IN: Howard W. Sams & Co.

Brattain, W. H. and Bardeen J. (1948). Nature of the forward current in germanium point contacts (letter to the editor). *Physical Review*, **74**, 2 (July 15), 231–2.

Bromberg, J. L. (1988). The birth of the laser. *Physics Today*, **41**, 10, 26–33.

Chapin, D. M., Fuller, C. S., and Pearson, G. L. (1954). A new silicon p–n junction photocell for converting solar radiation into electrical power. *Journal of Applied Physics*, **25**, 5, 676–7.

Fink, D. G. and Rockett, F. H. (1948). The transistor – A crystal triode. *Electronics* (September), 68–71.

Gordon, J. P., Ziegler, H. J., and Townes, C. H. (1954). Molecular microwave oscillator and new hyperfine structure in the microwave spectrum of NH_3. *Physical Review*, **95**, 1, 282–4.

Gray, P. E., DeWitt, D., Boothroyd, A. R., and Gibbons, J. F. (1964). *Physical Electronics and Circuit Models of Transistors*. New York: John Wiley & Sons.

Hall, R. N. (1976). Injection lasers. *IEEE Transactions on Electron Devices*, **ED-23**, 7, 700-4.

Hall, R. N., Fenner, G. E., Kingsley, J. D., Soltys, T. J., and Carlson, R. O. (1962). Coherent light emission from GaAs junctions. *Physical Review Letters*, **9**, 9 (November 1), 366–8.

Hecht, J. (1987). *Understanding Fiber Optics*. Carmel, IN: Howard W. Sams & Co.

Hubbard, H. M. (1989). Photovoltaics today and tomorrow. *Science*, **244**, 4902, 297–304.

Kittel, Charles (1996). *Introduction to Solid State Physics* (7th ed.). New York: John Wiley & Sons.

Lehovec, K., Accardo, C. A., and Jamgochian, E. (1951). Injected light emission of silicon carbide crystals. *Physical Review*, **83**, 3 (August 1), 603–7.

Loebner, E. E. (1976). Subhistories of the light emitting diode. *IEEE Transactions on Electron Devices*, **ED-23**, 7, 675–99.

McWhorter, R. W. (1976). The small electronic calculator. *Scientific American*, **234**, 3, 88–98.

Milnes, A. G. (1980). *Semiconductor Devices and Integrated Electronics*. New York: Van Nostrand Reinhold Co.

Muller, R. S. and Kamins, T. I. (1986). *Device Electronics for Integrated Circuits* (2d ed.). New York: John Wiley & Sons.

Nathan, M. I., Dumke, W. P., Burns, G., Dill, F. H. Jr., and Lasher, G. (1962). Stimulated emission of radiation from GaAs *p–n* junctions. *Applied Physics Letters*, **1**, 3 (1 November), 42–62.

Neudeck, G. W. (1989). *The PN Junction Diode* (2d ed.). Reading, MA: Addison–Wesley Publishing.

The news of radio. (1948). *The New York Times* (July 1), 46.

Palais, J. C. (1992). *Fiber Optic Communications* (3d. ed.). Englewood Cliffs, NJ: Prentice Hall.

Pierret, R. F. (1988). *Semiconductor Fundamentals* (2d ed.). Reading, MA: Addison–Wesley Publishing.

Quist, T. M., Rediker, R. H., Keyes, R. J., Krag, W. E., Lax, B., McWhorter, A. L., and Zeigler, H. J. (1962). Semiconductor maser of GaAs. *Applied Physics Letters*, **1**, 4 (1 December), 91–2.

Raisbeck, G. (1955). The solar battery. *Scientific American*, **193**, 6, 102–10.

Round, H. J. (1907). A note on carborundum (letter to the editor). *Electrical World*, **49**, 6 (February 9), 309.

Schawlow, A. L. (1976). Masers and lasers. *IEEE Transactions on Electron Devices*, **ED-23**, 7, 773–9.

Schawlow, A. L. and Townes, C. H. (1958). Infrared and optical masers. *Physical Review* **112**, 6, 1940–9.

Shockley, W. (1950). *Electrons and Holes in Semiconductors*. New York: D. Van Nostrand Co.

Shockley, W. (1976). The path to the conception of the junction transistor. *IEEE Transactions on Electron Devices*, **ED-23**, 7, 597–620.

Shockley, W. and Pearson, G. L. (1948). Modulation of conductance of thin films of semiconductors by surface charges (letter to the editor). *Physical Review*, **74**, 2 (July 15), 232–3.

Smits, F. M. (1976). History of silicon solar cells. *IEEE Transactions on Electron Devices*, **ED-23**, 7, 640–3.

Streetman, B. G. (1990). *Solid State Electronic Devices* (3d ed.). Englewood Cliffs, NJ: Prentice Hall.

Sze, S. M. (1981). *Physics of Semiconductor Devices* (2d ed.). New York: John Wiley & Sons.

Tuinenga, P. W. (1995). *SPICE: A Guide to Circuit Simulation and Analysis Using PSPICE* (3d ed.). Englewood Cliffs, NJ: Prentice-Hall.

Weaver, J. H. (1986). Metal-semiconductor interfaces. *Physics Today*, **39**, 1, 24–33.

Weiner, C. (1973). How the transistor emerged. *IEEE Spectrum*, **10** 1, 23–33.

Wilson, A. H. (1931). The theory of electronic semiconductors. *Proceedings of the Royal Society of London* **A133** (October), 458–91, and **A134** (November), 277–87.

Wolf, M. (1981) Photovoltaic solar energy conversion systems, in *Solar Energy Handbook*. Kreider, J. F. and Kreith, F. (Eds.). New York: McGraw-Hill, 24-1–24-35.

Wolfe, C. M., Holonyak, N., and Stillman, G. E. (1989). *Physical Properties of Semiconductors*. Englewood Cliffs, N J: Prentice Hall.

PROBLEMS

2.1 A current of 1 A is used to charge a 12-V battery. What is the number of electrons transferred to (and from) the battery over a time interval of 20 h?

2.2 AWG 12 (American Wire Gauge) copper wire is common for residential wiring (AWG 12 has a diameter of 2.05 mm). Consider the case for a current of 10 A.

a) What is the current density of the wire?
b) What is the drift velocity of the electrons?
c) What is the resistance per meter length of wire?

2.3 Repeat Problem 2.2 for aluminum wire. Assume three free electrons are available for each atom (aluminum has a density of 2.70 g/cm^3).

2.4 A two-conductor extension cord with a length of 25 m is used to supply a current of 12 A for an electric heater. The cord has AWG 12 copper wire and the supply voltage is 120 V.

a) What is the overall series resistance of the cord?
b) What is the resultant voltage across the heater?
c) What is the efficiency of the system, that is, the fraction of the input power supplied to the electric heater?

2.5 Repeat Problem 2.4 for extension cords having AWG 14, AWG 16, and AWG 18 (diameters of 1.63, 1.29, and 1.02 mm, respectively) copper wire.

2.6 An inductor is wound with AWG 24 enamel-covered (negligible thickness) copper wire (diameter of 0.51 mm). The inductor has an inner diameter of 1 cm, an outer diameter of 2 cm, and a length of 2 cm. Assume a packing factor of 0.90 for the round wire.

a) What is the number of turns of the inductor?
b) What is the resistance (static, dc) of the inductor?

2.7 Aluminum is used to interconnect components of a silicon integrated circuit. Assume a connecting strip with a width of 25 μm and a thickness of 2 μm. What is the resistance per centimeter length of the strip?

2.8 A particular 12-V automobile electric starter motor requires a current of 100 A. Suppose that a 1-m length of copper wire is used to connect the battery to the starter and that the maximum allowable voltage drop across the wire is 1 V. What is the minimum diameter of wire that can be used?

2.9 Aluminum wire is generally used for outdoor electric power lines. Consider the case for a 50-m length feed and 50-m return wire from the power pole to a house. What is the minimum diameter wire that can be used if the total voltage drop is not to exceed 5 V for a current of 100 A?

2.10 A sample of intrinsic silicon has a cross-sectional area of 0.1 mm^2 and a length of 1 cm. What is the resistance of the sample for a temperature of 27 °C?

2.11 What is the resistance of the silicon sample of Problem 2.10 for temperatures of 50, 100, and 150 °C?

2.12 Repeat Problem 2.10 for intrinsic germanium ($n_i = 2.4 \times 10^{13}/cm^3$ at 27 °C, $\mu_n = 3900$ cm^2/V · s, $\mu_p = 1900$ cm^2/V·s).

2.13 The intrinsic density of germanium doubles for each 16 °C increase in temperature. Repeat Problem 2.12 for temperatures of 50 and 100 °C.

2.14 Gallium arsenide, a III–IV semiconductor compound, is used to fabricate specialized electronic devices ($n_i = 1.8 \times 10^6/cm^3$, $\mu_n = 8500$ cm^2/V · s, $\mu_p = 400$ cm^2/V·s). What is the resistance of a sample with a 1-mm diameter cylindrical cross section and a length of 1 cm?

2.15 Assume that for a particular application an acceptor atom concentration in a silicon semiconductor that results in a conductivity that is 50 percent greater than that of intrinsic material is acceptable. What is the acceptable concentration of acceptor atoms? What is the ratio of acceptor atom density to the silicon atom density?

2.16 Repeat Problem 2.15 for a donor atom impurity.

2.17 Repeat Problem 2.15 for a germanium semiconductor.

2.18 A silicon semiconductor with a resistivity of $1.0 \ \Omega \cdot m$ is desired.

a) What is the donor atom density required for this resistivity?
b) What is the acceptor atom density required?

2.19 Repeat Problem 2.18 for a resistivity of $0.15 \ \Omega \cdot m$.

2.20 A silicon semiconductor with a resistivity of $1.0 \ \Omega \cdot m$ is desired. Owing to a prior doping it has a donor density of 10^{14} atoms/cm^3.

a) What is the acceptor atom density that would result in n-type material with a $1.0 \ \Omega \cdot m$ resistivity?
b) What is the acceptor atom density that would result in a p-type material with a $1.0 \ \Omega \cdot m$ resistivity?

2.21 A silicon semiconductor has a donor atom concentration of $100 \ n_i$ at a temperature of $27 \, ^\circ C$.

a) What is the ratio of free electrons to holes (n/p) of the semiconductor at $27 \, ^\circ C$?
b) At what temperature is $n/p = 10$?

2.22 Repeat Problem 2.21 for an initial donor concentration of $10^4 \ n_i$.

2.23 Repeat Problem 2.21 for a germanium semiconductor (n_i doubles for a temperature change of approximately $16 \ C^\circ$).

2.24 An n-type semiconductor with a free-electron concentration of 10^{15}/cm^3 is desired.

a) Compare n/p for silicon and germanium semiconductors for a temperature of $27 \, ^\circ C$.
b) What are the temperatures at which $n/p = 10$ for the silicon and germanium semiconductors? (Assume doubling temperatures of $11 \ C^\circ$ and $16 \ C^\circ$ for n_i of silicon and germanium, respectively.)

2.25 A junction diode has the charge density distribution of Figure 2.11(a). Assume that the two regions (marked $+$ and $-$) are rectangular and that the negative region has a width of x_p ($0.052 \ \mu m$) and the positive region has a width of x_n ($0.13 \ \mu m$). The diode has an acceptor doping of 10^{17}/cm^3 and a donor doping of 4×10^{16}/cm^3.

a) Use Gauss's law to determine E_x at $x = 0$ ($\epsilon_r = 11.5$, consult a physics text if necessary).
b) Write expressions for E_x within the depletion region.
c) Using the expressions for E_x, determine $V_{\text{built-in}}$ of the diode.

2.26 For a given set of doping densities, the size of I_s depends on the cross-sectional area of a junction diode. The area of a diode is generally chosen according to the current it may be required to conduct. Assume that $v_D = 0.75$ V for different diodes having currents of 1 μA, 1 mA, 1 A, and 100 A. What are the corresponding values of I_s for these diodes ($n = 1$, $V_T = 25$ mV)?

2.27 Repeat Problem 2.26 for diodes with $n = 1.4$.

2.28 For a given silicon junction diode it is found that $i_D = 10$ mA for $v_D = 0.75$ V and that $i_D = 1$ mA for $v_D = 0.65$ V. What are n and I_s of the diode ($T = 27°C$, $V_T = 25$ mV)?

2.29 Suppose that the data of Problem 2.28 were obtained at a temperature of 60 °C. Determine n and I_s for the diode at a temperature of 60 °C. What would I_s be if the temperature of the diode were reduced to 27 °C?

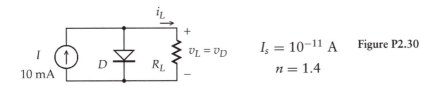

$I_s = 10^{-11}$ A   **Figure P2.30**

$n = 1.4$

2.30 A silicon junction diode with a current source is used in the circuit of Figure P2.30. Determine v_L for $i_L = 0$ and for $i_L = 9$ mA. What is the variation in v_L for this current range?

2.31 Repeat Problem 2.30 for a circuit in which the single diode is replaced by five series-connected diodes.

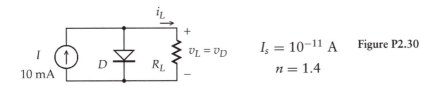

$I_s = 10^{-14}$ A   **Figure P2.32**

$n = 1.0$

2.32 A silicon junction diode with a voltage source is used in the circuit of Figure P2.32. What is v_L for $i_L = 0$ and for $i_L = 4$ mA? What is the variation in v_L for this current range?

2.33 Replace the diode of Problem 2.32 with four identical series diodes and increase V_A to 8 V. What is v_L for $i_L = 0$ and for $i_L = 4$ mA? What is the variation in v_L for this current range?

$I_s = 10^{-13}$ A   **Figure P2.34**

$n = 1.2$

2.34 A series silicon junction diode is used for the circuit of Figure P2.34. The input voltage v_{IN} has a sinusoidal time dependence as follows:

$$v_{IN} = V_m \sin \omega t \qquad V_m = 1.0\,\text{V}$$

a) What is v_{OUT} for $\omega t = 0$, $\pi/4$, $\pi/2$, and $3\pi/4$?
b) Over what fraction of a period is v_{OUT} essentially zero ($v_L \leq 1.0\,\text{mV}$)?

2.35 Repeat Problem 2.34 for $V_m = 1.5$ V.

2.36 Repeat Problem 2.34 for $V_m = 10$ V. Suggest an approximation for v_D that would yield a reasonably accurate result for v_{OUT} while simplifying the calculations.

2.37 Suppose V_m of Problem 2.34 is 100 V. For this condition, the approximation that $v_D = 0$ for $i_D > 0$ is reasonable.

a) Sketch v_{OUT} for this condition. What is its peak value?
b) What is the time average value of v_{OUT}?
c) What is the rms value of v_{OUT}?
d) What is the average power dissipated by R_L?

2.38 Repeat Problem 2.37 for $V_m = 50$ V.

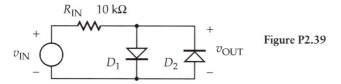

Figure P2.39

2.39 In the circuit of Figure P2.39, the diode voltage may be assumed to be 0.7 V ($v_{D(on)}$) for a current that is greater than zero.

a) Sketch v_{OUT} -versus-v_{IN} for the circuit.
b) Suppose v_{IN} is a sinusoidal function of time.

$$v_{IN} = V_m \sin 2\pi f t, \qquad V_m = 10\,\text{V}, \qquad f = 60\,\text{Hz}$$

Determine the rise and fall times of the approximate square waveform of v_{OUT}.

2.40 Rise and fall times of 100 μs are desired for v_{OUT} of Problem 2.39. Determine the value of V_m required.

2.41 Suppose that v_{IN} of Problem 2.39 is a symmetrical triangular wave with a peak amplitude of 10 V (V_m). Repeat Problem 2.39 for this input voltage.

2.42 Repeat Problem 2.39 for each diode replaced by two series diodes.

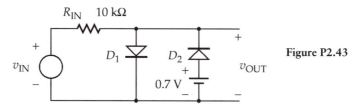

Figure P2.43

2.43 In the circuit of Figure P2.43, the diode voltages may be assumed to be 0.7 V ($v_{D(on)}$) if the diode current is greater than zero.

a) Sketch v_{OUT}-versus-v_{IN} for the circuit.
b) Suppose v_{IN} is a sinusoidal function of time.

$$v_{IN} = V_m \sin 2\pi ft, \qquad V_m = 10 \text{ V}, \qquad f = 60 \text{ Hz}$$

Determine the rise and fall times of the approximate square waveform of v_{OUT}.

2.44 Repeat Problem 2.43 for $V_m = 5$ V and $f = 1$ kHz.

2.45 Repeat Problem 2.43 for D_1 replaced by two series diodes.

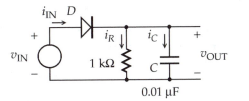

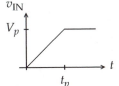

Figure P2.46

2.46 Consider the diode circuit of Figure P2.46 with the input voltage indicated.

$$V_p = 10 \text{ V}, \qquad t_p = 10 \text{ } \mu s$$

Assume a diode voltage $v_{D(on)}$ of 0.7 V and that $v_{OUT} = 0$ at $t = 0$.

a) Determine and sketch v_{OUT} as a function of time.
b) Determine and sketch i_R as a function of time. What is its peak value?
c) Determine and sketch i_C as a function of time. What is its peak value and over what time interval does it occur?
d) Determine and sketch i_{IN} as a function of time.

2.47 Assume that the increasing voltage segment of Problem 2.46 ($0 < t < t_p$) consists of a portion of a sinusoidal signal.

$$v_{IN} = V_p \sin(\pi t/2t_p), \qquad V_p = 10 \text{ V}, \quad t_p = 10 \text{ } \mu s$$

Repeat Problem 2.46 for this input voltage.

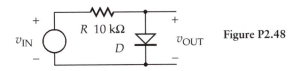

Figure P2.48

2.48 Consider the diode circuit of Figure P2.48.

a) Determine v_{OUT} versus v_{IN} (-5 to $+5$ V) with ideal behavior of the diodes assumed.
b) Determine v_{OUT} versus v_{IN} with $v_{D(on)} = 0.7$ V assumed.

2.49 Repeat Problem 2.48 with the polarity of the diode reversed.

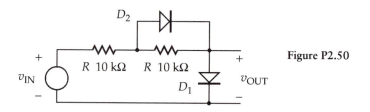

Figure P2.50

2.50 Consider the diode circuit of Figure P2.50.

a) Determine v_{OUT} versus v_{IN} (-5 to $+5$ V) with ideal behavior of the diodes.

b) Determine v_{OUT} versus v_{IN} with $v_{D(on)} = 0.7$ V assumed.

2.51 Repeat Problem 2.50 with the polarity of D_2 reversed.

2.52 Suppose that diode D_2 of Figure P2.50 is connected across both resistors. Discuss the behavior of the circuit, with $v_{D(on)}$ assumed to be 0.7 V. What would be the effect of $v_{IN} = 3$ V on the circuit?

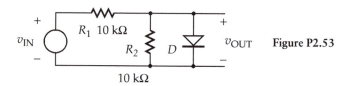

Figure P2.53

2.53 Consider the diode circuit of Figure P2.53.

a) Determine v_{OUT} versus v_{IN} (-5 to $+5$ V) with ideal behavior of the diodes assumed.

b) Determine v_{OUT} versus v_{IN} with $v_{D(on)}$ assumed to be 0.7 V.

2.54 Repeat Problem 2.53 with the polarity of the diode reversed.

2.55 Repeat Problem 2.53 for a second diode connected in parallel with D but with the opposite polarity.

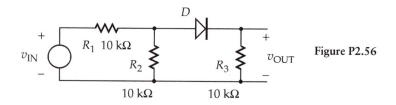

Figure P2.56

2.56 Consider the diode circuit of Figure P2.56.

a) Determine v_{OUT} versus v_{IN} (-5 to $+5$ V) with ideal behavior of the diodes assumed.

b) Determine v_{OUT} versus v_{IN} with $v_{D(on)}$ assumed to be 0.7 V.

2.57 Repeat Problem 2.56 with the polarity of the diode reversed.

2.58 Repeat Problem 2.56 for a second diode connected in parallel with D but with the opposite polarity.

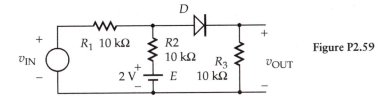

Figure P2.59

2.59 Consider the diode circuit of Figure P2.59.

    a) Determine v_{OUT} versus v_{IN} (-5 to $+5$ V) with ideal behavior of the diode assumed.

    b) Determine v_{OUT} versus v_{IN} with $v_{D(on)}$ assumed to be 0.7 V.

2.60 Determine the voltage of the battery E of Figure P2.59 that results in the following for v_{OUT}:

$$v_{OUT} = \begin{cases} 0 & \text{for } v_{IN} < 0 \\ kv_{IN} & \text{for } v_{IN} > 0 \end{cases}$$

Assume $v_{D(on)} = 0.7$ V. What is the value of k?

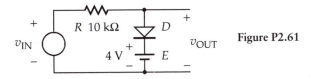

Figure P2.61

2.61 Consider the diode circuit of Figure P2.61.

    a) Determine v_{OUT} versus v_{IN} (-10 to $+10$ V) with ideal behavior of the diodes assumed.

    b) Determine v_{OUT} versus v_{IN} with $v_{D(on)}$ assumed to be 0.7 V.

    c) Determine v_{OUT} versus v_{IN} with a constant diode equivalent resistance r_d corresponding to a current of 0.25 mA assumed. Assume $v_{D(on)} = 0.7$ V.

2.62 Repeat Problem 2.61 for a 1 kΩ resistor in series with the diode. Is the effect of the diode's resistance of part (c) significant for this circuit?

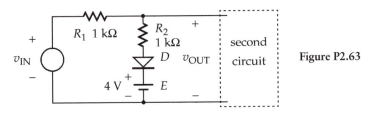

Figure P2.63

2.63 Suppose that the output of the circuit of Figure P2.63 were to be used as the input of a second circuit.

    a) Determine v_{OUT} versus v_{IN} (-10 to $+10$ V) with an open circuit for the second circuit assumed. Assume $v_{D(on)} = 0.7$ V.

    b) Determine two Thévenin equivalent circuits that can be used to predict v_{OUT} when the second circuit is in place. One circuit is to apply for the diode having zero current whereas the other is for the diode forward biased.

c) What is the range of v_{OUT} over which each of these circuits applies?

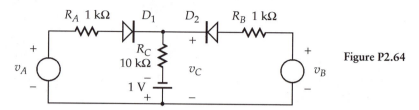

Figure P2.64

2.64 Consider the modification (Figure P2.64) of the diode **OR** gate discussed in the text. With ideal behavior of the diodes assumed, determine v_{OUT} for the following input voltages:

a) $v_A = 0$ V $v_B = 0$ V
b) $v_A = 5$ V $v_B = 0$ V
c) $v_A = 5$ V $v_B = 2.5$ V
d) $v_A = 5$ V $v_B = 5$ V

2.65 Suppose that in building the circuit of Figure P2.64 the polarity of D_2 was accidentally reversed. Repeat Problem 2.64 for this condition.

2.66 Repeat Problem 2.64 with $v_{D(on)}$ assumed to be 0.6 V for the diodes.

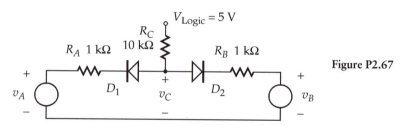

Figure P2.67

2.67 Consider the diode **AND** gate of Figure P2.67. Assume ideal behavior of the diodes. Determine v_C versus v_A for the following values of v_B:

a) $v_B = 0$ V.
b) $v_B = 2.5$ V.
c) $v_B = 5$ V.

2.68 Repeat Problem 2.67 with $v_{D(on)}$ assumed to be 0.6 V for the diodes.

2.69 Assume that a 10 kΩ resistor is connected in parallel with the output of the diode **AND** gate of Problem 2.67. Repeat Problem 2.67 for this condition.

2.70 Assume that a 50 kΩ resistor is connected in parallel with the output of the diode **AND** gate of Problem 2.67. Repeat Problem 2.67 for this condition.

2.71 Assume $v_A = v_B = 0$ for the circuit of Figure P2.67. Determine a set of Thévenin equivalent circuits for the output terminals of the circuit (v_C). Note: Owing to the symmetry of the circuit, only two equivalent circuits are needed: one for the diode currents being zero and the other for the diodes being forward biased. Assume $v_{D(on)} = 0.6$ V and determine the range of v_C over which each circuit is valid.

2.72 Repeat Problem 2.71 for $v_A = v_B = 5$ V.

2.73 Consider the diode voltage regulator of Example 2.14 (Figure 2.47). Determine the variation in v_{Load} for a variation of R_L from 1.5 kΩ to 4 kΩ. Assume V_{Supply} remains equal to 10 V.

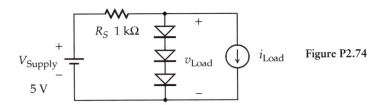

Figure P2.74

2.74 Three junction diodes are used in the voltage regulator circuit of Figure P2.74, which supplies a load current of i_{Load}. Assume that for the diodes $v_{D(on)} = 0.7$ V, $n = 1$, and $V_T = 25$ mV.

   a) What is the maximum load current for which the diode voltages remain equal to 0.7 V?

   b) What is the maximum load current for which the current of the diodes remains 1 mA or greater?

   c) Determine the variation in v_{Load} for a variation of load current from zero to that determined in part (b).

2.75 Assume $i_{Load} = 1.5$ mA for the circuit of Figure P2.74. Determine the variation in v_{Load} for a ± 1 V variation in V_{Supply}.

2.76 Repeat Problem 2.74 for a circuit with four series diodes.

2.77 A silicon photovoltaic cell in bright sunlight produces a short-circuit current I_{sc} of 1 A and an open-circuit voltage V_{oc} of 0.60 V. At a reduced illumination level (cloud cover), $I_{sc} = 0.1$ A and $V_{oc} = 0.50$ V.

   a) What are the parameters of the diode I_s and n ($V_T = 25$ mV)?

   b) Assume that the cell is in bright sunlight. What will be the electrical output power of the cell for $R_L = V_{oc}/I_{sc}$?

2.78 The open- and short-circuit measurements of Problem 2.77 are for a temperature of 27°C. What would I_{sc} and V_{oc} be for the same illumination levels but for a temperature of 60°C?

2.79 A silicon photovoltaic cell at a temperature of 27 °C has parameters of $I_s = 10^{-7}$ A and $n = 2$. For a particular illumination level $I_{photon} = 25$ mA.

   a) What are the open-circuit voltage and short-circuit current of the cell?

   b) What is the output power for $R_L = V_{oc}/I_{sc}$?

   c) To ascertain whether R_L is close to an optimal value, determine the electrical output power for values of R_L that are 10 percent greater and smaller than that used in part (b).

2.80 A series connection of three photovoltaic cells results in a short-circuit current I_{sc} of 100 mA and an open-circuit voltage V_{oc} of 1.55 V when the cells are in bright sunlight. For overcast conditions $I_{sc} = 15$ mA and $V_{oc} = 1.46$ V.

a) What are the individual diode parameters I_s and n ($V_T = 25$ mV)?
b) What is the power supplied to a load resistor of $R_L = V_{oc}/I_{sc}$ for bright sunlight conditions?

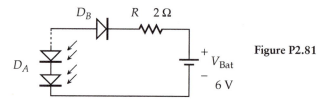

Figure P2.81

2.81 Sixteen series-connected photovoltaic cells are to be used to charge a nickel–cadmium battery with a nominal voltage of 6 V (Figure P2.81). A series silicon junction diode is used to prevent the battery from being discharged by the photovoltaic cell when the cell is not illuminated.

$$D_A : \quad I_s = 10^{-10} \text{ A}, \qquad n = 1$$
$$D_B : \quad I_s = 10^{-13} \text{ A}, \qquad n = 1$$

The series resistor R accounts for the resistance of the battery and circuit components.

a) Estimate the battery charging current for $I_{photon} = 0.1$ A. (Hint: Assume for an initial iteration that the current is equal to I_{photon}.)
b) Estimate the battery current for $I_{photon} = 0.5$ A.

2.82 Consider a multiplexed seven-segment display of Figure 2.63 with eight digits ($N = 7$). Assume that the output voltages of the logic drivers are either 0 or 5 V and that the LED segments may be treated as diodes with $v_{D(on)} = 1.75$ V. For static conditions, a current of 5 mA is required for a desired brightness level, and for pulsed operation it is found that the brightness of a segment is proportional to its average current.

a) Determine the diode current required if a segment is to be on one-eighth of the time.
b) What is the value required for the series resistors?
c) What is the current that the individual inverters must be designed to sink?
d) What is the power utilized by the circuit to display all zeroes? All ones? All eights?

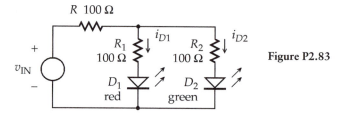

Figure P2.83

2.83 A red and a green LED of Figure 2.61 are used in the circuit of Figure P2.83. For convenience, assume a constant forward-biased voltage $v_{D(on)}$ for the diodes corresponding to a current of 2.5 mA. Determine and sketch the dependence of the diode currents i_{D1} and i_{D2}, on the input voltage v_{IN} (0 to 5 V). Hint: First determine the current of D_1 for $i_{D2} = 0$. When D_1 is conducting, a Thévenin equivalent circuit for it along with the resistances and v_{IN} may be used to determine the current of D_2.

2.84 Consider the photodiode of Example 2.18. Assume $C_L = 20$ pF and R_L is such as to result in a time constant of $T/2$, that is, 12.5 μs. Determine $v_{OUT}(t)$ for this condition.

2.85 Consider the photodiode of Example 2.18. Determine the fundamental Fourier component of $v_{OUT}(t)$ for both parts of the example. Hint: Utilize the fundamental Fourier component of $i_s(t)$ for the input of the $R_L C_L$ circuit rather than attempting to extract the fundamental component from $v'_{OUT}(t)$. Compare the peak-to-peak values of the sinusoids with those obtained in the example.

COMPUTER SIMULATIONS

C2.1 Determine the temperature dependence of a diode with $n = 1.4$, obtaining a set of curves similar to those of Figure 2.19 (0 to 10 mA for i_D). Assume $I_s = 2 \times 10^{-12}$ A. What is the voltage sensitivity for currents of 1 mA and 10 mA?

C2.2 Repeat Simulation C2.1 for $n = 2.0$ and $I_s = 10^{-9}$ A.

C2.3 Consider the diode circuit of Figure 2.22. Use your favorite programming langauge to write a program for solving the circuit using an iterative process. The element values, V_A and R and the diode parameters I_s and n are input quantities ($V_T = 25$ mV). A diode voltage within 1 mV of its true value is desired. Obtain numerical values for the following:

  a) $V_A = 2.5$ V, $R = 1$ kΩ, $I_s = 10^{-14}$ A, $n = 1$.
  b) $V_A = 5$ V, $R = 100$ Ω, $I_s = 10^{-10}$ A, $n = 1$.
  c) $V_A = 5$ V, $R = 10$ Ω, $I_s = 10^{-10}$ A, $n = 1.4$.
  d) $V_A = 1$ V, $R = 10$ kΩ, $I_s = 10^{-16}$ A, $n = 1$.
  e) $V_A = 0.7$ V, $R = 10$ kΩ, $I_s = 10^{-16}$ A, $n = 1$.

C2.4 Use SPICE to determine i_D and v_D for the circuit of Figure 2.22 for the conditions of Simulation C2.3. If Simulation C2.3 was performed, compare results.

C2.5 Repeat Simulation C2.3 for the case of two parallel diodes (same orientation as original diode). Assume diode paramters of I_{s1} and I_{s2} and that $n_1 = n_2 = 1$. For this circuit, an iterative process in which the currents are obtained after assuming a diode voltage will be required. A sum of currents at the node of the resistors that has a magnitude that is less than 10^{-4} of V_A/R is desired. Obtain solutions for the following conditions:

a) $V_A = 2.5$ V, $R = 1$ kΩ, $I_{s1} = I_{s2} = 10^{-14}$ A.
b) $V_A = 5$ V, $R = 100$ Ω, $I_{s1} = 10^{-10}$ A, $I_{s2} = 2 \times 10^{-10}$ A.
c) $V_A = 5$ V, $R = 10$ kΩ, $I_{s1} = 10^{-14}$ A, $I_{s2} = 10^{-15}$ A.
d) $V_A = 0.7$ V, $R = 10$ kΩ, $I_{s1} = 10^{-14}$ A, $I_{s2} = 10^{-15}$ A.

C2.6 Use SPICE to determine v_D ($v_D = v_{D1} = v_{D2}$), i_{D1} and i_{D2} for the conditions of simultion C2.5. If Simulation C2.5 was done, compare the results.

C2.7 Repeat Simulation C2.3 for the case of series diodes (same orientation as original diode). Assume $n_1 = n_2 = 1$. For this circuit it will be necessary to assume a diode current and then sum the voltages across the two diodes. A sum of diode voltages that is within 10^{-4} V_A is desired. Obtain v_{D1}, v_{D2} and i_D ($= i_{D1} = i_{D2}$) for the conditions of Simulation 2.5.

C2.8 Use SPICE to determine v_{D1}, v_{D2} and i_D ($= i_{D1} = i_{D2}$) for the conditions of Simulation C2.5.

C2.9 Determine the behavior of the circuit of Figure 2.34 using a SPICE simulation. Assume the diodes have an ideality factor n of 1.0 and that I_s is such as to result in a diode voltage of 0.7 V for a diode current of 1.0 mA ($V_T = 25$ mV). Obtain a plot of v_{OUT} for $-5 \leq v_{IN} \leq 5$ V. Also obtain plots of the diode voltages.

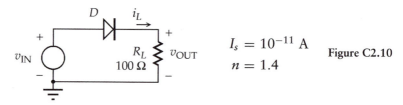

Figure C2.10

C2.10 The series diode circuit of Figure C2.10 is frequently used to rectify a voltage v_{IN}. As a result of the diode, the current of the circuit and hence the load voltage will never be negative.

a) Using a .DC solution obtain a plot of v_{OUT} versus v_{IN} for $-10 \leq v_{IN} \leq 10$ V.

b) Consider the case for which v_{IN} is a sinusoidal voltage with a peak amplitude of 10 V and a frequency of 60 Hz (a voltage that could be obtained from a 60-Hz power line). Obtain a transient solution for v_{OUT} and the diode voltage for two periods of the input voltage. What are the peak and average values of v_{OUT} during the second period of the simulation?

C2.11 A capacitor $C_L = 100$ μF is connected in parallel with R_L of Figure C2.10. Repeat Simulation C2.10 for this circuit. Also obtain plots of the currents of the diode and the capacitor.

C2.12 Use the circuit of simulation C2.11 to determine the effect of the value of C_L on the ripple output voltage, that is, the difference of the maximum and minimum output voltages after the first period. Use a range of 20 to 500 μF. From the data of the simulations, plot the ripple voltage versus C_L.

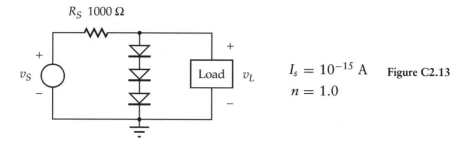

$I_s = 10^{-15}$ A   **Figure C2.13**

$n = 1.0$

C2.13 The three silicon diodes used in the voltage regulator of Figure C2.13 result in a load voltage, when the diodes are conducting, of approximately 2 V. Assume the load is a resistance R_L of 1 kΩ.

    a) Determine the dependence of v_L on the input voltage, v_S ($0 \le v_S \le 10$ V). What are the values of v_L for $v_S = 5$ and 10 V?

    b) The current-versus-voltage characteristic of the diodes depends on their temperature. Repeat part (a) for a temperature of 100 °C. What is the variation of v_L for the temperature increase ($v_S = 5$ and 10 V)?

C2.14 Consider the situation for which the input voltage of the regulator of Figure C2.13 is constant, $v_S = 10$ V. A variation in the load current results in a variation in the load voltage – an effect that may be simulated by using a current source for the load.

    a) Determine the dependence of v_L on the load current ($0 \le i_L \le 10$ mA). What is the maximum load current for which v_L remains within 0.1 V of its value for $i_L = 0$?

    b) Repeat part (a) for a temperature of 100°C.

C2.15 A SPICE simulation of the diode circuit of Figure 2.44 for ideal diodes as well as for actual silicon diodes is desired.

    a) The behavior of ideal diodes may be simulated by using an extremely small value of ideality factor for the diodes. Assume $n = 0.001$ and $I_s = 10^{-15}$ A for the diodes. Obtain curves of v_C versus v_A for $v_B = 0$, 2.5, and 5 V. A nested voltage sweep may be used to obtain this result with a single simulation run. Also obtain plots of the diode currents.

    b) Assume $n = 1$ and $I_s = 10^{-15}$ A for the diodes, a set of parameters corresponding to a silicon diode. Repeat part (a) for these diodes.

C2.16 A particular photovoltaic cell has $I_{sc} = 1$ A, $V_{oc} = 0.6$ V, and $n = 1$. Determine the load resistance for which the power supplied to the load is a maximum.

C2.17 Repeat Simulation C2.16 for $n = 2$.

C2.18 A SPICE simulation of the circuit with LEDs (Figure 2.71 of Example 2.17) is desired.

a) Estimate the diode voltages of the yellow LED for currents of 1 and 10 mA (Figure 2.61). From these quantities, values of I_s and nV_T may be obtained for the diode. For the default SPICE temperature of 27 °C, $V_T = 0.0259$ V. Obtain the value of n and perform a simulation to obtain the current-versus-voltage characteristic of the diode (check to verify that it is indeed approximately that of Figure 2.61).

b) Obtain a plot of the LED currents similar to that of Figure 2.71. If the ciruit is functioning properly, a current limiting should occur. Assume $I_s = 5 \times 10^{-15}$ A and $n = 1.0$ for the limiting diodes.

CHAPTER THREE

THE BIPOLAR JUNCTION TRANSISTOR: AN ACTIVE ELECTRONIC DEVICE

Several types of transistors are used in modern electronic systems, both individually as discrete devices and in conjunction with other transistors in integrated circuits. Transistors have three or more terminals and, as is the case for junction diodes, transistors are nonlinear elements. The bipolar junction transistor (BJT) that will be discussed in this chapter, as well as the field-effect transistor of the next chapter, are *active* devices. Electronic amplifying circuits in which a small input voltage, current, or both, produces a larger output voltage, current, or both depend on active devices. Amplification is required for nearly all electronic systems. The analysis of transistor circuits is considerably more difficult (a greater challenge) than that of circuits with two-terminal passive elements – resistors, capacitors, and inductors.

The history of active electronic circuits dates from the invention of the vacuum tube (the audion) by Lee De Forest in 1906 (De Forest 1906). From the very beginning the challenge was to develop circuits to utilize this new device. Edwin H. Armstrong was foremost among the early designers of electronic circuits that were initially used to improve wireless communication (Armstrong 1915). The junction transistor, developed in 1950 (following the invention of its predecessor, the point-contact transistor, in 1948), and other transistors, have replaced vacuum tubes for most (but not all!) applications. These applications, however, tend to rely on electronic circuits similar to those initially used with vacuum tubes (*Electronics* 1980).

A bipolar junction transistor consists of two junction diodes fabricated from a single semiconductor crystal (Figure 3.1). The n-type regions, the emitter and collector, are separated by a very thin p-type base region (an *NPN*-type transistor). Normal operation of this device depends primarily on free electrons of the emitter region crossing the forward-biased base–emitter junction (the effect of other carriers will be considered in the next section). These free electrons diffuse across the very thin base region and cross a normally reverse-biased base–collector junction. Because the base–collector junction is reverse biased, the collector current depends primarily on these free electrons from the emitter. Hence, for the

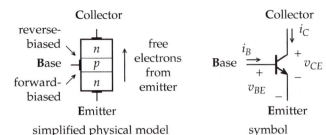

Collector

reverse-
biased

free
electrons
from
emitter

Base

forward-
biased

Emitter

simplified physical model

Collector

Base

Emitter

symbol

Figure 3.1: Physical model and symbol of a bipolar junction transistor in the normal mode of operation.

device of Figure 3.1, the base–emitter junction's forward bias tends to control the collector current. For a properly fabricated transistor, a small current controls a much larger collector current.

The static terminal properties of a transistor may be specified by a set of graphical characteristics such as those of Figure 3.2. It will be noted that the collector current i_C depends on the collector–emitter voltage v_{CE} and the base current i_B, and the base characteristic is essentially that of a forward-biased diode (the base–emitter junction diode of the transistor). As long as the base–collector junction diode remains reverse biased, the effect of the collector–emitter voltage on the base characteristic tends to be extremely small and can, for most applications, be neglected. Hence, the base current of the transistor depends only on the external circuit connected to the base.

To gain an appreciation for the concept of amplification, consider the transistor with the characteristic of Figure 3.2 that is used in the circuit of Figure 3.3. Individual load lines, as were used for junction diode circuits, may be drawn for both the input and output circuits of the transistor (Figure 3.4). For a particular value of v_{IN}, a base current i_B may be determined from the intersection of the load line and the base characteristic. This current, in turn, determines the particular curve (or intermediate "interpolated" curve) of the collector characteristic. The intersection of this curve with the load line determines the collector current i_C and collector–emitter voltage v_{CE}. From a set of closely spaced input voltages, a transfer characteristic $v_{OUT} (= v_{CE})$ versus v_{IN} may be obtained (Figure 3.5).

Two important points may be drawn from the transfer characteristic. First, if circuit components are properly chosen, a small change in input voltage v_{IN} can

Figure 3.2: Base and collector characteristic of a typical low-power bipolar junction transistor.

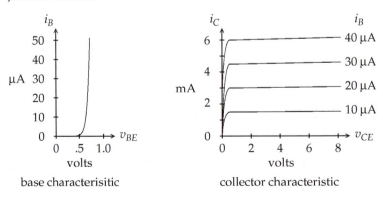

base characterisitic

collector characteristic

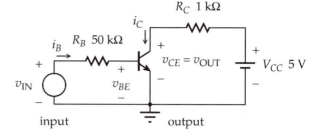

Figure 3.3: A bipolar junction transistor amplifier circuit.

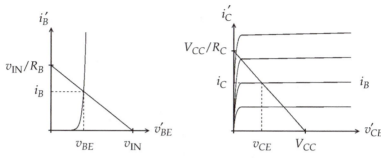

Figure 3.4: Base and collector characteristic with load lines for a transistor amplifier.

Figure 3.5: Transfer characteristic of a basic transistor amplifier.

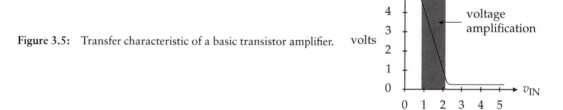

result in a much larger magnitude of change in output voltage v_{OUT}. Although the ratio of these changes is negative (an increase in v_{IN} results in a decrease of v_{OUT}), this inverting of the input signal is not important for many applications (such as amplifying an audio signal). Furthermore, if a second amplifier is connected to the output of the first, it will produce an output that follows the input signal. The second important point relates to the input and output currents of the circuit i_B and i_C, respectively. For the transistor of Figure 3.2, the base current is much smaller than the collector current. Hence, the power supplied by v_{IN} could be much smaller than that which might be extracted from the amplifier. This is possible because power is supplied to the circuit by the battery V_{CC}.

The transfer characteristic of Figure 3.5 is typical of an elementary basic amplifier. For a signal $v_s(t)$ such as that of Figure 3.6, an offset voltage is necessary for amplification because, for input voltages near zero, v_{OUT} does not change (v_{OUT} remains equal to 5 V in Figure 3.5). An offset voltage of V_{BB} ($V_{BB} \approx 1.5$ V would work fairly well for the circuit of Figure 3.3) is included for the circuit

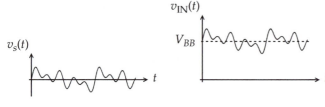

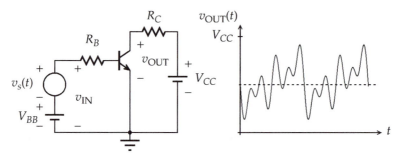 is
the signal voltage that is to be
amplified.

The output signal of the amplifier is the variation of v_{OUT}
about its midvalue. For a properly functioning amplifier, the
output signal is an inverted and
enlarged version of the input signal. Several circuit schemes are available for
eliminating the output offset voltage (a coupling capacitor will work for many
applications). Large voltage and current gains, such as needed for many electronic
applications (radio, TV, etc.) are achieved with several cascaded amplifiers, each
producing an amplified version of its input signal, which then serves as the input
of the next amplifier.

Figure 3.6: Shifting the level of an input signal $v_s(t)$.

The circuit of Figure 3.3 with the transfer characteristic of Figure 3.5 may also
be used as an active logic inverter gate. For this gate, $v_{OUT} \approx 5$ V for $v_{IN} < V_{Low}$
and v_{OUT} is small but not quite zero for $v_{IN} > V_{High}$ (Figure 3.8). Of importance
is that, for $v_{IN} < V_{Low}$, the output corresponds to a logic low input, and, for
$v_{IN} < V_{High}$, the output corresponds to a logic high input. Precise logic levels of the
input voltage are not necessary; it needs only to be less than V_{Low} and greater than
V_{High} to be interpreted properly by the gate. It will be noted that the magnitude
of the slope of the characteristic for the intermediate region of the characteristic
is greater than 1. Both analog and logic circuits generally require amplification.

To analyze and design circuits using bipolar junction transistors, it will be necessary to develop a better understanding of the physical operation of a transistor

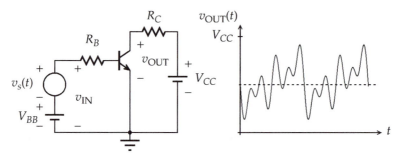

Figure 3.7: Amplification of a signal.

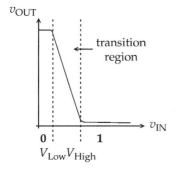

transition
region

Figure 3.8: A transistor logic inverter.

and to devise equivalent circuit models that can be used for analytic solutions and computer simulations. This will be followed by a consideration of a few basic bipolar junction transistor circuits used in analog and digital electronic systems.

3.1 THE COMMON–BASE CONFIGURATION: A PHYSICAL DESCRIPTION

A three-terminal device such as the bipolar junction transistor of Figure 3.1 has three currents – one associated with each terminal. Only two currents, however, are required to describe the behavior of the device; the third current depends on the other two. As a consequence, one may arbitrarily choose which two to use. In the introduction to the chapter, the base and collector currents were used, leaving the emitter to be treated as the common terminal of the device. The voltages of the base and collector were referred to the emitter.

The currents and voltages of a bipolar junction transistor are indicated in Figure 3.9. By convention, currents are defined as being directed into the terminals of the device; if the physical current is in the opposite direction, the numerical value of the current will be negative. When Kirchhoff's current law is applied to the device, the currents must sum to zero as follows:

$$i_B + i_C + i_E = 0, \qquad v_{BE} + v_{CB} = v_{CE} \tag{3.1}$$

If this did not occur for the currents, the device, over time, would accumulate or be depleted of charge. Summing the voltages around the device yields the voltage relationship of Eq. (3.1).

Although the common–emitter configuration of a bipolar junction transistor is generally (but not always) employed to describe a transistor's behavior using an appropriate equivalent circuit model, the common–base configuration lends itself to developing an understanding of the behavior of the device based on physical considerations (Figure 3.10).

For normal operation, the collector–base junction of a transistor is reverse biased. In Figure 3.10, an external potential V_{CC} is used to achieve this. A forward-biased base–emitter junction is necessary to produce the free electrons responsible for the operation of the transistor. The free electrons of the emitter that cross the junction are injected into the base region. In a conventional diode, the injected free electrons would eventually recombine with the plentiful holes of the p-type region. Recombination, however, is not an instantaneous process; a finite time is required for the thermal wandering of a free electron before it comes within the vicinity of a hole and recombines. While this is occurring, the injected free electrons diffuse across the very thin base region of a transistor. That the diffusion of minority carriers in a semiconductor could indeed result in a current was established by the Haynes–Shockley experiment. A reading of this classic paper describing the experiment (Shockley, Pearson, and Haynes 1949) is highly recommended. The free electrons that reach the base–collector junction readily cross this junction. It will be recalled that, although the

Figure 3.9: Current and voltages of a bipolar junction transistor.

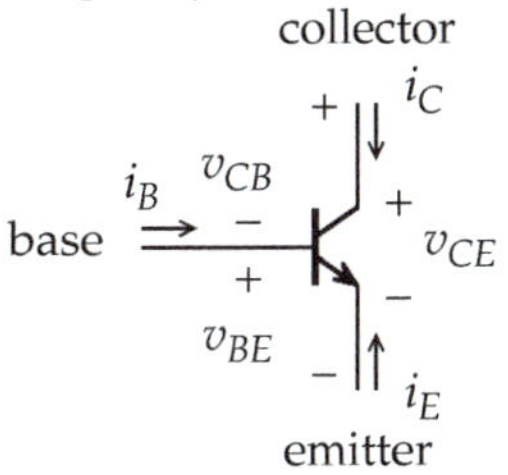

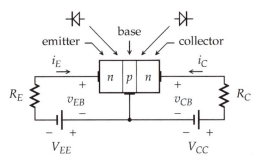

Figure 3.10: Common–base configuration for determining a transistor's behavior based on physical considerations.

built-in electric field of a junction diode inhibits majority carriers from crossing the junction, the field has an attractive effect on minority carriers. Therefore, the free electrons that have diffused across the base may be thought of as "falling down" the potential difference of the base–collector junction.

A few of the injected electrons, regardless of how thin the base region might be, are lost through recombination. This is indicated in Figure 3.11(a) as an equivalent flow of electrons out of the base. Simultaneously, holes of the base region of the transistor cross the base–emitter junction to the emitter. Because these carriers do not contribute to the collector current, the current due to those holes is minimized in a well-designed transistor. This is achieved with a heavily doped emitter region, a doping that is much larger than the base doping. Hence, many more free electrons from the emitter cross the junction than holes from the base with a lower doping. Other carriers, the conventional minority carriers, also cross the junctions and contribute to the terminal currents of the device. At nominal temperatures, the currents due to these carriers tend to be negligible and can generally be ignored, but at elevated temperatures they may be significant.

Because currents are defined in terms of positive charges, the conventional current is in the opposite direction of the electron flow. Hence, the base and collector currents of Figure 3.11(b) are positive, and the emitter current is negative. For a first approximation, the collector current of a transistor with a reverse-biased base–collector junction ($v_{CB} > 0$) is proportional to the emitter current as given by

$$i_C = -\alpha_F i_E \qquad (3.2)$$

The quantity α_F is the common–base current gain, which is a positive quantity because the relationship of Eq. (3.2) has a negative sign. The F subscript implies the forward direction of the transistor; a subscript R is used for the reverse direction when the roles of the emitter and collector are interchanged. If recombination of the injected free electrons did not occur, and if the hole current of the

Figure 3.11: The role of majority carriers of a bipolar junction transistor.

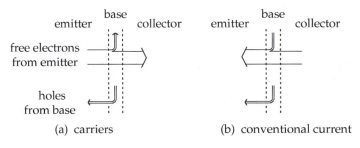

(a) carriers   (b) conventional current

base–emitter junction were zero, then α_F would equal 1(negligible effect of minority carriers). For high-quality transistors, α_F is very close to, but less than 1, and values greater than 0.98 are common.

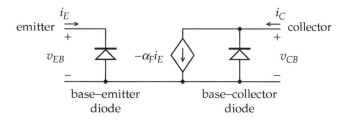

The current relationship of Eq. (3.2) suggests an equivalent circuit with a dependent current source. By taking into account the junction diodes of the device, the equivalent circuit of Figure 3.12 is obtained. Only when v_{EB} is negative will the base–emitter diode conduct ($i_E < 0$). When v_{CB} is positive, the base–collector diode is re-

Figure 3.12: Equivalent circuit of common–base transistor configuration.

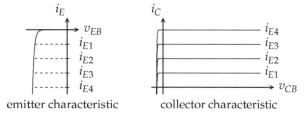

Figure 3.13: Emitter and collector characteristic of a common–base transistor configuration.

verse biased, that is, its current is negligible. However, when v_{CB} is negative, the base–collector diode is forward biased, thus reducing the collector current (Figure 3.13). The point at which $i_C = 0$ (an open collector circuit condition), implies that the current due to the injected free electrons from the emitter, $-\alpha_F i_E$, is equal to the diode current.

The equivalent circuit model of Figure 3.12 will be used to predict the behavior of the common–base amplifier circuit of Figure 3.14 that uses a silicon junction transistor. The first step is to draw the overall circuit with the transistor replaced by its equivalent circuit (Figure 3.15). It is convenient initially to consider the case for no signal, that is, for $v_s(t) = 0$. For the biasing voltage V_{EE}, the base–emitter diode will be forward biased. A constant-voltage diode model having a forward voltage of $v_{EB(\text{on})}$ (≈ -0.7 V) will be assumed. Summing the voltages around the emitter loop yields the following:

$$V_{EE} + v_{EB(\text{on})} + i_E R_E = 0$$
$$i_E = -(V_{EE} + v_{EB(\text{on})})/R_E \qquad (3.3)$$
$$= -(2 - 0.7)/(0.22 \text{ k}\Omega) = -5.909 \text{ mA}$$

From the emitter current, a numerical value of the collector current, if it is assumed that the base–collector diode is reverse biased, will be obtained. Let $\alpha_F = 0.99$, a typical value.

$$i_C = -\alpha_F i_E = 5.850 \text{ mA} \qquad (3.4)$$

The collector–base voltage v_{CB} can now be determined because the collector current is known:

$$v_{CB} = V_{CC} - i_C R_C = 9.15 \text{ V} \qquad (3.5)$$

The voltages and currents for no signal are often referred to as quiescent values (quiet being associated with no signal).

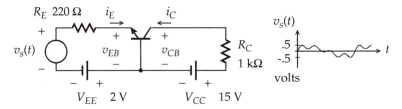

Figure 3.14: Common–base amplifier.

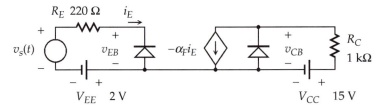

Figure 3.15: Equivalent circuit for the common–base amplifier.

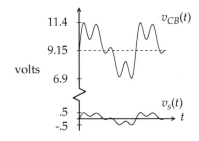

Figure 3.16: Collector–base voltage of the common–base amplifier.

The presence of a signal may readily be taken into account.

$$V_{EE} + v_{EB(\text{on})} + i_E R_E - v_s(t) = 0$$
$$i_E = -(V_{EE} + v_{EB(\text{on})} - v_s(t))/R_E \tag{3.6}$$

When $v_s(t)$ is equal to its maximum value of 0.5 V, the following is obtained:

$$i_E = -(2 - 0.7 - 0.5)/(0.22 \text{ k}\Omega) = 3.636 \text{ mA}$$
$$v_{CB} = V_{CC} + \alpha_F i_E R_C = 11.4 \text{ V} \tag{3.7}$$

In a like manner, a solution can be obtained for $v_s(t)$ being equal to its minimum value of -0.5 V:

$$i_E = -8.182 \text{ mA}, \qquad v_{CB} = 6.9 \text{ V} \tag{3.8}$$

This implies that the peak-to-peak value of v_{CB} is 4.5 V compared with a peak-to-peak value of 1 V for $v_s(t)$. The voltage gain is thus 4.5. The instantaneous values of v_{CB} and $v_s(t)$ are shown in Figure 3.16.

EXAMPLE 3.1

Consider the common–base transistor circuit of Figure 3.14 having an input signal $v_s(t)$ with a peak-to-peak amplitude variation of 1 V. Estimate the

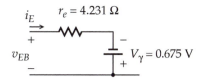

Figure 3.17: A series resistance equivalent circuit for the forward-biased base–emitter diode of a transistor.

peak-to-peak variation in the emitter–base voltage v_{EB} for this signal. Assume $nV_T = 25$ mV for the diode of the transistor equivalent circuit. What is the emitter-to-collector voltage gain of the circuit?

SOLUTION The quiescent emitter current $(v_s(t) = 0)$ was found to be -5.909 mA (Eq. (3.3)). Because this current corresponds to a forward-biased diode current of 5.909 mA, the equivalent resistance of the diode, r_e, is $nV_T/(5.909$ mA) or 4.231 Ω. For the forward-biased diode voltage of 0.7 V, the equivalent diode circuit has a parameter V_γ that is nV_T (25 mV) less than 0.7 V. Hence, $V_\gamma = 0.675$ V. From the circuit of Figure 3.17, the following is obtained for the emitter current:

$$i_E = -(V_{EE} - V_\gamma - v_s(t))/(R_E + r_e)$$

When $v_s(t) = 0.5$ V, $i_E = -3.679$ mA, and when $v_s(t) = -0.5$ V, $i_E = -8.139$ mA. These currents differ but slightly from those obtained using a constant emitter–base voltage of $v_{EB(on)}$ (-0.7 V). The emitter–base voltages corresponding to these currents may now be obtained as follows:

$$v_{BE} = i_E r_e - V_\gamma$$

For $v_s(t) = 0.5$ V, $v_{EB} = -0.691$ V, and for $v_s(t) = -0.5$ V, $v_{EB} = -0.709$ V. The peak-to-peak variation of v_{EB} is thus 18 mV, a very small quantity. For a peak-to-peak collector-base variation of 4.5 V, this implies an emitter-to-collector voltage gain of 250, a rather large quantity. (It may readily be verified that the new emitter currents result in essentially the same peak-to-peak variation in v_{CB}.)

EXAMPLE 3.2

The circuit of Figure 3.14 requires a "floating" input signal source, that is, a source that does not have one of its terminals connected to the common or ground point of the amplifier. From a practical perspective, an amplifier that utilizes an input signal source with a common terminal is preferable. Consider the circuit of Figure 3.18, a modification of the basic common–base circuit.

Figure 3.18: A modified common–base amplifier circuit.

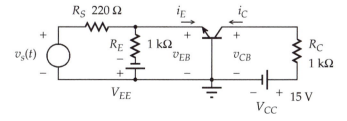

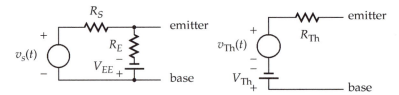

Figure 3.19: A Thévenin equivalent input circuit.

a. Determine the biasing voltage V_{EE} required to achieve the same quiescent emitter current as in Figure 3.14.

b. Determine the voltage gain of this circuit using the input signal of Figure 3.14.

SOLUTION

a. To determine the behavior of the circuit, a Thévenin equivalent circuit will be used for the elements connected to the emitter (Figure 3.19). By superposition, a Thévenin equivalent source dependent on the signal $v_{\mathrm{Th}}(t)$ and a nonvarying voltage dependent on the biasing battery V_{Th} may be obtained as follows:

$$R_{\mathrm{Th}} = R_S \| R_E = 180 \ \Omega$$

$$v_{\mathrm{Th}}(t) = \frac{R_E v_s(t)}{R_E + R_S} = 0.820 v_s(t)$$

$$V_{\mathrm{Th}} = \frac{R_S V_{EE}}{R_E + R_S} = 0.180 V_{EE}$$

The voltage V_{Th} of Figure 3.19 may be associated with the quiescent emitter current by

$$V_{\mathrm{Th}} = -\left(i_E R_{\mathrm{Th}} + v_{EB(\mathrm{on})}\right)$$

For a current of -5.909 mA (Eq. (3.3)) a voltage (V_{Th}) of 1.764 V is required. Hence, $V_{EE} = 9.798$ V.

b. The Thévenin equivalent circuit of Figure 3.19 may be used to determine the emitter current when a signal is present as follows:

$$-V_{\mathrm{Th}} + v_{\mathrm{Th}}(t) = i_E R_{\mathrm{Th}} + v_{EB(\mathrm{on})}$$
$$i_E = -\left(V_{\mathrm{Th}} + v_{EB(\mathrm{on})} - v_{\mathrm{Th}}(t)\right)/R_{\mathrm{Th}}$$

The following is obtained for $v_s(t) = \pm 0.5$ V:

$v_s(t)$	$v_{\mathrm{Th}}(t)$	i_E	i_C	v_{CB}
0.5 V	0.41 V	-3.633 mA	3.597 mA	11.4 V
-0.5 V	-0.41 V	-8.189 mA	8.107 mA	6.9 V

The peak-to-peak collector–base voltage remains 4.5 V, resulting in a voltage gain of 4.5.

3.2 THE COMMON-EMITTER CONFIGURATION: SAME DEVICE, DIFFERENT PERSPECTIVE

The physical operation of an *NPN* bipolar junction transistor was discussed in the previous section. For normal operation, the collector current of the device was found to be approximately equal to the current out of its emitter, that is $-i_E$ or $|i_E|$. Therefore, the base current is much smaller than the collector current or the magnitude of the emitter current. This suggests using the base terminal with its small current for the input of a transistor circuit and the emitter terminal as the common terminal. A common–emitter transistor configuration will have, if used in a properly designed circuit, both a voltage and a current gain.

EQUIVALENT CIRCUIT

If the collector–base voltage of a transistor remains positive, the base–collector diode of the equivalent circuit will be reverse biased, and its current will tend to be negligible (except at elevated temperatures). Hence, the simplified equivalent circuit of Figure 3.20(a) applies if $v_{CB} > 0$. To obtain a common–emitter configuration, it is simply necessary to redraw the equivalent circuit of Figure 3.20(a) using the emitter as the common terminal. Although the corresponding common–emitter equivalent circuit (Figure 3.20(b)) has a new set of terminal voltages, it is not convenient to have a current generator that depends on the emitter current i_E, the current of the common terminal. An explicit dependence on i_B, however, is readily obtained as follows:

$$i_B + i_C + i_E = 0, \qquad i_E = -i_C/\alpha_F$$
$$i_B + (1 - 1/\alpha_F)i_C = 0, \qquad i_C = \left(\frac{\alpha_F}{1 - \alpha_F}\right) i_B = \beta_F i_B \tag{3.9}$$

The quantity β_F is the common–emitter current gain of the transistor given by

$$\beta_F = \frac{\alpha_F}{1 - \alpha_F} \tag{3.10}$$

Because α_F is close to unity, the denominator of the right-hand side of Eq. (3.10) is small, and β_F is large (Table 3.1). Values of 100 and larger for β_F are not uncommon for modern bipolar junction transistors.

Figure 3.20: Common–base and common–emitter transistor equivalent circuits.

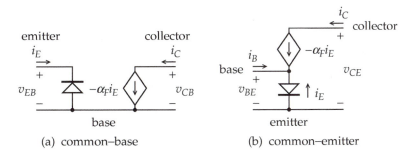

(a) common–base   (b) common–emitter

TABLE 3.1	COMMON–BASE AND COMMON–EMITTER CURRENT GAINS					
α_F	0.96	0.97	0.98	0.985	0.99	0.995
β_F	24	32.3	49	65.7	99	199

From the result of Eq. (3.9), the common–emitter equivalent circuit of Figure 3.21(a) is obtained. It will be noted that the base–emitter voltage v_{BE} depends on the current of the diode, namely $(1 + \beta_F)i_B$. If this diode is replaced by a new diode that has a smaller value of reverse saturation current, namely a value that is $1/(1 + \beta_F)$ of its original value, the equivalent circuit of Figure 3.21(b) yields the correct base–emitter voltage. The current of the diode of the simplified circuit of Figure 3.21(b) is only i_B; the current of the dependent current source $\beta_F i_B$ does not go through the diode of this circuit. Concurrently, the new connection for the lower end of the dependent current source will have no effect on either i_C or v_{CE} (an ideal current source results in a current that is independent of its voltage difference).

It is the circuit of Figure 3.21, or modification of this circuit, that will generally be used for analyzing bipolar junction transistor circuits. When using this circuit, however, one must keep in mind that it is valid only for $v_{CE} \geq v_{BE}$, the condition for which the base–collector diode is reverse biased. A further refinement of this circuit has additional capacitances between the terminals of the transistor to account for the behavior of the charges associated with the junction diodes. In addition, there are small parasitic resistances in series with each terminal, the equivalent resistance of the metal–semiconductor connection. Furthermore, for an integrated circuit transistor the junction capacitance between the transistor's collector and its substrate would also be included.

TRANSFER CHARACTERISTIC

The common–emitter equivalent circuit will be used to determine the behavior of the transistor circuit of Figure 3.22, which is essentially the amplifier circuit of Figure 3.3 considered in the introduction to the chapter. The first step of a solution is redrawing the circuit by inserting the equivalent circuit model for the transistor.

Figure 3.21: Common–emitter equivalent circuits.

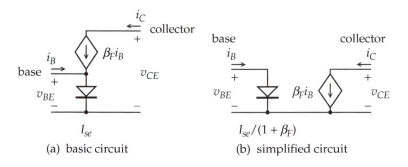

(a) basic circuit   (b) simplified circuit

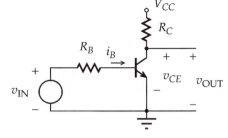

Figure 3.22:   Common–emitter transistor circuit.

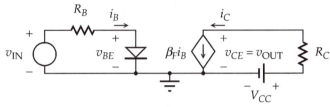

Figure 3.23: Equivalent circuit of a common–emitter transistor circuit.

The circuit of Figure 3.23 may readily be analyzed using a constant forward-biased voltage diode model $v_{BE(\text{on})} \approx 0.7$ V. This results in the following expression for the base current:

$$i_B = \begin{cases} 0 & \text{for } v_{\text{IN}} \leq v_{BE(\text{on})} \\ (v_{\text{IN}} - v_{BE(\text{on})})/R_B & \text{for } v_{\text{IN}} > v_{BE(\text{on})} \end{cases} \tag{3.11}$$

The right-hand side of the equivalent circuit may now be used to determine v_{OUT} as follows:

$$v_{\text{OUT}} = V_{CC} - \beta_F i_B R_C \tag{3.12}$$

From the expressions for i_B, the following is obtained:

$$v_{\text{OUT}} = \begin{cases} V_{CC} & \text{for } v_{\text{IN}} \leq v_{BE(\text{on})} \\ V_{CC} - (v_{\text{IN}} - v_{BE(\text{on})})\beta_F R_C/R_B & \text{for } v_{\text{IN}} > v_{BE(\text{on})} \end{cases} \tag{3.13}$$

The solution is indicated in Figure 3.24.

On the basis of the result of Eq. (3.13), negative values of v_{OUT} result for large values of v_{IN} (the dashed line of Figure 3.24). This invalid response is a result of ignoring the base–collector diode of the common–base equivalent circuit. As a result, the common–emitter circuit of Figure 3.21(b) is strictly valid only for

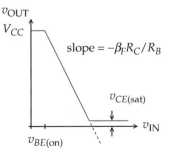

Figure 3.24: Transfer characteristic of a common-emitter circuit.

$v_{CE} > v_{BE}$. In practice, however, it is found that this circuit tends to be valid for v_{CE} down to a few tenths of a volt because the current of the equivalent base–collector diode remains negligible. A useful transistor approximation is that the circuit is valid for v_{CE} greater than a saturation voltage $v_{CE(\text{sat})}$ and that $v_{CE} = v_{CE(\text{sat})}$ if the circuit predicts a voltage less than $v_{CE(\text{sat})}$. This approximation was used for the characteristic of Figure 3.24. Because the behavior of many circuits is not critically dependent on the precise value of v_{CE} for the saturated condition, a voltage of 0.3 V is frequently assumed for $v_{CE(\text{sat})}$.

SPICE SIMULATION MODEL

Computer simulation programs include algorithms for active electronic devices. The bipolar junction transistor algorithm is based on a modification of the equivalent circuit of Figure 3.21(b). A forward-biased base-emitter diode of the model results in the following for the base current:

$$i_B = \left(\frac{I_{se}}{1 + \beta_F} \right) e^{v_{BE}/n_F V_T} \tag{3.14}$$

The collector current for $v_{CE} > v_{BE}$ is $\beta_F i_B$.

$$i_C = \left(\frac{\beta_F I_{se}}{1 + \beta_F} \right) e^{v_{BE}/n_F V_T} = I_s e^{v_{BE}/n_F V_T} \tag{3.15}$$

A new parameter, known as the transport saturation current I_s has been introduced. This quantity is similar to the reverse saturation current of a junction diode. Equation (3.15), however, yields the collector current of the transistor, not the diode current. An expression may be obtained for the base current i_B in terms of I_s and the common–emitter current gain β_F as follows:

$$i_B = (I_s/\beta_F)e^{v_{BE}/n_F V_T} \tag{3.16}$$

Both I_s and β_F are generally specified for a SPICE simulation (unless one is willing to accept built-in default values).

Simulation algorithms account for a forward-biased base–collector junction that occurs for small values of collector–emitter voltage. A built-in set of default parameters for this effect is generally acceptable for most applications (a discussion of these, the reverse parameters of a transistor, will be deferred to a later time). It should be stressed that simulation programs utilize a set of numerically related quantities to determine currents and voltages of a device rather than an explicit equivalent circuit model such as used for analytic solutions.

A close examination of the collector characteristic of Figure 3.2 or an experimentally determined collector characteristic of a transistor reveals that the collector current for a particular value of base current tends to increase by a small amount as the collector–emitter voltage increases. This is known as the Early effect, a phenomenon first explained by J. Early (1952). As v_{CE} is increased, the reverse bias of the base–collector junction of the transistor increases. This results in an increased charge separation of the diode, that is, an increased depletion width. Hence, the effective width of the very thin base region is decreased, thus

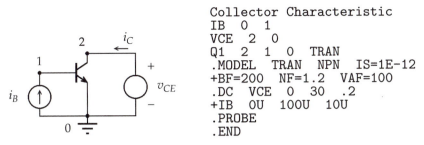

```
                        Collector Characteristic
                        IB   0   1
                        VCE  2   0
                        Q1   2   1   0    TRAN
                        .MODEL   TRAN   NPN   IS=1E-12
                        +BF=200   NF=1.2   VAF=100
                        .DC   VCE   0   30   .2
                        +IB   0U   100U   10U
                        .PROBE
                        .END
```

Figure 3.25: Circuit and SPICE file for determining a collector characteristic.

tending to increase the rate at which injected free electrons diffuse across the base. Although the Early effect is usually ignored for analytic solutions, it is readily included for computer simulations through an Early voltage parameter V_{AF} as follows:

$$i_C = I_s e^{v_{BE}/n_F V_T}(1 + v_{CE}/V_{AF}) \tag{3.17}$$

If, for example, $V_{AF} = 100$ V, the collector current will be 10 percent larger for $v_{CE} = 10$ V than if the Early effect were ignored. Ignoring the Early effect is equivalent to setting V_{AF} equal to infinity – the default value used by SPICE if V_{AF} is not specified.

To illustrate the use of a SPICE simulation of a bipolar junction transistor circuit, a collector and base characteristic will be obtained for a transistor with the following parameters:

$$\beta_F = 200, \quad I_s = 10^{-12} \text{ A}, \quad n_F = 1.2, \quad V_{AF} = 100 \text{ V} \tag{3.18}$$

Values of base currents up to 100 μA and collector–emitter voltages up to 30 V are desired for the characteristic. The SPICE circuit and file of Figure 3.25 will be used for the collector characteristic. The bipolar junction transistor has an element label of Q, and three nodes are specified; collector, base, and emitter (in this order), followed by an arbitrary model name. The corresponding model statement has an NPN specification (in contrast to an alternative PNP transistor type). If no other parameters were to be specified, a SPICE simulation would use a set of built-in default values rather than those of Eq. (3.18). The .DC statement results in a set of base currents of 0 to 100 μA in steps of 10 μA. With the nested specification, the collector–emitter voltage VCE is swept from 0 to 30 V (0.2 V increment) for each value of base current. The collector characteristic of Figure 3.26 is obtained where the effect of the Early voltage is readily apparent.

The circuit and SPICE file of Figure 3.27 will be used to determine the base characteristic of the transistor for collector–emitter voltages of 0, 1, 5, and 20 V. Although this could be achieved with a nested sweep specification, it is not possible to distinguish readily on a PROBE graph which base current corresponds to which collector–emitter voltage. A common base–emitter voltage source VBE is used for the four circuits.

It should be noted that large values of base–emitter voltage will result in excessive device currents. The circuits of Figure 3.27 would not be used for a

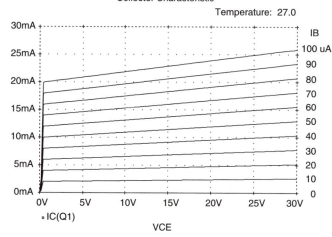

Collector Characteristic
Temperature: 27.0

Figure 3.26: SPICE-generated collector characteristic for the circuit of Figure 3.25.

laboratory experimental determination of the base characteristic because the excessive currents would most likely burn out the device. A resistor in series with the base connection would be used to limit the base current. The justification for this circuit is that, with a voltage source connected directly between the base and emitter of the device, the simulation will produce a set of characteristics with a common voltage scale (Figure 3.28). Although devices are not destroyed in simulation programs, regardless of how excessive the voltages and currents, it would be of value if simulation programs had a "burn out" algorithm that would remind one of necessary practical laboratory precautions. It will be noted that, except for $v_{CE} = 0$, the base characteristic is essentially independent of the collector–emitter voltage of the transistor.

Figure 3.27: Circuit and SPICE file for determining a base characteristic.

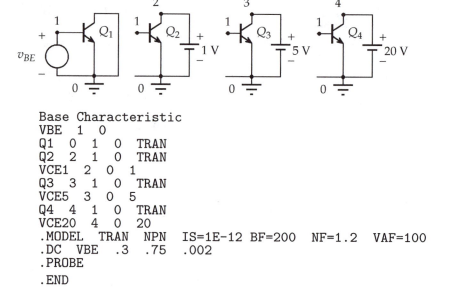

```
Base Characteristic
VBE   1   0
Q1    0   1   0   TRAN
Q2    2   1   0   TRAN
VCE1  2   0   1
Q3    3   1   0   TRAN
VCE5  3   0   5
Q4    4   1   0   TRAN
VCE20 4   0   20
.MODEL   TRAN   NPN   IS=1E-12 BF=200   NF=1.2   VAF=100
.DC   VBE   .3   .75   .002
.PROBE
.END
```

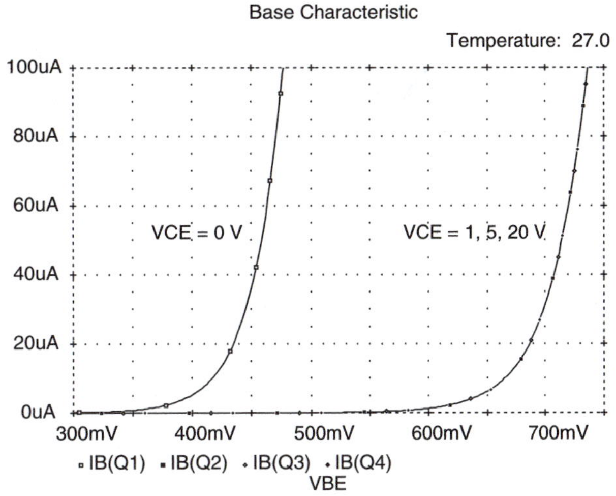

Base Characteristic
Temperature: 27.0
100uA
80uA
60uA
40uA
20uA
0uA
VCE = 0 V
VCE = 1, 5, 20 V
300mV
400mV
500mV
600mV
700mV
IB(Q1) IB(Q2) IB(Q3) IB(Q4)
VBE

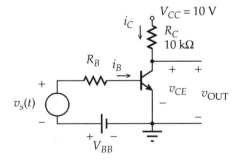

Figure 3.28: SPICE-generated base characteristic for the circuit of Figure 3.21.

EXAMPLE 3.3

Consider the common–emitter transistor circuit of Figure 3.29.

The input source is a voltage with a sinusoidal time dependence as follows:

$$v_s(t) = V_m \sin 2\pi f t, \qquad V_m = 0.5 \text{ V}$$

A transistor with a common–emitter current gain β_F of 150 and a base–emitter voltage $v_{BE(\text{on})}$ of 0.7 V is used in the circuit.

a. Determine values for V_{BB} and R_B that result in an output voltage with a quiescent value of 5 V and a peak-to-peak value of 8 V.

b. Determine and sketch $v_{\text{OUT}}(t)$.

c. Suppose that the transistor of the circuit with the values of V_{BB} and R_B determined in part (a) is replaced with a transistor having a current gain β_F of 100. Determine and sketch $v_{\text{OUT}}(t)$ for this condition.

d. Repeat part (c) for a transistor with $\beta_F = 200$.

SOLUTION The circuit of Figure 3.30 is obtained when the transistor is replaced with its common–emitter equivalent circuit. For $V_{BB} + v_s(t) \geq v_{BE(\text{on})}$ the

V_{CC} = 10 V
i_C
R_C
10 kΩ
R_B
i_B
+
+
+
v_{CE}
v_{OUT}
$v_s(t)$
−
−
−
+
−
V_{BB}

Figure 3.29: Common–emitter transistor amplifier of Example 3.3.

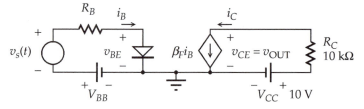

Figure 3.30: Equivalent circuit for Example 3.3.

following is obtained:

$$i_B = (V_{BB} + v_s(t) - v_{BE(on)})/R_B$$

$$v_{OUT} = V_{CC} - i_C R_C = V_{CC} - (V_{BB} + v_s(t) - v_{BE(on)})\beta_F R_C/R_B$$

$$= V_{CC} - (V_{BB} - v_{BE(on)})\beta_F R_C/R_B - (\beta_F R_C/R_B)V_m \sin 2\pi f t$$

a. Because the input voltage has a peak-to-peak value of 1 V, an output voltage with a peak-to-peak value requires that $\beta_F R_C/R_B = 8$.

$$R_B = \beta_F R_C/8 = 187.5 \text{ k}\Omega$$

A quiescent collector voltage of 5 V implies a collector current of 0.5 mA and a base current of 0.5 mA/150 = 3.33 μA.

$$V_{BB} = i_B R_B + v_{BE(on)} = 1.325 \text{ V} \quad \text{(quiescent value)}$$

b. The expression for $v_{OUT}(t)$ yields the following:

$$v_{OUT} = 5 - 4 \sin 2\pi f t \quad \text{V}$$

This is shown in Figure 3.31.

c. $\beta_F = 100$. The base current of the circuit will not be affected if $v_{BE(on)}$ of the new transistor remains equal to 0.7 V. Hence, $i_C = \beta_F i_B = 0.333$ mA, and $v_{CE} = 6.67$ V for quiescent conditions. The following is obtained for $v_{OUT}(t)$ (Figure 3.31):

$$v_{OUT} = 6.67 - 2.67 \sin 2\pi f t \quad \text{V}$$

d. $\beta_F = 200$. The transistor will have a quiescent collector current of 0.677 mA and a collector–emitter voltage of 3.33 V. The expression for $v_{OUT}(t)$ yields

Figure 3.31: Output voltages of Example 3.3.

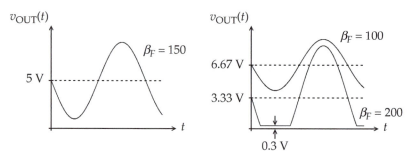

the following:

$$v_{OUT} = 3.33 - 5.33 \sin 2\pi ft \quad \text{V}$$

This relationship is obviously not valid because it results in values of collector–emitter voltage that are less than $v_{CE(sat)}$. This is a result of the equivalent base–collector diode being ignored. Given that $v_{CE} \geq v_{CE(sat)} \approx 0.3$ V is required, the following is obtained for $v_{OUT}(t)$ (Figure 3.31):

$$v_{OUT} = \begin{cases} 3.33 - 5.33 \sin 2\pi ft & \text{for } 3.33 - 5.33 \sin 2\pi ft \geq 0.3 \text{ V} \\ 0.3 \text{ V} & \text{otherwise} \end{cases}$$

EXAMPLE 3.4

It is often necessary to cascade transistor circuits to achieve a desired response. Identical transistors with $\beta_F = 50$ and $v_{BE(on)} = 0.7$ V are used in the two-stage logic buffer of Figure 3.32. Determine the overall transfer characteristic v_{OUT2} versus v_{IN} of the circuit.

SOLUTION The transfer characteristic of stage 2, v_{OUT2} versus v_{OUT1}, is that of Figure 3.24 with $V_{CC} = 5$ V and a transition with a slope of $-\beta_F R_C/R_B = -5$. This is indicated in Figure 3.33 with $v_{CE(sat)} = 0.3$ V. The input current of stage 2, that is, i_{B2}, must be taken into account when determining the transfer characteristic of stage 1. For $v_{CE1} \geq 0.7$ V, the equivalent circuit diode of Q_2

Figure 3.32: Two-stage transistor circuit of Example 3.4.

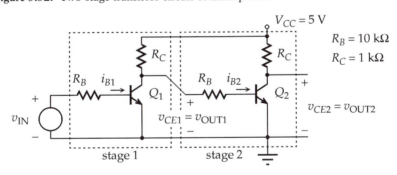

Figure 3.33: Transfer characteristic of the individual transistor stages.

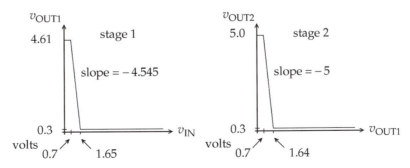

can be replaced with a voltage source of $v_{BE(on)}$. The circuit external to the collector of Q_1 may be replaced by a Thévenin equivalent circuit (Figure 3.34):

$$R_{Th} = R_B \| R_C = 0.909 \text{ k}\Omega$$

$$V_{Th} = \frac{R_B V_{CC}}{R_B + R_C} + \frac{R_C v_{BE(on)}}{R_B + R_C} = 4.61 \text{ V}$$

The slope of the transition is $-\beta_F R_{Th}/R_B = -4.55$, and the maximum value of v_{OUT1} is 4.61 V (Figure 3.33). The response of stage 1 for the Thévenin equivalent circuit is valid for $v_{CE1} \geq 0.7$ V, that is, while the base–emitter voltage of Q_2 is $v_{BE(on)}$. For $v_{CE1} \leq 0.7$ V, $i_{B2} = 0$, and the external collector circuit consists of only R_C and V_{CC}. However, from a practical perspective, the impact of this effect on the transfer characteristic for $0.3 \leq v_{CE1} \leq 0.7$ V is negligible – it cannot be seen on the graph with the resolution of Figure 3.33.

Figure 3.34: Collector circuit of stage 1.

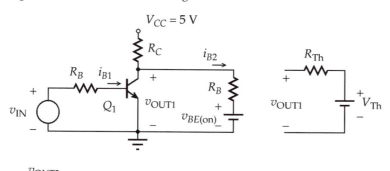

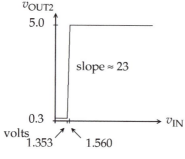

slope ≈ 23

Figure 3.35: Overall transfer characteristic of Example 3.4.

TABLE 3.2 TRANSFER CHARACTERISTIC OF A TWO-TRANSISTOR CIRCUIT

v_{IN} V	v_{OUT1} V	v_{OUT2} V
0	4.61	0.3
0.7	4.61	0.3
1.353	1.64	0.3
1.560	0.7	5.0
5.0	0.3	5.0

A point-by-point procedure may be used to determine the overall transfer characteristic. For $v_{IN} < 0.7$ V, $v_{OUT1} = 4.61$ V, and because this is greater than 1.64 V, $v_{OUT2} = 0.3$ V. For $0.7 < v_{IN} < 1.65$ V, v_{OUT1} will linearly decrease from 4.61 V to 0.3 V. However, v_{OUT2} will remain equal to 0.3 V until v_{OUT1} falls below 1.64 V. On the basis of the transfer characteristic of stage 1, $v_{IN} = 1.353$ V for this condition. The output v_{OUT2} will reach 5 V when $v_{OUT1} = 0.7$ V. This occurs when $v_{IN} = 1.560$ V. These data are summarized in Table 3.2 and shown in Figure 3.35.

3.3 THE COMMON–EMITTER EQUIVALENT CIRCUIT: SOLVING TRANSISTOR CIRCUITS

The transfer characteristic, that is, the dependence of an output voltage on an input voltage of bipolar junction transistor circuits has been discussed. A single transistor can serve as a logic inverter or, if a suitable input offset voltage is provided, a linear amplifier. Only the most basic of circuits has been considered. Numerous modifications of a single-transistor circuit may be used to achieve a desired characteristic. Alternatively, additional transistors may be used to achieve results not obtainable with a single device. From an analysis and design perspective, a transistor is considerably more difficult to treat than a resistor with a linear voltage-versus-current relationship. However, a transistor generally requires a smaller chip area than a resistor when an integrated circuit is fabricated. As a consequence, circuit configurations that tend to rely primarily on transistors are preferred for integrated circuits. Fortunately, computer simulation tools are available that make the design and analysis of circuits with large numbers of transistors manageable.

A voltage transfer characteristic provides only a partial description of the behavior of a circuit. A knowledge of the input current and the ability of the circuit to supply or sink an output current is also of importance. These considerations are generally perceived in terms of impedance levels (for a static condition, only resistance quantities are required). Because a transistor is a nonlinear device, a degree of caution is necessary in applying an impedance concept. For example, consider the basic common–emitter circuit of Figure 3.22 with the equivalent circuit of Figure 3.23. The current of v_{IN} is zero for v_{IN} less than $v_{BE(on)}$ (an infinite input resistance for this condition). For v_{IN} greater than $v_{BE(on)}$, the input consists of the resistor R_B in series with a voltage of $v_{BE(on)}$. The input Thévenin equivalent resistance is thus R_B. From these considerations, one can conclude that, if the resistance of an input source is small compared with R_B, it will have only a small effect on the transfer characteristic of the circuit. On the other hand, a source with a resistance comparable to R_B will have significant effect on the transfer characteristic. A similar situation exists for the output of the transistor circuit. The Thévenin equivalent output resistance of the circuit is R_C, provided the output is not saturated ($v_{CE} > v_{CE(sat)}$). When saturation occurs, the equivalent output circuit is a constant voltage of $v_{CE(sat)}$.

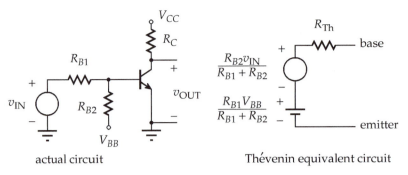

Figure 3.36: Common–emitter circuit with an input offset voltage.

Several modifications of the basic common–emitter circuit will be considered in this section. An understanding of these circuits is necessary to appreciate the operation of more complex logic and amplifier circuits that will be treated in the following sections. Concurrently, the circuits will provide opportunities to increase one's ability in using the common–emitter equivalent circuit for analyzing circuits. Only the static behavior of transistor circuits will be considered; a treatment of the dynamic effects of charge storage by transistors and circuit capacitances will be deferred.

AN EXTERNAL BASE BIAS

An external voltage source V_{BB} is frequently used to shift the transfer characteristic of a transistor circuit (Figure 3.36). Because v_{IN}, R_{B1}, R_{B2}, and V_{BB} are linear elements, they may be replaced by the Thévenin equivalent circuit indicated in the figure. This results in the equivalent circuit of Figure 3.37, which yields the following:

$$R_{Th} = R_{B1} \| R_{B2} = \frac{R_{B1} R_{B2}}{R_{B1} + R_{B2}} \tag{3.19}$$

If $\dfrac{R_{B2} v_{IN} + R_{B1} V_{BB}}{R_{B1} + R_{B2}} < v_{BE(on)}$,

then $i_B = 0$, $i_C = 0$, $v_{OUT} = V_{CC}$,

otherwise $i_B = \left(\dfrac{R_{B2} v_{IN} + R_{B1} V_{BB}}{R_{B1} + R_{B2}} - v_{BE(on)} \right) \Big/ R_{Th}$. $\tag{3.20}$

When the output circuit is not saturated ($v_{CE} > v_{CE(sat)}$), the following is obtained

Figure 3.37: Equivalent circuit of Figure 3.36.

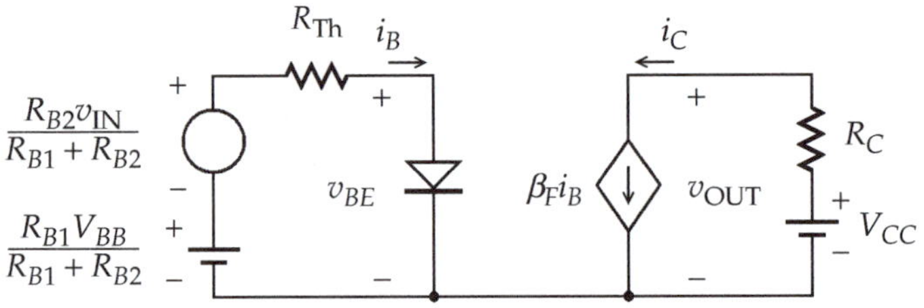

for the output voltage:

$$v_{\text{OUT}} = v_{CE} = V_{CC} - \beta_F i_B R_C$$
$$= V_{CC} - \left(\frac{R_{B2}v_{\text{IN}} + R_{B1}V_{BB}}{R_{B1} + R_{B2}} - v_{BE(\text{on})}\right)\frac{\beta_F R_C}{R_{\text{Th}}} \tag{3.21}$$

The slope of the characteristic during the transition is readily obtained from Eq. (3.21) as follows:

$$\text{slope} = -\frac{R_{B2}\beta_F R_C}{(R_{B1} + R_{B2})R_{\text{Th}}} = -\beta_F R_C/R_{B1} \tag{3.22}$$

This is the same result as that obtained without a biasing voltage V_{BB} and a second resistor R_{B2}.

The voltage V_{BB} and the resistance R_{B2} may be used to change the input voltage over which the output voltage transition occurs. For convenience, let $v_{\text{IN on}}$ be the input voltage that corresponds to the onset of the transition of the output voltage ($v_{\text{OUT}} = V_{CC}$). This voltage, $v_{\text{IN on}}$, is the largest input voltage for which the inequality of Eq. (3.20) applies.

$$\frac{R_{B2}v_{\text{IN on}} + R_{B1}V_{BB}}{R_{B1} + R_{B2}} = v_{BE(\text{on})} \tag{3.23}$$
$$v_{\text{IN on}} = (1 + R_{B1}/R_{B2})v_{BE(\text{on})} - (R_{B1}/R_{B2})V_{BB}$$

It is instructive to consider the special case of $R_{B1} = R_{B2}$.

$$v_{\text{IN on}} = 2v_{BE(\text{on})} - V_{BB} \tag{3.24}$$

This results in the transfer characteristics of Figure 3.38 (it has been assumed that $v_{BE(\text{on})} = 0.7$ V).

Equal values of R_{B1} and R_{B2} are not necessary. When designing a circuit of this type, the value of R_{B1} may be chosen to achieve a desired slope for the output voltage transition. Both R_{B2} and V_{BB} can then be used to establish the input voltage at which the transition occurs. Very likely, a circuit voltage source from some other part of the circuit will be used for V_{BB} (some characteristics can be achieved with $V_{BB} = 0$).

The equivalent input circuit when the transistor is not conducting is R_{B1} and R_{B2} in series with V_{BB}. When the transistor is conducting ($i_B > 0$), the input is a resistance of R_{B1} in series with a voltage of $v_{BE(\text{on})}$. Hence, an input source with a series-equivalent resistance that is small compared with R_{B1} is required if the transfer characteristic is not to deviate appreciably from that predicted with only R_{B1} in the circuit. When the transistor is not saturated ($v_{CE} > v_{CE(\text{sat})}$), the output resistance is R_C.

AN EMITTER RESISTOR

Another modification of the common-emitter circuit is the circuit of Figure 3.39

Figure 3.38: Transfer characteristic of the circuit of Figure 3.36.

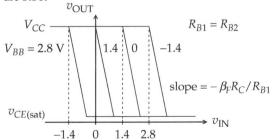

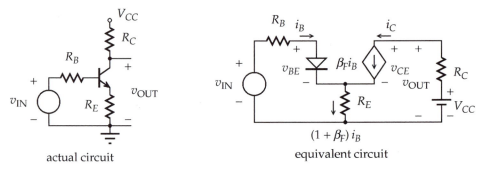

Figure 3.39: Common–emitter circuit with an emitter resistor.

with an emitter resistor R_E. The addition of this resistor (generally much smaller than R_C) to the circuit could have a large effect on the circuit's transfer characteristic and its input impedance. The voltage across R_E depends on the collector current of the transistor. This voltage, however, is in series with the input source and therefore affects the base current of the transistor – a mechanism known as feedback. Feedback of this type is used extensively in electronic circuits.

From the equivalent circuit of Figure 3.39, it may be seen that the current of R_E is $(1 + \beta_F)i_B$ for $v_{CE} > v_{CE(\text{sat})}$. This results in the following for the base current:

$$v_{\text{IN}} = i_B R_B + v_{BE(\text{on})} + (1 + \beta_F)i_B R_E \quad \text{for } v_{\text{IN}} > v_{BE(\text{on})}$$
$$i_B = \frac{v_{\text{IN}} - v_{BE(\text{on})}}{R_B + (1 + \beta_F)R_E} \tag{3.25}$$

The output voltage is readily obtained for these conditions by

$$v_{\text{OUT}} = V_{CC} - \beta_F i_B R_C$$
$$= V_{CC} - \frac{\beta_F \left(v_{\text{IN}} - v_{BE(\text{on})} \right) R_C}{R_B + (1 + \beta_F)R_E} \tag{3.26}$$

The slope of response depends on R_E:

$$\text{slope of } v_{\text{OUT}} = -\frac{\beta_F R_C}{R_B + (1 + \beta_F)R_E} \tag{3.27}$$

As a result of the emitter resistor, the collector–emitter voltage and the output voltage are no longer equal.

$$v_{CE} = v_{\text{OUT}} - (1 + \beta_F)i_B R_E \tag{3.28}$$

This expression is valid only for $v_{CE} > v_{CE(\text{sat})}$. From the expressions for i_B and v_{OUT}, the following is obtained:

$$v_{CE} = V_{CC} - \left(\frac{\beta_F(R_C + R_E) + R_E}{R_B + (1 + \beta_F)R_E} \right) \left(v_{\text{IN}} - v_{BE(\text{on})} \right)$$
$$\text{slope of } v_{CE} = -\frac{\beta_F(R_C + R_E) + R_E}{R_B + (1 + \beta_F)R_E} \tag{3.29}$$

The expression for v_{CE} may be used to determine the onset of saturation, the value of v_{IN} for which $v_{CE} = v_{CE(sat)}$. The response of the circuit, on the basis of the above results, is indicated in Figure 3.40.

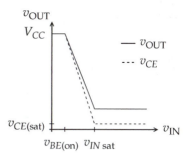

Figure 3.40: Transfer characteristic of the circuit of Figure 3.39.

For an input voltage greater than that corresponding to the onset of saturation, $v_{IN\,sat}$, the equivalent circuit of Figure 3.39 is no longer valid because the collector voltage remains equal to $v_{CE(sat)}$. This suggests the equivalent circuit of Figure 3.41. Because this circuit is valid only for $i_B > 0$, the equivalent base–emitter diode has been replaced by a voltage source of $v_{BE(on)}$. This circuit is valid, provided the collector current that it predicts, i_C, is less than $\beta_F i_B$.

From the equivalent circuit of Figure 3.41, an expression may be obtained for v_{OUT} when v_{IN} is greater than $v_{IN\,sat}$. However, for typical component values that are used in a circuit of this type, the dependence of v_{OUT} on v_{IN} is found to be very slight. Hence, the approximation that v_{OUT} is equal to its value determined for $v_{IN\,sat}$ for values of v_{IN} greater than $v_{IN\,sat}$ is generally justified.

The voltage across the emitter resistor of this circuit affects the current that is supplied by v_{IN}. For the transition region of the response, that is, for $v_{IN} > v_{BE(on)}$ and $v_{CE} > v_{CE(sat)}$, the following is obtained from Eq. (3.25):

$$v_{IN} = v_{BE(on)} + i_B[R_B + (1 + \beta_F)R_E] = v_{BE(on)} + i_B R_{eq}$$
$$R_{eq} = R_B + (1 + \beta_F)R_E \tag{3.30}$$

The input circuit is thus a battery $v_{BE(on)}$ in series with an equivalent resistance of R_{eq}. Because β_F for most transistors is large (50 or greater), this equivalent resistance, for typical values of R_E, is much larger than R_B.

A zero value of R_B is acceptable for this circuit; the input source may be connected directly to the base of the transistor. On the basis of Eq. (3.27), the following is obtained for the slope of v_{OUT} versus v_{IN} in the transition region:

$$\text{slope of } v_{OUT} = -\frac{\beta_F R_C}{(1 + \beta_F)R_E} \approx -\frac{R_C}{R_E} \quad \text{for } \beta_F \gg 1 \tag{3.31}$$

This approximate result is particularly interesting in that it does not depend on the common–emitter current gain of the transistor β_F. Stated in pragmatic terms, the response of this circuit does not depend on the particular transistor that might happen to be used. This is an important consideration when designing circuits because a large variation in β_F generally occurs between "like" transistors (same part number for discrete devices; same fabrication procedure for integrated circuits).

Figure 3.41: Equivalent circuit for the transistor being saturated.

The behavior of this circuit may be modified by connecting R_E to an emitter offset potential V_{EE}. Although the slope of the resultant characteristic in the transition region will remain

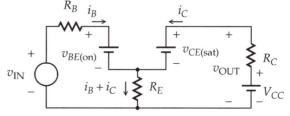

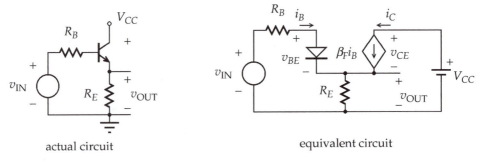

actual circuit   equivalent circuit

Figure 3.42: Emitter–follower configuration and equivalent circuit.

unchanged, the input voltage over which the transition occurs will depend on V_{EE}. As for the circuit with a base voltage V_{BB}, the voltage V_{EE} may be used in place of an input offset voltage for an amplifier.

AN EMITTER–FOLLOWER CIRCUIT

The output terminal of the previous circuits has been the collector of the transistor. Alternatively, the voltage across the emitter resistor of Figure 3.39 could also be used for the output – a collector resistor is not necessary for this configuration (Figure 3.42). If the input and output voltages were specified relative to the positive terminal of V_{CC} rather than relative to its negative terminal, the collector of the transistor of this circuit could be considered as the common terminal (common–collector configuration). This configuration is generally designated an emitter–follower circuit – a designation that is derived from the tendency of the output voltage to follow the input voltage.

The procedure for obtaining a solution for this circuit tends to follow that for a circuit with an emitter and collector resistor. For $v_{IN} < v_{BE(on)}$, $i_B = 0$ and $v_{OUT} = 0$. For $v_{IN} > v_{BE(on)}$, the base current of the transistor is the same as that given by Eq. (3.25) ($v_{CE} > v_{CE(sat)}$):

$$i_B = \frac{v_{IN} - v_{BE(on)}}{R_B + (1 + \beta_F) R_E} \tag{3.32}$$

Because the current of R_E is $i_B + i_C$, the following is obtained for v_{OUT}:

$$v_{OUT} = (1 + \beta_F) i_B R_E = \frac{\beta_F \left(v_{IN} - v_{BE(on)}\right) R_E}{R_B + (1 + \beta_F) R_E} \tag{3.33}$$

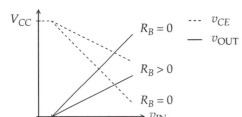

Figure 3.43: Transfer characteristic of emitter–follower circuit of Figure 3.42

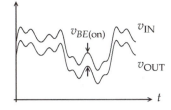

Figure 3.44: Input and output voltages of an emitter follower ($R_B = 0$).

By noting that $v_{CE} = V_{CC} - v_{OUT}$, an expression is readily obtained for v_{CE} as follows:

$$v_{CE} = V_{CC} - \frac{\beta_F \left(v_{IN} - v_{BE(on)} \right) R_E}{R_B + (1 + \beta_F) R_E} \tag{3.34}$$

The transfer characteristic of this circuit, v_{OUT} versus v_{IN}, along with the dependence of v_{CE} on v_{IN} is indicated in Figure 3.43. This solution is valid for $v_{CE} > v_{CE(sat)}$; and even for $R_B = 0$, this condition is valid for $v_{IN} \leq V_{CC}$.

A particularly simple result is obtained for $R_B = 0$.

$$\begin{aligned} v_{OUT} &= \frac{\beta_F R_E}{(1 + \beta_F) R_E} \left(v_{IN} - v_{BE(on)} \right) \\ &\approx v_{IN} - v_{BE(on)} \qquad \text{for } \beta_F \gg 1 \end{aligned} \tag{3.35}$$

For this condition, the output voltage follows the input voltage, that is, it is always $v_{BE(on)}$ less than the input voltage. This is indicated in Figure 3.44 for a time-dependent input voltage with an appropriate offset voltage.

Because the slope of v_{OUT} versus v_{IN} is unity, it is reasonable to ask what is the value of this circuit or what possible use could it serve? The answer lies not with the voltage transfer characteristic but with the current transfer characteristic of the circuit. Although the current supplied by v_{IN} is i_B, the current of R_E is $(1 + \beta_F) i_B$. If R_E is considered as the load resistance of the circuit, the current of the load is $(1 + \beta_F)$ times the current supplied by v_{IN}. Even though voltage gain of an emitter–follower is unity or less, its current gain is generally large. Alternatively, an emitter–follower circuit may be viewed as an active impedance transformer. The equivalent resistance of the input of the circuit R_{eq} is $(1 + \beta_F) R_E$ for $R_B = 0$. Hence, a source with a much larger resistance can be used with this circuit than if the source were to be connected directly to the load R_E.

EXAMPLE 3.5

Consider the common–emitter transistor circuit of Figure 3.45. With an appropriate input offset voltage, this circuit could be used as an amplifier. The slope of the transfer characteristic for the transition region of the circuit is $-\beta_F R_C / R_B$. To increase the magnitude of the slope (that is, to increase the magnitude of voltage gain), one is tempted to reduce the resistance of the base resistor R_B. On the basis of the preceding expression for the slope, $-\infty$ is obtained for $R_B = 0$. This result, however, is not valid because it is premised

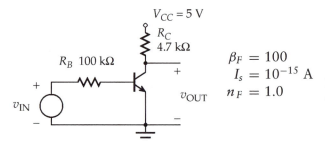

$$\beta_F = 100$$
$$I_s = 10^{-15} \, \text{A}$$
$$n_F = 1.0$$

Figure 3.45: Common–emitter transistor circuit of Example 3.5.

on the base–emitter voltage v_{BE} remaining constant and equal to $v_{BE(on)}$. For small values of R_B, it is necessary to account for the dependence of the base current on the base–emitter voltage.

$$i_B = (I_s/\beta_F)e^{v_{BE}/n_F V_T}$$

This relationship is extremely temperature sensitive; both I_s and V_T depend on temperature. Although small values of R_B will result in a very abrupt transition in v_{OUT}, these circuits tend to be impractical because of the temperature dependence of the response. To show this, consider the circuit of Figure 3.45 along with a circuit in which $R_B \approx 0$. Determine, using SPICE, the dependence of v_{OUT} on v_{IN} for temperatures of 27 °C (normal default temperature) along with elevated temperatures of 50 and 100 °C.

SOLUTION If the input voltage v_{IN} is swept from 0 to 5 V, a zero value for R_B would result in inordinate base currents. Therefore, to limit the current

Figure 3.46: SPICE circuit and file for Example 3.5.

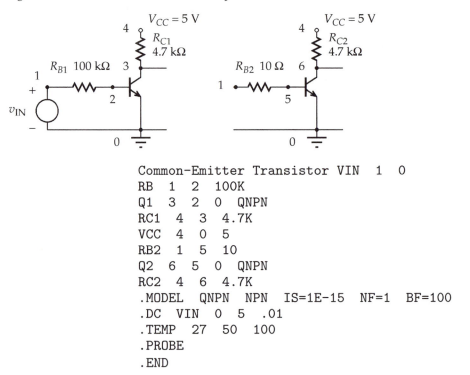

```
Common-Emitter Transistor  VIN   1   0
RB    1   2   100K
Q1    3   2   0   QNPN
RC1   4   3   4.7K
VCC   4   0   5
RB2   1   5   10
Q2    6   5   0   QNPN
RC2   4   6   4.7K
.MODEL  QNPN  NPN  IS=1E-15  NF=1  BF=100
.DC  VIN  0  5  .01
.TEMP  27  50  100
.PROBE
.END
```

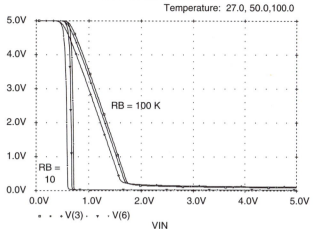

Figure 3.47: SPICE solution for Example 3.5. For each value of R_B, the traces, from left to right correspond to temperatures of 100, 50, and 27 °C.

Common-Emitter Transistor
Temperature: 27.0, 50.0,100.0

RB = 100 K

RB = 10

◦ · ·V(3)· ▾ ·V(6)

VIN

without an appreciable effect on the results, a value of $R_B = 10\ \Omega$ will be used for the "zero" solution (Figure 3.46). The result of the simulation is indicated in Figure 3.47. Consider the case for $R_B = 100\ \text{k}\Omega$. An offset voltage of 1.23 V results in $v_{\text{OUT}} = 2.5$ V for a temperature of 27 °C. At a temperature of 50 °C, $v_{\text{OUT}} = 2.33$ V, and at a temperature of 100 °C, $v_{\text{OUT}} = 1.96$ V for this offset voltage. These small changes in v_{OUT} would be, for most applications, acceptable. This is not the case for $R_B = 10\ \Omega$. For a temperature of 27 °C, an input offset voltage of 0.698 V results in $v_{\text{OUT}} = 2.5$ V. At the elevated temperatures, the offset voltage of 0.698 V results in $v_{\text{OUT}} \approx 0$ V. Only a slight increase in temperature ($T = 40$ °C) results in a saturation of the circuit $v_{\text{OUT}} \approx 0$. To utilize the steep transition obtained for $R_B \approx 0$ for amplification, a different, more sophisticated offset biasing scheme is required.

EXAMPLE 3.6

In the transistor amplifier circuit of Figure 3.48, an emitter voltage source V_{EE} is used in place of an input offset voltage. Assume $v_s(t)$ is a sinusoidal signal ($V_m \sin \omega t$) with a peak-to-peak amplitude of 1.0 V.

a. Determine the value of V_{EE} required to obtain an output voltage of 5 V for $v_s(t) = 0$.

b. Determine the maximum and minimum values of $v_{\text{OUT}}(t)$. Sketch the resultant voltage.

c. The rationale for an emitter resistor is that the dependence of the voltages and currents on β_F is reduced. Use the value of V_{EE} determined in part (a) and repeat part (b) for a transistor with $\beta_F = 100$.

SOLUTION The equivalent circuit of Figure 3.49 applies for an emitter offset voltage of V_{EE} ($i_B > 0$). The following is obtained for this circuit:

$$v_s(t) = i_B R_B + v_{BE(\text{on})} + (1 + \beta_F)i_B R_E + V_{EE}$$

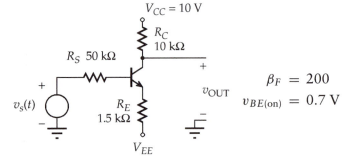

$$\beta_F = 200$$
$$v_{BE(\text{on})} = 0.7 \text{ V}$$

Figure 3.48: Common–emitter transistor amplifier with an emitter offset voltage.

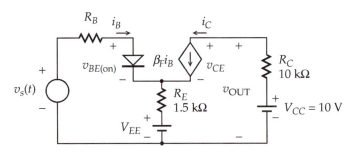

Figure 3.49: Equivalent circuit for Example 3.6.

$$i_B = \frac{v_s(t) - V_{EE} - v_{BE(\text{on})}}{R_B + (1 + \beta_F)R_E}$$

$$v_{\text{OUT}} = V_{CC} - i_C R_C = V_{CC} - \beta_F i_B R_C$$

a. A quiescent value of $v_{\text{OUT}} = 5$ V requires a collector current of 0.5 mA. Because $i_B = i_C/\beta_F$, a quiescent base current of 2.5 μA is required.

$$V_{EE} = -v_{BE(\text{on})} - i_B[R_B + (1 + \beta_F)R_E] = -1.58 \text{ V}$$

b. Because the peak-to-peak value of $v_s(t)$ is 1.0 V, $V_m = 0.5$ V. Hence, for $v_s(t) = 0.5$ V, $i_B = 3.93$ μA, $i_C = 0.785$ mA, and $v_{\text{OUT}} = 2.15$ V. For $v_s(t) = -0.5$ V, $i_B = 1.08$ μA, $i_C = 0.216$ mA, and $v_{\text{OUT}} = 7.84$ V. It should be noted that Eq. (3.27) yields a slope of -5.69, a result that may be shown to be independent of V_{EE}. The calculated values of v_{OUT} for $v_s(t) = \pm 0.5$ V yield a peak-to-peak amplitude of 5.69 V, which is precisely that expected for a slope of -5.69 (Figure 3.50(a)).

c. Current-gain $\beta_F = 100$. For $v_s(t) = 0$, $i_B = 4.37$ μA, $i_C = 4.37$ mA, and

Figure 3.50: Output voltage of Example 3.6.

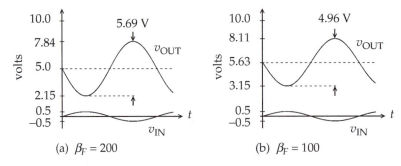

(a) $\beta_F = 200$ (b) $\beta_F = 100$

$v_{OUT} = 5.63$ V (a shift of 0.63 V in the quiescent value of v_{OUT}). For $v_s(t) = 0.5$ V, $i_B = 6.85$ μA, $i_C = 6.85$ mA, and $v_{OUT} = 3.15$ V. For $v_s(t) = -0.5$ V, $i_B = 1.89$ μA, $i_C = 0.189$ mA, and $v_{OUT} = 8.11$ V (Figure 3.50(b)).

EXAMPLE 3.7

The circuit of Figure 3.51 with an LED is used to indicate when the input voltage v_{IN} exceeds a prescribed value. Because the equivalent series resistance of the input voltage source R_{IN} is very large (1 MΩ), a two-transistor circuit is required. Transistor Q_1 is used in an emitter–follower configuration, whereas Q_2 is used in a common–emitter configuration. The transistors have identical parameters.

a. Determine the dependence of i_{C2} on v_{IN} ($0 \leq v_{IN} \leq 5$ V).

b. The LED requires a current of 1.0 mA to produce a noticeable light emission. What is the value of v_{IN} required to achieve this diode current?

c. Suppose Q_1 is omitted from the circuit, that is, R_{IN} is connected directly to the base of Q_2. What is the value of v_{IN} required for a noticeable light emission ($i_D = 1.0$ mA)?

SOLUTION

a. Both Q_1 and Q_2 will be replaced by their equivalent circuits (Figure 3.52). For $v_{IN} < 2v_{BE(on)}$, $i_{B1} = 0$, $i_{B2} = 0$, and $i_{C2} = 0$. An input voltage, v_{IN}, greater than $2v_{BE(on)}$ is required for nonzero transistor currents. For $v_{IN} > 2v_{BE(on)}$ and with neither transistor saturated, the following is

Figure 3.51: Circuit of Example 3.7 with an LED.

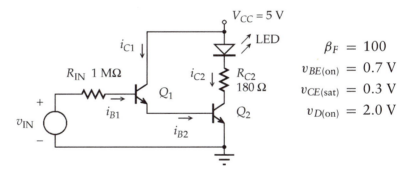

$$\beta_F = 100$$
$$v_{BE(on)} = 0.7 \text{ V}$$
$$v_{CE(sat)} = 0.3 \text{ V}$$
$$v_{D(on)} = 2.0 \text{ V}$$

Figure 3.52: Equivalent circuit for Example 3.7.

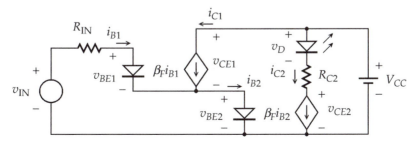

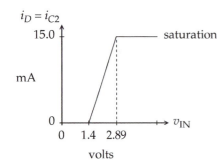

$i_D = i_{C2}$

15.0

mA

saturation

0

0   1.4   2.89

volts

v_{IN}

Figure 3.53: The dependence of diode current on v_{IN} for Example 3.7.

obtained:

$$v_{IN} = i_{B1} R_{IN} + 2v_{BE(on)}, \qquad i_{B1} = (v_{IN} - 2v_{BE(on)})/R_{IN}$$

$$i_{B2} = (1 + \beta_F)i_{B1}$$

$$i_{C2} = \beta_F(1 + \beta_F)i_{B1} = \beta_F(1 + \beta_F)(v_{IN} - 2v_{BE(on)})/R_{IN}$$

$$= 10.1(v_{IN} - 1.4) \text{ mA}$$

From the collector current of Q_2, i_{C2}, its collector–emitter voltage can be determined ($v_D = v_{D(on)}$ for the LED).

$$V_{CC} = v_{D(on)} + i_{C2} R_{C2} + v_{CE2}$$

$$v_{CE2} = 3.0 - 1.818(v_{IN} - 1.4 \text{ V}) \quad \text{V}$$

This expression is valid if $v_{CE2} > v_{CE(sat)} = 0.3$ V. A value of $v_{IN} = 2.89$ V (or greater) results in a saturation of Q_2. The collector current of Q_2 is 15.0 mA when the transistor is saturated (Figure 3.53). It should be noted that the collector–emitter voltage of the emitter–follower transistor Q_1 remains equal to $V_{CC} - v_{BE(on)} = 4.3$ V regardless of i_{C1} ($i_{C1} > 0$).

b. For $i_{C2} = 1.0$ mA, $v_{IN} = 1.50$ V.

c. With R_{IN} connected directly to the base of Q_2, the following is obtained for $v_{IN} > v_{BE(on)}$.

$$i_D = i_{C2} = \beta_F i_{B2} = \beta_F(v_{IN} - v_{BE(on)})/R_{IN}$$

$$= 0.1(v_{IN} - 0.7 \text{ V}) \quad \text{mA}$$

To obtain a noticeable light emission, an input voltage of 10.7 V is required.

3.4 DIGITAL LOGIC CIRCUITS: STATIC AND DYNAMIC CHARACTERISTICS

The earliest integrated-circuit digital logic gates utilized bipolar junction transistors in circuits that were modifications of the basic common–emitter configuration that has previously been considered. These circuits evolved into standardized integrated circuits containing several transistors and resistors that performed numerous logic functions. An advanced family of integrated circuits introduced in the 1960s, the transistor-transistor-logic (TTL) family, is not only still widely used, but its terminal characteristics have dictated the design of more advanced gates (TTL compatibility). The earliest IBM personal computer (1981) was designed to use off-the-shelf standardized TTL integrated circuits and other components (Cringely 1993). Through the development of very-large-scale integrated circuits using metal-oxide field-effect transistors (MOSFETs), the logic functions of arrays of TTL gates were achieved with a single integrated circuit. A modern personal computer using these large-scale integrated circuits appears "empty" compared with the earliest personal computers that were literally packed full of integrated circuits.

Two basic considerations must be taken into account when designing logic circuits. One is the circuit's static transfer characteristic, and the other is the dynamic behavior of the circuit. Achieving a desired static voltage (or current) transfer characteristic is usually not difficult. The challenge is obtaining the short response time required by modern systems. This is generally referred to as the "speed" of a gate, which is the rate at which a gate will respond to a set of changing input signals.

There is an old addage, "haste makes waste," which is closely related to the thermodynamic concept of reversibility. The "waste" of an electronic circuit is the electrical energy dissipated by the components and devices of the circuit. To increase the rate at which a particular circuit will operate, it is generally necessary to modify circuit components and devices, that is, decrease resistances and increase the currents of devices. This increases the electrical energy dissipated by the circuit. Electrical energy is converted to thermal energy, which is removed from the integrated circuit by conduction, thus increasing the temperature of the components and devices. Because the characteristics of semiconductors depend on temperature, an upper operating temperature limit restricts the rate at which electrical energy (power) can be dissipated.

Both capacitive and inductive effects influence the dynamic behavior of circuits. For most logic circuits, however, it is capacitive effects, that is, charge storage, that is the main factor. Capacitances are unavoidable and, although they can be reduced with improved designs, they can never be reduced to zero. It should be noted that charge storage, for example, by an ideal capacitor, is not dissipative in that electrical energy is conserved. However, to change the voltage of a capacitor, its stored charge must be changed ($v = Q/C$). This requires a current ($i = \frac{dQ}{dt}$), and the faster the voltage is changed, the larger the current that is required. Larger currents necessarily increase the electrical power dissipated by a circuit. It is also desirable to minimize the difference in the voltages corresponding to high and low logic levels. Voltage differences smaller than the standard "0-to-5 V" TTL voltage change are therefore used for very rapidly responding circuits.

TRANSISTOR OPERATING REGIONS

Three distinct regions of operation, namely, cutoff, active, and saturation, can be identified for the basic transistor circuit of Figure 3.54. Cutoff is the region for which the device currents are zero. This occurs for negative values of base–emitter voltage as well as for positive values (forward bias) that are insufficient to result in a significant junction current. For silicon devices, this is usually the case for base–emitter voltages that are less than approximately 0.5 V. In the

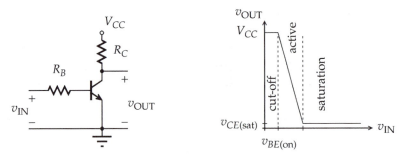

Figure 3.54: Regions of operation for a basic common–emitter transistor configuration.

TABLE 3.3 BIPOLAR JUNCTION TRANSISTOR – REGIONS OF OPERATION

Region	Currents and voltages	Base–emitter junction bias	Base–collector junction bias
Cutoff	$i_B = 0, i_C = 0$	[a]	Reverse
Active	$i_C = \beta_F i_B{}^b, v_{CE} > v_{CE(sat)}$	Forward	Reverse
Saturated	$i_C < \beta_F i_B, v_{CE} = v_{CE(sat)}$	Forward	Forward

[a] Reverse or insufficient forward bias for significant curent.
[b] or modified by Early effect.

---

active region, the base current controls the collector current, that is, $i_C \approx \beta_F i_B$. The base–emitter junction is forward biased, whereas the base–collector junction is reverse biased. This is the region of operation in which amplification can be obtained. Finally, there is the saturated region in which the collector voltage remains very small, $v_{CE(sat)}$, regardless of the base current ($i_C < \beta_F i_B$). Both junctions of the transistor are forward biased for these conditions. The regions of operation are summarized in Table 3.3.

For saturated transistor logic circuits (these include TTL integrated circuits), the cutoff and saturation regions correspond to static logic levels. It is, however, necessary for the device to make a transition through its active region when changing its logic levels. It is the change from one region to another, as well as the transition through the active region, that limits the rate at which logic levels of a circuit can be changed.

CAPACITIVE LOAD

The behavior of the circuit of Figure 3.55 with an output load capacitance C_L will be determined for abrupt changes in the input voltage. The load capacitance includes the capacitance to which the circuit is connected (e.g., a data bus). It also includes the equivalent capacitances of the devices along with the parasitic capacitance of the leads and devices to the substrate of an integrated circuits. To determine the dynamic behavior of this circuit fully, other capacitances, voltage-dependent capacitances of the transistor, need to be taken into account. In addition, the charge carriers of the transistor result in a charge storage

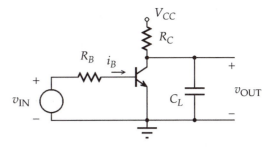

Figure 3.55: A basic common–emitter circuit with a capacitive load.

within the device that must also be accounted for. To include all these effects, it is generally necessary to use a computer simulation to obtain an accurate dynamic result, albeit reasonable estimates can be obtained with a "charge-controlled" transistor model and a judicious set of analytic approximations.

Despite the foregoing considerations, it is of value to consider the elementary circuit of Figure 3.55 with a single capacitor. For some circuits, the charge storage of a single capacitor may be the dominant effect; other charge storage effects may have a negligible impact on the results. Although the effects of multiple charge storage elements cannot be obtained by considering these elements individually, useful insights can be gained through such a process.

Two different input voltages will be considered, one with a high-to-low transition, and the other with a low-to-high transition. Assume that $v_{\text{IN}}(t)$ has been equal to V_p for a very long time and that it is suddenly switched to zero at $t = 0$.

$$v_{\text{IN}}(t) = V_p(1 - u(t)) \tag{3.36}$$

If $V_p < v_{BE(\text{on})}$, then the transistor will remain cut off and $v_{\text{OUT}} = V_{CC}$ for all times. The interesting case is for $V_p > v_{BE(\text{on})}$, which results in the following for the initial base current ($t < 0$):

$$i_B = (V_p - v_{BE(\text{on})})/R_B \quad t < 0 \tag{3.37}$$

If $\beta_F i_B R_C < V_{CC} - v_{CE(\text{sat})}$, then the transistor is in its active region of operation as given by

$$v_{\text{OUT}} = v_{CE} = V_{CC} - \beta_F i_B R_C \quad \text{active, } t < 0 \tag{3.38}$$

On the other hand, if the base current is sufficiently large, the transistor will be saturated, $\beta_F i_B R_C > V_{CC} - v_{CE(\text{sat})}$.

$$v_{\text{OUT}} = v_{CE(\text{sat})} \quad \text{saturated, } t < 0 \tag{3.39}$$

Because it is this case that is generally of interest, this condition will be assumed for the solution that follows.

For $t > 0$, the base current is reduced to zero, resulting in a cutoff condition for the transistor, $i_C = 0$. The equivalent circuit of Figure 3.56, in which R_C charges the load capacitor applies.

$$(V_{CC} - v_{\text{OUT}})/R_C = C_L \frac{dv_{\text{OUT}}}{dt}$$

$$\frac{dv_{\text{OUT}}}{dt} + \frac{v_{\text{OUT}}}{R_C C_L} = \frac{V_{CC}}{R_C C_L} \quad t > 0 \tag{3.40}$$

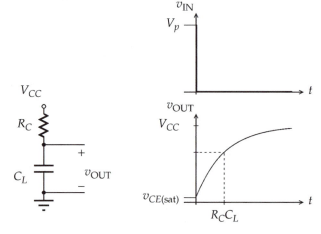

Figure 3.56: Equivalent circuit for charging of a load capacitor.

A solution for $v_{OUT}(t)$ having an arbitrary constant of integration A is obtained for a solution of the differential equation

$$v_{OUT}(t) = Ae^{-t/R_C C_L} + V_{CC} \qquad t > 0. \tag{3.41}$$

As a result of the capacitor, the output voltage will not change abruptly at $t = 0$. Hence, A must be such that $v_{OUT}(0) = v_{CE(sat)}$.

$$A = v_{CE(sat)} - V_{CC}$$
$$v_{OUT}(t) = V_{CC} - (V_{CC} - v_{CE(sat)})e^{-t/R_C C_L} \quad t > 0 \tag{3.42}$$

When t is equal to the time constant of the circuit $R_C C_L$, the exponential is equal to 0.37 (e^{-1}), that is, the output voltage has changed 63 percent of its way toward reaching V_{CC} (Figure 3.56).

The solution for an upward transition of $v_{IN}(t)$ is somewhat more complex. It will be assumed that $v_{IN}(t) = 0$ for $t < 0$, resulting in a cutoff condition for the transistor ($i_C = 0$). At $t = 0$, the input voltage is suddenly increased to V_p, resulting in a base and collector for the transistor ($V_p > v_{BE(on)}$) as follows:

$$v_{IN}(t) = V_p u(t)$$
$$i_B = (V_p - v_{BE(on)})/R_B \tag{3.43}$$
$$i_C = \beta_F i_B = \beta_F(V_p - v_{BE(on)})/R_B \quad t > 0$$

The initial value of v_{CE} (= v_{OUT}) is V_{CC}. Because this voltage will not change abruptly (owing to C_L), the transistor immediately following the change in input voltage will be in its active mode of operation. The equivalent circuit of Figure 3.57 is therefore valid as long as $v_{OUT} > v_{CE(sat)}$.

On the basis of the circuit of Figure 3.57, the following is obtained for the time-dependent behavior of v_{OUT}:

$$(V_{CC} - v_{OUT})/R_C = C_L \frac{dv_{OUT}}{dt} + \beta_F i_B \quad t > 0 \tag{3.44}$$

For many circuits, $\beta_F i_B$ is very much larger than the left-hand side of Eq. (3.44). If

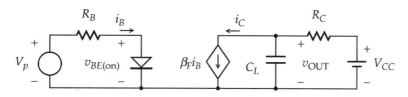

Figure 3.57: Transistor equivalent circuit for $v_{OUT} > v_{CE(sat)}$.

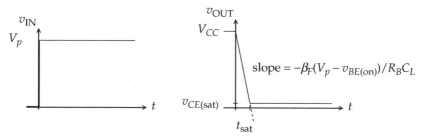

Figure 3.58: Input and output voltage for a downward transition of v_{OUT}.

this is the case, a rather simple result is obtained for v_{OUT} (Figure 3.58) as follows:

$$\frac{dv_{OUT}}{dt} = -\beta_F i_B / C_L = -(V_p - v_{BE(on)})\beta_F / R_B C_L$$

$$v_{OUT}(t) = V_{CC} - (V_p - v_{BE(on)})\beta_F t / R_B C_L \quad t > 0$$

(3.45)

This solution is valid until v_{OUT} falls to $v_{CE(sat)}$. The time required for this to occur, t_{sat}, is readily obtained by

$$(V_p - v_{BE(on)})\beta_F t_{sat} / R_B C_L = V_{CC} - v_{CE(sat)}$$

$$t_{sat} = \frac{(V_{CC} - v_{CE(sat)}) R_B C_L}{(V_p - v_{BE(on)})\beta_F}$$

(3.46)

This time will generally be considerably less than $R_C C_L$, the time constant associated with an upward transition of v_{OUT}.

The response of this circuit for pulsed input voltage is given in Figure 3.59. It will be noted that the pulse length t_p needs to be larger than t_{sat} for saturation to occur. Otherwise, the minimum value of v_{OUT} will be greater than $v_{CE(sat)}$. The time constant for the upward transition is equal to $R_C C_L$. Therefore, a time

Figure 3.59: Response of the circuit of Figure 3.55 for an input pulse ($t_p > t_{sat}$).

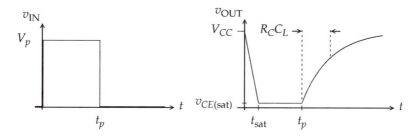

interval of three (or more) times $R_C C_L$ is required for v_{OUT} to reach approximately V_{CC}.

Both the time constant associated with the upward transition of v_{OUT}, $R_C C_L$, and the time required for a downward transition of v_{OUT}, t_{sat}, are linearly dependent on the load capacitance C_L. Hence, to the extent that C_L can be reduced, the response times of the gate can be reduced. It is often the case that C_L is due primarily to an external load over which one has little control. To decrease the upward response time of v_{OUT}, it is then necessary to reduce the collector resistor R_C. Similarly, t_{sat}, can be reduced by decreasing the base resistor R_B.

For a given set of voltage levels, the response times of the circuit are directly proportional to R_B and R_C. Reducing both these quantities by 50 percent, for example, reduces the response times by 50 percent. Reducing the resistance values, however, increases the currents of a circuit (a doubling in currents for a 50-percent reduction in resistances). Consider the static collector current of the basic current of Figure 3.54:

$$i_C = \begin{cases} (V_{CC} - v_{CE(sat)})/R_C \approx V_{CC}/R_C & \text{for } v_{OUT} = v_{CE(sat)} \\ 0 & \text{for } v_{OUT} = V_{CC} \end{cases} \qquad (3.47)$$

For a low-output voltage, the power supplied by V_{CC} is approximately V_{CC}^2/R_C, whereas for a high output voltage it is zero (assuming no load current). If, on the average, the output of the logic circuit is high as often as it is low, the average power supplied by V_{CC} is the average of the values for a high-and low-output voltage.

$$P_{CC(av)} = V_{CC}^2/2R_C \qquad (3.48)$$

Therefore, if R_C is reduced to decrease transition times, the average power supplied by V_{CC} is increased. The price "paid" to achieve a more rapid response is an increased average power that must be dissipated by the gate circuit.

To perform basic combinational logic functions, the common–emitter transistor circuit could be used in conjunction with the diode logic circuits discussed in Sections 1.7 and 2.6. Alternatively, transistor circuits can be directly used to perform combinational logic functions.

LOGIC FAMILIES

Three early families of transistor logic circuits are indicated in Figure 3.60 (Garret 1970; Glaser and Subak-Sharpe 1977; Gray, and Searle 1966; Gray and Searle 1969; Harris, Gray, and Searle 1966; Haznedar 1991; Hodges and Jackson 1988; Millman and Taub 1965; Taub and Schilling 1977). If one input of the resistor-transistor circuit is zero, that is, $v_B = 0$, the resultant circuit is essentially that of Figure 3.36 with $V_{BB} = 0$. As a result of the voltage divider found by the two base resistors, an input voltage of $2v_{BE(on)}$ is required to turn the transistor on. For a properly designed circuit, the transistor will be saturated for an input voltage that is less than V_{CC}. Therefore, if $v_B = V_{CC}$, the transistor will be saturated regardless of the value of v_A ($v_A \geq 0$). The condition of saturation occurs if either (or both) input(s) is(are) high, that is, it corresponds to the logic **OR** operation on the two inputs. Because saturation results in a low output voltage,

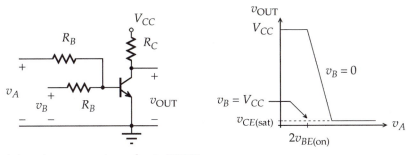

(a) resistor-transistor logic **NOR** gate

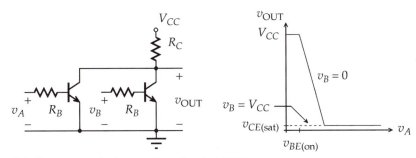

(b) direct-coupled transistor logic **NOR** gate

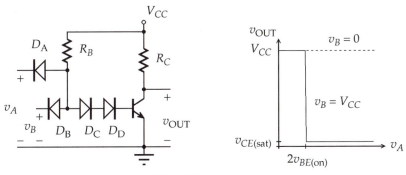

(c) diode-transistor logic **NAND** gate

Figure 3.60: Early bipolar-junction-transistor logic gates.

the gate performs the logical **NOR** operation (positive logic). Additional base resistors may be used for more inputs; this increases the input voltage required for a single input (for n inputs, it may be shown to be $nv_{BE(on)}$). A base biasing resistor and a negative biasing supply can be used to reduce this potential.

An alternative logic **NOR** gate configuration is the direct-coupled transistor gate in which each input is connected to a separate transistor (Figure 3.60(b)). As for the previous circuit, component values are chosen such that a transistor is saturated if its input is high. A transistor with a low input, however, is cut off, that is, its collector current is zero. If one or both transistors are satured, the output of the circuit v_{OUT} is low. Additional transistors may be used for more inputs.

Logic gates are generally designed to function as a family, that is, the output of a gate must serve as the input of one or more other gates of that family. For the gates

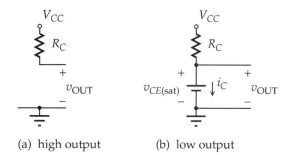

(a) high output (b) low output

Figure 3.61: Equivalent output circuits of a transistor logic gate.

considered, the equivalent output circuit for a high output has a Thévenin equivalent resistance of R_C. A low output is a voltage of $v_{CE(sat)}$ as long as the collector current of the transistor is less than $\beta_F i_B$ (Figure 3.61). The circuits of Figures 3.60(a) and 3.60(b) result in an input current for a high input signal. The gate providing the input signal must therefore supply the input currents of the gates to which it is connected. Because these currents must be supplied by the circuit of Figure 3.61(a) with the series resistance of R_C, the currents will be less than for an input voltage of V_{CC}. This sets a limit to the number of gates that can be connected to the output of a gate. Furthermore, a reduced base current tends to result in a slower response of the circuit.

The output characteristics of the resistor–transistor and direct-coupled transistor logic gates are ill-matched to provide the input current that these gates require. When the transistor's output is high, it is least able to supply the base currents of the gates to which it is connected. On the other hand, when a transistor is saturated, it can readily sink additional current. The diode-transistor-logic gate of Figure 3.60(c) provides a much better match between the output characteristic of the gate and its input current requirement. The two inputs, their diodes, and R_B connected to V_{CC} will be recognized as a diode logic **AND** gate. Because the common–emitter transistor behaves as an inverter, a logic **NAND** operation would be expected for the gate.

To gain an understanding of the operation of the gate, consider the case for which both inputs are floating, that is, the currents of the input diodes D_A and D_B are zero. The other two diodes D_C and D_D will be conducting, resulting in the following for the base current of the transistor:

$$V_{CC} = i_B R_B + 2v_{D(on)} + v_{BE(on)}$$
$$i_B = \left(V_{CC} - 2v_{D(on)} - v_{BE(on)}\right)/R_B \tag{3.49}$$

The circuit is designed such that i_B is sufficiently large to saturate the transistor, resulting in $v_{OUT} = v_{CE(sat)}$. For $v_{D(on)} = v_{BE(on)} = 0.7$ V, the voltage at the junction of the diodes v_J will be equal to 2.1 V. Hence, if v_A and v_B are greater than 2.1 V, the input diodes will be reverse biased, and the same solution is obtained. Therefore, if both inputs are high, or floating, the output of the gate is low.

Consider the case for which $v_A = 0$ and the other input v_B is high. This results in a current through D_A and a voltage of $v_{D(on)}$ for the junction voltage v_J. This voltage, approximately 0.7 V, is, because of the series diodes D_C and D_D, insufficient to result in a significant base current of the transistor. Hence, the transistor is cut off for this condition, and its output is high ($v_{OUT} = V_{CC}$). The corresponding current of the input i_A is readily determined:

$$V_{CC} = -i_A R_B + v_{D(on)}$$
$$i_A = -(V_{CC} - v_{D(on)})/R_B \tag{3.50}$$

The current is negative, that is, it is out of the input terminal. A gate serving as the input must therefore be capable of sinking this current when its output is low, which is a condition that coincides with the capability of a saturated transistor. As indicated in Figure 3.60(c), an abrupt transition of the output voltage occurs for an input voltage of about 1.4 V. If either or both inputs are low, the output of the gate is high. Only if both inputs are high is the output low – the logic **NAND** function. A floating input results in the same behavior as a high logic input.

TRANSISTOR-TRANSISTOR LOGIC

The transistor–transistor logic circuit is a major improvement over the diode–transistor logic circuit. This circuit, developed by J. L. Buie in 1961 (Buie 1966), is the basis of the widely used TTL integrated circuit logic gates (Elmasry 1983, 1985). A switching limitation of bipolar junction transistors arises from the charge storage within the device associated with its charge carriers. To minimize switching delays, in particular those associated with leaving the saturated region, external circuits that can rapidly remove stored charges are necessary. The transistors of a TTL circuit, in effect, work together to achieve this.

The basic two-input logic **NAND** gate circuit indicated in Figure 3.62(a) is similar to the diode–transistor circuit. The input protective diodes, for normal

Figure 3.62: Transistor–transistor two-input logic NAND gates. Approximate resistance values are given for the standard 7400 series gates.

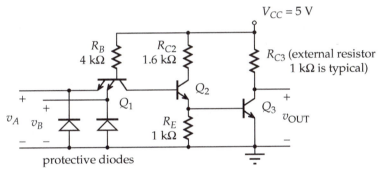

(a) open-collector output circuit

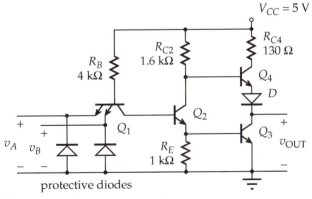

(b) totem-pole output circuit

3.4 DIGITAL LOGIC CIRCUITS: STATIC AND DYNAMIC CHARACTERISTICS 173

operation in which the input voltages are zero or greater, are reverse biased and thus do not affect the behavior of the circuit. The input transistor Q_1 has two emitters that share a common base and collector. For static conditions, the base–emitter and base–collector junction diodes tend to behave as diodes D_A, D_B, and D_C of the diode–transistor circuit of Figure 3.60. The advantage of the transistor circuit over the diode circuit is that it responds much faster. Transistor Q_2 will be recognized as forming an emitter–follower circuit. When this transistor is active, or saturated, its base–emitter voltage $v_{BE(on)}$ is essentially the same as the voltage across diode D_D of the diode–transistor circuit when it is conducting. As a result of the current gain of Q_2, the base current of Q_3 is much larger than if the circuit had only a diode. This circuit has nearly the same transfer characteristic as the diode–transistor circuit.

A further enhancement of the TTL circuit is the totem-pole output circuit of Figure 3.62(b). When both inputs of the gate are high, the current through the base–collector junction of Q_1 (the base current of Q_2) is sufficient to saturate Q_2. This, in turn, results in a base current that saturates Q_3. Because the collector–emitter voltage of Q_2 is small, $v_{CE(sat)}$, the base–emitter voltage of Q_4 is insufficient for it to conduct, that is, the currents of Q_4 are essentially zero. Hence, the equivalent output circuit is simply the saturated transistor Q_3.

A high-output state occurs if either input is low. For this condition, the base current of Q_2 is zero – it is cut off. The output transistor Q_3 is also cut off. This results in an equivalent output circuit that consists of V_{CC}, R_{C2}, R_{C4}, Q_4, and D. This circuit has a much lower equivalent resistance than a circuit with a collector resistor. However, to determine the behavior of this circuit, two regions of operation for Q_4 must be considered; Q_4 may be either in its active or in its saturated region.

The transfer characteristics of the two TTL circuits are indicated in Figure 3.63. As a result of the totem-pole output circuit, the high output voltage of this gate is only 3.6 to 3.8 V. Furthermore, when the input voltage reaches approximately 0.7 V, Q_2 starts to conduct, that is, its collector voltage decreases. As a result of Q_4, which functions as an emitter-follower, an initial small downward slope of the characteristic results (the telltale mark of a TTL gate). When the input voltage reaches approximately 1.4 V, the output transistor is turned on, and an abrupt output voltage transition occurs.

Figure 3.63: Transfer characteristics of transistor–transistor logic circuits.

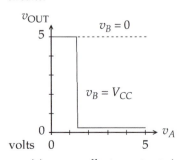

(a) open-collector output circuit

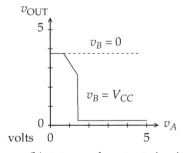

(b) totem-pole output circuit

Integrated logic circuits with numerous modifications of the basic TTL circuit are common. Both low- and a high-power versions of the circuit are available. A version with Schottky-type transistors, in which saturation does not occur, results in very short response times. Standard integrated circuits with up to 100 or more transistors that perform numerous logic functions are available (Lancaster 1974; Morris and Miller 1971; Texas Instruments 1976).

EXAMPLE 3.8

The common-emitter logic inverter gate of Figure 3.64 has a load capacitance of 50 pF. Assume that a step function with a peak voltage of 5 V is used for the input of the gate.

a. Determine t_{sat} for the downward transition of the output voltage.

b. What are the times constant associated with an upward transition of the output voltage?

c. What is the time required for the transitions of the output voltage to reach their midvoltage value, that is, $(V_{CC} + v_{CE(sat)})/2$?

SOLUTION It is necessary to determine if the transistor is saturated for $v_{IN} = 5$ V.

$$i_B = (v_{IN} - v_{BE(on)})/R_B = 0.43 \text{ mA}$$

If i_C were equal to $\beta_F i_B$, 21.5 mA, the collector–emitter voltage would be -16.5 V. This is not the case; the transistor is obviously saturated and $v_{OUT} = v_{CE(sat)} = 0.2$ V. The collector current is a current considerably less than $\beta_F i_B$, $(V_{CC} - v_{CE(sat)})/R_C = 4.8$ mA.

a. The downward transition time t_{sat} is given by Eq. (3.46).

$$t_{sat} = \frac{(V_{CC} - v_{CE(sat)}) R_B C_L}{(V_p - v_{BE(on)})\beta_F} = 11.2 \text{ ns}$$

b. The time constant for an upward transition is $R_C C_L$, 50 ns.

c. The midpoint for the downward transition occurs at $t_{sat}/2$, 5.6 ns. The following is obtained for the upward transition of the output voltage (Eq. (3.42)):

$$v_{OUT}(t) = V_{CC} - (V_{CC} - v_{CE(sat)})e^{-t/R_C C_L}$$
$$e^{-t_{mid}/R_C C_L} = 0.5, \quad t_{mid} = R_C C_L \ln 2 = 34.7 \text{ ns}$$

Figure 3.64: Logic inverter gate of Example 3.8.

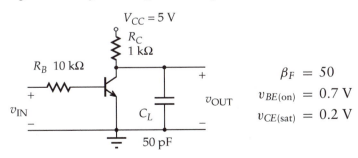

EXAMPLE 3.9

The dynamic performance of an inverter gate is usually determined using an input voltage with the same time-dependent waveform as the output voltage of the gate. To achieve this, several gates can be connected in cascade, as indicated in Figure 3.65. The last gate (Gate 5) is the output load of the gate being tested (Gate 4). Assume that the circuit of Figure 3.64 applies for each of the gates. A SPICE simulation is desired for this circuit ($I_s = 10^{-15}$ A, $n_F = 1.0$). Use an input voltage pulse with a peak amplitude of 5 V and a duration of 200 ns. Except for a time displacement, the waveforms of the input and output voltages of Gate 4 should be nearly identical.

a. Determine the output voltages of the first four gates and verify the previous statement.

b. For Gate 4, determine the 10- to 90-percent transition times for the upward and downward transitions of its output voltage.

Figure 3.65: Cascade inverter gates of Example 3.9.

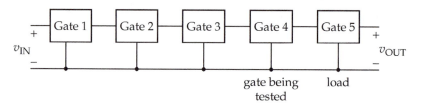

Figure 3.66: Circuit and SPICE file for Example 3.9.

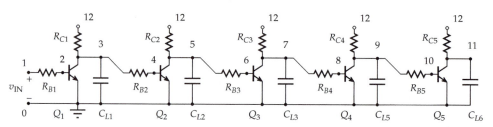

```
Inverter Gates                  RB4   7   8    10K
VIN 1 0 PULSE(0 5 0 1N 1N 200N) Q4    9   8  0   QNPN
RB1   1   2   10K               RC4   12  9    1K
Q1    3   2   0   QNPN          CL4   9   0    50P
RC1   12  3   1K                RB    9   10   10K
CL1   3   0   50P               Q5    11  10  0   QNPN
RB2   3   4   10K               RC5   12  11   1K
Q2    5   4   0   QNPN          CL5   11  0    50P
RC2   12  5   1K                VCC   12  0    DC   5
CL2   5   0   50P               .MODEL QNPN NPN IS=1E-15 NF=1
RB3   5   6   10K               .TRAN  1N   400N
Q3    7   6   0   QNPN          .PROBE
RC3   12  7   1K                .END
CL3   7   0   50P
```

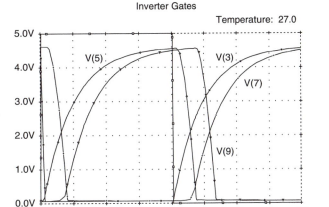

Figure 3.67: SPICE solution for Example 3.9.

c. Determine the propagation delay times of Gate 4. These times, the delay in the gate's output voltage response to an input voltage change, are measured between the midvoltage point of the input and the midvoltage point of the output.

SOLUTION The circuit and SPICE file of Figure 3.66 will be used.

a. The PROBE response of Figure 3.67 was obtained for the circuit. Except for the time difference, the voltage V(9) is essentially the same as V(7).

b. The maximum output voltage is 4.48 V, and the minimum is 73 mV (saturation). The 10- and 90-percent points are thus 0.52 and 4.11 V. An upward transition time of 95 ns and a downward time of 19 ns are obtained.

c. For a downward transition of the output voltage, the propagation delay time is 41 ns. A time of −10.7 ns is obtained for the upward transition. This does not mean that the change in the output occurs before an input (a "predictive" circuit). As expected, the output voltage of Stage 4, V(9), does not begin to change until after the input begins to change.

3.5 AMPLIFIER CIRCUITS: SMALL-SIGNAL BEHAVIOR

It is often necessary to increase the voltage or current levels of a signal, or both. A common example is the amplification of the low-level audio signal produced by a microphone or radio detector in which the signal may have a power level of only a few microwatts (or less) and needs to be amplified to a power level of a few watts (or more) to drive a loudspeaker. This requires active devices (such as bipolar junction transistors) that are powered by an external power supply (V_{CC}). Although amplification is most frequently thought of in terms of analog signals, it is frequently necessary to amplify digital signals. For example, a "receiver" at the end of a digital transmission line is used to amplify and restore the logic levels of the low-level digital signal. Another example is the amplification of the low-level signal produced by a photodetector of a fiber-optic system.

ANALOG SIGNALS

Signals may generally be classified into two categories. One type, an alternating current (ac) signal, has a zero average value.

$$\lim_{T \to \infty} \frac{1}{T} \int_0^T v_s(t)\, dt = 0 \qquad \text{ac signal} \tag{3.51}$$

Signals of this type, for example an audio signal, are generally such that the integral term of Eq. (3.51) tends to be very small for relatively small values of T (a dc voltmeter would read zero). The other type of signal has a direct current (dc) component, that is, it does not have a zero average value. The output voltage of a TTL logic gate, for example, has an average value that falls between its low- and high-level output voltages – midway between these levels if the output is low as frequently as it is high.

It is generally easier to amplify ac signals faithfully than signals with a dc component. As a result, most communications systems are designed to process and transmit ac signals. Although many analog signals fall naturally into this category, other signals can be transformed into an ac signal. The standard serial communications port of a computer, for example, uses binary levels of approximately -10 and $+10$ V, which result, if the binary levels occur with equal frequency, in an output voltage with a zero average value. Alternatively, a special encoding might be used for digital signals. For example, a logic **0** might be transmitted by a zero voltage and a logic **1** by either a positive or negative pulse, the polarity alternating from pulse to pulse (bipolar signaling). This signal has a zero average value regardless of the relative frequency of 0s and 1s. Alternatively, a split-phase pulse can be used in which the pulse signal has a zero average value (Figure 3.68).

A practical advantage of an amplifier of ac signals is that capacitors can be used to couple an ac signal from one node of the circuit to another node. To illustrate this, consider a signal that is composed of two components, an average value and a signal component without an average value, that is, an ac signal.

$$\underbrace{v_S(t)}_{\text{physical quantity}} = \underbrace{V_S}_{\text{average value}} + \underbrace{v_s(t)}_{\text{ac component}} \tag{3.52}$$

The notation of Eq. (3.52) is important. A lowercase symbol with an uppercase subscript is used for the physical quantity, which is the notation that has generally been used for currents and voltages up to this point. An uppercase symbol and subscript are used for the average value, and a lowercase symbol and subscript are used for the ac component. This is the generally accepted notation used for electronic circuits. The term *ac component*, however, is not quite correct because ac is an abbreviation for alternating current, a current that is periodically reversed in time (it also tends to carry a sinusoidal connotation). Although the polarity of communications signals is frequently reversed, these signals are generally not periodic. A periodic

Figure 3.68: Split-phase pulses with a zero average value.

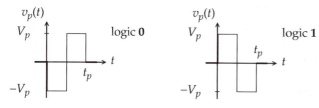

signal conveys little information – seeing one period is equivalent to seeing all periods. Therefore, the term *varying component* will generally be used in the discussions that follow. It is to be understood that the varying component has a zero average value. (This also eliminates the misnomer of ac voltage, that is, alternating current voltage; or ac current, alternating current current.)

CAPACITIVE COUPLING

Suppose a voltage $v_S(t)$ is coupled to a resistor R by means of a capacitor (Figure 3.69). Superposition may be used to solve the circuit. Consider the case for the voltage source V_S, that is, the condition for $v_s(t) = 0$. The steady-state solution is $v_C(t) = V_S$ and $v_R(t) = 0$. For the other case, $V_S = 0$, the following is obtained:

$$v_s(t) = v_C(t) + v_R(t) = \frac{1}{C} \int i \, dt + i R \tag{3.53}$$

The average value of $v_s(t)$ is zero. For a linear circuit, it would therefore be expected that the average value of the current would also be zero. Hence, for a sufficiently large value of capacitance, the integral term of Eq. (3.53) becomes negligible. It should be noted that a large value of C is insufficient to assure that the integral term is negligible; it is necessary that the integral of the current over time be bounded – a condition that is satisfied for a current with a zero average value.

$$v_s(t) = i R = v_R(t) \qquad C \to \infty \tag{3.54}$$

The result of Eq. (3.54) is the "infinite capacitance" approximation. If both superposition results are taken into account, the following is obtained:

$$v_R(t) = v_s(t), \quad v_C(t) = V_S \qquad C \to \infty \tag{3.55}$$

The average value of the signal V_S is "lost," that is, it does not appear across the resistor R. On the other hand, the varying component of the signal appears across the resistor of the circuit. This component is coupled to the resistor by the capacitor (hence, the label of coupling capacitor).

The circuit of Figure 3.69 is used for the ac input of an oscilloscope in which, typically, a 1-μF capacitor is connected in series with the 1-MΩ input resistance of the oscilloscope deflection amplifier. Only the varying component (ac component) of the input signal results in an oscilloscope deflection – the average value of the input signal is lost. For the dc input of the oscilloscope, the capacitor is shorted out.

Capacitor coupling may be used to extract the output signal from a transistor amplifier that has an input signal $v_s(t)$ with a zero average value (Figure 3.70). The behavior of the base circuit will not be affected by the capacitor of the output

Figure 3.69: Capacitor coupling.

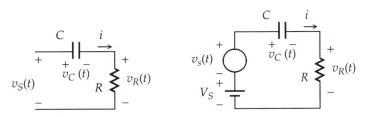

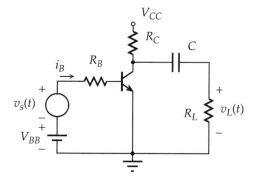

Figure 3.70: A common–emitter amplifier with an output coupling capacitor.

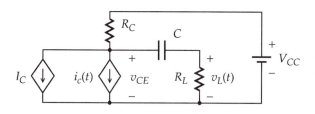

Figure 3.71: Equivalent-circuit for the collector of the common–emitter amplifier of Figure 3.70.

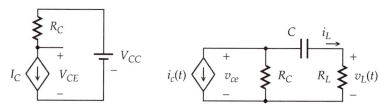

(a) dc equivalent circuit   (b) varying component equivalent circuit

Figure 3.72: A solution of the circuit of Figure 3.71 using superposition.

circuit. If $V_{BB} + v_s(t)$ exceeds $v_{BE(\text{on})}$, the following is obtained:

$$i_B = (V_{BB} + v_s(t) - v_{BE(\text{on})})/R_B$$
$$i_C = \beta_F i_B = \beta_F(V_{BB} - v_{BE(\text{on})})/R_B + \beta_F v_s(t)/R_B$$

(3.56)

In obtaining the collector current i_C it has been assumed that a saturation of the transistor does not occur. This current consists of two components, a dc component I_C and a varying component $i_c(t)$:

$$I_C = \beta_F(V_{BB} - v_{BE(\text{on})})/R_B, \quad i_c(t) = \beta_F v_s(t)/R_B$$

(3.57)

Separate dependent-current sources may be used for these currents (Figure 3.71).

Superposition may be used to solve the circuit of Figure 3.71. First, consider the case for the dc sources, namely I_C and V_{CC}. For this case, the steady-state current of C is zero, and the capacitor may be treated as an open circuit (Figure 3.72(a)). This circuit yields $V_{CC} - I_C R_C$ for the dc collector–emitter voltage. This voltage (no signal) is the quiescent collector–emitter voltage V_{CE}, whereas I_C is the quiescent collector current. To obtain a solution for the varying component of collector current, the sources I_C and V_{CC} are removed by being replaced with an open- and a short-circuit, respectively (Figure 3.72(b)). The parallel combination of the current source $i_c(t)$ and the collector resistor R_C may be replaced by a Thévenin equivalent circuit of a voltage source $-i_c(t)R_C$ and a series resistance of R_C as follows:

$$-i_c(t)R_C = i_L R_C + \frac{1}{C} \int i_L \, dt + i_L R_L \tag{3.58}$$

For an infinite C approximation, the integral term can be treated as negligible.

$$i_L = -\frac{i_c(t)R_C}{R_C + R_L}$$

$$v_L(t) = i_L R_L = -\frac{i_c(t)R_C R_L}{R_C + R_L} = -i_c(t)R_C \| R_L \tag{3.59}$$

The solution of Eq. (3.59) is the same as that obtained by replacing C of Figure 3.72(b) with a short circuit.

Using superposition, these results yield the following for the output voltage $v_L(t)$:

$$v_L(t) = -\beta_F v_s(t) R_C \| R_L / R_B \tag{3.60}$$

The effect produced by V_{BB} is removed from the output of the circuit. The collector–emitter voltage, however, has both a dc and a varying component as follows:

$$v_{CE} = \underbrace{V_{CC} - I_C R_C}_{\text{dc (quiescent) value}} - \underbrace{\beta_F v_s(t) R_C \| R_L / R_B}_{\text{varying component}} \tag{3.61}$$

$$= \quad V_{CE} \quad + \quad v_{ce}(t)$$

The dc component is the quiescent value of voltage (no signal). Following the notation previously adopted, an uppercase symbol and subscript are used for the quiescent value, and a lowercase symbol and subscript are used for the signal component.

SMALL-SIGNAL EQUIVALENT CIRCUIT

A set of equivalent circuits similar to those used for the collector current could have been employed to obtain the quiescent and varying component of the transistor's base current. For the solution obtained (Eq. (3.56)), it was assumed that the diode of the equivalent circuit was conducting and that it had a constant voltage of $v_{BE(\text{on})}$. Often the signal voltage of the amplifier is very small. For this case, an improved model for the equivalent base–emitter diode is appropriate – a diode with an equivalent series resistance r_{be}. This resistance depends on the current of the diode, that is, the quiescent base current of the transistor. Two equivalent circuits, one for quiescent quantities and the other for varying components, are then obtained (Figure 3.73).

If $V_{BB} > V_\gamma$, a condition required for conduction of the ideal diode of Figure 3.73(a), the following is obtained for quiescent quantities:

$$V_{BB} = I_B R_B + I_B r_{be} + V_\gamma = I_B R_B + v_{BE(\text{on})}$$

$$I_B = \left(V_{BB} - v_{BE(\text{on})} \right) / R_B \tag{3.62}$$

This previously obtained solution for the quiescent base current may now be

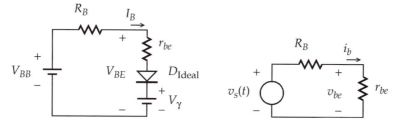

(a) quiescent equivalent circuit   (b) small-signal equivalent circuit

Figure 3.73: Quiescent and small-signal equivalent base circuits. The voltage V_{BB} must be sufficient for D_{Ideal} to be conducting for the small-signal equivalent circuit to apply.

used to determine the equivalent resistance r_{be} as follows:

$$r_{be} = n_F V_T / I_B \tag{3.63}$$

The equivalent series resistance of the diode appears in the small-signal equivalent circuit and results in the following for the varying components of the base current (Figure 3.73(b)):

$$v_s(t) = i_b R_B + i_b r_{be}, \qquad i_b = v_s(t)/(R_B + r_{be}) \tag{3.64}$$

The varying component of collector current is $\beta_F i_b$, which is a quantity that differs from that previously obtained. The difference, however, is small if $r_{be} \ll R_B$.

Before proceeding, a summary of the previous results is in order. The first step in solving a circuit such as that of Figure 3.70 is determining the quiescent currents and voltages. For $v_s(t) = 0$, the quiescent condition, this results in the equivalent circuit of Figure 3.74 in which it is assumed that steady-state conditions prevail – the current of the capacitor is zero.

$$I_B = \left(V_{BB} - v_{BE(\text{on})}\right)/R_B$$
$$I_C = \beta_F I_B, \qquad V_{CE} = V_{CC} - I_C R_C \tag{3.65}$$

If I_B is known, the equivalent resistance r_{be} may be obtained as follows:

$$r_{be} = n_F V_T / I_B \tag{3.66}$$

The small-signal equivalent circuit may now be used to determine the varying components of currents and voltages.

Figure 3.74: Quiescent equivalent circuit for $V_{BB} > v_{BE(\text{on})}$.

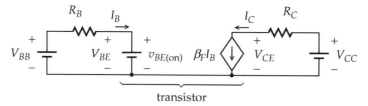

transistor

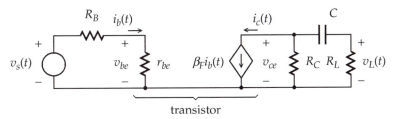

Figure 3.75: Small-signal equivalent circuit.

As a result of the capacitor, the load voltage has only a varying component defined by

$$i_b(t) = v_s(t)/(R_B + r_{be}), \quad i_c(t) = \beta_F i_b(t)$$

$$v_L(t) = -i_c(t) R_C \| R_L = -\frac{\beta_F v_s(t) R_C \| R_L}{R_B + r_{be}} \tag{3.67}$$

Although this result assumes an infinite capacitance, such an approximation is not necessary to obtain a solution for $v_L(t)$. The small-signal equivalent circuit is linear. Therefore, if the input voltage were a sinusoidal quantity, a phasor solution in which the capacitor is replaced by a complex impedance of $1/j\omega C$ could be used. For other input voltage functions, a step-by-step numerical integration could be used to obtain $v_L(t)$.

The procedure for determining the behavior of an amplifier for low-level signals involves the solution, in succession, of two equivalent circuits. A current of the quiescent equivalent circuit I_B is used to determine the base–emitter equivalent resistance r_{be} of the small-signal equivalent circuit. This diode linear resistance approximation sets a limit for the validity of the small-signal analysis. The magnitude of the varying component of the base current, $|i_b(t)|$, needs to remain small compared with the quiescent base current. However, for most applications, sufficiently accurate results are obtained for varying components that are as large as 10 or 20 percent of the quiescent value.

HYBRID-π TRANSISTOR MODEL

A slightly different form of the small-signal transistor equivalent circuit indicated in Figure 3.76 is commonly utilized in which a voltage-dependent source is used in place of the current-dependent source for the collector circuit. In addition, a small-signal equivalent resistance designation of r_π is used in place of r_{be} (for the level of coverage up to this point, they have equal values). Because

Figure 3.76: Small-signal equivalent circuits.

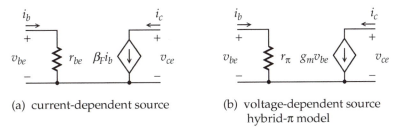

(a) current-dependent source

(b) voltage-dependent source
hybrid-π model

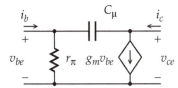

Figure 3.77: The small-signal hybrid-π equivalent circuit.

both circuits must result in the same varying component of collector current, the following is valid:

$$\beta_F i_b = g_m v_{be} = g_m i_b r_\pi$$
$$\beta_F = g_m r_\pi, \quad g_m = \beta_F / r_\pi$$

(3.68)

The quantity g_m is the small-signal mutual conductance of the transistor. When the quiescent value of I_B is used to determine r_π ($= r_{be}$), the following is obtained for the mutual conductance:

$$g_m = \frac{\beta_F}{r_\pi} = \frac{\beta_F}{n_F V_T / I_B} = \frac{\beta_F I_B}{n_F V_T} = \frac{I_C}{n_F V_T}$$

(3.69)

Hence, the mutual conductance is directly dependent on the quiescent collector current of the transistor.

The small-signal equivalent circuit of Figure 3.76(b) is generally referred to as the transistor hybrid-π equivalent circuit. This description comes about from the equivalent circuit that is obtained when the base–collector capacitance C_μ is included (Figure 3.77). It will be noted that the three elements form what might be described as the Greek letter π (vertical elements on each side and one across the top). Although this is the basic hybrid-π circuit, other elements can be incorporated to account for effects that have not been covered in this brief introductory treatment of the bipolar junction transistor.

There is a class of amplifiers that do not utilize coupling capacitors but rely on direct coupling. To achieve this, considerably more complex transistor circuits are required. These amplifiers amplify signals with a dc component; an amplified dc and a varying component appear at the output of the amplifier. This type of response is required for operational amplifiers that use external feedback circuits.

Only the smallest of capacitors, having capacitances no larger than a few tens of picofarads, can be fabricated on an integrated circuit. Hence, if larger capacitances, such as used for coupling (measured in microfarads), are required, they need to be externally connected to the integrated circuit. Direct-coupled circuits eliminate the need for these capacitors, but they require considerably more transistors to achieve the same amplification. A problem with a direct-coupled amplifier is that a small internally generated dc output offset voltage is unavoidable. Furthermore, this voltage is usually dependent on temperature and on the supply voltage. Hence, if integrated amplifier circuits are connected so that one amplifies the output of another, a coupling capacitor is generally used to eliminate the effect of the internally generated offset voltage.

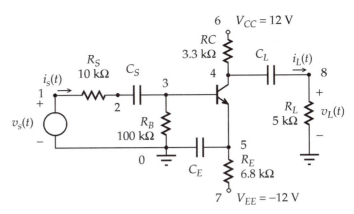

Figure 3.78: Transistor amplifier of Example 3.10. (The node numbers are for the SPICE solution of Example 3.11.)

EXAMPLE 3.10

The behavior of the transistor amplifier circuit of Figure 3.78 is to be determined. The input signal has a zero average value, and infinite values of capacitance may be assumed for the capacitors.

a. Determine the quiescent base and collector currents and collector–emitter voltage of the transistor.

b. Determine the small-signal behavior of the circuit, that is, $v_L(t)$ as a function of $v_s(t)$.

c. What is the small-signal voltage gain of the amplifier, $v_L(t)/v_s(t)$?

d. What is the small-signal current gain of the amplifier, $i_L(t)/i_s(t)$?

e. What is the small-signal power gain of the amplifier?

SOLUTION

a. The quiescent equivalent circuit of the transistor will be used to obtain the quiescent currents (Figure 3.79). The following is obtained for the quiescent base current:

$$0 = I_B R_B + v_{BE(on)} + (I_B + I_C)R_E + V_{EE}$$

$$I_B = \frac{-V_{EE} - v_{BE(on)}}{R_B + (1 + \beta_F)R_E} = 14.4 \ \mu A$$

Figure 3.79: Quiescent equivalent circuit of Example 3.10.

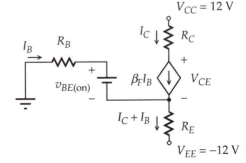

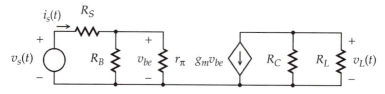

Figure 3.80: Small-signal equivalent circuit – an infinite C approximation.

A value of -12 V was used for V_{EE}. Hence, $I_C = \beta_F I_B = 1.44$ mA. If I_C is known, the quiescent collector–emitter voltage may be determined as follows:

$$V_{CC} = I_C R_C + V_{CE} + (1 + \beta_F) I_B R_E + V_{EE}$$

$$V_{CE} = V_{CC} - V_{EE} - I_C R_C - (1/\beta_F + 1) I_C R_E = 9.40 \text{ V}$$

This solution is valid – the transistor is not saturated.

b. If I_B is known, the small-signal equivalent resistance r_π may be determined by

$$r_\pi = n_F V_T / I_B = 1.74 \text{ k}\Omega$$

The small-signal equivalent circuit of Figure 3.80 applies for an infinite capacitance approximation. The input elements of $v_s(t)$ and R_S may be converted to a Norton equivalent circuit that has a source current of $v_s(t)/R_S$:

$$v_{be} = \frac{v_s(t)}{(1/R_S + 1/R_B + 1/r_\pi) R_S}$$

This voltage may be used to obtain $v_L(t)$ as follows:

$$v_L(t) = -g_m v_{be} R_C \| R_L = -\frac{g_m R_C \| R_L v_s(t)}{1 + R_S/R_B + R_S/r_\pi}$$

A numerical evaluation of the preceding is readily obtained as follows:

$$g_m = I_C / n_F V_T = 57.4 \text{ mS}$$

$$R_C \| R_L = 1.99 \text{ k}\Omega, \quad v_L(t) = -16.7 v_s(t)$$

c. The voltage gain of the amplifier is -16.7. The significance of the minus sign is that a positive input voltage results in a negative output voltage.

d. The input and output resistances may be used to determine the current gain of the amplifier as follows:

$$R_{in} = v_s(t)/i_s(t) = R_S + R_B \| r_\pi = 11.7 \text{ k}\Omega$$

$$R_L = v_L(t)/i_L(t) = 5 \text{ k}\Omega$$

$$i_s(t) = v_s(t)/R_{in}, \qquad i_L(t) = v_L(t)/R_L$$

$$\frac{i_L(t)}{i_s(t)} = \frac{v_L(t)}{v_s(t)} \frac{R_{in}}{R_L} = -39.0$$

e. The power gain is the product of the voltage and current gains, namely, 650.

EXAMPLE 3.11

A SPICE solution of Example 3.10 is desired using the .AC analysis mode. The program automatically determines quiescent quantities, that is, an .OP statement results in a listing for each transistor. A phasor-type solution is then obtained for the corresponding small-signal equivalent circuit in which the effects of all capacitances are included:

$$C_S = 1\ \mu\text{F}, \quad C_L = 1\ \mu\text{F}, \quad C_E = 100\ \mu\text{F}$$

An .AC statement causes the frequency of the phasor source (or sources) to be swept. As a result of the finite capacitance values (C_S, C_L, and C_E), the response of the circuit for low-frequency signals will be affected. For higher-frequency signals, the infinite C approximation applies. In an actual circuit, the response of the transistor will be a limiting factor for very-high-frequency signals. An important high-frequency effect is the base–collector capacitance of the reverse-biased base–collector junction diode (C_μ of the small-signal hybrid-π equivalent circuit). To account for this, include an external capacitance C_{bc} of 5 pF connected from the base to the collector of the transistor. Use a value of I_s for the transistor model that results in the quiescent collector current determined in Example 3.10 for a base–emitter voltage of 0.7 V. For the 27 °C SPICE default temperature, $V_T = 25.9$ mV.

SOLUTION   A quiescent base–emitter voltage V_{BE} of 0.7 V is to result in a quiescent collector current I_C of 1.44 mA.

$$I_C = I_s e^{V_{BE}/V_T}, \quad I_s = I_C e^{-V_{BE}/V_T} = 2.63 \times 10^{-15}\ \text{A}$$

It will be noted that an input phasor signal with an amplitude of 1.0 V and zero phase angle has been specified (Figure 3.81). This amplitude is obviously much

```
Small-Signal Amplifier
VS 1 0 AC 1
RS 1 2 10K
CS 2 3 1U
RB 3 0 100K
Q1 4 3 5 QNPN
CBC 4 3 5P
VCC 6 0 DC 12
RC 6 4 3.3K
RE 5 7 6.8K
CE 5 0 100U
VEE 7 0 DC -12
CL 4 8 1U
RL 8 0 5K
.MODEL QNPN NPN IS=2.63E-15 NF=1 BF=100
.OP
.AC DEC 10 1 1MEG
.PROBE
.END
```

Figure 3.81:   SPICE circuit file for Example 3.11. The node numbers of the circuit are indicated on Figure 3.78.

**** BIPOLAR JUNCTION TRANSISTORS

NAME Q1	RPI 1.80E+03
MODEL QNPN	RX 0.00E+00
IB 1.44E-05	RO 1.00E+12
IC 1.44E-03	CBE 0.00E+00
VBE 6.99E-01	CBC 0.00E+00
VBC -8.70E+00	CBX 0.00E+00
VCE 9.40E+00	CJS 0.00E+00
BETADC 1.00E+02	BETAAC 1.00E+02
GM 5.55E-02	FT 8.84E+17

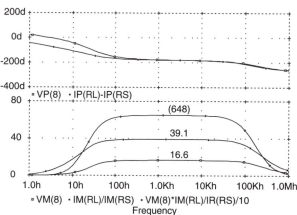

Figure 3.82: SPICE solution for Example 3.11.

larger than that of a small signal. In solving the linear small-signal equivalent circuit, the SPICE program does not take into account signal amplitudes (it "blindly" obtains a solution for the circuit). For the 1-V input signal specified, the output voltage will be numerically equal to the voltage gain of the amplifier. The .OP statement produces an output table for the transistor (Figure 3.82). Quiescent voltages and currents as well as calculated values of g_m and r_π (GM and RPI) are included. The very slight numerical differences are the result of a SPICE value of 25.9 mV being used for V_T. The PROBE response includes plots of both the amplitude and phase of the voltage and current gains. It should be noted that for the midfrequency range the gains agree with those obtained in Example 3.10, and the phase difference of about $-180°$ corresponds to the minus sign of the analytic expressions.

EXAMPLE 3.12

Consider the amplifier of Figure 3.83 in which the quiescent base curent is derived from the collector circuit of the transistor. Assume the average value of $v_s(t)$ is zero and that the infinite capacitance approximation applies.

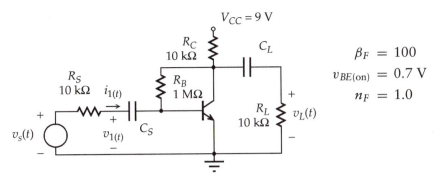

Figure 3.83: Transistor amplifier of Example 3.12

a. Determine the quiescent base and collector currents and the collector–emitter voltage of the transistor.
b. Determine the small-signal voltage gain $v_L/v_s(t)$ of the amplifier.
c. Determine the input resistance $v_1(t)/i_1(t)$ of the amplifier.

SOLUTION

a. The quiescent equivalent circuit of Figure 3.84 will be used.

$$V_{CC} = (I_B + I_C)R_C + I_B R_B + v_{BE(\text{on})}$$
$$= [(1 + \beta_F)R_C + R_B]I_B + v_{BE(\text{on})}$$
$$I_B = \frac{V_{CC} - v_{BE(\text{on})}}{R_B + (1 + \beta_F)R_C} = 4.13 \ \mu A$$
$$I_C = \beta_F I_B = 413 \ \mu A, \qquad V_{CE} = V_{CC} - (I_C + I_B)R_C = 4.83 \ V$$

The transistor is not saturated, and the solution is valid. It should be noted that for this circuit V_{CE} must be greater than V_{BE} for a positive quiescent base current. Hence, the transistor will not saturate regardless of the resistance values of the circuit.

b. The small-signal equivalent circuit must be solved to determine the voltage gain of the amplifier (Figure 3.85). The circuit of Figure 3.85(b) is designed by obtaining a Norton equivalent circuit for $v_s(t)$ and R_S.

$$r_\pi = n_F V_T/I_B = 6.05 \ k\Omega, \qquad g_m = I_C/n_F V_T = 16.5 \ mS$$
$$R_1 = R_S\|r_\pi = 3.77 \ k\Omega, \qquad R_2 = R_C\|R_L = 5.0 \ k\Omega$$

Figure 3.84: Quiescent equivalent circuit of Example 3.12.

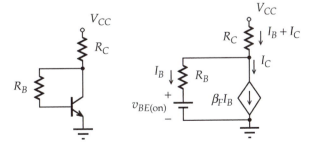

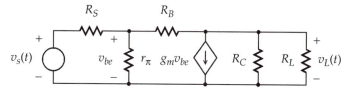

(a) actual small-signal equivalent circuit

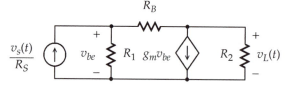

(b) simplified small-signal equivalent circuit

Figure 3.85: Small-signal equivalent circuit for the amplifier of Example 3.12.

A set of simultaneous node-voltage equations needs to be solved.

$$\frac{v_s}{R_S} + \frac{v_L - v_{be}}{R_B} = \frac{v_{be}}{R_1}, \qquad \frac{v_{be} - v_L}{R_B} = g_m v_{be} + \frac{v_L}{R_L}$$

These equations may be written as follows:

$$v_{be} - v_L \frac{R_1}{R_1 + R_B} = v_s \frac{R_1 R_B}{(R_1 + R_B) R_S}, \qquad v_{be} + v_L \frac{R_2 + R_B}{(g_m R_B - 1) R_2} = 0$$

If v_{be} is eliminated, the following relation between v_L and v_s is obtained (the time dependence, though not written, is assumed):

$$v_L \left(\frac{R_1}{R_1 + R_B} + \frac{R_2 + R_B}{(g_m R_B - 1) R_2} \right) = -v_s \frac{R_1 R_B}{(R_1 + R_B) R_S}$$

$$0.0159\, v_L = -0.376\, v_s, \qquad v_L/v_s = -23.6$$

c. The input voltage v_1 is equal to v_{be}. The second of the simultaneous equations for v_{be} and v_L yields the following:

$$v_{be} + 0.0122\, v_L = 0, \qquad v_1 = v_{be} = -0.0122\, v_L = 0.287\, v_s$$

From Figure 3.85(a), the current i_1 is readily determined as follows:

$$i_1 = (v_s - v_1)/R_S = 0.713\, v_s/R_S$$

$$v_1/i_1 = \frac{0.287\, v_s}{0.713\, v_s/R_S} = 0.403\, R_S = 4.03\ \text{k}\Omega$$

This resistance, less than r_π (6.05 kΩ), is the result of R_B connected from the base to the collector of the transistor.

3.6 THE *PNP* TRANSISTOR: A COMPLEMENTARY DEVICE

Up to this point, only *NPN* bipolar junction transistors, devices with a *p*-type base sandwiched between an *n*-type emitter and collector region, have been considered. An interchange of the semiconductor doping, as indicated in Figure 3.86, results in a *PNP* transistor, a complementary device with particularly useful circuit properties. Circuits that simultaneously use *NPN* and *PNP* transistors can often accomplish tasks that are not possible with circuits using only *NPN* or *PNP* transistors.

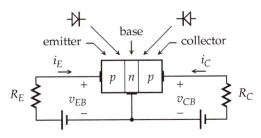

Figure 3.86: A *PNP* transistor in a common–base circuit.

To forward bias the emitter–base junction of a *PNP* transistor, a positive emitter–base potential v_{EB} is required. For this biasing, holes of the emitter cross the emitter–base junction and diffuse across the thin base region of the transistor. These holes readily cross a reverse-biased collector–base junction ($v_{CB} < 0$). For normal operation, it is primarily holes originating in the emitter region that account for the behavior of a *PNP* transistor.

The roles of the holes and free electrons in a *PNP* transistor are reversed from their roles in an *NPN* transistor (Figure 3.87). For normal operation of a *PNP* transistor, the emitter current i_E is positive, whereas the base and collector currents are negative. Furthermore, all voltages have opposite polarities. Nevertheless, the same relationship applies between the collector and emitter current as for an *NPN* transistor:

$$i_C = -\alpha_F i_E \qquad (3.70)$$

The common–base current gain α_F remains a positive quantity.

As a result of the interchange of the *n*- and *p*-type regions, the polarity of the diodes of the equivalent circuit of a *PNP* transistor are reversed from those of the equivalent circuit of an *NPN* transistor (Figure 3.88). Concurrently, a set of emitter and collector characteristics that are the mirror image (mirrored about both axes) of those for an *NPN* transistor result (Figure 3.89).

As for the *NPN* transistor, the equivalent-circuit model for the common–emitter configuration of a *PNP* transistor is generally used to analyze circuits. A simplified equivalent circuit model is indicated in Figure 3.90 (the polarity of the diode is reversed). It will be noted that the emitter arrow of the transistor symbol is reversed from that of an *NPN* transistor. The arrow of the emitter is in the direction of the physical emitter current that occurs for a forward-biased base–emitter junction.

Figure 3.87: The role of majority carriers in a *PNP* transistor.

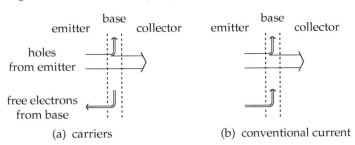

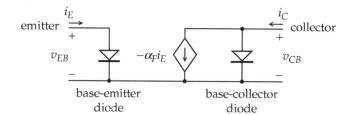

Figure 3.88: Equivalent circuit of a common–base *PNP* transistor.

base-emitter diode   base-collector diode

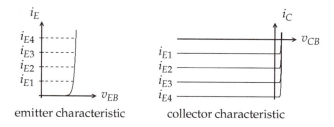

emitter characteristic   collector characteristic

Figure 3.89: Emitter and collector characteristics of a *PNP* transistor.

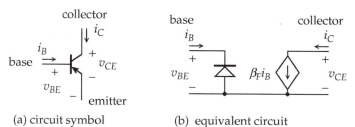

(a) circuit symbol   (b) equivalent circuit

Figure 3.90: Common–emitter equivalent circuit of a *PNP* transistor.

The collector and base currents for normal operation of a *PNP* transistor are related by the common–emitter current gain β_F by

$$i_C = \beta_F i_B, \qquad \beta_F = \frac{\alpha_F}{1 - \alpha_F} \tag{3.71}$$

These are the same relations as for an *NPN* transistor. Expressions for the dependence of the base and collector currents of a *PNP* transistor on the base–emitter voltage require a set of minus signs:

$$i_C = -I_s e^{-v_{BE}/n_F V_T}, \qquad i_B = -(I_s/\beta_F) e^{-v_{BE}/n_F V_T} \tag{3.72}$$

A negative base–emitter voltage is required, and the parameter I_s is a positive quantity. When using SPICE, a PNP .MODEL specification causes the program to utilize the relationships appropriate for a *PNP* transistor.

Consider the situation in which a *PNP* transistor is used in a basic common–emitter circuit. As indicated in Figure 3.91, a negative supply voltage V_{CC} is required for normal operation of the transistor, and the diode of the equivalent circuit is reversed from what it would be for an *NPN* transistor. For positive values of v_{IN}, the diode is reverse biased, resulting in a zero base current ($i_C = 0$ and $v_{OUT} = V_{CC}$, a negative quantity). Negative values of v_{IN} are necessary to forward bias the diode ($v_{BE(on)} \approx -0.7$ V for a silicon *PNP* transistor).

An analysis, similar to that for the *NPN* transistor, results in the transfer characteristic of Figure 3.92 (the algebra of the analysis is the same). Although

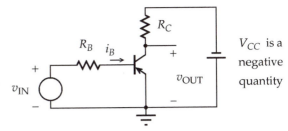

Figure 3.91: Common–emitter *PNP* transistor circuit.

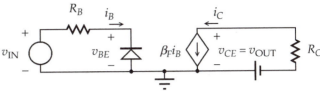

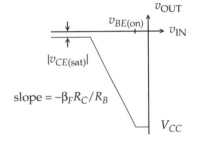

Figure 3.92: Transfer characteristic of a common–emitter *PNP* transistor circuit.

the transition region occurs when v_{IN} and v_{OUT} are negative, the slope for the transition region, namely $-\beta_F R_C/R_B$, is the same as that for an *NPN* transistor.

COMPLEMENTARY SYMMETRY

Frequently, a *PNP* transistor is used in conjunction with an *NPN* transistor with one device supplying current when a positive quantity is required and the other when a negative quantity is required. Often an emitter–follower-type of circuit is used. The response of the *NPN* transistor of Figure 3.93(a) is that previously obtained for a circuit with $R_B = 0$. A subscript of *N* has been used to distinguish these results from those that will be obtained for the *PNP* device. For the *PNP* transistor circuit of Figure 3.93(b), a value of $v_{IN} < v_{BEP(\text{on})}$ is necessary to achieve a base and collector current ($v_{BEP(\text{on})} \approx -0.7$ V for a silicon device). When $v_{IN} < v_{BEP(\text{on})}$, $v_L = v_{IN} - v_{BEP(\text{on})}$. For these individual circuits, the *NPN* transistor supplies current for v_{IN} positive and the *PNP* transistor for v_{IN} negative.

Suppose that the separate emitter–follower circuits are combined to provide current for a single load resistor (Figure 3.94). The resultant transfer characteristic is the sum of the individual transfer characteristics. This apparent superposition of results, it should be emphasized, does not follow (at least directly) from the superposition theorem of circuits – transistors are nonlinear circuit elements! An analysis of this circuit is relatively straightforward. For simplicity, it will be assumed that the magnitude of the base–emitter on voltages is 0.7 V ($v_{BEN(\text{on})} = -v_{BEP(\text{on})} = 0.7$ V). For $|v_{IN}| < 0.7$ V, neither device will be

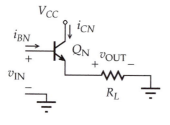
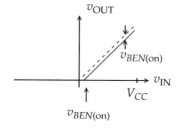

(a) NPN emitter–follower amplifer

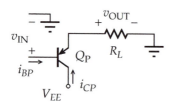
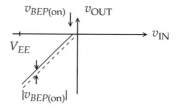

(b) PNP emitter–follower amplifer

Figure 3.93: *NPN* and *PNP* transistor emitter-follower circuits.

conducting.

For $-0.7\,\text{V} < v_{\text{IN}} < 0.7$,
$$v_{BEN} = v_{BEP} = v_{\text{IN}}, \quad i_{BN} = i_{BP} = 0, \quad v_{\text{OUT}} = 0. \tag{3.73}$$

When v_{IN} is greater than 0.7 V, the *NPN* transistor will be conducting.

For $v_{\text{IN}} > 0.7\,\text{V}$,
$$v_{BEN} = v_{BEP} = 0.7\,\text{V}, \quad i_{BP} = 0, \quad v_{\text{OUT}} = v_{\text{IN}} - 0.7\,\text{V}. \tag{3.74}$$

As a result of their common emitter and base connections, both devices have the same base–emitter voltage. A base–emitter voltage of 0.7 V for the *PNP* transistor implies that its base–emitter equivalent-circuit diode is reverse biased – its base and collector currents are therefore zero. For $v_{\text{IN}} < -0.7\,\text{V}$, the *PNP* device will conduct.

For $v_{\text{IN}} < -0.7\,\text{V}$,
$$v_{BEN} = v_{BEP} = -0.7\,\text{V}, \quad i_{BN} = 0, \quad v_{\text{OUT}} = v_{\text{IN}} + 0.7\,\text{V}. \tag{3.75}$$

Figure 3.94: An emitter–follower circuit using both an *NPN* and *PNP* transistor – complementary symmetry.

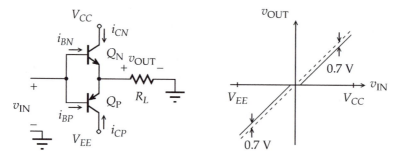

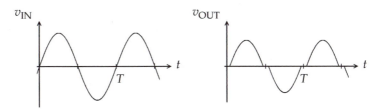

Figure 3.95: Distortion of a sinusoidal input signal.

The base–emitter equivalent-circuit diode of the *NPN* transistor is reverse biased for this condition.

The behavior of the emitter–follower amplifier of Figure 3.94 relies on the complementary symmetry of its *NPN* and *PNP* transistors. For this basic circuit, a distortion of the output signal occurs around $v_{IN} = 0$ (Figure 3.95). This crossover distortion can be reduced with appropriate circuit modifications. Suppose, for example, that batteries with potentials slightly less than 0.7 V, the potential assumed for the magnitude of the base–emitter voltage of the preceeding analysis, are inserted in series with the base of each transistor. For a battery polarity opposite to the base–emitter voltage, the magnitude of the voltage required to turn on a transistor would be reduced to a very small value. This is equivalent to a small value of $v_{BEN(on)}$ or $|v_{BEP(on)}|$. Hence, the width of the crossover region, the range of v_{IN} for which $v_{OUT} = 0$, would be very small.

A circuit requiring a set of batteries is not acceptable even if batteries with the precise potential required were available. Furthermore, the base–emitter voltages of the transistors do not remain constant, as assumed for the model utilized. The effect of the batteries can be approximated with a set of diodes and current sources (Figure 3.96). As long as the diodes are forward biased, their voltage is nearly equal to that required to keep the transistors in their active region of operation. This circuit is generally designed so that, for a zero input voltage, there is a small collector current of Q_N and an equal magnitude collector current of Q_P. Owing to the complementary symmetry of the circuit, the load voltage and current will be zero for $v_{IN} = 0$. For positive values of v_{IN}, the collector current of Q_N increases, whereas the magnitude of the collector current of Q_P decreases to essentially zero for fairly small positive values of v_{IN}. For negative values of v_{IN}, the magnitude of the current of Q_P increases (providing a negative load current), whereas the collector current of Q_N decreases. The net effect is that crossover distortion is essentially eliminated. By using diodes with appropriate parameters (in particular I_s), the no-signal collector currents will be fairly small compared with the currents that occur for normal levels of the signal voltages.

Figure 3.96: A complementary emitter-follower circuit using diode biasing.

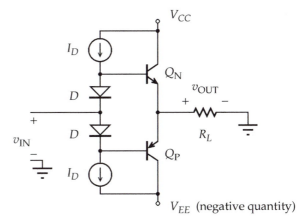

Emitter-follower circuits using complementary symmetry are generally used when an output load current must be provided. This is the case for an operational amplifier – an amplifier for which the output voltage, ideally, is independent of the load current. Typical low-power integrated-circuit operational amplifiers will generally supply at least 10 mA of current with either polarity. A circuit employing complementary emitter-follower transistors is generally used for integrated circuits incorporating bipolar junction transistors. Another application is the output circuit of an audio power amplifier. The load resistor for this case is the loudspeaker; two amplifiers are required for a stereo system.

EXAMPLE 3.13

A *PNP* transistor is used in the circuit of Figure 3.97, which uses a positive supply voltage. The output voltage is the voltage across the collector resistor. Determine the voltage transfer characteristic of the circuit.

SOLUTION The transistor needs to be replaced with its equivalent circuit (Figure 3.98). A forward-biased base–emitter junction has been assumed, that is, $i_B < 0$ for a *PNP* transistor.

$$V_{CC} = -v_{BE(on)} - i_B R_B + v_{IN}, \qquad i_B = -(V_{CC} + v_{BE(on)} - v_{IN})/R_B$$

For $i_B < 0$, $v_{IN} < 4.3$ V is required.

$$v_{OUT} = -i_C R_C = (V_{CC} + v_{BE(on)} - v_{IN})\beta_F R_C/R_B$$

This solution is valid only if the transistor is not saturated, that is, only if

Figure 3.97: Transistor circuit of Example 3.13.

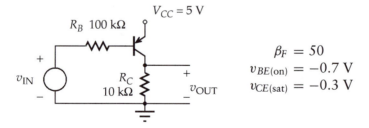

$$\beta_F = 50$$
$$v_{BE(on)} = -0.7 \text{ V}$$
$$v_{CE(sat)} = -0.3 \text{ V}$$

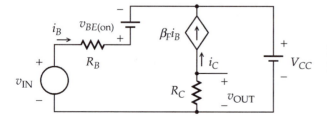

Figure 3.98: Equivalent circuit of Example 3.13.

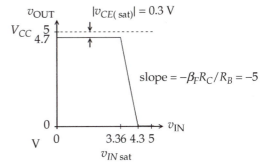

Figure 3.99: Transfer characteristic of Example 3.13.

$v_{CE} < v_{CE(\text{sat})}$. The following is obtained at the edge of saturation, $v_{IN} = v_{IN\,\text{sat}}$:

$$V_{CC} = -v_{CE(\text{sat})} + v_{OUT}$$
$$= -v_{CE(\text{sat})} + (V_{CC} + v_{BE(\text{on})} - v_{IN\,\text{sat}})\beta_F R_C / R_B$$
$$v_{IN\,\text{sat}} = V_{CC} + v_{BE(\text{on})} - (V_{CC} + v_{CE(\text{sat})}) R_B / \beta_F R_C = 3.36 \text{ V}$$

The resultant voltage transfer characteristic is indicated in Figure 3.99.

EXAMPLE 3.14

The circuit of Figure 3.100 is typical of that used for the output stage of an audio amplifier connected directly to a loudspeaker having an 8 Ω coil. Consider the case for which $v_L(t)$ is a sinusoidal voltage:

$$v_L(t) = V_m \sin \omega t, \quad V_m = 20 \text{ V}$$

A feedback circuit is generally used to obtain an input voltage that will produce a sinusoidal output voltage. The base currents of the devices may be assumed to be negligible, that is, the current out of an emitter of a transistor may be assumed equal to its collector current.

a. Determine the average power delivered to the loudspeaker.
b. What are the average values of i_{CN} and i_{CP}?
c. What is the average power supplied by each of the voltage sources V_{CC} and V_{EE}?
d. What is the average power dissipated by each of the transistors?

SOLUTION

a. The average loudspeaker power depends on the rms value of the load voltage, that is, $V_m / \sqrt{2}$.

$$P_{L\,\text{av}} = V_m^2 / 2 R_L = 25 \text{ W}$$

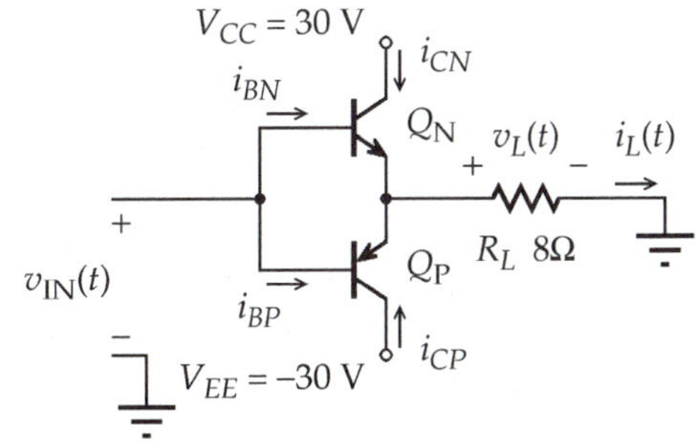

Figure 3.100: Transistor circuit of Example 3.14.

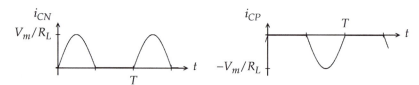

Figure 3.101: Collector currents of the transistors of Example 3.14.

b. The following is obtained if the base currents of the transistors are ignored:

$$i_L(t) = i_{CN}(t) + i_{CP}(t)$$

For $i_L(t) > 0$, only the *NPN* transistor is conducting, $i_{CP}(t) = 0$.

$$i_{CN}(t) = i_L(t) \quad \text{for} \quad i_L(t) > 0$$

A similar situation occurs for the *PNP* transistor:

$$i_{CP}(t) = i_L(t) \quad \text{for} \quad i_L(t) < 0$$

Each of these currents is a half-wave rectified sinusoidal current (Figure 3.101). The average value of each current is $1/\pi$ of its peak value.

$$i_{CN\,av} = V_m/\pi R_L = 0.796 \text{ A}$$
$$i_{CP\,av} = -V_m/\pi R_L = -0.796 \text{ A}$$

c. The average power supplied by V_{CC} or V_{EE} is equal to the product of its voltage, a constant, and its average current:

$$P_{CC\,av} = i_{CN\,av} V_{CC} = 23.9 \text{ W}$$
$$P_{EE\,av} = i_{CP\,av} V_{EE} = 23.9 \text{ W}$$

d. The total average power supplied to the circuit is $P_{CC\,av} + P_{EE\,av}$, 47.8 W. Therefore, $47.8 - 20 = 27.8$ W is dissipated by the two transistors, 13.9 W by each transistor.

REFERENCES

Armstrong, E. H. (1915). Some recent developments in the audion receiver. *Proceedings of the Institute of Radio Engineers*, 3, 3, 215–47.

Buie, J. L. (1966). U. S. Patent 3,233,125.

Cringely, R. X. (1992). *Accidental Empires*. New York: HarperCollins Publishers.

De Forest, L. (1914). The audion – detector and amplifier. *Proceedings of the Institute of Radio Engineeers*, 2, 1, 15–36.

Early, J. M. (1952). Effects of space-charge layer widening in junction transistors. *Proceedings of the Institute of Radio Engineers*, 40, 11, 1401–6.

Electronics (1980). Entire issue, 53, 9 (17 April).

Elmasry, M. I. (1983). *Digital Bipolar Integrated Circuits*. New York: John Wiley & Sons.

Elmasry, M. I. (1985). Digital bipolar integrated circuits: A tutorial. In *Digital VLSI Systems*, M.I. Elmasry, ed., 38–46. New York: IEEE Press.

Garrett, L. S. (1970). Integrated-circuit digital logic families. Part I: Requirements and features of a logic family; RTL, DTL, and HTL devices. *IEEE Spectrum*, 7, 10, 46–58. Part II: TTL devices; *IEEE Spectrum*, 7, 11, 63–72. Part III: ECL and MOS devices. *IEEE Spectrum*, 7, 12, 30–42.

Glazer, A. B. and Subak-Sharpe, G. E. (1977). *Integrated Circuit Engineering: Design, Fabrication, and Applications*. Reading, MA: Addison-Wesley Publishing Co.

Gray, P. E. and Searle, C. L. (1969). *Electronic Principles: Physics, Models, and Circuits*. New York: John Wiley & Sons.

Harris, J. N., Gray, P. E., and Searle, C. L. (1966). *Digital Transistor Circuits*. Semiconductor Electronics Education Committee. New York: John Wiley & Sons.

Haznedar, H. (1991). *Digital Microelectronics*. Redwood City, CA: The Benjamin/Cummings Publishing Co.

Hodges, D. A. and Jackson, H. G. (1988). *Analysis and Design of Digital Integrated Circuits*. New York: McGraw-Hill.

Lancaster, D. (1974). *TTL Cookbook*. Indianapolis, IN: Howard W. Sams & Co.

Millman, J. and Taub, H. (1965). *Pulse, Digital, and Switching Waveforms*. New York: McGraw-Hill.

Morris, R. L. and Miller, J. R. (eds.) (1971). *Designing with TTL Integrated Circuits*. Texas Instruments Electronics Series. New York: McGraw-Hill.

Shockley, W., Pearson, G. L., and Haynes, J. R. (1949). Hole injection in germanium – Quantitative studies and filamentary transistors. *The Bell System Technical Journal*, **28**, 3, 344–66.

Taub, H. and Schilling, D. (1977). *Digital Integrated Electronics*. New York: McGraw-Hill.

Texas Instruments (1976). *The TTL Data Book for Design Engineers* (2d ed.). Dallas, TX: Texas Instruments, Inc.

PROBLEMS

3.1 Consider the common–base circuit of Figure P3.1 that uses a silicon *NPN* transistor with $\alpha_F = 0.995$ and $v_{EB(on)} = -0.7$ V.

   a) Suppose $V_{EE} = 2.0$ V. Determine i_E, i_C, and v_{CB} for this condition.
   b) Determine the maximum value of V_{EE} for which $i_C = 0$.
   c) What is the value of V_{EE} that results in $v_{CB} = 0$?
   d) Repeat the previous parts for $V_{CC} = 20$ V.

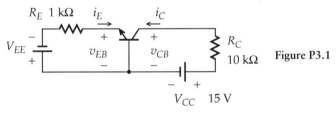

Figure P3.1

3.2 Repeat parts (a), (b), and (c) of Problem 3.1 for $V_{CC} = 10$ V.

3.3 Repeat Problem 3.1 with a 1-kΩ resistor R_{EB} connected between the emitter and base of the transistor.

3.4 The transistor of Problem 3.1 is replaced by another transistor having parameters of $\alpha_F = 0.98$ and $v_{EB(on)} = -0.7$ V. Repeat Problem 3.1 for this transistor.

3.5 A silicon transistor with $\alpha_F = 0.99$ and $v_{EB(on)} = -0.7$ V is used in the circuit of Figure P3.5. Determine the transfer characteristic of the circuit, that is, v_{CB} versus v_{IN}, for $-8 < v_{IN} < 0$ V.

3.6 Repeat Problem 3.5 for $R_C = 1$ kΩ.

3.7   Repeat Problem 3.5 for $R_C = 10$ kΩ.

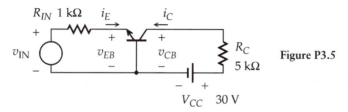

Figure P3.5

3.8   The silicon transistor circuit of Figure P3.8 ($\alpha_F = 0.99$, $v_{EB(on)} = -0.7$ V) has an input signal source of $v_s(t)$. An emitter biasing source V_{EE} of 4 V is used to achieve a desired operating point.

a) Assume quiescent conditions prevail, $v_s(t) = 0$. Determine i_E, i_C, and v_{CB}.

b) The periodic input signal $v_s(t)$ has a symmetrical triangular waveform with a peak-to-peak amplitude of 1.0 V. What is the peak-to-peak value of v_{CB}?

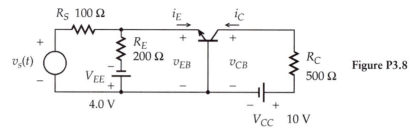

Figure P3.8

3.9   Repeat Problem 3.8 assuming $v_s(t)$ has a sinusoidal waveform with a peak-to-peak value of 0.5 V.

3.10   Repeat Problem 3.8 assuming $v_s(t)$ has a symmetrical square waveform with a peak-to-peak value of 1.0 V.

3.11   The circuit of Figure P3.11 has a silicon diode in series with the base of the transistor.

a) Determine the emitter and collector currents of the transistor.

b) What are the voltages v_{EG} and v_{CG}?

c) Suppose the diode of the base circuit is reversed. Determine i_E and i_C for this condition.

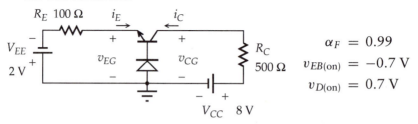

Figure P3.11

3.12   A base biasing voltage V_{BB} of 3 V is used for the silicon transistor of the circuit of Figure P3.12 ($\alpha_F = 0.995$, $v_{EB(on)} = -0.7$ V). Use the transistor common–base equivalent circuit to determine i_E and i_C of the circuit. What are the transistor voltages v_{EB} and v_{CB}?

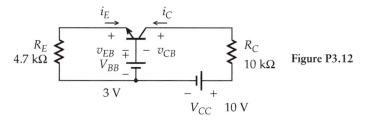

Figure P3.12

3.13 Repeat Problem 3.12 with the polarity of the battery V_{BB} reversed.

3.14 Because the common–base current gain of a transistor is very close to unity, small changes in this gain have a marked effect on the common–emitter current gain. Suppose that a nominal common–emitter current gain β_F of 150 is required.

a) What is the nominal value of α_F required?
b) Suppose a tolerance of ±50 percent is required for β_F. What would be the corresponding tolerance of α_F?
c) What would be the tolerance required for α_F if that of β_F were ±10 percent?

3.15 Repeat Problem 3.14 for a nominal value of 100 for β_F.

3.16 Repeat Problem 3.14 for a nominal value of 200 for β_F.

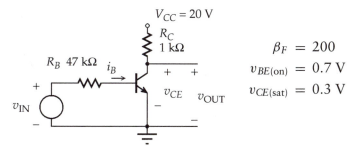

Figure P3.17

3.17 A silicon *NPN* transistor is used in the circuit of Figure P3.17.

a) Determine the transfer characteristic of this circuit, that is, v_{OUT} versus v_{IN} ($0 < v_{IN} < 20$ V).
b) What is the minimum value of v_{IN} for which $v_{OUT} = v_{CE(sat)}$?
c) What is the slope of the transfer characteristic for v_{IN} less than that required for saturation ($v_{IN} > 0.7$ V)?

3.18 Repeat Problem 3.17 for $R_C = 5$ kΩ.

3.19 Repeat Problem 3.17 for $R_B = 100$ kΩ.

3.20 Repeat Problem 3.17 for $\beta_F = 100$.

3.21 A silicon diode is connected in series with the base of the transistor of Figure P3.21.

a) Determine the transfer characteristic of the circuit.
b) What is the slope of the characteristic for the region over which v_{IN} affects the output voltage?
c) Suppose the diode of the circuit is reversed. Determine the transfer characteristic for this condition.

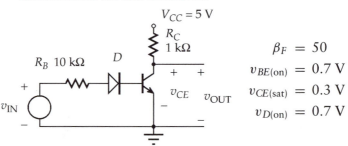

$$\beta_F = 50$$
$$v_{BE(on)} = 0.7 \text{ V}$$
$$v_{CE(sat)} = 0.3 \text{ V}$$
$$v_{D(on)} = 0.7 \text{ V}$$

Figure P3.21

3.22 Repeat Problem 3.21 for two diodes in series with the base of the transition (same polarity as the diode of Figure P3.21).

3.23 Repeat Problem 3.21 for three diodes in series with the base of the transition (same polarity as the diode of Figure P3.21).

3.24 Repeat Problem 3.21 with the diode removed from the base circuit and placed in series with the emitter of the transistor (downward forward-biased current).

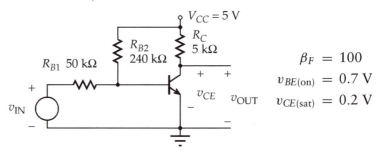

$$\beta_F = 100$$
$$v_{BE(on)} = 0.7 \text{ V}$$
$$v_{CE(sat)} = 0.2 \text{ V}$$

Figure P3.25

3.25 In the transistor circuit of Figure P3.25, a base biasing resistor is connected directly to V_{CC} ($V_{BB} = V_{CC}$).

a) Determine i_C and v_{OUT} for $v_{IN} = 0$.
b) What is the slope of the voltage transfer characteristic for $v_{IN} = 0$?
c) What is the minimum value of v_{IN} for which $v_{OUT} = v_{CE(sat)}$?
d) What is the maximum value of v_{IN} for $v_{OUT} = V_{CC}$?

3.26 Consider the circuit of Figure P3.25. What is the equivalent input circuit for $v_{IN} = 0$? Over what range of v_{IN} is this circuit valid? What is the equivalent output circuit for this condition?

3.27 Repeat Problem 3.25 for $R_{B1} = 100 \text{ k}\Omega$.

3.28 Repeat Problem 3.25 for $R_C = 10 \text{ k}\Omega$.

3.29 Repeat Problem 3.25 for $\beta_F = 150$.

3.30 A load resistor R_L of 10 kΩ is connected to the output of the circuit of Figure P3.25 (across v_{OUT}). Repeat parts (a), (b), and (c) of Problem 3.25 for this condition.

    d) What is the maximum value of v_{OUT} and the minimum value of v_{IN} for which it is obtained?

3.31 Suppose input voltage source v_{IN} of Figure P3.25 is replaced by a sinusoidal signal source $v_s(t) = V_p \sin 2\pi f t$.

    a) Determine $v_{OUT}(t)$ for $V_p = 0.2$ V.
    b) Determine $v_{OUT}(t)$ for $V_p = 1.0$ V.
    c) What is the largest value of V_p for which the output voltage is not distorted?

3.32 Repeat Problem 3.31 for a 10-kΩ load resistor connected to the output of the circuit (across v_{OUT}).

3.33 A common–emitter transistor circuit is used in conjunction with a light-sensitive diode to indicate the presence of a light signal (Figure P3.33).

    a) Consider the condition for no light – the reverse-biased current of the diode may be treated as zero. Determine the maximum value of R_B for which saturation of the transistor occurs.
    b) Determine the light-generated current of the diode required to result in $v_{OUT} = 5$ V for the value of R_B determined in the previous part.
    c) What is the minimum value of light-generated diode current required for $v_{OUT} = V_{CC}$?

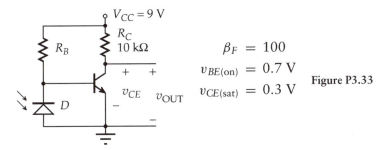

$\beta_F = 100$

$v_{BE(on)} = 0.7$ V

$v_{CE(sat)} = 0.3$ V

Figure P3.33

3.34 Repeat Problem 3.33 for a 1-kΩ resistor in series with the emitter of the transistor.

3.35 Repeat Problem 3.33 for a 50-kΩ resistor connected in series with the base of the transistor.

3.36 The transistor circuit of Figure P3.36 is used to amplify an audio signal source $v_s(t)$. It is found that the audio signal can be simulated with a sinusoidal voltage as follows:

$$v_s(t) = V_p \sin 2\pi f t, \qquad f = 600 \text{ Hz}$$

    a) Suppose $V_p = 0.5$ V. Determine the peak-to-peak value of $v_{OUT}(t)$ and the amplification of the circuit.
    b) What is the largest value of V_p that may be used without resulting in a distorted output voltage?

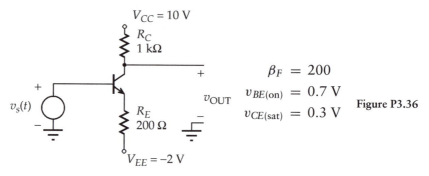

Figure P3.36

$\beta_F = 200$

$v_{BE(on)} = 0.7\ \text{V}$

$v_{CE(sat)} = 0.3\ \text{V}$

c) What is the equivalent input circuit for $v_{IN} = 0$?

3.37 Repeat Problem 3.36 for $R_E = 500\ \Omega$.

3.38 Repeat Problem 3.36 for a 10-kΩ resistor in series with the base of the transistor.

3.39 Repeat Problem 3.36 for a 1-kΩ load resistor connected to the output of the circuit (across v_{OUT}).

3.40 The circuit of Figure P3.36 is to be redesigned to respond to an input signal $v_S(t)$ that is a negative pulse.

   a) Determine a new value of R_E that results in $v_{OUT}(t) = 0$ for $v_S(t) = 0$ (no change in other circuit values).
   b) What is v_{OUT} for a -1.0 V input pulse?
   c) What is the pulse amplitude required for $v_{OUT} = V_{CC}$?

3.41 Repeat parts (a) and (b) of Problem 3.40 for a load resistor of 1 kΩ connected to the output of the circuit. What is the maximum value of v_{OUT} and for what input pulse voltage does it occur?

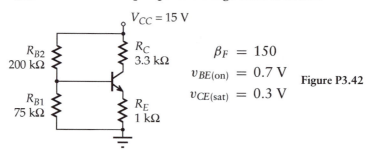

Figure P3.42

$\beta_F = 150$

$v_{BE(on)} = 0.7\ \text{V}$

$v_{CE(sat)} = 0.3\ \text{V}$

3.42 A silicon transistor is used in the biasing circuit of Figure P3.42. With appropriate capacitors, this basic circuit could be converted to a high-gain signal amplifier. Although a direct solution of this circuit may be obtained by writing the appropriate circuit equations, a simpler approach is to utilize a Thévenin equivalent circuit for the base resistors.

   a) Determine an equivalent circuit for R_{B1}, R_{B2}, and V_{CC}.
   b) Using the circuit of part (a), determine the base and collector currents of the transistor.
   c) Determine v_{CE} of the transistor.

3.43 Repeat Problem 3.42 for $\beta_F = 50$ and $\beta_F = 300$. Explain why the collector current of the transistor tends to have only a minimal dependence on β_F.

3.44 Repeat Problem 3.42 for $v_{BE(on)} = 0.6$ and 0.8 V. Explain why the dependence of i_C on $v_{BE(on)}$ is not very great.

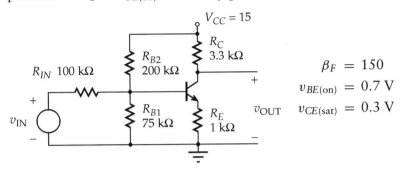

$\beta_F = 150$

$v_{BE(on)} = 0.7$ V

$v_{CE(sat)} = 0.3$ V

Figure P3.45

3.45 The circuit of Problem 3.42 is to be used with an input voltage v_{IN} and resistance R_{IN} (Figure P3.45).

a) Determine the values of i_B, i_C, and v_{CE} for $v_{IN} = 0$.
b) What is the change in v_{OUT} for $v_{IN} = 1$ V?
c) What is the change in v_{OUT} for $v_{IN} = -1$ V?
d) What is the voltage gain of the circuit, $\Delta v_{OUT}/\Delta v_{IN}$?

3.46 Suppose that the emitter resistor R_E of Figure P3.45 is replaced with an ideal battery V_{EE}. The potential of V_{EE} is such as to result in the same quiescent currents as obtained in Problem 3.45.

a) What is the value of V_{EE}?
b) What is the change in v_{OUT} for $v_{IN} = -10$ mV? For input signals that vary in time, a large capacitor connected in parallel with R_E will often achieve the same result.

3.47 Repeat Problem 3.45 for R_{B1} removed from the circuit (an infinite resistance).

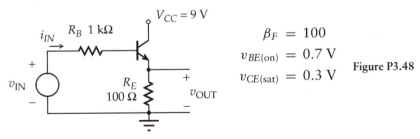

$\beta_F = 100$

$v_{BE(on)} = 0.7$ V

$v_{CE(sat)} = 0.3$ V

Figure P3.48

3.48 A silicon transistor is used in the emitter–follower circuit of Figure P3.48.

a) Determine the voltage transfer characteristic of the circuit v_{OUT} versus v_{IN} $(0 < v_{IN} < V_{CC})$.
b) Determine the current transfer characteristic of the circuit i_{OUT} versus i_{IN} $(i_{OUT} = v_{OUT}/R_E)$.

3.49   Repeat Problem 3.48 for $R_B = 0$.

3.50   Consider the emitter-follower circuit of Figure P3.48 with $v_{IN} = 3$ V.

a) Determine v_{OUT}.
b) Determine the output short-circuit current of the circuit.
c) What is the Thévenin equivalent resistance of the circuit?
d) Repeat the previous parts for $v_{IN} = 6$ V.

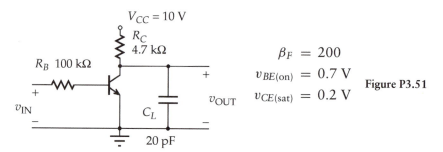

$V_{CC} = 10$ V

R_C
4.7 kΩ

R_B 100 kΩ

v_{IN}

C_L

20 pF

$+$

v_{OUT}

$\beta_F = 200$

$v_{BE(on)} = 0.7$ V

$v_{CE(sat)} = 0.2$ V

Figure P3.51

3.51   The transistor logic inverter of Figure P3.51 has a load capacitance C_L of 20 pF.

a) Suppose v_{IN} has been equal to V_{CC} for a very long time and that it is suddenly switched to zero at $t = 0$. Determine the times necessary for v_{OUT} to reach 5 V and 9 V.
b) The input voltage v_{IN} is zero and is suddenly switched to V_{CC} at $t = 0$. What are the times necessary for v_{OUT} to fall to 5 V and to 1 V?

3.52   Repeat Problem 3.51 for an input signal that has a high logic level of only 5 V.

3.53   Repeat Problem 3.51 for $R_C = 10$ kΩ.

3.54   Repeat Problem 3.51 for $R_B = 47$ kΩ.

3.55   Repeat Problem 3.51 for transistors with $\beta_F = 50$ and $\beta_F = 200$. Why is only one transition of the output affected?

3.56   Suppose that the circuit of Figure P3.51 has two inputs as the logic **NOR** gate of Figure 3.60(a) (both base resistors are 100 kΩ). The input voltage of the second input is zero. Repeat Problem 3.51 for this condition.

3.57   The supply voltage of the logic inverter circuit of Figure P3.51 is reduced to 5 V.

a) Determine the times necessary for v_{OUT} to reach 2.5 V and 4.5 V.
b) The input voltage v_{IN} is zero and is suddenly switched to V_{CC} at $t = 0$. What are the times necessary for v_{OUT} to fall to 2.5 V and to 0.5 V?

3.58   A silicon *NPN* transistor is used in the logic inverter of Figure P3.58 in which the load capacitance C_L is that of a data bus.

a) The input has been high (V_{CC}) and is suddenly switched to zero. Determine the 10-to 90-percent rise time of v_{OUT}.
b) The input is suddenly switched from zero to V_{CC}. Determine the 90-to 10-percent fall time of v_{OUT}.

c) The maximum 10- to 90-percent rise or fall time is not to exceed 200 ns. Redesign the circuit to achieve this.

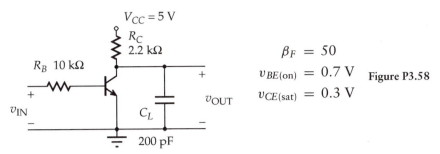

$\beta_F = 50$

$v_{BE(on)} = 0.7\ \text{V}$

$v_{CE(sat)} = 0.3\ \text{V}$

Figure P3.58

3.59 A voltage with a periodic square waveform is often used to test logic circuits. Suppose the input voltage of the circuit of Figure P3.58 is the periodic square voltage of Figure P3.59. Determine the maximum frequency f for which the output voltage is a reasonable response for the input voltage. It will be necessary to arrive at a quantitative definition of a reasonable response.

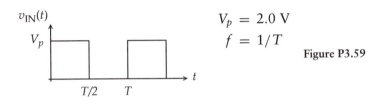

$V_p = 2.0\ \text{V}$

$f = 1/T$

Figure P3.59

3.60 The resistor–transistor logic **NOR** gate of Figure P3.60 has four input voltages. Determine the transfer characteristic of the circuit v_{OUT} versus v_A for $v_B = v_C = v_D = 0$. What is the slope of the characteristic for the transition region of the output voltage?

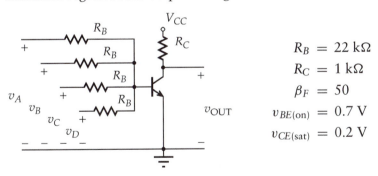

$R_B = 22\ \text{k}\Omega$

$R_C = 1\ \text{k}\Omega$

$\beta_F = 50$

$v_{BE(on)} = 0.7\ \text{V}$

$v_{CE(sat)} = 0.2\ \text{V}$

Figure P3.60

3.61 Repeat Problem 3.60 for $R_B = 10\ \text{k}\Omega$.

3.62 Consider the direct-coupled transistor logic **NOR** gate of Figure 3.60(b) with $R_C = 500\ \Omega$. The transition of v_{OUT} is to occur in a 0.5 V range of v_A with $v_B = 0$. Determine the maximum value of R_B for a circuit with a transistor having a common–emitter current gain β_F of 50 to 200 ($v_{BE(on)} = 0.7\ \text{V}$, $v_{CE(sat)} = 0.2\ \text{V}$).

3.63 Repeat Problem 3.62 for a logic **NOR** gate with five inputs. What are the limiting factors related to the number of inputs for a gate with this configuration?

3.64 A sinusoidal voltage is used for the input signal $v_S(t)$ of the RC circuit of Figure P3.68. Assume the following:

$$v_S(t) = V_s \cos 2\pi ft, \quad v_R(t) = V_r \cos (2\pi ft + \theta)$$

Determine V_r / V_s and θ in terms of f and the element values of the circuit. Show that the results depend on the product RC and not on either individual value.

3.65 Evaluate the result of Problem 3.64 for $f = 1$ kHz, $R = 10$ kΩ, and $C = 1$ μF.

3.66 Use the results of Problem 3.64 to determine the frequency f for which $V_r / V_s = 1/\sqrt{2}$. What is θ for this frequency? What is the frequency for which $V_r / V_s = 0.9$? What is the angle for this frequency?

3.67 The input voltage of the RC circuit of Figure 3.69 is a symmetrical square wave with a periodic frequency of 50 Hz and peak amplitudes of ± 1 V. Component values are $R = 10$ kΩ and $C = 10$ μF. Determine and sketch $v_R(t)$. What is the amount by which the output voltage changes over the interval for which $v_S(t) = 1$ V?

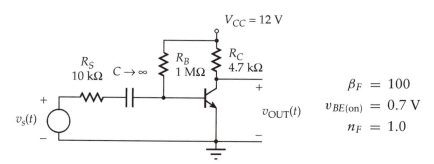

Figure P3.68

3.68 A silicon *NPN* transistor is used in the amplifier circuit of Figure P3.68.

a) Draw the quiescent equivalent circuit for the amplifier. Determine I_C and V_{CE}.

b) Draw the small-signal equivalent circuit.

c) Determine the small-signal voltage gain $v_{out}(t)/v_s(t)$.

d) Suppose $v_s(t) = V_m \cos \omega t$. Determine the value of V_m that results in a peak-to-peak value of 1 V for $v_{OUT}(t)$.

3.69 Repeat Problem 3.68 for a transistor with $\beta_F = 150$.

3.70 Repeat Problem 3.68 for $R_S = 0$.

3.71 Suppose that a transistor with $\beta_F = 200$ is to be used in the circuit of Figure P3.68. Determine a new value of R_B that results in a quiescent value of 5 V for $v_{OUT}(t)$. Determine $v_{out}(t)/v_s(t)$ for this circuit.

3.72 Repeat Problem 3.68 for a circuit with a 270 Ω resistor in series with the emitter of the transistor.

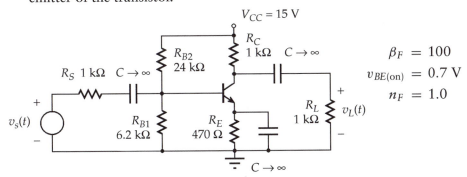

$V_{CC} = 15$ V

R_C
1 kΩ $C \to \infty$

R_{B2}
24 kΩ

R_S 1 kΩ $C \to \infty$

$\beta_F = 100$
$v_{BE(\text{on})} = 0.7$ V
$n_F = 1.0$

$v_s(t)$

R_{B1}
6.2 kΩ

R_E
470 Ω

R_L
1 kΩ $v_L(t)$

$C \to \infty$

Figure P3.73

3.73 An *NPN* silicon transistor is used in the small-signal common–emitter amplifier of Figure P3.73.

  a) Draw the quiescent equivalent circuit for the amplifier. Determine I_C and V_{CE}.
  b) Draw the small-signal equivalent circuit.
  c) Determine the small-signal voltage gain $v_L(t)/v_s(t)$.
  d) A peak-to-peak value of 1 V is required for $v_L(t)$. What is the minimum peak-to-peak value of $v_s(t)$ necessary?

3.74 It is desired to determine the small-signal output resistance of the amplifier circuit of Figure P3.73. To do this, find $v_L(t)$ for $R_L \to \infty$ and the load current of R_L for $R_L \to 0$.

3.75 Repeat Problem 3.73 for a transistor with $\beta_F = 50$.

3.76 Repeat Problem 3.73 for a transistor with $\beta_F = 200$.

3.77 Determine the quiescent power supplied by V_{CC} of the circuit of Figure P3.73. What is the quiescent power dissipated by the transistor?

3.78 To reduce the quiescent power consumed by the circuit of Figure P3.73, an engineer decided to double all resistance values.

  a) Determine the quiescent power supplied by V_{CC} and that dissipated by the transistor.
  b) Determine the small-signal voltage gain $v_L(t)/v_s(t)$ for this condition ($R_S = 1$ kΩ, $R_L = 1$ kΩ).

3.79 Repeat Problem 3.78 for a circuit in which all resistance values are multiplied by a factor of 10.

3.80 A silicon transistor is used in the emitter-follower amplifier of Figure P3.80.

  a) Draw the quiescent equivalent circuit of the amplifier. Determine I_C and V_{CE}.
  b) Draw the small-signal equivalent circuit.
  c) Determine the small-signal voltage gain $v_L(t)/v_s(t)$.
  d) Determine the small-signal current gain $i_L(t)/i_s(t)$ of the amplifier.

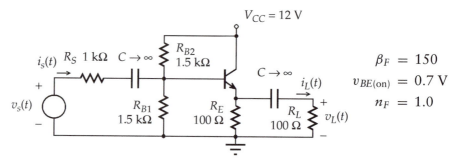

Figure P3.80

3.81 Repeat Problem 3.80 for $R_S = 100\ \Omega$.

3.82 Repeat Problem 3.80 for $R_S = 0$.

3.83 The logic circuits of Figure P3.83 utilize silicon *PNP* transistors. The logic input voltage levels are 0 and V_{CC}. Determine a voltage truth table for each of the circuits. What are the logic functions of the gates?

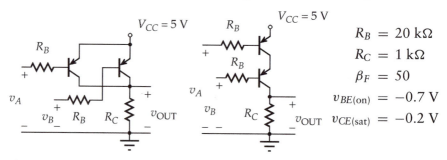

Figure P3.83

3.84 The output of the *PNP* transistor logic circuit of Example 3.13 is connected to a 50-pF capacitor. Determine the 10- to 90-percent rise and fall times of $v_{OUT}(t)$. Assume input step functions with voltage levels of 0 and V_{CC}.

COMPUTER SIMULATIONS

C3.1 The emitter and collector characteristics of a common–base transistor, as indicated in Figure 3.13, are to be obtained using a SPICE simulation. Assume the transistor has parameters of $\beta_F = 100$, $I_s = 10^{-12}$ A, and $n_F = 1.2$. Using an appropriate circuit, obtain the characteristics for a range of -10 to 0 mA for i_E (steps of 2 mA) and for v_{CB} up to 10 V. Be sure to use a negative v_{CB} to obtain the collector characteristic for all positive values of collector current.

C3.2 A SPICE simulation of the common–base amplifier of Figure 3.14 is desired. Assume the transistor has parameters of $\beta_F = 100$, $I_s = 10^{-12}$ A, and $n_F = 1.2$.

a) Use a .DC analysis mode to obtain a plot of v_{CB} versus $v_s (-3 \le v_s \le 3 \text{ V})$. What is the slope of this characteristic for v_s small (the voltage gain)?

b) Use a .TRAN analysis mode to determine the behavior of the circuit for a sinusoidal input voltage $v_s(t)$ that has an amplitude of 0.5 V and a frequency of 1 kHz. On the basis of the peak-to-peak amplitude of these curves, what is the voltage gain of the amplifier? This quantity should be the same as that obtained in part (a).

c) Repeat part (b) for the largest input voltage for which an undistorted collector–base voltage can be obtained.

C3.3 A simulation of the circuits of Problems 3.21, 3.22, and 3.23 is desired. For the diodes, assume $I_s = 10^{-13}$ A and $n = 1.0$, whereas for the transistors, assume $\beta_F = 50$, $I_s = 10^{-14}$ A, and $n_F = 1.0$. If the three circuits are included in a single circuit file and run simultaneously, their transfer characteristics can be obtained on a single graph. A single input voltage source v_{IN} can be used for the three circuits. The derivative of a trace can be obtained by requesting DV(N) for a Probe display. On a separate graph, obtain curves of $\frac{dv_{OUT}}{dv_{IN}}$ for each of the circuits. What is the value of the derivative of each curve for $v_{OUT} = 2.5$ V?

C3.4 Solve the common–emitter transistor amplifier circuit of Example 3.3 using a SPICE simulation. Use the values of R_B and V_{BB} determined in part (a) and, for the transistor, assume that $n_F = 1.4$ and that I_s has a value that yields $i_C = 1$ mA for $v_{BE} = 0.7$ V. Use a .TRAN mode of analysis and a frequency of 1 kHz for the sinusoid. By using three circuits, curves for all three values of β_F may be obtained on a single graph (similar to those of Figure 3.31).

C3.5 Solve the cascade amplifier circuit of Example 3.4 using a SPICE simulation. Assume the transistor has parameters of $\beta_F = 50$, $I_s = 10^{-15}$ A, and $n_F = 1.0$. Obtain curves of v_{OUT1} and v_{OUT2} versus v_{IN}. Compare the resultant response with that of Table 3.2. Determine the slopes of the output voltages for $v_{OUT2} = 2.5$ V and compare these values with the analytically obtained values.

C3.6 A simulation of the transistor circuit of Problem 3.45 (Figure P3.45) is desired. Assume the transistor has parameters of $\beta_F = 150$, $I_s = 10^{-15}$ A, $n_F = 1.0$, and $V_{AF} = 100$ V.

a) Using a .DC sweep, determine the dependence of v_{OUT} on v_{IN} ($-10 < v_{IN} < 10$ V). What is v_{OUT} and the slope of the characteristic for $v_{IN} = 0$? What are the maximum and minimum values of v_{OUT}?

b) A transient (.TRAN) solution for an input sinusoidal signal $V_m \sin 2\pi f t$, is desired ($f = 500$ Hz). Obtain v_{OUT} for $V_m = 4.0$ V. What are the maximum and minimum values of v_{OUT}? What is the ratio of $v_{OUT\,p\text{-}p}$ to $v_{IN\,p\text{-}p}$? This ratio should be nearly the same as the magnitude of the slope obtained in part (a).

c) Repeat part (b) for $V_m = 8$ V. What are the maximum and minimum values of v_{OUT}?

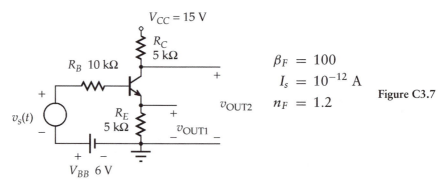

Figure C3.7

C3.7   The transistor circuit of Figure C3.7 has identical collector and emitter resistors.

a) From a static solution (.DC mode) obtain plots of the output voltages and the collector–emitter voltage of the transistor. A range of ±6 V is needed for v_s.

b) Consider the case for $v_s(t)$ being a sinusoidal signal $V_m \sin 2\pi ft$ with an amplitude of 1 V and a frequency of 1 kHz. Using a .TRANS solution, obtain plots of the input voltage and the two output voltages. What are the gains for varying voltage components? What is the input resistance for the varying component of the input voltage, that is, the ratio of the peak-to-peak input voltage divided by the peak-to-peak value of the input current? This circuit is known as a phase splitter because, for a sinusoidal input voltage, the varying components of its output voltages are 180° out of phase.

C3.8   Repeat Simulation C3.7 using the maximum input signal amplitude V_m for which undistorted outputs are achieved. Determine the cause of the distortion by observing v_{CE}.

C3.9   A SPICE simulation of the logic inverter of Figure 3.64 is desired. Assume the transistor has parameters of $\beta_F = 50$, $I_s = 10^{-15}$ A, $n_F = 1.0$, and $V_{AF} = 50$ V.

a) Determine the static transfer characteristic of the gate.

b) Determine, using a .TRAN solution, the dynamic behavior of the gate. Use an input voltage that has a 0- to 5-V transition at $t = 0$ and a downward transition at $t = 50$ ns. Compare the simulation results with the analytic results of Example 3.8.

C3.10   Repeat Simulation C3.9 for a transistor with a base-to-collector capacitor of 5 pF. This capacitor corresponds to the junction capacitance of the transistor.

C3.11   A diode–transistor logic **NAND** gate is indicated in Figure 3.60(c). Assume the transistor has parameters of $\beta_F = 50$, $I_s = 10^{-15}$ A, $n_F = 1.0$, and $V_{AF} = 50$ V and that each diode is a diode-connected transistor (base and collector tied together). The parameters of the circuit are $V_{CC} = 5$ V, $R_C = 1$ kΩ, and $R_B = 20$ kΩ. Use SPICE to determine the static transfer characteristic of the gate v_{OUT} versus v_A for $v_B = 0$ and 5 V. For $v_B = 5$ V, determine $\frac{dv_{OUT}}{dv_{IN}}$ for $v_{OUT} = 2.5$ V.

C3.12 An open-collector **TTL NAND** gate is indicated in Figure 3.62(a). Assume that the gate has only a single input; it therefore behaves as a logic **NOT** gate. For normal input voltages, the input protective diode is reverse biased (it is therefore not required for a simulation). Assume the transistor has parameters of $\beta_F = 50$, $I_s = 10^{-15}$ A, $n_F = 1.0$, and $V_{AF} = 50$ V.

    a) Determine the static transfer characteristic of this circuit ($0 \leq v_A \leq 5$ V). Determine $\frac{dv_{OUT}}{dv_{IN}}$ for $v_{OUT} = 2.5$ V.

    b) Determine the input characteristic of the gate, that is, obtain a plot of the input current versus the input voltage.

    c) Assume that the output of the gate is connected to a 50-pF load capacitor. Determine the transient behavior of v_{OUT} for an input voltage that has a 0- to 5-V transition at $t = 0$ and a 5- to 0-V transition at $t = 50$ ns. What are the delay times of the output voltage, that is, the time required for the output to reach its midvalue?

C3.13 Repeat Simulation C3.12 for a **TTL** gate with a totem-pole output circuit (Figure 3.62(b)). Assume that the diode of the output circuit is a diode-connected transistor (collector and base tied together).

C3.14 A SPICE simulation of the transistor circuit of Figure P3.73 using the .AC analysis mode is desired.

    $C_S = 10\ \mu F$ (in series with R_S)

    $C_L = 10\ \mu F$ (in series with R_L)

    $C_E = 100\ \mu F$ (in parallel with R_E)

    $C_{bc} = 5\ pF$ (base-to-collector)

Follow the approach of Example 3.11 to obtain voltage, current, and power gains for a frequency range of 1 Hz to 1 MHz.

C3.15 A SPICE simulation is desired to verify the analytic solution of Example 3.12. Assume that the transistor has parameters of $I_s = 5 \times 10^{-15}$ A, $n_F = 1.0$, and $V_{AF} = 100$ V, and that $C_S = 0.1\ \mu F$ and $C_L = 0.2\ \mu F$.

    a) Compare the quiescent currents and voltages of the devices (produced by an .OP statement) with the analytically obtained quantities.

    b) Obtain plots of the magnitude and phase of the small-signal voltage gain of the amplifier ($1\ Hz \leq f \leq 100$ kHz). At what frequency is the response 0.707 of its high-frequency value? This is the lower half-power frequency because power is proportional to the square of the voltage. At what frequency is the response 90 percent of its high-frequency value?

    c) Obtain plots of the magnitude and phase of the input impedance of the amplifier. Compare the high-frequency value (it should be real) with the analytic result.

C3.16 The transistor amplifier of Example 3.13 (Figure 3.97) uses a *PNP* transistor. For a SPICE simulation, a PNP designation in the .MODEL statement is required. By convention, the parameters I_s and V_{AF} are positive – appropriate minus signs are introduced within the program. Assume the transistor has parameters of $\beta_F = 50$, $I_s = 10^{-13}$ A, $n_F = 1.2$ and $V_{AF} = 50$ V.

  a) Obtain a static solution (.DC) and a plot of v_{OUT} versus v_{IN}. Compare the slope of this characteristic with the analytic result.
  b) Consider the case for which a 30-pF load capacitor C_L is connected in parallel with the output of the circuit. Assume v_{IN} has an upward transition, 0- to 5V, at $t = 0$ and a downward transition at $t = 1$ μs. What are the rise and fall times of the output voltage (10- to 90-percent change)? Obtain a plot of the time-dependent collector current of the transistor. Why is the upward transition of v_{OUT} much smaller than its downward transition?

C3.17 A SPICE simulation to determine the behavior of the amplifier with complementary symmetry of Example 3.14 is desired. (See Simulation C3.16 for modeling the *PNP* transistor.) Assume $I_s = 3 \times 10^{-10}$ A, $n_F = 1.4$, and $V_{AF} = 100$ V for the "power" transistors of the circuit.

  a) Obtain a static transfer characteristic of the circuit, that is, v_{OUT} versus v_{IN}. Also obtain curves of the currents out of the emitters of the transistors (in the direction of i_L) and i_L. Does QN provide the load current when v_{IN} is positive and QP when v_{IN} is negative, as expected?
  b) Assume v_{IN} is a sinusoidal voltage $V_m \sin 2\pi ft$ with $V_m = 25$ V and $f = 1$ kHz. Using a transient solution (.TRANS), obtain a plot giving time dependence of v_{OUT}. On a second graph, obtain plots of the instantaneous powers dissipated by the load resistor and by the collectors of each of the transistors ($i_C v_{CE}$).

DESIGN EXERCISES

D3.1 A common–emitter transistor amplifier is shown in Figure D3.1. A value of $v_{OUT} = 5$ V for $v_{IN} = 0$ V and $\frac{dv_{OUT}}{dv_{IN}} = -5$ for $v_{OUT} = 5$ V is desired. Determine values for R_{B1} and R_{B2} that achieve these conditions. What is the effect of using a transistor with $\beta_F = 150$ in the circuit? What would be the effect of designing the circuit for a transistor with $\beta_F = 125$? Would the resultant circuit work reasonably well for $100 \le \beta_F \le 150$?

D3.2 In the common–emitter amplifier indicated in Figure D3.1, a 1-kΩ resistor is connected in parallel with v_{OUT}. A value of $v_{OUT} = 2.5$ V for $v_{IN} = 0$ V and $\frac{dv_{OUT}}{dv_{IN}} = -5$ for $v_{OUT} = 2.5$ V is desired. Determine values for R_{B1} and R_{B2} that achieve these conditions. What is the effect of using a transistor with $\beta_F = 150$ in the circuit? What would be the effect of designing the circuit for a transistor with $\beta_F = 125$? Would the resultant circuit work reasonably well for $100 \le \beta_F \le 150$?

D3.3 Suppose that a 100 Ω resistor R_E is connected in series with the emitter of the transistor of Figure D3.1. Repeat Design D3.1 for this circuit.

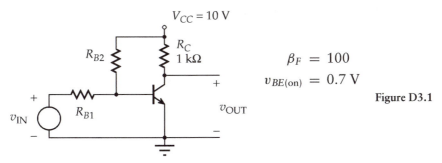

$$V_{CC} = 10 \text{ V}$$

R_C
1 kΩ

$\beta_F = 100$

$v_{BE(on)} = 0.7 \text{ V}$

Figure D3.1

D3.4 In the transistor circuit of Figure 3.36, $V_{CC} = 5$ V, $R_C = 1$ kΩ, $\beta_F = 50$, and $v_{BE(on)} = 0.7$ V. Determine values for the base resistors that result in an output transition at $v_{IN} \approx 2.5$ V and an output transition that has a slope of -5.

D3.5 Repeat Design D3.4 for a circuit with the configuration of Figure 3.48 (component values and transistor parameters of D3.4). Values for R_E and V_{EE} are to be determined.

D3.6 Consider the transistor amplifier circuit of Figure 3.51 that has an input voltage source with an equivalent series resistance R_{IN} of 1 MΩ. Modify the circuit so that the LED turns on at $v_{IN} = 0$ V.

D3.7 Modify the transistor amplifier circuit of Figure 3.51 so that the LED is on for a negative value of v_{IN} and turns off when v_{IN} becomes positive. An additional transistor will be required.

D3.8 Design a resistor–transistor logic **NOT** gate with the configuration of Figure 3.55 ($V_{CC} = 5$ V). Assume the transistor has parameterrs of $\beta_F = 50$ and $v_{BE(on)} = 0.7$ V. The load capacitance is 50 pF and, for static conditions, the transistor is to be saturated for $v_{IN} \geq 2$ V. The delay time is to be no more than 50 ns (50-percent change in the output). To minimize power dissipation, use maximum resistance values.

D3.9 Consider the direct-coupled logic **NOR** gate of Figure 3.60(b) ($V_{CC} = 5$ V). The output of the gate is to be capable of providing the input signal of at least four similar gates (a fan-out of 4). It is desired that $v_{OUT} \geq 4.5$ V for a high output. It is also desired that $v_{OUT} = v_{CE(sat)}$ for a single input voltage of 2 V or greater. Assume $R_C = 1$ kΩ. Determine R_B for a transistor with $\beta_F = 50$ and $v_{BE(on)} = 0.7$ V. Suppose that a transistor with $\beta_F = 100$ in the circuit designed for the transistor with $\beta_F = 50$. What would be the effect of this change?

D3.10 Consider the transistor biasing circuit of Figure D3.10. A nominal collector–emitter voltage of 5 V and a collector current of 5 mA is desired. Determine values of R_C and R_B that achieve this for a transistor with $v_{BE(on)} = 0.7$ V and $100 \leq \beta_F \leq 200$. Minimize the variation in the collector–emitter voltage for different transistors.

D3.11 Repeat Design D3.10 for the circuit of Figure D3.11. Assume $R_E = 0.5 R_C$ and use resistance values for the base resistors that are large compared with r_π of the transistor.

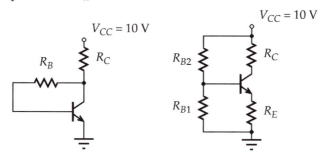

Figure D3.10   **Figure D3.11**

D3.12 A transistor amplifier circuit with the configuration of Figure 3.83 (but with different component values) is to be designed. The circuit is to work with transistors with $v_{BE(on)} = 0.7$ V and $100 \leq \beta_F \leq 200$. The source resistance R_S is 10 kΩ, the load resistance R_L is 1 kΩ, and $V_{CC} = 10$ V. The amplifier is to have a voltage gain with a magnitude of at least 20. The peak-to-peak undistorted output voltage is to be at least 1 V. Assume an infinite capacitance. Verify, using SPICE, that your design is indeed valid for transistors with $\beta_F = 100$ and 200.

CHAPTER FOUR

THE METAL-OXIDE FIELD-EFFECT TRANSISTOR: ANOTHER ACTIVE DEVICE

The idea of a field-effect transistor predates that of the junction transistor by two decades. In the late 1920s, Julius Edgar Lilienfeld proposed using an electric field to control the conductance of a semiconductor crystal (Sah 1988). Although Lilienfeld was granted three patents for proposed devices, there is no evidence that he was able to build an actual working transistor, probably because the required semiconductor technologies were not available at the time. It was Shockley's 1939 consideration of a related field-effect process, the "Schottky gate," that initiated his thought processes and ultimately led to the invention of the point-contact transistor in 1948 (Shockley 1976). Shockley recognized that a surface field-effect played a role in the operation of this device (Shockley and Pearson 1948). Not only did the invention of the bipolar junction transistor follow this device, but so too the junction field-effect transistor (Shockley 1952).

Semiconductor field-effect devices rely on a single type of carrier for conduction, that is, either free electrons or holes. Hence, these devices are frequently referred to as unipolar transistors (single-polarity charges). The earliest commercially produced field-effect device is the junction field-effect transistor (JFET) in which conduction is controlled by a reverse-biased junction diode. This device, therefore, is characterized by a very high input resistance (negligible diode current for static conditions). Junction field-effect transistors are utilized both as discrete devices and, most frequently, in conjunction with bipolar junction transistors in integrated circuits.

It is a more recent development, however, that of the metal-oxide semiconductor field-effect transistor (MOSFET), that has had perhaps the most profound effect on the design of electronic systems. This transistor was first proposed by Kahng in 1960 (Kahng 1976), and its theory of operation was published 3 years later (Hofstein and Heiman 1963). Very-large-scale integrated circuits rely on this device, which is, structurally, considerably simpler than the bipolar junction transistor. In addition, MOSFET devices require fewer processing steps and can be made considerably smaller than bipolar junction transistors. As a result,

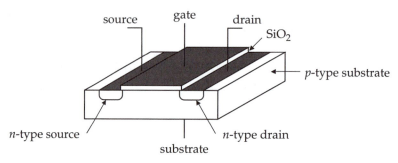

source · gate · drain · SiO₂ · p-type substrate · n-type source · substrate · n-type drain

Figure 4.1: Metal-oxide field-effect transistor.

extremely high device (and therefore logic) densities can be achieved (Hittinger 1973).

The basic structure of a MOSFET device with a p-type substrate is illustrated in Figure 4.1. This device has two heavily doped n-type semiconductor wells, labeled source and drain, embedded in the p-type substrate. A metallic gate extends between the wells and is insulated from the substrate by a thin silicon dioxide layer. Because silicon dioxide is a dielectric, the gate and substrate form a capacitor in which a gate-to-substrate potential results in induced surface charges at the boundary of the dielectric and the substrate.

To gain an appreciation of the operation of this device, suppose that the gate is floating (no connection) and that it has no residual charge. An equivalent circuit consisting of two junction diodes applies (Figure 4.2). For no connection to the substrate, the current between the source and drain will be negligible regardless of the polarity of an externally applied voltage difference (one diode will be reverse biased). This is also the case for a substrate connection if the source and drain potentials relative to those of the substrate are zero or greater.

The capacitive effect of the gate is utilized to induce surface charges on the substrate that, in turn, provide a current path between the source and drain of the device. Consider a positive gate-to-substrate potential that induces negative surface charges on the substrate between the n-type source and drain of the device. Because the surface charge density is proportional to the electric field of the dielectric, extremely thin oxide thicknesses, no more than a fraction of a micron, are required to achieve useful charge densities for a reasonable potential difference.

For small gate voltages, the negative substrate charges are the result of mobile holes moving away from the dielectric–substrate boundary, thus leaving behind unneutralized acceptor atoms. A further increase in the gate volt-

Figure 4.2: Equivalent circuit for a MOSFET device with a floating gate.

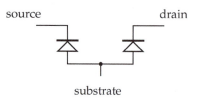

source · drain · substrate

age results in the generation of free electrons (mobile carriers) as a result of a shifting of the internal energy levels of the semiconductor. These free electrons provide a current path between the drain and the source of the device. For a positive drain-to-source potential, free electrons originate at the n-type source and are collected by the drain. It should be noted that MOSFET devices generally have a symmetrical physical structure – the source and drain are not physically distinguishable. It

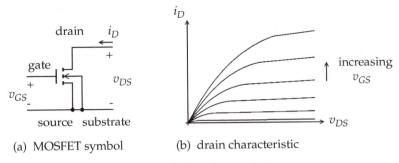

(a) MOSFET symbol   (b) drain characteristic

Figure 4.3: MOSFET symbol and drain characteristic.

is the external potentials that determine which n-type region functions as a source or a drain. Hence, the device has a bidirectional property – a characteristic that is extremely useful.

A MOSFET device is a four-terminal element, a characteristic that complicates determining its behavior in a circuit. However, if the source and substrate are connected together, as in Figure 4.3, the equivalent of a three-terminal device is obtained. The drain current depends on both the gate-to-source voltage (the same as the gate-to-substrate voltage) and the drain-to-source voltage (Figure 4.3(b)). A gate-to-source voltage v_{GS} greater than a threshold value V_T is required for a drain current, and for $v_{GS} > V_T$ the drain current increases for an increasing value of v_{GS}. Because the metal gate is one terminal of a capacitor, the gate current for static conditions is essentially zero. Therefore, to the extent that the current of this capacitor is negligible, the power provided by an input signal connected to the gate is likewise negligible.

To illustrate the operation of a MOSFET device, consider the basic circuit of Figure 4.4, a circuit analogous to that of a bipolar junction transistor (Figure 3.3). A transfer characteristic is readily obtained by drawing a load line on the drain characteristic of the device. To construct an amplifier, an input biasing voltage would be inserted in series with an input signal source. Alternatively, the transfer characteristic (Figure 4.4(b)) is essentially that desired for a logic **NOT** gate. For an integrated circuit, a second MOSFET device is used in place of the drain resistor R_D. As a result, logic circuits are fabricated entirely from MOSFET devices – a

Figure 4.4: Common–source circuit and transfer characteristic.

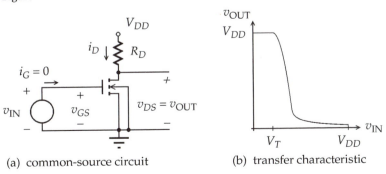

(a) common-source circuit   (b) transfer characteristic

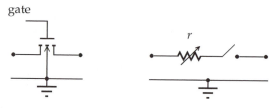

gate

Figure 4.5: MOSFET switch.

decided fabrication advantage because MOSFET devices tend to require a much smaller chip area than a resistor. Both **NAND** and **NOR** operations may be obtained using additional devices.

In addition to being used for conventional logic gates, MOSFET devices are frequently used as logic switches. When an integrated circuit is fabricated, it is generally desirable that a common substrate be utilized for all devices. If a *p*-type substrate is used, as for the device of Figure 4.1, the substrate is connected to the lowest potential of the circuit, normally the common ground of the circuit. This ensures that neither the source nor the drain diode of the device will be forward biased. For a sufficient gate-to-source voltage, the device behaves as a relatively low, albeit nonlinear, resistance. However, for a zero gate-to-substrate voltage, the device behaves as an open circuit. This property of MOSFET devices provides an additional logic function – a bidirectional switch (Figure 4.5).

A memory cell of a dynamic random-access memory (DRAM) uses a MOSFET device as a bidirectional switch. It has been pointed out by Sah that this one-transistor memory cell is probably the most abundant man-made [*sic*] object on the planet earth (Sah 1988, p. 1301). The single transistor memory cell was invented by Robert H. Dennard in 1966 (Dennard 1984). Rather than using an electronic flip-flop requiring at least four MOSFET devices for the storage of a binary bit, Dennard proposed an ingenious circuit that requires only a single MOSFET switch and capacitor. The memory state (logic **1** or **0**) is associated with the charge of the capacitor (charged or uncharged).

Although Dennard is credited with inventing the one-transistor semiconductor memory cell, the concept of using a capacitor as a memory element goes back to the 1940s computer of John V. Atanasoff (Mackintosh 1988). His early computer (some historians argue precomputer) used a rotating memory disk of 50 capacitors for data storage. A capacitor was either charged (it had a potential difference) or was uncharged to represent a binary **1** or **0**. A contact on the periphery of the disk was used to read or write an individual capacitor as the connection to the capacitor rotated past the contact. As is true for today's semiconductor memories, a refresh circuit was required to compensate periodically for charge leaking from the capacitors, which is the consequence of unavoidable dielectric losses.

The configuration of a modern dynamic memory is illustrated in Figure 4.6 – additional rows and columns are used for typical memory chips (1024 rows × 1024 columns for a 1-Mbit memory). To access a particular memory cell, the appropriate column and row needs to be addressed. A high voltage (logic **1**) applied to a row line turns on all the MOSFET switches of that row, connecting each memory capacitor to its column line. If the capacitor was initially charged, its charge will increase the potential of an initially uncharged column line. This increase is sensed by the column circuit, and the result is transferred to the

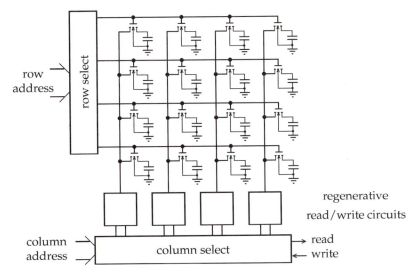

Figure 4.6: An elementary 16-bit dynamic memory. Each memory element requires but a single MOSFET device with the capacitor being the substrate to source–drain capacitance of the device.

read-output line through the column select circuit. Connecting the memory capacitor to the column line and circuit, which has a much larger capacitance to ground, depletes the charge of the memory capacitor. A regenerative-type circuit is utilized to restore the voltage and hence the charge of the memory capacitor. To write to a memory cell, either a high or low voltage is applied to the appropriate column line.

Periodically the memory cells must be read and their voltage for a charged condition restored. Because each column has its own regenerative circuit, an entire row can be refreshed simultaneously. It is, however, necessary to sequence through each row to refresh the entire memory. Depending on the memory characteristic, it must be refreshed every few milliseconds or less.

An understanding of the physical operation of a MOSFET device is necessary to devise suitable equivalent circuit models. These models will then be used to develop an understanding of several commonly used MOSFET circuits. It is these circuits, when combined in very-large-scale integrated circuits, that have revolutionized the diversity and complexity of modern electronic systems.

4.1 FIELD-INDUCED CARRIERS: THE PHYSICS OF A MOSFET DEVICE

Although a quantitative description of a MOSFET device premised on a detailed understanding of semiconductor physics is beyond the scope of an introductory text, a qualitative description of the key physical mechanisms involved is desirable. The device of Figure 4.7, a device with a *p*-type substrate in which free electrons are the current carriers, will be considered. Because the carriers of the MOSFET device are free electrons that form a channel between the source and drain, an *n*-channel designation is used. To simplify the analysis, a common

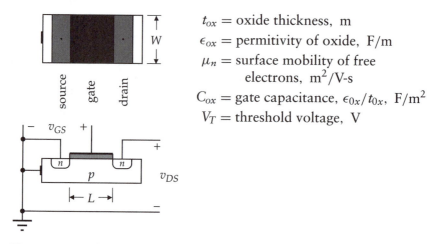

t_{ox} = oxide thickness, m

ϵ_{ox} = permitivity of oxide, F/m

μ_n = surface mobility of free electrons, m²/V-s

C_{ox} = gate capacitance, ϵ_{0x}/t_{ox}, F/m²

V_T = threshold voltage, V

Figure 4.7: Metal-oxide field-effect transistor configuration.

connection for the source and substrate will be assumed; the effect of a source-to-substrate bias will be treated later. Complementary devices with an *n*-type substrate in which holes are the current carriers, that is, a *p*-channel MOSFET device, are also widely used.

It is the gate-to-substrate capacitance, the result of the thin silicon dioxide dielectric that separates the gate and substrate, that plays a pivotal role in the behavior of the device. For a zero gate-to-substrate bias (Figure 4.8(a)), the substrate, except in the vicinity of the *n*-type regions, tends to have a uniform distribution of holes, and the hole density is approximately equal to the acceptor doping density. For each hole, however, there is an acceptor atom tending to have an extra valence electron that completes its valence bonds. Hence, charge neutrality prevails.

To begin, consider the case for a small gate-to-substrate voltage (Figure 4.8(b)). This results in a downward-directed electric field within the dielectric approximately equal to the voltage divided by the dielectric thickness v_{GS}/t_{ox}. A positive surface charge therefore resides on the gate and an equal negative charge on (or slightly within) the *p*-type substrate (surface charge densities of $\pm\epsilon_{ox}v_{GS}/t_{ox}$). Within the substrate, the negative surface charge is the result of the mobile

Figure 4.8: The effect of a gate voltage.

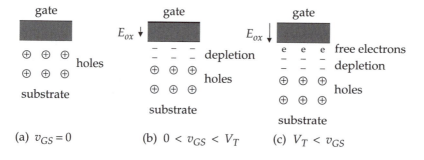

holes' moving away from the surface of the substrate – an effect of the electric field's penetrating the substrate. However, acceptor atoms that are locked into the semiconductor crystalline structure are not mobile. Hence, the surface region is depleted of holes, leaving it with a net negative charge (a depletion region).

The thickness of the depletion region tends to increase as the gate-to-substrate voltage is increased until a critical gate-to-substrate voltage is reached. This is the threshold, voltage V_T.* For larger gate-to-substrate voltages, a new phenomenon occurs: some valence electrons near the surface of the substrate gain sufficient energy to escape their valence bonds and become free electrons (Figure 4.8(c)). This is the result of a bending of the semiconductor energy levels caused by the electric field produced by the gate-to-substrate voltage. It is the free (and hence mobile) electrons of the very thin inversion layer that provide the current path between the drain and source of a MOSFET device.

With these considerations in mind, let us return our attention to the MOSFET device of Figure 4.7 having a common substrate–source connection. Assume, initially, that the drain-to-source voltage is zero ($v_{DS} = 0$) and that a gate-to-source voltage less than the threshold voltage is applied ($0 < v_{GS} < V_T$). This results in a substrate depletion region, as illustrated in Figure 4.8(b). Because the negative charges of the depletion region are not mobile, they do not provide a current path between the drain and the source. Hence, the drain current i_D is zero if a small voltage, v_{DS}, is applied. Furthermore, the drain current remains zero for any positive value of v_{DS} (providing breakdown does not occur).

$$\text{for } v_{GS} < V_T \quad \text{and} \quad v_{DS} > 0, \quad i_D = 0 \quad \text{cutoff condition} \tag{4.1}$$

This is the cutoff region of operation.

Suppose that v_{GS} is now increased so that it is greater than the threshold voltage, whereas v_{DS} remains equal to zero. The substrate condition of Figure 4.8(c) with a free electron inversion layer now prevails. Free electrons provide a current path for the device – in essence, they "connect" the source to the drain of the device. The stored charge of the gate–substrate capacitor is its voltage multiplied by its capacitance. This charge resides on the gate surface and in the vicinity of the substrate–oxide boundary (a positive quantity on the gate and an equal but negative quantity on the substrate). The surface charge density (q_s expressed in C/m^2), is the capacitance per unit area (C_{0x} expressed in F/m^2) times the voltage difference:

$$q_s = -C_{ox}v_{GS} \quad \text{C/m}^2 \tag{4.2}$$

In Eq. (4.2) it is assumed that the gate-to-source potential is nearly entirely across the dielectric – a generally valid assumption. Only that portion of the

---

* It should be noted that the symbol V_T has already been used in conjunction with diodes and bipolar junction transistors ($V_T = kT/e \approx 25$ mV for room temperature). Ideally, a different symbol should be used for these two totally unrelated voltages. Unfortunately, the same symbol is used in most of the published literature. Because the behavior of both devices is seldom considered simultaneously, a confusion of terms rarely arises.

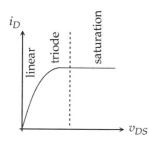

Figure 4.9: Drain current versus drain-to-source voltage.

surface charge due to a gate-to-source voltage in excess of the threshold voltage V_T gives rise to free-electron charges:

$$q_s' = -C_{ox}(v_{GS} - V_T) \quad \text{C/m}^2 \text{ free electrons} \qquad (4.3)$$

The conductivity of the channel formed by the free electrons may now be obtained. The width of the channel is W, its length is L, and the surface mobility of free electrons is μ_n.

$$G = \mu_n q_s' W/L = \mu_n C_{ox} W/L(v_{GS} - V_T) = k(v_{GS} - V_T) \quad \text{S}$$

where $k = \mu_n C_{ox} W/L \quad \text{A/V}^2$

$$\qquad (4.4)$$

The quantity k is known as the transconductance of the device. For a very small drain-to-source voltage, the conductivity calculated for $v_{DS} = 0$ would be expected to yield the drain current of the device as follows:

$$i_D \approx Gv_{DS} \approx k(v_{GS} - V_T)v_{DS}, \quad v_{DS} \text{ small} \qquad (4.5)$$

This results in a linear dependence of drain current on drain-to-source voltage for a given gate-to-source voltage (Figure 4.9).

The basis for the naming of the *n*-type regions should now be clear. Free electrons originate at the source, traverse the channel, and leave the device at the drain. Although the electron flow is from source to drain, the direction of conventional current is in the opposite direction, that is, from drain to source.

A linear current-versus-voltage relationship of Eq. (4.5) would not be expected to prevail as v_{DS} is increased. A potential difference will exist from one end to the other of the channel, which, in turn, implies that the voltage across the oxide varies along the length of the channel. At the source end, the voltage is v_{GS}, but at the drain end it will be smaller, namely $v_{GS} - v_{DS}$. Therefore, the surface charge density at the drain end is reduced and the current is less than that for a uniform channel. A detailed analysis predicts the following:

$$\text{for } v_{GS} > V_T \quad \text{and} \quad 0 < v_{DS} < v_{GS} - V_T$$

$$i_D = k[(v_{GS} - V_T)v_{DS} - \tfrac{1}{2}v_{DS}^2] \qquad (4.6)$$

This expression is obtained by accounting for the variation of gate-to-substrate potential that occurs between the source and drain of the device. It is valid only for a gate-to-source potential greater than the threshold voltage, that is, $v_{GS} - v_{DS} > V_T$. Hence, Eq. (4.6) is valid only for $v_{DS} < v_{GS} - V_T$.

The condition for which $v_{DS} = v_{GS} - V_T$ is known as pinch-off. The surface charge density of free electrons at the drain end is zero because the gate-to-drain voltage is only V_T. Does this imply that for a further increase in v_{DS} the drain current will fall to zero? For this condition, the conducting channel of free electrons will no longer contact the drain region. Abrupt discontinuities in the behavior of electron devices are not common. The current of a MOSFET device does not fall to zero when v_{DS} exceeds $v_{GS} - V_T$, rather it remains equal to a saturation value predicted by Eq. (4.6) for $v_{DS} = v_{GS} - V_T$.

for $v_{GS} \geq V_T$ and $v_{DS} > v_{GS} - V_T$

$$i_D = \frac{k}{2}(v_{GS} - V_T)^2 \tag{4.7}$$

For $v_{DS} > v_{GS} - V_T$, current carriers, in effect, cross the depletion region between the end of the channel and the drain, and the drain current is independent of the drain-to-source voltage of the device.

To summarize, the MOSFET device has three distinct regions of operation. For small gate-to-source voltages, voltages less than the threshold voltage, $v_{GS} < V_T$, the device is cut off – its drain current is zero. For larger gate-to-source voltages, $v_{GS} > V_T$, and for small drain-to-source voltages, namely $v_{DS} < v_{GS} - V_T$, the drain current depends on v_{GS} and v_{DS}. This region of operation is designated by various terms such as linear, resistance, below pinch-off, and triode. The triode designation is the result of a recognition that the behavior of this device is similar to the behavior of the triode vacuum tube (a device familiar to the engineers and scientists who developed field-effect devices). The triode designation appears to have the widest acceptance and will thus be used in this text. Large values of drain-to-source voltage result in a saturation of the drain current. This region of operation is referred to as *above pinch-off* or *saturation*. Saturation will be used in this text.* These results are summarized in Table 4.1.

On the basis of the preceding considerations, the behavior of a MOSFET device may be described by two parameters: its transconductance k and its threshold voltage V_T. The MOSFET characteristic of Figure 4.10 was obtained for $k = 0.8$ mA/V^2 and $V_T = 2.0$ V. For gate-to-source voltages of 2 V or less, the drain current is zero. For a gate-to-source voltage of 3 V, the expression for the triode

Table 4.1 Drain Current of an n-channel MOSFET

Region		Drain-Source Voltage	Drain Current
Cutoff $v_{GS} < V_T$		$v_{DS} \geq 0$	$i_D = 0$
Conduction $v_{GS} \geq V_T$	Triode	$0 < v_{DS} < v_{GS} - V_T$	$i_D = k\left[(v_{GS} - V_T)v_{DS} - \frac{1}{2}v_{DS}^2\right]$
	Saturation	$v_{DS} \geq v_{GS} - V_T$	$i_D = \frac{k}{2}(v_{GS} - V_T)^2$

* Unfortunately saturation is also used for the region of operation of the bipolar junction transistor (BJT) that occurs for small collector-to-emitter voltages corresponding to a forward biasing of the base–collector junction of the device. For a BJT, saturation is the result of an external circuit limitation, namely, insufficient voltage. On the other hand, the saturation of a MOSFET occurs for large drain-to-source voltages; in essence, the opposite condition to that for a BJT. Unless one is dealing with a circuit having both types of devices, it is generally clear which type of behavior is being described. It is important, however, to remember that the term *saturation* is used to describe rather dissimilar circuit characteristics of BJT and MOSFET devices.

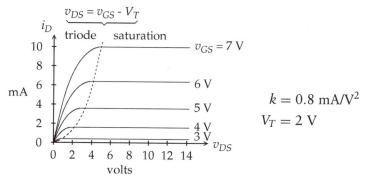

Figure 4.10: Drain characteristic of a MOSFET device obtained using the equations of Table 4.1.

region applies when drain-to-source voltages are less than 1 V. Larger drain-to-source voltages result in a saturation of the drain current $i_D = 0.4$ mA (last line of Table 4.1 and Eq. (4.7)). Higher gate-to-source voltages result in higher drain-to-source voltages that correspond to the dividing line between the triode and saturation regions (the dotted line on Figure 4.10).

The current expressions of Table 4.1 are for static conditions. Capacitive currents must also be considered for rapid voltage changes. Equivalent capacitances exist between the gate and the source and the gate and the drain of the device. These capacitances are the result of the close proximity of the edges of the gate and the source and drain regions (a small overlapping of the gate and these regions is common). There is also an equivalent gate-to-substrate capacitance, the result of charges due to the electric field of the oxide layer. Between the source and the substrate and the drain and the substrate there may also be capacitive currents resulting from the junction diodes between these regions. In addition, the dynamic response of the mobile charges of the channel must also be taken into account for rapid voltage changes. Numerous texts are available that provide a comprehensive quantitative treatment of relevant solid-state physics principles and also discuss the operation of other types of MOSFET devices (Milnes 1980; Pierret 1990; Pulfrey and Tarr 1989; Schroder 1987; Streetman 1990; Sze 1981; Taur and Ning 1998; Tsividis 1987).

SPICE MODEL

The SPICE computer program includes a set of algorithms to simulate the behavior of a MOSFET device. An indirect specification is used for the transconductance parameter k, that is, a parameter known as the transconductance process parameter k' (KP in SPICE).

$$k' = KP = \mu_n C_{ox} = \mu_n \epsilon_{ox}/t_{ox}, \qquad k = k' W/L \tag{4.8}$$

The process parameter KP is specificied in a .MODEL statement, whereas values of W and L are specified in the device statement. Default values of 100 μm for both W and L are incorporated in the MicroSim PSPICE simulation program (other versions of SPICE may have different default values). If W and L are not explicitly specified, the default values are automatically used – an .OPTIONS statement

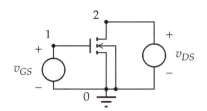

```
MOSFET Drain Characteristic
M1   2  1  0  0   MOSTRANS
VGS  1  0
VDS  2  0
.MODEL MOSTRANS NMOS KP=50U
+LAMBDA=.02 VTO=1
.DC  VDS 0 6 .1   VGS 1 5 .5
.PROBE
.END
```

Figure 4.11: SPICE circuit and file for determining the drain characteristic of a MOSFET device.

may be used to change the default values. It is not infrequent that W and L are not known, that is, k is either specified or has been determined experimentally. Because the default values of W and L are equal, KP is numerically equal to k if W and L are not specified.

To illustrate the computer simulation of a MOSFET device, the circuit and SPICE file of Figure 4.11 will be used to obtain a drain characteristic. Device parameters of $k = 50$ μA/V^2 and $V_T = 1.0$ V have been assumed. An M is used for the MOSFET label, and four terminals are specified: drain, gate, source, and substrate (in this order). This is followed by an arbitrary device name (MOSTRAN). If W and L were to be specified, they would follow the device name (for example, W=800U L=400U). With no specification, the default values result in $W/L = 1$. The .MODEL statement includes a device-type NMOS for an n-channel device (PMOS for a p-channel device). The numerical value of KP is equal to k for $W/L = 1$, and the threshold voltage VTO is also specified.

An additional parameter LAMBDA (λ) of 0.02 is also included in the .MODEL statement. This parameter accounts for the small increase in drain current that generally occurs in the saturation region. In Figure 4.10, the drain current lines are horizontal for saturation, that is, the drain circuit of the MOSFET device behaves as an ideal current source. For an actual device, the drain current tends to increase somewhat as the drain-to-source voltage is increased (Figure 4.3). This phenomenon, which is the result of several effects, is generally accounted for by a channel-length modulation parameter λ. As the drain-to-source voltage is increased, the channel pinch-off position tends to move away from the drain, thus reducing the effective channel length. Taking this into account, one obtains the following equations for the drain current (conduction):

$$i_D = \begin{cases} k\left[(v_{GS} - V_T)v_{DS} - \frac{1}{2}v_{DS}^2\right](1 + \lambda v_{DS}) & \text{triode region} \\ \frac{k}{2}(v_{GS} - V_T)^2(1 + \lambda v_{DS}) & \text{saturation} \end{cases} \tag{4.9}$$

The factor $1 + \lambda v_{DS}$ is used in both expressions to avoid a discontinuity in current at the transition between the expressions.

Although λ is usually assumed to be zero for analytic calculations, a nonzero value is readily handled by simulation programs. The similarity to the Early effect for bipolar junction transistors will be noted – λ plays the same role as $1/V_A$. Values of 0.005 to 0.04 V^{-1} for λ are typical for MOSFET devices, and the largest values occur for small channel lengths L. A value of 0.02 V^{-1} was

MOSFET Drain Characteristic

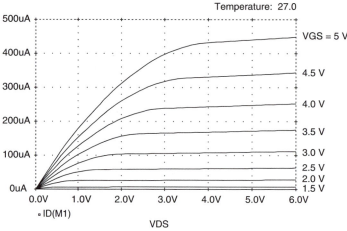

Figure 4.12: SPICE-generated MOSFET drain characteristic.

used for the simulation. A nested sweep was specified that yields 0.5-V increments for the gate-to-source voltages. The resultant drain characteristic is given in Figure 4.12.

EXAMPLE 4.1

The MOSFET device of Figure 4.7 has the following parameters:

$$V_T = 1.0 \text{ V} \quad t_{ox} = 0.1 \text{ } \mu\text{m} \quad \epsilon_{ox} = 3.9 \, \epsilon_0$$
$$W = 80 \text{ } \mu\text{m} \quad L = 20 \text{ } \mu\text{m} \quad \mu_n = 800 \text{ cm}^2/\text{V} \cdot \text{s}$$

The source and substrate are connected.
a. What is the overall gate–substrate capacitance of the device?
b. What is the transconductance k?
c. The gate-to-source voltage is 2 V. What is i_D for $v_{DS} = 0.5, 1.0, 1.5,$ and 2.0 V?

SOLUTION
a. The overall gate-to-source capacitance is the capacitance per unit area C_{ox} multiplied by the area of the gate:

$$C_{ox} = \epsilon_{ox}/t_{ox} = 3.45 \times 10^{-4} \text{ F/m}^2$$
$$C_{gs} = C_{ox} WL = 0.55 \text{ pF}$$

b. The transconductance depends on C_{ox} as follows:

$$k = \mu_n C_{ox} W/L = 0.11 \text{ mA/V}^2$$

c. $v_{GS} = 2.0$ V. For $v_{DS} = 0.5$ V, the MOSFET device will be in its triode region of operation ($v_{DS} < v_{GS} - V_T$).

$$i_D = k[(v_{GS} - V_T)v_{DS} - \tfrac{1}{2}v_{DS}^2] = 41.3 \ \mu A$$

For $v_{DS} = 1.0$ V the device is at pinch-off ($v_{DS} = v_{GS} - V_T$). Either expression may be used to obtain the drain current:

$$i_D = \frac{k}{2}(v_{GS} - V_T)^2 = 55 \ \mu A$$

This will also be the current for $v_{DS} = 1.5$ and 2.0 V because the current is independent of v_{DS} for saturation.

EXAMPLE 4.2

In MOSFET integrated circuits, MOSFET devices are generally used in place of resistors. This simplifies the fabrication process – the circuit then consists entirely of devices that differ only in the width, length, (W, or both L, or both) of the channel. Furthermore, much smaller areas are required for MOSFET "resistors" than for true ohmic resistors. A nonlinear resistor can be obtained by connecting the drain and gate of the device of Example 4.1 (Figure 4.13).

a. Determine a set of expressions for the current as a function of v_{DS}.
b. What is the equivalent resistance v_{DS}/i_D for $v_{DS} = 2$ V?
c. What is v_{DS}/i_D for $v_{DS} = 3$ V?
d. An equivalent resistance of 10 kΩ is required for $v_{DS} = 3$ V. What is the channel width W that could be used to obtain this resistance (no change in the other parameters)?

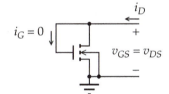

Figure 4.13: MOSFET circuit of Example 4.2.

SOLUTION

a. Because $v_{DS} > v_{GS} - V_T$ for all values of v_{DS}, the device will be either cut off or saturated – it will never be in its triode region of operation.

$$\text{for } v_{DS} < 1 \text{ V}, \qquad i_D = 0 \quad \text{cutoff}$$
$$\text{for } v_{DS} \geq 1 \text{ V}, \qquad i_D = \frac{k}{2}(v_{DS} - V_T)^2 \ \text{mA}$$

b. $v_{DS} = 2$ V, $i_D = 55 \ \mu A$, $R_{eq} = 36.4$ kΩ
c. $v_{DS} = 3$ V, $i_D = 0.22$ mA, $R_{eq} = 13.6$ kΩ
d. $R_{eq} = 10$ kΩ for $v_{DS} = 3$ V, $i_D = 0.3$ mA, $k = 37.5 \ \mu A/V^2$.

Because k is proportional to the width of the channel, $W = (37.5/55)(80 \ \mu m) = 54.5 \ \mu m$.

EXAMPLE 4.3

A MOSFET device is used in the circuit of Figure 4.14.

a. Determine the value of v_{IN} that results in $v_{OUT} = 6$ V.

b. Determine the variation in v_{IN} required to produce a variation in v_{OUT} of ± 0.1 V about 6 V.

c. Determine the variation in v_{IN} required to produce a variation in v_{OUT} of ± 1 V about 6 V.

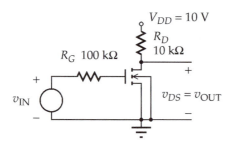

Figure 4.14: MOSFET circuit of Example 4.3.

SOLUTION

a. The gate resistor R_G will have no effect on the behavior of the circuit because for static conditions the gate current is zero $v_{GS} = v_{IN}$. A value of v_{IN} greater than 2 V is required for conduction ($i_D > 0$). Because it is not known if the device is in its triode or saturated region of operation, it is necessary to guess. The corresponding value of v_{IN} may be determined and compared with v_{DS} to determine if v_{DS} is greater or less than $v_{GS} - V_T$. If the guess was incorrect, the device is in the other region of operation. Because the expression for the saturated drain current is the simplest, it is reasonable to use this for the initial guess:

$$V_{DD} = i_D R_D + v_{OUT}, \qquad i_D = 0.4 \text{ mA}$$

$$v_{GS} - V_T = \sqrt{2i_D/k} = 2.828, \qquad v_{GS} = 4.828 \text{ V}$$

Because $v_{GS} - V_T = 2.828$ V, the initial guess of saturation is valid.

b. For $v_{OUT} = 6.1$ V, $i_D = 0.39$ mA, $v_{IN} = V_T + \sqrt{2i_D/k} = 4.793$ V.
For $v_{OUT} = 5.9$ V, $i_D = 0.41$ mA, $v_{IN} = V_T + \sqrt{2i_D/k} = 4.864$ V
A variation of about ∓ 0.035 V is required for a variation of ± 0.10 V in v_{OUT} (a voltage gain of -2.86).

c. For $v_{OUT} = 7$ V, $i_D = 0.3$ mA, $v_{IN} = V_T + \sqrt{2i_D/k} = 4.449$ V (-0.379 V)
For $v_{OUT} = 5$ V, $i_D = 0.5$ mA, $v_{IN} = V_T + \sqrt{2i_D/k} = 5.162$ V ($+0.334$ V)
The variation in v_{IN} for a ± 1 V variation in v_{OUT} is not symmetrical as a result of the nonlinear dependence of the drain current on $v_{GS} - V_T$.

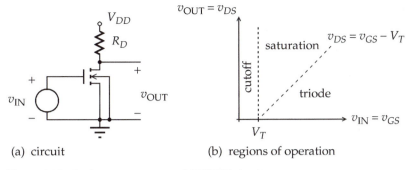

(a) circuit   (b) regions of operation

Figure 4.15: Basic common–source MOSFET circuit.

4.2 THE COMMON–SOURCE EQUIVALENT CIRCUIT: APPLICATIONS

The transfer characteristic of the basic common–source MOSFET circuit of Figure 4.15 will be determined using the analytic expressions summarized in Table 4.1 to calculate the drain current of the device. Because $v_{IN} = v_{GS}$, the MOSFET device is cut off ($i_D = 0$) for v_{IN} less than the threshold voltage. When the device is conducting, the dividing line between the triode and saturated region occurs for $v_{DS} = v_{GS} - V_T$, the 45° upward-sloping line of Figure 4.15(b) that extends upward from the $v_{IN} = V_T$ point on the input voltage axis. To be valid, a solution must fall in the region corresponding to the expression that was used to calculate the drain current.

To obtain a transfer characteristic of v_{OUT} versus v_{IN}, it will be assumed that v_{IN} is increased from zero. For $v_{IN} < V_T$, the device is cut off, yielding $i_D = 0$ and $v_{OUT} = V_{DD}$. However, as v_{IN} is increased above V_T, the saturation region of operation is entered, and the following is obtained:

$$i_D = \frac{k}{2}(v_{IN} - V_T)^2 \quad \text{saturation}$$
$$v_{OUT} = V_{DD} - i_D R_D = V_{DD} - \frac{kR_D}{2}(v_{IN} - V_T)^2$$

(4.10)

The solution of Eq. (4.10), indicated in Figure 4.16, is valid until $v_{OUT} = v_{IN} - V_T$, which corresponds to the boundary between the saturation and triode regions. Setting v_{OUT} equal to $v_{IN} - V_T$ in Eq. (4.10) will yield the value of v_{IN} at which

Figure 4.16: Transfer characteristic for the cutoff and saturation regions.

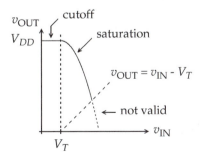

the device enters the triode region:

$$v_{IN} - V_T = V_{DD} - \frac{kR_D}{2}(v_{IN} - V_T)^2$$

$$(v_{IN} - V_T)^2 + \frac{2}{kR_D}(v_{IN} - V_T) - \frac{2V_{DD}}{kR_D} = 0 \tag{4.11}$$

$$v_{IN} = V_T - \frac{1}{kR_D} + \sqrt{\left(\frac{1}{kR_D}\right)^2 + \frac{2V_{DD}}{kR_D}}$$

Only the positive square-root term of the solution that arises in solving the preceding quadratic equation yields a valid solution.

For values of v_{IN} larger than that of Eq. (4.11), the device is in the triode region of operation.

$$i_D = k\left[(v_{IN} - V_T)v_{OUT} - \tfrac{1}{2}v_{OUT}^2\right]$$

$$v_{OUT} = V_{DD} - kR_D\left[(v_{IN} - V_T)v_{OUT} - \tfrac{1}{2}v_{OUT}^2\right] \tag{4.12}$$

A quadratic equation must be solved to obtain an explicit functional dependence of v_{OUT} on v_{IN}. This dependence corresponds to causality in which an input voltage is applied and the output is observed. However, if one is merely interested in obtaining a curve of v_{OUT} versus v_{IN}, a solution of v_{IN} as a function of v_{OUT} is equally useful. This corresponds to v_{OUT} being treated as the independent variable.

$$(v_{IN} - V_T)v_{OUT} - \frac{1}{2}v_{OUT}^2 + \frac{v_{OUT}}{kR_D} - \frac{V_{DD}}{kR_D} = 0$$

$$v_{IN} = V_T + \frac{1}{2}v_{OUT} + \frac{V_{DD}}{kR_D v_{OUT}} - \frac{1}{kR_D} \tag{4.13}$$

A point-by-point determination of v_{IN} for a range of values of v_{OUT} is readily obtained (Figure 4.17).

The transfer characteristic of Figure 4.17 could be used either for a logic **NOT** gate or for an amplifier circuit. For a logic **NOT** gate, a threshold voltage considerably less that the supply voltage V_{DD} is necessary. Furthermore, for $v_{IN} = V_{DD}$ an output voltage less than V_T is desirable. These conditions ensure that the

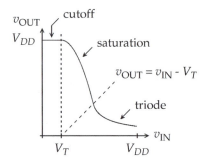

Figure 4.17: Transfer characteristic of a common–source MOSFET circuit.

output of one logic gate is properly interpreted when it is used for the input of a second logic gate.

To illustrate these requirements, suppose that a particular MOSFET device has a threshold voltage V_T of 1.0 V and is used in a circuit with a supply voltage V_{DD} of 5.0 V. A reasonable value of v_{OUT} for $v_{IN} = V_{DD}$ would be 0.5 V ($V_T/2$). From Eq. (4.12), a numerical value for the design parameter kR_D is obtained; $kR_D = 2.40$ V. What does this result imply? If, for example, $R_D = 1$ kΩ, a device with a transconductance parameter k of 2.4 mA/V^2 is required. On the other hand, if $R_D = 10$ kΩ, $k = 0.24$ mA/V^2. Because a device with a larger value of k results in a smaller value of v_{OUT} for $v_{IN} = V_{DD}$, these values are, in effect, minimum transconductance values required for a particular drain resistor R_D. If $kR_D = 2.40$ V is assumed, the point at which the transition from the saturation region to the triode region occurs may be determined using Eq. (4.11):

$$v_{IN} = 2.667 \text{ V}, \qquad v_{OUT} = 1.667 \text{ V} \tag{4.14}$$

At this value of input voltage, it may be noted, the second derivative of the transfer characteristic changes sign.

A COMMON–SOURCE AMPLIFIER

Although the transfer characteristic of the basic MOSFET circuit is well suited for a logic gate, its utility as a basic amplifier circuit is more limited. Unlike the transfer characteristic of the basic common–emitter bipolar junction transistor circuit (e.g., Figure 3.24), the characteristic of the basic MOSFET circuit does not have a region over which v_{OUT} tends to have a linear dependence on v_{IN}. Suppose that an input biasing voltage V_{GG} is used with a sinusoidal input signal voltage that is to be amplified (Figure 4.18). If the device remains in its saturated region of operation, the following is obtained:

$$v_{IN} = V_{GG} + V_m \sin \omega t$$

$$v_{OUT} = V_{DD} - \frac{kR_D}{2}(V_{GG} + V_m \sin \omega t - V_T)^2 \tag{4.15}$$

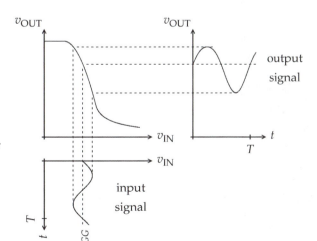

Figure 4.18: Distortion caused by the transfer characteristic of the basic MOSFET circuit.

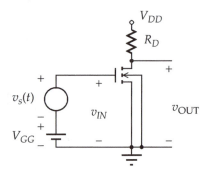

Figure 4.19: MOSFET small-signal amplifier circuit.

An alternative to determining v_{OUT} using Eq. (4.15) is a point-by-point procedure employing the transfer characteristic. For the sinusoidal input voltage indicated in Figure 4.18, the output voltage is significantly distorted. Although alternative circuits may be used to reduce the distortion (for example, a MOSFET load), this circuit is useful for amplifying small-amplitude signals.

Consider the small-signal amplifier circuit of Figure 4.19 and assume that the input voltage v_{IN} is such that the device is always conducting ($v_{IN} > V_T$) and that the device remains in its saturated region of operation. The quiescent value of drain current corresponds to a zero value of a signal.

$$I_D = \frac{k}{2}(V_{GG} - V_T)^2 \tag{4.16}$$

If the magnitude of the input signal $v_s(t)$ is sufficiently small, an approximate value for the drain current may be obtained using the derivative of the drain current as follows:

$$i_D = \frac{k}{2}(v_{GS} - V_T)^2, \qquad \frac{di_D}{dv_{GS}} = k(v_{GS} - V_T)$$

$$\left.\frac{di_D}{dv_{GS}}\right|_{V_{GG}} = k(V_{GG} - V_T) = g_m \tag{4.17}$$

The derivative has been evaluated for the quiescent condition ($v_{IN} = V_{GG}$). The quantity g_m, the mutual conductance of the MOSFET, depends on the quiescent gate-to-source voltage. This results in the following:

$$i_D = I_D + \left.\frac{di_D}{dv_{GS}}\right|_{V_{GG}} v_s(t) = I_D + g_m v_s(t) \tag{4.18}$$

This implies that two current sources may be used for the drain current: one a quiescent value given by Eq. (4.16), and the other, a signal value of $g_m v_s(t)$.

The two current sources of Eq. (4.18) result in the equivalent circuit of Figure 4.20. Although the gate-to-source voltage v_{GS} determines the drain current

Figure 4.20: Overall equivalent circuit of MOSFET amplifier.

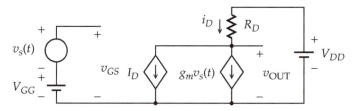

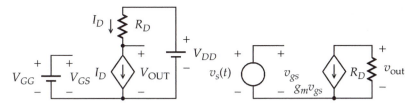

(a) quiescent equivalent circuit   (b) small-signal equivalent circuit

Figure 4.21: Equivalent circuits of a MOSFET amplifier.

of the MOSFET device, the gate behaves as an open circuit because $i_G = 0$. As for the bipolar junction transistor circuit, superposition may be used to obtain a solution for v_{OUT}.

The quiescent circuit of Figure 4.21(a) is obtained for $v_s(t) = 0$. All other voltage sources are included, and I_D (for saturation) is given by Eq. (4.16). The small-signal circuit has a single voltage source $v_s(t)$. A notation similar to that already used for BJT circuits is employed for all quantities. For example, the following applies for the gate-to-source voltage:

$$v_{GS} = \underbrace{V_{GS}}_{\text{dc (quiescent) value}} + \underbrace{v_{gs}}_{\text{varying component}} \tag{4.19}$$

The solution for v_{OUT} is the sum of the solutions of the individual circuits of Figure 4.21.

$$V_{OUT} = V_{DD} - \frac{kR_D}{2}(V_{GG} - V_T)^2 \qquad \text{quiescent solution}$$

$$v_{out} = -g_m R_D v_s(t) = -kR_D(V_{GG} - V_T)v_s(t) \quad \text{varying component} \tag{4.20}$$

$$v_{OUT} = V_{OUT} + v_{out}$$

Although the magnitude of the voltage gain of a MOSFET amplifier tends to be less than that of a BJT amplifier, a MOSFET amplifier tends to have a much higher input impedance. To the extent that the effect of the gate-to-source capacitance can be ignored (Example 4.1), the input current is negligible. This implies an extremely large power gain. An advantage of MOSFET amplifiers is that they can be used to amplify signals from sources that have very large internal impedances.

Figure 4.22: A MOSFET source–follower circuit.

A SOURCE–FOLLOWER AMPLIFIER

An alternative to the MOSFET common–source configuration having a resistor in the drain circuit is a source–follower configuration having a resistor in the source circuit (Figure 4.22). This circuit will be recognized as being the MOSFET analogy of the BJT emitter–follower circuit (Figure 3.42). As the names of these circuits imply, their output voltage tends to follow their input voltage.

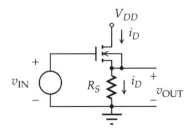

An advantage of the MOSFET source–follower over a MOSFET common–source amplifier is that its output resistance tends to be much smaller.

It will be noted that the current out of the source of the MOSFET device is equal to its drain current – the result of a zero gate current. For $v_{IN} < V_T$, the gate-to-source voltage is less than V_T, resulting in a zero drain current and zero output voltage. As v_{IN} is increased so that it exceeds V_T, the MOSFET device will start to conduct. The drain-to-source voltage v_{DS} depends on V_{DD} and v_{OUT} as follows:

$$v_{DS} = V_{DD} - v_{OUT} \tag{4.21}$$

Hence, for $v_{IN} = V_T$ and $v_{OUT} = 0$, $v_{DS} = V_{DD}$. The drain-to-source voltage is therefore initially large, considerably larger than $v_{GS} - V_T$, and the device is in its saturated region of operation.

$$i_D = \frac{k}{2}(v_{GS} - V_T)^2 \tag{4.22}$$

The gate-to-source voltage depends on the output voltage v_{OUT} as well as the input voltage v_{IN} as given by

$$v_{IN} = v_{GS} + v_{OUT}, \qquad v_{GS} = v_{IN} - v_{OUT} \tag{4.23}$$

Given that $v_{OUT} = i_D R_S$, the following is obtained:

$$i_D = \frac{k}{2}(v_{IN} - v_{OUT} - V_T)^2$$
$$v_{OUT} = i_D R_S = \frac{kR_S}{2}\left[(v_{IN} - V_T)^2 - 2(v_{IN} - V_T)v_{OUT} + v_{OUT}^2\right] \tag{4.24}$$

If a numerical value is available for kR_S, the preceding quadratic equation may readily be solved to determine the dependence of v_{OUT} on v_{IN}.

As for the common–source circuit, a somewhat simpler expression is obtained if v_{OUT} is treated as the independent variable:

$$\sqrt{v_{OUT}} = \sqrt{kR_S/2}(v_{IN} - V_T - v_{OUT})$$
$$v_{IN} = V_T + \sqrt{\frac{2v_{OUT}}{kR_S}} + v_{OUT} \tag{4.25}$$

Using this expression or solving the quadratic expression of Eq. (4.24) for v_{OUT} results in the transfer characteristic of Figure 4.23. Before proceeding, it should be verified that the MOSFET does indeed remain in its saturation region of operation. When $v_{IN} = V_{DD}$, the gate-to-source and drain-to-source voltages are equal, $v_{GS} = v_{DS}$. If it is assumed that V_T is a positive quantity, the condition for saturation applies because v_{DS} is obviously greater than $v_{GS} - V_T$. The input voltage needs to be greater than V_{DD} for the device to enter its triode region of operation.

The slope of the transfer characteristic of Figure 4.23 is less than unity. This is invariably the case for a MOSFET source–follower circuit – its small-signal voltage

Figure 4.23: Transfer characteristic of a MOSFET source–follower circuit.

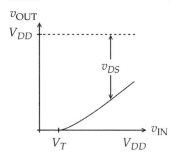

gain, which depends on the slope of the characteristic, will be less than unity (or, at best, close to unity). The utility of this circuit is that its input current, the current of the gate of the MOSFET, is for many applications negligible, and, therefore, its power gain is large. A small-signal analysis will be used to determine the voltage gain.

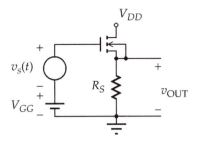

Figure 4.24: MOSFET source–follower small-signal amplifier.

Consider the MOSFET source–follower circuit of Figure 4.24 in which the amplitude of the input signal $v_s(t)$ is small. For the circuit to function as an amplifier (i.e., for $v_{OUT} > 0$), the device must be conducting. This requires that the biasing voltage V_{GG} be greater than V_T. For $V_{GG} < V_{DD}$, the device is in its saturated region of operation. A quiescent solution is obtained using Eq. (4.24) or 4.25 ($v_{IN} = V_{GG}$). From the quiescent value of drain current I_D, the following is obtained:

$$V_{OUT} = I_D R_S, \qquad V_{GS} = V_{GG} - V_{OUT}$$
$$g_m = k(V_{GS} - V_T) = k(V_{GG} - V_{OUT} - V_T)$$

$$(4.26)$$

For a circuit in which the biasing and supply voltages along with the device parameters are specified, Eq. (4.26) yields a numerical value of the mutual conductance g_m. A small-signal equivalent circuit may now be used to determine the small-signal behavior of the MOSFET source–follower (Figure 4.25).

The voltage sources V_{GG} and V_{DD} are not included in this circuit – their effect has already been taken into account in determining the quiescent solution. The following is obtained for the small-signal quantities in which it will be noted that $i_d = g_m v_{gs}$:

$$v_s(t) = v_{gs} + i_d R_S = v_{gs} + g_m v_{gs} R_S, \qquad v_{out} = i_d R_S = g_m v_{gs} R_S$$
$$v_{out} = \frac{g_m R_S v_s(t)}{1 + g_m R_S}, \qquad \frac{v_{out}}{v_s} = \frac{g_m R_S}{1 + g_m R_S}$$

$$(4.27)$$

The voltage gain v_{out}/v_s is less than unity.

As a final consideration, the small-signal equivalent output resistance of the circuit will be determined. This is the resistance seen looking into the output terminal of the small-signal equivalent circuit when all *independent* sources are properly removed. The mutual conductance current source $g_m v_{gs}$ is a dependent source – it cannot be removed from the circuit unless its controlling voltage v_{gs} happens to be zero. The output resistance is essentially the resistance that would be measured with an ohmmeter. Therefore, it will be imagined that a measuring voltage v_x is applied to the output, and the corresponding current will be determined (as an ohmmeter does). From Figure 4.26, in which the independent source $v_s(t)$ has been replaced with

Figure 4.25: Small-signal equivalent circuit of a MOSFET source–follower amplifier.

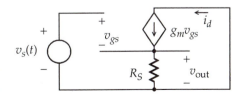

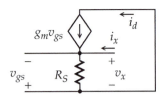

Figure 4.26: Equivalent circuit used for determining the output resistance of a source–follower amplifier.

a short-circuit, the following is obtained:

$$i_x = v_x/R_S - g_m V_{gs} = v_x/R_S + g_m v_x$$

$$i_x/v_x = 1/R_S + g_m \tag{4.28}$$

It will be noted that the output conductance is greater than the conductance of the resistor R_S. The output resistance is, in effect, the resistance R_S in parallel with an equivalent resistance of $1/g_m$.

EXAMPLE 4.4

The common–source MOSFET circuit of Figure 4.15 has the following circuit values and device parameters:

$$V_{DD} = 5 \text{ V}, \qquad R_D = 10 \text{ k}\Omega, \qquad k = 0.5 \text{ mA/V}^2, \qquad V_T = 1 \text{ V}$$

Determine the following points of the circuit's transfer characteristic:
a. The input voltage for $v_{OUT} = 2.5$ V.
b. The values of v_{IN} and v_{OUT} corresponding to the transition between the saturation and triode region of operation (pinch-off).
c. The output voltage for $v_{IN} = 5$ V.
d. The input and output voltages corresponding to the points of the characteristic that have a slope of -1.

SOLUTION

a. Equation (4.10) is valid for the saturation region ($kR_D = 5 \text{ V}^{-1}$).

$$v_{OUT} = V_{DD} - \frac{kR_D}{2}(v_{IN} - V_T)^2$$

$$v_{IN} = V_T + \sqrt{\frac{2(V_{DD} - v_{OUT})}{kR_D}} = 2 \text{ V} \quad \text{for } v_{OUT} = 2.5 \text{ V}$$

b. Equation (4.11) may be used to determine the input voltage at which the transition between the regions occurs.

$$v_{IN} = V_T - \frac{1}{kR_D} + \sqrt{\left(\frac{1}{kR_D}\right)^2 + \frac{2V_{DD}}{kR_D}} = 2.228 \text{ V}$$

$$v_{OUT} = v_{IN} - V_T = 1.228 \text{ V}$$

c. Equation (4.12) can be used to determine v_{OUT} for $v_{IN} = 5$ V.

$$v_{OUT} = V_{DD} - kR_D\left[(v_{IN} - V_T)v_{OUT} - \tfrac{1}{2}v_{OUT}^2\right]$$
$$= 5 - 5\left(4\,v_{OUT} - \tfrac{1}{2}v_{OUT}^2\right)$$
$$v_{OUT}^2 - 8.4\,v_{OUT} + 2 = 0, \qquad v_{OUT} = 0.245, \quad 8.155 \text{ V}$$

Only the first solution, $v_{OUT} = 0.245$ V, is valid. A simpler alternative approach generally yields an acceptable result when v_{IN} is large and v_{OUT} is small. This involves using an approximate expression for the drain current as follows:

$$i_D = k\left[(v_{IN} - V_T) - \tfrac{1}{2}v_{DS}\right]v_{DS}$$
$$\approx k(v_{IN} - V_T)v_{DS} \quad \text{if } v_{DS} \ll 2(v_{IN} - V_T)$$

It will be noted that $2(v_{IN} - V_T) = 8$ V, which is a value considerably larger than the expected value of v_{DS} ($v_{OUT} = v_{DS}$). This result corresponds to a linear dependence of i_D on v_{DS}, that is, the device's behavior is approximately that of a resistor r_n.

$$v_{DS}/i_D = r_n = \frac{1}{k(v_{IN} - V_T)} = 0.5 \text{ k}\Omega$$

By using this resistance in place of the MOSFET, the following is obtained for v_{OUT}:

$$v_{OUT} = \frac{r_n V_{DD}}{r_n + R_D} = 0.238 \text{ V}$$

This approximate result is only 7 mV less than the exact solution.

d. The transfer characteristic will have two points at which it has a slope of -1. Consider initially the point near $v_{IN} = V_T$ in which v_{OUT} is large. The device is in its saturated region of operation, and Eq. (4.10) applies.

$$\frac{dv_{OUT}}{dv_{IN}} = -kR_D(v_{IN} - V_T)$$

With the derivative equal to -1, the following is obtained for v_{IN} and v_{OUT}:

$$1 = kR_D(v_{IN} - V_T), \qquad v_{IN} = 1.20 \text{ V}, \qquad v_{OUT} = 4.90 \text{ V}$$

The other point for which the slope is equal to -1 occurs for v_{IN} large and v_{OUT} small corresponding to the triode region of operation. Differentiating Eq. (4.13) with respect to v_{IN} yields the following:

$$1 = \frac{1}{2}\frac{dv_{OUT}}{dv_{IN}} - \frac{V_{DD}}{kR_Dv_{OUT}^2}\frac{dv_{OUT}}{dv_{IN}}$$

By substituting -1 for the derivative, a value for v_{OUT} is obtained.

$$v_{OUT}^2 = \frac{2V_{DD}}{3kR_D}, \qquad v_{OUT} = 0.816 \text{ V}$$

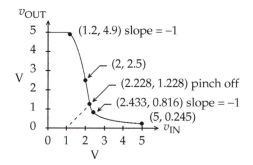

v_{OUT}

$(1.2, 4.9)$ slope $= -1$

$(2, 2.5)$

$(2.228, 1.228)$ pinch off

$(2.433, 0.816)$ slope $= -1$

$(5, 0.245)$

v_{IN}

Figure 4.27: Transfer characteristic for MOS-FET circuit of Example 4.4.

This value of v_{OUT} may now be substituted into Eq. (4.13) to obtain a value for v_{IN}; $v_{IN} = 2.433$ V. These points, as well as those of the previous parts, are indicated on the transfer characteristic of Figure 4.27.

EXAMPLE 4.5

Determine the small-signal voltage gain of the common–source MOSFET amplifier of Figure 4.28.

SOLUTION The quiescent conditions need to be determined using the circuit of Figure 4.29. Because the static gate current is zero, a voltage divider relationship may be used to determine V_{GS}.

$$V_{GS} = \frac{R_{G1}V_{DD}}{R_{G1} + R_{G2}} = 3.056 \text{ V}$$

If the device is assumed to be in its saturated region of operation, the following is obtained:

$$I_D = \frac{k}{2}(V_{GS} - V_T)^2 = 0.223 \text{ mA}$$
$$V_{DS} = V_{DD} - I_D R_D = 5.54 \text{ V}$$

Because V_{DS} of 5.54 V is greater than $V_{GS} - V_T$ (1.056 V), the initial assumption of saturation is valid. The mutual conductance may now be determined as

Figure 4.28: MOSFET amplifier circuit of Example 4.5. (Node numbers are those used for the SPICE simulation of Example 4.6.)

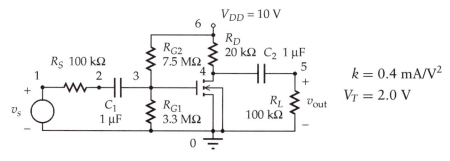

$V_{DD} = 10$ V

R_{G2} 7.5 MΩ

R_D 20 kΩ C_2 1 μF

R_S 100 kΩ

C_1 1 μF

R_{G1} 3.3 MΩ

R_L 100 kΩ

v_{out}

v_s

$k = 0.4 \text{ mA/V}^2$

$V_T = 2.0$ V

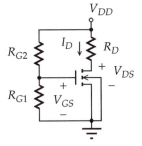

Figure 4.29: Quiescent circuit of Example 4.5.

follows:

$$g_m = k(V_{GS} - V_T) = 0.422 \text{ mS}$$

The small-signal equivalent circuit will now be solved (Figure 4.30). It is assumed that the capacitors behave as short circuits for the signal components. Numerical resistance values are included on the circuit for the parallel combination of resistors.

$$v_{gs} = \frac{R_{G1} \| R_{G2} v_s}{R_{G1} \| R_{G2} + R_S} = 0.958 v_s$$

$$v_{out} = -g_m v_{gs} R_D \| R_L = -7.035 v_{gs}$$

$$v_{out} = -6.740 v_s, \qquad v_{out}/v_s = -6.74$$

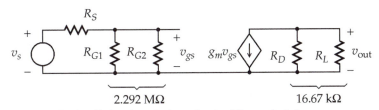

Figure 4.30: Small-signal equivalent circuit of Example 4.5.

EXAMPLE 4.6

Determine, using a SPICE simulation, the large-signal behavior of the amplifier of Example 4.5 for a sinusoidal input voltage with a frequency of 1 kHz. To show the distortion that occurs, assume amplitudes of the input signal that would result in, on the basis of the small-signal voltage gain determined in Example 4.5, peak-to-peak output voltages of 1, 2, 3, and 4 V. Obtain graphs of v_{out} and numerical values for the harmonic distortion.

SOLUTION For a small-signal voltage gain of 6.74, a peak-to-peak input voltage of 0.148 V will produce a peak-to-peak output voltage of 1 V. Hence, an amplitude of 0.074 V is required for an input signal, and amplitudes of 0.148, 0.222, and 0.296 V correspond to peak-to-peak voltages of 2, 3, and 4 V. If the results were to be obtained experimentally, a single circuit would be constructed. The input amplitude would then be successively set to the preceding values while the output voltage is observed. For a SPICE simulation, however,

```
MOSFET Amplifier                      VS3   31  0   SIN(0 .222 1000)
VS1   1   0   SIN(0 .074 1000)        RS3   31  32  100K
RS1   1   2   100K                    CS3   32  33  1U
CS1   2   3   1U                      RG13  33  0   3.3MEG
RG1   3   0   3.3MEG                  RG23  6   33  7.5MEG
RG2   6   3   7.5MEG                  M3    34  33  0   0  MOST
M1    4   3   0   0  MOST             RD3   6   34  20K
RD1   6   4   20K                     CL3   34  35  1U
CL1   4   5   1U                      RL3   35  0   100K
RL1   5   0   100K
VDD   6   0   DC   10                 VS4   41  0   SIN(0 .296 1000)
                                      RS4   41  42  100K
VS2   21  0   SIN(0 .148 1000)        CS4   42  43  1U
RS2   21  22  100K                    RG14  43  0   3.3MEG
CS2   22  23  1U                      RG24  6   43  7.5MEG
RG12  23  0   3.3MEG                  M4    44  43  0   0  MOST
RG22  6   23  7.5MEG                  RD4   6   44  20K
M2    24  23  0   0  MOST             CL4   44  45  1U
RD2   6   24  20K                     RL4   45  0   100K
CL2   24  25  1U
RL2   25  0   100K                    .MODEL   MOST NMOS KP=.4M VTO=2
                                      .TRAN   .01M 2M  0M   .01M
                                      .FOUR 1000 V(45)
                                      .PROBE
                                      .END
```

Figure 4.31: SPICE simulation circuit file of Example 4.6.

it is more convenient to determine the behavior of four circuits simultaneously, each having a different input voltage. This is accomplished with the circuit file of Figure 4.31, the node numbers of the first circuit being indicated in Figure 4.28. By running the circuits simultaneously, a plot of output voltages on a common graph can be obtained.

Even though the four circuits are identical (except for their input voltages), a unique labeling of each element and device is required. However, only a single .MODEL statement is required for the MOSFET devices (MOST). The last quantity of the transient statement (.TRAN), .01M, limits the internal integration step size to 0.01 ms, that is, one hundredth of the period of the 1-kHz signal. Without such a step-size limit, the SPICE program tends to use a larger step size, resulting in a poorly defined sinusoidal voltage – one that is noticeably distorted. It is the author's experience that the step size should be no greater than one hundredth of the period of a sinusoid. Although a smaller value produces still better results, a longer computational time is required. It should be noted that the first value of the .TRAN statement, the step size for the output data file, does not constrain the internal integration step size. Output data are generated through an interpolation process.

The result of the simulation is given in Figure 4.32. It will be noted that for the larger amplitudes, the output voltage is appreciably distorted. Although the peak-to-peak amplitudes are very close to that expected, the resultant signals are no longer symmetrical, that is, their positive excursions are considerably

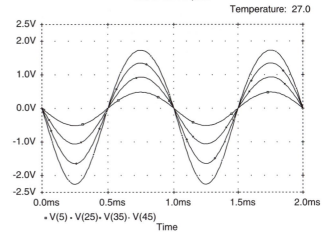

Figure 4.32: SPICE solution for Example 4.6.

less than their negative excursions. A .Four command produces a Fourier analysis of the voltage specified V(45) if a fundamental frequency of 1000 Hz is assumed. The program uses the last period of the signal, that is the last 1.0 ms of the transient result. The data of Figure 4.33 appeared in the output file. It will be noted that the fundamental component's having an amplitude of nearly 2 V would be the case for no distortion. The distortion is essentially that due to the dc component of −0.13 V and a second harmonic with an amplitude of 0.13 V. The contribution of the other harmonics is negligible. To obtain the harmonic distortion of the other components, separate runs of the program were required – the version of SPICE used by the author did not allow more than one .Four statement.

Figure 4.33: Harmonic distortion.

```
FOURIER COMPONENTS OF TRANSIENT RESPONSE V(45)

DC COMPONENT =  -1.296614E-01
```

HARMONIC NO	FREQUENCY (HZ)	FOURIER COMPONENT	NORMALIZED COMPONENT	PHASE (DEG)	NORMALIZED PHASE (DEG)
1	1.000E+03	1.995E+00	1.000E+00	-1.799E+02	0.000E+00
2	2.000E+03	1.339E-01	6.708E-02	9.012E+01	2.700E+02
3	3.000E+03	1.143E-04	5.730E-05	1.800E+02	3.599E+02
4	4.000E+03	8.552E-05	4.286E-05	1.799E+02	3.598E+02
5	5.000E+03	6.824E-05	3.420E-05	1.799E+02	3.599E+02
6	6.000E+03	5.664E-05	2.838E-05	1.799E+02	3.598E+02
7	7.000E+03	4.831E-05	2.421E-05	1.798E+02	3.597E+02
8	8.000E+03	4.208E-05	2.109E-05	1.798E+02	3.597E+02
9	9.000E+03	3.717E-05	1.863E-05	1.797E+02	3.596E+02

```
   TOTAL HARMONIC DISTORTION =   6.708376E+00 PERCENT
```

One might be wondering why an .AC analysis was not used. Is this not the option generally used for sinusoidal signals? The .AC option is based on a phasor analysis (an $e^{j\omega t}$ time dependence) that is valid for only a linear circuit. SPICE, when doing an .AC analysis, uses the small-signal linear circuit and treats this circuit as being valid regardless of the signal amplitudes. This option predicts the same voltage gain (with no distortion because the circuit is linear) regardless of the signal amplitudes. A 1-mV input signal is treated in the same fashion as a 1-MV signal.

4.3 MOSFET LOGIC GATES: BASIC CONSIDERATIONS

To change the voltage of a device or circuit capacitance, a current is required. Furthermore, the more rapidly this change in the voltage of a capacitance needs to be achieved, the larger the current required. Circuit limitations of devices therefore impose a limit on the speed at which these circuits can respond. Capacitances may generally be treated as being lossless, that is, the electrical energy they dissipate tends to be negligible. However, as a result of capacitive currents, electrical power is dissipated by the circuit associated with the capacitance. Therefore, limits on the power that may be safely dissipated limit the response time of a logic circuit.

A capacitor stores electric charge, and as a result of its charge storage q a capacitor may also be viewed as storing a quantity of electrical energy E_C as follows:

$$q = Cv \quad \text{C, coulombs;} \qquad E_C = \tfrac{1}{2}Cv^2 \quad \text{J, joules} \tag{4.29}$$

For a periodic voltage with a period of T, the stored charge and energy are also periodic.

$$q(t + T) = q(t); \qquad E_C(t + T) = E_C(t) \tag{4.30}$$

Hence, at the conclusion of one period, the capacitor's stored energy is the same as at the beginning of the period. This implies that the net energy supplied over a period by $v_S(t)$ in Figure 4.34(a), the circuit without a resistor, is zero. This is obviously not the case for the circuit with a series resistor (Figure 4.34(b)). To

Figure 4.34: Periodic voltage source and a capacitor.

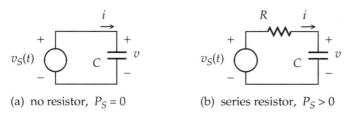

(a) no resistor, $P_S = 0$ (b) series resistor, $P_S > 0$

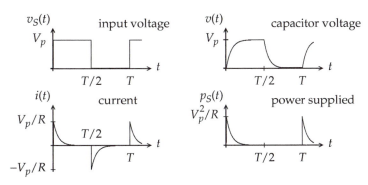

Figure 4.35: Switching the voltage of a capacitor–resistor circuit with a periodic square-wave voltage.

change a capacitor's voltage, a current is required:

$$i = \frac{dq}{dt} = C\frac{dv}{dt} \tag{4.31}$$

As a consequence, electrical power $i^2 R$ is dissipated by the resistor.

Consider the case of a periodic square-wave input voltage $v_S(t)$, which could represent a logic signal with voltage levels of 0 and V_p (Figure 4.35). As a result of the series resistance of the circuit of Figure 4.34(b), the capacitor's voltage $v(t)$ does not respond instantaneously to changes in the input voltage. The capacitor's voltage has finite rise and fall times – a result of the current required to charge and discharge the capacitor. When $v_S(t)$ increases from 0 to V_p (for example, at $t = 0$), the energy it supplies is equal to the integral over time of its instantaneous power $p(t)$ as follows:

$$p(t) = v_S(t)i(t)$$
$$E_S = \int_0^{T/2} p(t)\,dt = \int_0^{T/2} v_S(t)i(t)\,dt = V_p \int_0^{T/2} i(t)\,dt \tag{4.32}$$

The time $T/2$ is assumed to be sufficient for $v(t)$ to reach approximately V_p (theoretically, the voltage will never quite reach V_p). From the expression for the capacitor's current, Eq. (4.31), a relatively simple expression, is obtained for E_S as follows:

$$E_S = V_p \int_0^{T/2} C\frac{dv}{dt}\,dt = CV_p \int_0^{V_p} dv = CV_p^2 \tag{4.33}$$

Because during this interval, the capacitor's voltage increases from zero to V_p, its increase in stored energy is $\frac{1}{2}CV_p^2$. Hence, half the energy supplied is stored by the capacitor while the other half is dissipated by the series resistor, that is, converted to thermal energy. It will be noted that the value of the resistance not only did not enter into the result, but the result is valid for a nonlinear resistive element. When $v_S(t)$ returns to zero (for example at $t = T/2$), a negative capacitor current results – charge leaves the capacitor. Because $v_S(t)$ is equal to zero when this occurs, the energy it supplies (or absorbs) is zero. At the conclusion of the discharge, the potential of the capacitor and its stored energy are approximately zero. Hence, an energy of $\frac{1}{2}CV_p^2$ is again dissipated by the resistor.

To summarize, an energy of CV_p^2 must be supplied by $v_S(t)$ to charge the capacitor; half of this energy is stored by the capacitor, whereas the other half is dissipated by the resistive series element. When the capacitor is discharged, the half of the energy stored by the capacitor is dissipated by the resistor. Therefore, over a complete cycle, the resistor dissipates an energy of CV_p^2. If $v_S(t)$ supplies an energy of E_S each period of T seconds, the average power supplied, P_{Sav}, is the following:

$$P_{S\,av} = E_C/T = E_C f = CV_p^2 f \ \text{W} \tag{4.34}$$

A periodic frequency f has been introduced. An average power P_{Sav}, is required to compensate for the resistive losses of the circuit (this result may be shown to remain valid in the limit as R goes to zero because its peak current becomes infinite). This is an important consideration for designing logic circuits – in particular circuits that may be driving data busses with an appreciable capacitance. Suppose $C = 100$ pF, $V_p = 5$ V, and $f = 10$ MHz. Then,

$$P_{S\,av} = CV_p^2 f = 25 \ \text{mW} \tag{4.35}$$

For this situation, an electronic circuit must be capable of dissipating at least 25 mW. A larger capacitance or a higher switching frequency increases the power.

AN ELEMENTARY LOGIC INVERTER

Both MOSFET and BJT devices may often be treated as switches that are either open circuits or closed circuits with small resistances. To the extent that the equivalent series resistance of the device can be treated as being zero, the basic logic inverter circuit of Figure 4.36 in which the device is simulated by a switch can be utilized. The switch will be assumed to be open if v_{IN} is less than an input transition voltage V_t and closed if it is above this voltage. (SPICE includes a switch model that requires a finite input voltage range for the switch to move from its open to its closed condition.)

The static output voltage levels of the inverter gate of Figure 4.36 are zero and V_{DD}. The output is high when the switch is open and the current supplied by V_{DD} is zero. For a low output, which occurs when the switch closed, a power of V_{DD}^2/R is supplied by V_{DD}. Hence, on the basis of static conditions and if it is assumed the output is low as often as it is high, the average power that must be supplied by V_{DD} is $V_{DD}^2/(2R)$. This power is dissipated by the resistor R. To determine the dynamic behavior of this basic circuit, suppose that the logic input voltage of the gate is simulated with a periodic square-wave having a frequency of f and an amplitude greater than V_t. When the switch is opened, an energy of CV_{DD}^2 must be supplied by V_{DD} to charge the capacitor. When the switch is closed,

Figure 4.36: A basic logic inverter using an ideal switch.

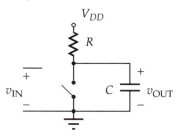

the energy stored by the capacitor, namely $\frac{1}{2}CV_{DD}^2$, is dissipated by the small but nevertheless finite resistance of the switch. The result is that an average dynamic power as well as a static power must be supplied by V_{DD} as follows:

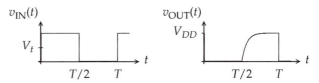

Figure 4.37: The dynamic behavior of a logic inverter using an ideal switch.

$$P_{S\,\text{dynamic}} = CV_{DD}^2 f, \qquad P_{S\,\text{static}} = V_{DD}^2/(2R) \qquad (4.36)$$

The ratio of these quantities is of value in judging their relative importance.

$$\frac{P_{S\,\text{dynamic}}}{P_{S\,\text{static}}} = 2RCf = \frac{RC}{T/2} \qquad (4.37)$$

For a useful circuit, the time constant of the circuit, RC, must be small compared with $T/2$; otherwise, the output voltage would not rise to approximately V_{DD}. Hence, the dynamic power tends to be much less than the static power for this particular circuit.

The rapidity of the response of the circuit is an important factor (Figure 4.37). A downward transition of v_{IN} (for example, at $t = T/2$) results in an exponential rise in the output voltage. On the other hand, an upward transition of v_{IN} (for example, at $t = 0$) results in a nearly instantaneous fall of v_{OUT}, the consequence of an assumed zero switch resistance. Although a zero fall time does not occur with an actual device such as a MOSFET, the fall time of a circuit of this type is much less than the rise time. A time equal to the time constant of the circuit, RC, is required for v_{OUT} to increase from 0 to 63.2 percent of its final value of V_{DD}. A longer time, namely $2.2RC$, is required for the 10- to 90-percent rise time of v_{OUT}. A quantity that may be utilized when designing logic circuits is the product of the rise time and the average static power supplied by V_{DD}:

$$t_{\text{rise time}} = 2.2RC, \qquad P_{S\,\text{static}}\, t_{\text{rise time}} = 1.1\, V_{DD}^2 C \qquad (4.38)$$

This is the static power – rise time product. It is a function of the supply voltage and the load capacitance, not the resistor of the circuit:

$$P_{S\,\text{static}} = \frac{1.1\, V_{DD}^2 C}{t_{\text{rise time}}} \qquad (4.39)$$

To decrease the rise time, that is, to increase the speed at which the gate will respond, the power supplied by V_{DD} and hence that dissipated by the circuit must be increased. A reduction in the circuit capacitance, if possible, will reduce the rise time. Furthermore, the circuit supply voltage V_{DD} can also be reduced – a technique frequently used for very high speed circuits.

A MOSFET INVERTER GATE

A basic MOSFET logic inverter gate, along with its static transfer characteristic obtained in the previous section, is indicated in Figure 4.38. Included in this

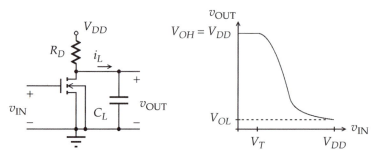

Figure 4.38: Basic MOSFET logic inverter gate.

circuit is an output capacitance C_L, which accounts not only for the capacitance of the external circuit connected to the inverter but also for the capacitance of the MOSFET device. For an input voltage that is less than the threshold voltage V_T, the output voltage is V_{DD}, the high-output voltage of the gate V_{OH}. For this to occur when a similar gate is used to produce v_{IN}, its output voltage for $v_{IN} = V_{DD}$ must be less than V_T. This is the low-output voltage V_{OL}. As a consequence, an input of V_{OL} results in an output voltage of V_{OH}, and an input of V_{OH} results in an output voltage of V_{OL}.

Of interest at this point is the dynamic behavior of the gate for an input voltage having upward and downward step functions with levels of V_{OL} and V_{OH}. Consider, initially, an input voltage with a downward step function (Figure 4.39). For $t < 0$, $v_{IN} = V_{OH}$, and the output voltage is V_{OL} (static behavior is assumed for $t < 0$). After $t = 0$, the gate-to-source voltage of the MOSFET is less than its threshold voltage, and the device is cut off. This is equivalent to the device being removed from the circuit. Hence, the equivalent circuit for $t > 0$ is simply a series resistor and capacitor R_D and C_L connected to the supply voltage V_{DD}. An increasing output voltage is obtained as defined by

$$
\begin{aligned}
v_{OUT}(t) &= V_{OL} + (V_{OH} - V_{OL})\left(1 - e^{-t/R_D C_L}\right) \\
&= V_{OH} - (V_{OH} - V_{OL})e^{-t/R_D C_L}
\end{aligned}
\tag{4.40}
$$

The 10- to-90-percent rise time of v_{OUT} is simply $2.2\,R_D C_L$. It will be noted that if $V_{OL} = 0$, the solution is the same as that obtained for the basic logic inverter with an ideal switch.

A more complex situation occurs for an upward transition of $v_{IN}(t)$ that results in a downward transition of $v_{OUT}(t)$ (Figure 4.40). For $t < 0$, the MOSFET is cut

Figure 4.39: A downward transition of $v_{IN}(t)$.

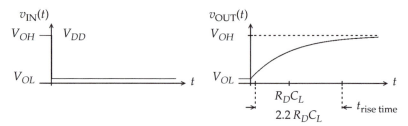

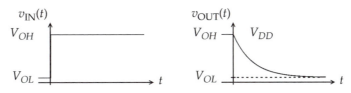

Figure 4.40: An upward transition of $v_{IN}(t)$.

off ($V_{OL} < V_T$) – its drain current is zero. After the upward transition of $v_{IN}(t)$, the device is turned on. As a result of the capacitor, $v_{OUT}(t)$ will not change instantaneously. Hence, initially $v_{DS} = V_{DD}$, and the device is in its saturated mode of operation with v_{GS} also being equal to V_{DD} (Figure 4.41).

An expression for the drain current i_D is readily obtained for saturation as follows:

$$i_D = \frac{k}{2}(v_{GS} - V_T)^2 = \frac{k}{2}(V_{DD} - V_T)^2 \tag{4.41}$$

On the basis of the circuit of Figure 4.41, the following is obtained by summing the currents at the drain node:

$$\frac{V_{DD} - v_{OUT}}{R_D} = i_D + C_L\frac{dv_{OUT}}{dt} = \frac{k}{2}(V_{DD} - V_T)^2 + C_L\frac{dv_{OUT}}{dt} \tag{4.42}$$

For t very small ($t = 0^+$), the output voltage is equal to approximately V_{DD}. The initial value of the time derivative of v_{OUT} is readily obtained from

$$\frac{dv_{OUT}}{dt} = -\frac{k(V_{DD} - V_T)^2}{2C_L} \quad \text{at } t = 0^+ \tag{4.43}$$

A line having this slope is indicated on the response of Figure 4.41. The intercept t_0 is readily determined as follows:

$$\frac{V_{DD}}{t_0} = \frac{k(V_{DD} - V_T)^2}{2C_L}, \qquad t_0 = \frac{2V_{DD}C_L}{k(V_{DD} - V_T)^2} \tag{4.44}$$

Although the actual transition time of v_{OUT} (10- to-90-percent) is larger than t_0, this readily calculated time provides a useful scale factor for the downward response time of v_{OUT}.

The drain current of Figure 4.41 remains equal to its saturated value as long as v_{OUT} exceeds $V_{DD} - V_T$. When v_{OUT} falls below $V_{DD} - V_T$, the device enters its triode region of operation, and hence, the drain current depends on $v_{DS}(= v_{OUT})$. While an analytic expression can be obtained for v_{OUT}, a simpler albeit approximate solution, is sufficient. To change the output voltage, the charge of the capacitor C_L must be changed. This requires a current that is out of the capacitor; that is, i_L of Figure 4.38 is a

Figure 4.41: The discharge of C_L.

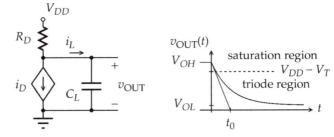

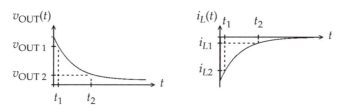

Figure 4.42: Determining a downward transition time of $v_{OUT}(t)$.

negative quantity:

$$i_L = C_L \frac{dv_{OUT}}{dt}, \qquad \langle i_L \rangle \approx C_L \frac{v_{OUT2} - v_{OUT1}}{t_2 - t_1} \tag{4.45}$$

For each output voltage v_{OUT1} and v_{OUT2}, the corresponding MOSFET drain currents i_{D1} and i_{D2} and capacitor currents i_{L1} and i_{L2} may be calculated (Figure 4.42). The average capacitor current $\langle i_L \rangle$ may be approximated as the average of its end values as follows:

$$\langle i_L \rangle = (i_{L2} + i_{L1})/2 \tag{4.46}$$

Although this may seem to be a rather crude approximation, the result is reasonably valid because i_L for an actual circuit tends to have a nearly linear time dependence (if i_L is linear in time, Eq. (4.46) is exact). Combining Eqs. (4.45) and (4.46) results in the following for $t_2 - t_1$:

$$t_2 - t_1 = \frac{2C_L(v_{OUT2} - v_{OUT1})}{i_{L2} + i_{L1}} \tag{4.47}$$

If v_{OUT1} and v_{OUT2} correspond to the 10- to-90-percent change in v_{OUT}, the time difference $t_2 - t_1$ is the corresponding response time. A numerical example will be used to illustrate this approximate solution more fully.

EXAMPLE 4.7

A basic logic inverter gate is to be approximated with a circuit using a switch that has an *on* resistance R_{on} of 1 kΩ (Figure 4.43).

a. What is V_{OL}, the static output voltage for a high-logic input voltage?

b. Suppose v_{IN} is high and that it is suddenly switched low. What is the time required for v_{OUT} to increase from V_{OL} to $V_{DD}/2$?

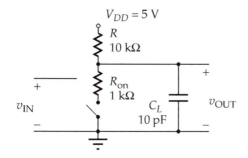

Figure 4.43: Switch-type logic inverter circuit of Example 4.7.

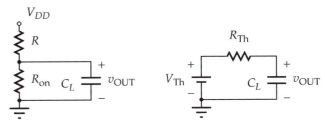

Figure 4.44: Thévenin equivalent circuit of Example 4.7.

c. Suppose v_{IN} is low and that it is suddenly switched high. What is the time required for v_{OUT} to fall from V_{DD} to $V_{DD}/2$?

SOLUTION

a. When the switch is closed, the output voltage for static conditions is determined by the voltage divider consisting of R_{on} and R:

$$V_{OL} = R_{on} V_{DD}/(R + R_{on}) = 0.455 \text{ V}$$

b. The capacitor is charged from its initial voltage of V_{OL} to V_{DD} by the resistor R. If $v_{OUT} = V_{OL}$ at $t = 0$, the following is obtained:

$$v_{OUT}(t) = V_{OL} + (V_{DD} - V_{OL})(1 - e^{-t/RC_L})$$
$$= V_{DD} - (V_{DD} - V_{OL}) e^{-t/RC_L}$$

Setting $v_{OUT}(t)$ equal to $V_{DD}/2$ yields the desired time $t_{mid\,up}$:

$$V_{DD}/2 = V_{DD} + (V_{DD} - V_{OL}) e^{-t/RC_L}$$

$$t_{mid\,up} = RC_L \ln 2(1 - V_{OL}/V_{DD}) = 0.598 \, RC_L = 59.8 \text{ ns}$$

c. When the switch is turned on, the circuit of Figure 4.44 applies as follows:

$$V_{Th} = R_{on} V_{DD}/(R + R_{on}) = V_{OL} = 0.455 \text{ V},$$
$$R_{Th} = R_{on} \| R = 0.909 \text{ k}\Omega$$

If it is assumed that the input transition again occurs at $t = 0$, the following is obtained for $v_{OUT}(t)$:

$$v_{OUT}(t) = V_{OL} + (V_{DD} - V_{OL}) e^{-t/R_{Th}C_L}$$

Setting $v_{OUT}(t)$ equal to $V_{DD}/2$ yields the desired time $t_{mid\,down}$:

$$V_{DD}/2 = V_{OL} + (V_{DD} - V_{OL}) e^{-t_{mid\,down}/R_{Th}C_L}$$

$$t_{mid\,down} = R_{Th} C_L \ln \left(\frac{V_{DD} - V_{OL}}{V_{DD}/2 - V_{OL}} \right) = 0.799 \, R_{Th} C_L = 7.26 \text{ ns}$$

EXAMPLE 4.8

An n-channel MOSFET device is used for the ouput circuit of an integrated circuit. The equivalent capacitance of the external circuit C_L is 50 pF (Figure 4.45).

a. What are the static low-and high-level output voltages V_{OL} and V_{OH}, respectively, of the logic gate?

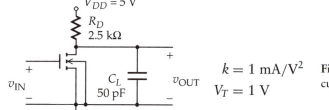

$k = 1$ mA/V^2 **Figure 4.45:** MOSFET logic inverter circuit of Example 4.8.

$V_T = 1$ V

b. Suppose that v_{IN} has been equal to V_{OH} for a very long time and that it is suddenly switched to V_{OL} at $t = 0$. Determine the rise time (10- to-90-percent) of v_{OUT}.

c. Consider the opposite situation in which v_{IN} has been equal to V_{OL} for a very long time and is suddenly switched to V_{OH} at $t = 0$. This results in the MOSFET device's being rapidly turned on. Determine the capacitor's current for a 10- and 90-percent change of v_{OUT}. What is the approximate fall time of v_{OUT}?

SOLUTION

a. $V_{OH} = V_{DD} = 5$ V. For $v_{IN} = V_{DD}$, the following approximate solution is obtained:

$$r_n = \frac{1}{k(V_{DD} - V_T)} = 0.25 \text{ k}\Omega, \qquad V_{OL} = \frac{r_n V_{DD}}{(R_D + r_n)} = 0.455 \text{ V}$$

b. The MOSFET device is cut off for this transition, and v_{OUT} increases with a time constant of $R_D C_L$. The 10- to-90-percent rise time is 2.2 $R_D C_L$, 0.275 μs.

c. The MOSFET is conducting for this transition. The total change in v_{OUT} is $V_{DD} - V_{OL} = 4.545$ V:

10-percent : $v_{OUT1} = 5 - .1(4.545) = 4.546$ V   (saturation region)

$$i_{D1} = \frac{k}{2}(V_{DD} - V_T)^2 = 8.0 \text{ mA}$$

$$i_{L1} = (V_{DD} - v_{OUT1})/R_D - i_{D1} = -7.82 \text{ mA}$$

90-percent : $v_{OUT2} = 5 - .9(4.545) = 0.910$ V   (triode region)

$$i_{D2} = k\left[(V_{DD} - V_T)v_{OUT2} - \frac{1}{2}v_{OUT2}^2\right] = 3.226 \text{ mA}$$

$$i_{L2} = (V_{DD} - v_{OUT2})/R_D - i_{D2} = -1.59 \text{ mA}$$

The average current $(i_{L1} + i_{L2})/2$ will be used to determine the time difference $t_2 - t_1$ as follows:

$$(i_{L1} + i_{L2})(t_2 - t_1)/2 = C_L(v_{OUT1} - v_{OUT2})$$

$$t_2 - t_1 = \frac{2C_L(v_{OUT1} - v_{OUT2})}{i_{L1} + i_{L2}} = 38.6 \text{ ns}$$

4.4 INTEGRATED-CIRCUIT LOGIC GATES: NO RESISTORS

The logic inverter gate of the previous section used a "pull-up" resistor to produce a high-level output voltage V_{DD} when the MOSFET device was cut off. A conventional carbon resistor would be employed if the logic gate were to be constructed using discrete components. For an integrated circuit, an n-type resistor could be diffused into the p-type substrate used for the n-channel devices. An alternative to a linear resistor is a second MOSFET device connected so as to behave as a resistor, albeit a nonlinear resistor. MOSFET pull-up resistors are universally used in modern integrated circuits. Not only is the processing step to form the resistor not required with a MOSFET device, but it generally requires a much smaller chip area – an important factor for very-large-scale integrated circuits (Elmasry 1992; Haznedar 1991; Hodges and Jackson 1988; Mukherjee 1986).

AN ENHANCEMENT-TYPE LOAD

Consider the MOSFET device of Figure 4.46(a) that has its gate and drain connected. Because $v_{DS} = v_{GS}$, v_{DS} is always greater than $v_{GS} - V_T$ for a device with a positive threshold voltage. Its triode region of operation is therefore excluded – the device is either cut off or saturated.

$$i_D = \begin{cases} 0 & \text{if } v_{DS} < V_T, \quad \text{cutoff} \\ \frac{k}{2}(v_{DS} - V_T)^2 & \text{for } v_{DS} \geq V_T, \quad \text{saturation} \end{cases} \tag{4.48}$$

Instead of a linear (straight-line) current-versus-voltage relationship, the MOSFET device has a nonlinear characteristic (Figure 4.46(b)).

In the inverter circuit of Figure 4.47, the pull-up resistor has been replaced by a MOSFET device. Because the gate voltage of this device is not controlled by v_{IN}, it is described as being a passive device as opposed to the inverter transistor M_1, which is an active device. As for a circuit with a resistor, a load line may be determined for the MOSFET load M_2 and V_{DD}.

$$i_{D1} = i_{D2}, \qquad V_{DD} = v_{DS2} + v_{DS1}, \qquad v_{DS1} = V_{DD} - v_{DS2} \tag{4.49}$$

These equations result in the load line of Figure 4.47(b), a line generated from the characteristic of Figure 4.46(b) (by reversing the voltage axis and shifting it to the right by V_{DD}).

Figure 4.46: A MOSFET device connected to form a two-terminal element.

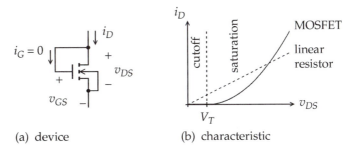

(a) device   (b) characteristic

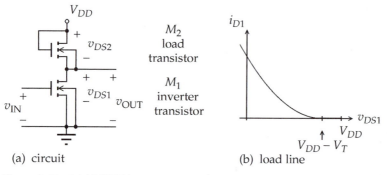

(a) circuit   (b) load line

Figure 4.47: A MOSFET inverter gate with a saturated MOSFET load.

Although a graphical analysis of the logic inverter gate with a MOSFET load device could be used to obtain its transfer characteristic, an analytic solution is readily obtained. For convenience, both devices will be assumed to have the same threshold voltage – a not uncommon situation when devices are fabricated simultaneously. However, different values of transconductance parameters will be assumed, k_1 and k_2 for M_1 and M_2, respectively. For v_{IN} less than the threshold voltage, M_1 is cut off and its drain current is zero. On the basis of the load line of Figure 4.47(b), the drain-to-source voltage of M_2, v_{DS2}, is V_T or less. If, however, one is attempting to measure v_{OUT}, for example, using an oscilloscope with a high but finite input resistance, the load device would need to supply the slight current required by the oscilloscope measurement. For a very small current, the drain-to-source voltage of M_2, v_{DS2}, would be approximately V_T, and v_{OUT} would be equal to $V_{DD} - V_T$. This portion of the circuit's transfer characteristic is indicated in Figure 4.48.

As v_{IN} is increased above the threshold voltage V_T, the active device M_1 begins to conduct. Because M_1 will be in its saturated region of operation, the following is obtained:

$$i_{D2} = \frac{k_2}{2}(V_{DD} - v_{OUT} - V_T)^2 \quad \text{because} \quad v_{DS2} = V_{DD} - v_{OUT}$$

$$\tag{4.50}$$

$$i_{D1} = \frac{k_1}{2}(v_{IN} - V_T)^2$$

Figure 4.48: Transfer characteristic of a MOSFET logic inverter gate with a MOSFET load.

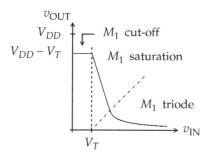

Recognizing that the drain currents must be equal yields the following for v_{OUT}:

$$(V_{DD} - v_{OUT} - V_T) = \sqrt{k_1/k_2}(v_{IN} - V_T)$$

$$v_{OUT} = V_{DD} - V_T - \sqrt{k_1/k_2}(v_{IN} - V_T)$$

$$\tag{4.51}$$

This linear transfer relationship having a slope of $-\sqrt{k_1/k_2}$ is valid as long as M_1 remains in its saturated region of operation. It will be noted that $\sqrt{k_1/k_2}$, the magnitude of the slope of the characteristic, needs to be larger than 1 for a useful logic inverter characteristic.

As v_{IN} is further increased, M_1 will enter its triode region of operation.

$$i_{D1} = k_1 \left[(v_{IN} - V_T)v_{OUT} - \frac{1}{2}v_{OUT}^2 \right]$$

$$\frac{k_2}{2}(V_{DD} - v_{OUT} - V_T)^2 = k_1 \left[(v_{IN} - V_T)v_{OUT} - \frac{1}{2}v_{OUT}^2 \right] \tag{4.52}$$

If v_{OUT} is desired for a particular value of v_{IN}, a quadratic equation must be solved. Alternatively, the transfer characteristic for this region may readily be obtained by treating v_{OUT} as the independent variable and solving for v_{IN}. If this gate is properly designed, $V_{OH} = V_{DD} - V_T$ and V_{OL} is the value of v_{OUT} obtained from Eq. (4.52) for $v_{IN} = V_{DD} - V_T$. The design criteria is that V_{OL} be less than the threshold voltage.

SUBSTRATE BIAS

In the logic inverter circuit of Figure 4.47, the substrate of the load transistor was connected directly to its source. This simplifies the analysis because one can treat the transistor as a three-terminal device. If this circuit were to be used within an integrated circuit, separate, electrically isolated, p-type substrates would be required for each device (Figure 4.49(a)). Although the inverter devices of an integrated circuit could share a common substrate, each load device would require a separate substrate. This complicates the fabrication of the integrated circuit. Furthermore, the load device has a relatively large substrate-to-base capacitance – a capacitance that appears across the output of the gate. This additional load capacitance increases the transition times of the circuit.

If a common substrate is used for load and inverter transistors (Figure 4.49(b)), not only is the fabrication process simplified, but a much smaller area is required for the circuit. Furthermore, a common n-type well can be used for the drain of the inverter and the source of the load. To analyze this circuit, however, it is necessary to treat the load transistor as a four-terminal device (Figure 4.50(a)) in which an additional voltage, the source-to-substrate voltage v_{SB} is taken into account. The second subscript of v_{SB} refers to "bulk," an alternative designation for the substrate. For static conditions, the substrate current is zero; the only currents of the device that remain are the drain and source currents ($i_D = -i_S$).

A substrate bias v_{SB}, for most situations, affects only the threshold voltage V_T of a MOSFET device (Figure 4.50(b)). For an n-channel device, the substrate

Figure 4.49: Physical structure of MOSFET logic inverter gates.

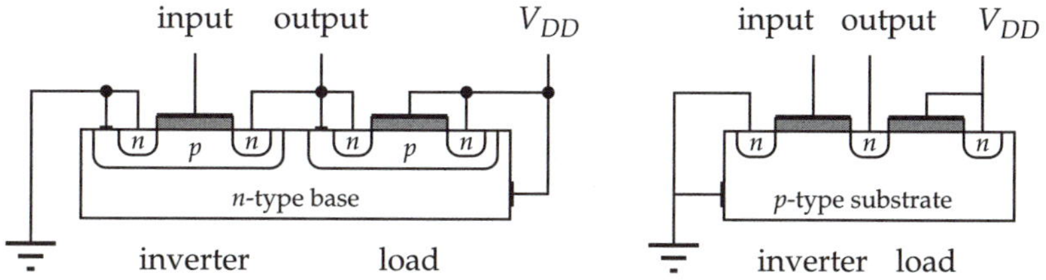

(a) substrate-to-source connection   (b) common substrate

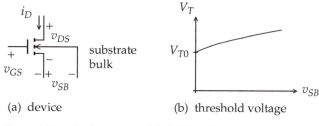

(a) device   (b) threshold voltage

Figure 4.50: The four-terminal MOSFET device.

must be at a potential less than that of either the source or drain of the device. If this were not the case, an *n*-type source or drain well would be forward biased. Hence, the substrate (bulk)-to-source potential will be zero or negative; conversely, v_{SB}, the source-to-substrate potential, is zero or positive for normal operation of the device. An analytic expression for the threshold voltage V_T based on theoretical considerations is generally utilized.

$$V_T = V_{T0} + \gamma \left(\sqrt{v_{SB} + 2\phi_F} - \sqrt{2\phi_F} \right) \tag{4.53}$$

where

V_{T0} = threshold voltage for zero substrate bias, V
γ = body-bias coefficient, $V^{1/2}$
$2\phi_F$ = surface potential, V

This is the relationship used by SPICE (the default value of γ is zero). A value of $0.4\ V^{1/2}$ is typical for the body-bias coefficient, and 0.6 V is typical for the surface potential. Precise values of these parameters for a particular device generally necessitate a set of experimental measurements.

A MOSFET logic inverter circuit with a load transistor sharing a common substrate is indicated in Figure 4.51 – the circuit corresponding to the physical structure of Figure 4.49(b). It will be noted from the following that the source-to-substrate voltage of the load device M_2 is equal to the output voltage:

$$v_{SB2} = v_{OUT}, \qquad V_{T2} = V_{T0} + \gamma \left(\sqrt{v_{OUT} + 2\phi_F} - \sqrt{2\phi_F} \right) \tag{4.54}$$

The high-level output voltage V_{OH} of this circuit, the voltage that occurs for $v_{IN} < V_T$, is reduced as a result of the dependence of the threshold voltage of M_2 on v_{OUT}.

$$V_{OH} = V_{DD} - V_{T2}, \qquad V_{T2} = V_{T0} + \gamma \left(\sqrt{V_{OH} + 2\phi_F} - \sqrt{2\phi_F} \right) \tag{4.55}$$

A simultaneous solution of the preceding equations can be obtained to find V_{OH}.

Figure 4.51: A MOSFET logic inverter circuit with a common substrate.

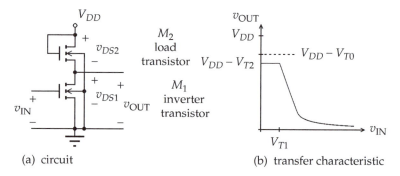

(a) circuit   (b) transfer characteristic

However, owing to the small sensitivity of the threshold voltage to the substrate bias, an iterative-type solution generally entails less effort.

To illustrate the effect of a common substrate, consider the case for $V_{DD} = 5$ V, $V_{T0} = 1$ V, $\gamma = 0.4$ V$^{1/2}$, and $2\phi_F = 0.6$ V. If the effect of the substrate bias is ignored ($V_T = V_{T0}$), the threshold voltage of M_2 is 1 V, and $V_{OH} = 4$ V. However, for a substrate bias of 4 V, the threshold voltage predicted by Eq. (4.55) is 1.548 V. This implies the output voltage V_{OH} is only 3.452 V ($V_{DD} - V_{T2}$). This value of output voltage yields a threshold voltage of 1.495 V and a corresponding output voltage V_{OH} of 3.505 V. The resultant output voltage may be shown, to an accuracy better than ± 10 mV, to be 3.50 V. The overall effect of the common substrate is to depress the entire transfer characteristic curve (Figure 4.51(b)).

A DEPLETION-TYPE LOAD

A disadvantage of the logic inverter gates with the MOSFET loads that have been considered is that their output voltage for a low-level input voltage is considerably less than V_{DD}. For the MOSFET devices that have been discussed, a positive gate-to-substrate voltage is required to form a channel of free electrons (V_T is positive). An alternative device can be fabricated with a surface layer of donor doping atoms that extends from the source to the drain of the device. The free electrons contributed by the donor atoms result in a "built-in" channel that exists in the absence of a gate-to-source bias. The net effect is that the threshold voltage is reduced, and, with sufficient doping, the threshold voltage is negative (Figure 4.52). The device symbol implies a built-in channel – the drain characteristic is unchanged except that a reduced gate-to-source voltage is required for a given drain current. A negative gate-to-source voltage is required to reduce the drain current to zero for the device of Figure 4.52(b). Because the channel must be depleted for the threshold condition, this is known as a depletion-type MOSFET device. A MOSFET device having a positive threshold voltage is known as an enhancement-type device – its channel must be enhanced to achieve a threshold condition.

Although enhancement-type devices are used for the inverter transistors of logic circuits, depletion-type devices are frequently used for the load transistors. Consider the logic inverter circuit of Figure 4.53 with a depletion-type load transistor. The gate-to-source connection of M_2 results in a current-versus-voltage

Figure 4.52: A depletion-type MOSFET device.

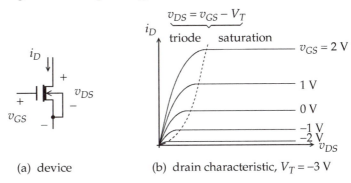

(a) device

(b) drain characteristic, $V_T = -3$ V

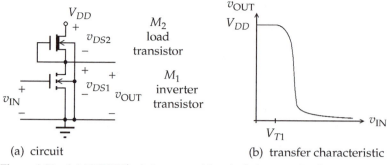

(a) circuit

(b) transfer characteristic

Figure 4.53: A MOSFET logic inverter with a depletion-type load.

characteristic of Figure 4.52(b) corresponding to $v_{GS} = 0$. For a negative threshold voltage V_{T2}, the device is never cut off ($v_{DS2} \geq 0$). Analytic expressions for the drain current of the load i_{D2}, are readily obtained ($v_{GS2} = 0$, $V_{T2} = -|V_{T2}|$).

$$i_{D2} = \begin{cases} k_2\left(|V_{T2}|v_{DS2} - \frac{1}{2}v_{DS2}^2\right) & \text{for } 0 \geq v_{DS2} < |V_{T2}|, \quad \text{triode} \\ \dfrac{k_2}{2}|V_{T2}|^2 & \text{for } v_{DS2} \geq |V_{T2}|, \quad \text{saturation} \end{cases} \quad (4.56)$$

With a common substrate connection, the threshold voltage of the load transistor V_{T2} depends on v_{OUT}. Equation (4.53) also applies for a depletion-type device, in which case V_{T0} is a negative quantity. An important advantage of this circuit is that when the inverter transistor is cut off and has a zero drain current ($v_{IN} < V_{T1}$), the drain-to-source voltage of the load transistor v_{DS2} is zero. Hence, the output voltage is V_{DD}. The typical transfer characteristic of a logic inverter with a depletion-type load is shown graphically in Figure 4.53(b).

The dynamic behavior of a MOSFET circuit depends on the load capacitance of the gate. Approximate analytic techniques, similar to those applied to circuits with a resistor-type load, may be used to obtain estimates of rise and fall times. Alternatively, a SPICE simulation can be used. The effect of a substrate bias ($v_{SB2} = v_{OUT}$) generally needs to be included when calculating the drain current of the load device.

Both **NOR** and **NAND** logic operations may readily be obtained with circuits using only MOSFET devices. For a two-input **NAND** gate, Figure 4.54(a), two

Figure 4.54: MOSFET logic gates.

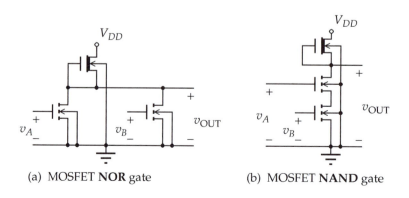

(a) MOSFET **NOR** gate

(b) MOSFET **NAND** gate

inverter transistors are connected in parallel. With a depletion-type load device the high-level output voltage (both v_A and v_B less than the threshold voltage of the inverter devices) is V_{DD}. If an enhancement-type load is used, the high-level output voltage is reduced. For a two-input **NAND** gate, Figure 4.54(b), the inverter transistors are connected in series. It should be noted that for both gates, the devices have a common substrate and that the logic functions are performed entirely with MOSFET devices – no resistors are required.

EXAMPLE 4.9

The devices of the MOSFET logic inverter gate of Figure 4.47 have the following parameters:

$$k_1 = 90 \ \mu A/V^2, \qquad k_2 = 10 \ \mu A/V^2, \qquad V_{T1} = V_{T2} = V_T = 1.0 \ V$$

The supply voltage V_{DD} is 5.0 V, and there is a load capacitance C_L of 1 pF.

a. Determine the static characteristic of the gate for the region in which M_1 is saturated. What is the slope of the characteristic, and what is the maximum value of v_{IN} for which M_1 is saturated?

b. Determine the static high- and low-output voltages V_{OH} and V_{OL} of the circuit.

c. The input voltage has been equal to V_{OH} for a very long time and is suddenly switched to V_{OL} at $t = 0$. What is $\frac{dv_{OUT}}{dt}$ for $t = 0^+$? On the basis of the derivative's value, estimate the time required for v_{OUT} to fall from V_{OH} to its midvalue of $(V_{OH} + V_{OL})/2$.

d. Consider the case for which v_{IN} has been equal to V_{OL} and is suddenly switched to V_{OH} at $t = 0$. What is $\frac{dv_{OUT}}{dt}$ for $t = 0^+$? Estimate the time required for v_{OUT} to reach its mid-value.

SOLUTION

a. For $v_{IN} < V_T$ (1.0 V), $v_{OUT} = V_{DD} - V_T = 4$ V, the high-level value of v_{OUT}, V_{OH}. For $v_{IN} > V_T$ and M_1 saturated, the following is obtained:

$$i_{D1} = \frac{k_1}{2}(v_{IN} - V_T)^2, \qquad i_{D2} = \frac{k_2}{2}(v_{DS2} - V_T)^2$$

The following results if these currents are equated and one recognizes that $v_{DS2} = V_{DD} - v_{OUT}$:

$$v_{OUT} = V_{DD} - V_T - \sqrt{k_1/k_2}(v_{IN} - V_T) = 4 - 3(v_{IN} - 1) \ V$$

The transition of M_1 to its triode region of operation can be obtained by setting $v_{OUT} = v_{IN} - 1$.

$$v_{IN} - 1 = 4 - 3(v_{IN} - 1), \qquad v_{IN} = 2 \ V$$

b. The output voltage is equal to V_{OL} for an input voltage of V_{OH}. For this condition, M_1 is in its triode region of operation.

$$i_{D1} = k_1 \left[(V_{OH} - V_T)V_{OL} - \tfrac{1}{2}V_{OL}^2 \right]$$

$$(V_{DD} - V_{OL} - V_T)^2 = \frac{2k_1}{k_2} \left[(V_{OH} - V_T)V_{OL} - \tfrac{1}{2}V_{OL}^2 \right]$$

Introducing numerical values results in the following quadratic equation for V_{OL}:

$$V_{OL}^2 - 6.2\,V_{OL} + 1.6 = 0, \qquad V_{OL} = 0.270, \quad 5.930 \text{ V}$$

Only the first solution, $V_{OL} = 0.27$ V, is valid. For $v_{IN} = 0.27$ V, M_1 is cut off and v_{OUT} is equal to $V_{DD} - V_T = 4$ V, the value of V_{OH}.

c. For this condition, M_1 is cut off for $t > 0$. The current of M_2 will charge the capacitor.

$$i_{D2} = \frac{k_2}{2}(V_{DD} - V_{OL} - V_T)^2 = 69.6 \ \mu\text{A} \quad \text{at } t = 0^+$$

$$C_L \frac{dv_{OUT}}{dt} = i_{D2}, \qquad \frac{dv_{OUT}}{dt} = \frac{i_{D2}}{C_L} = 69.6 \text{ V}/\mu\text{s}$$

To reach its midvalue, it is necessary for v_{OUT} to change by $(V_{OH} - V_{OL})/2$, 1.865 V.

$$\frac{\Delta v_{OUT}}{t_{\text{mid up}}} = 69.6 \text{ V}/\mu\text{s}, \qquad t_{\text{mid up}} = 26.8 \text{ ns}$$

d. For an upward transition of v_{IN}, M_1 is conducting at $t = 0^+$ and at the edge of its saturation–triode region. The current of M_2 is zero for $v_{OUT} = V_{OH}$.

$$i_{D1} = \frac{k_1}{2}(V_{OH} - V_T)^2 = 405 \ \mu\text{A}$$

$$\frac{dv_{OUT}}{dt} = \frac{i_{D1}}{C_L} = 405 \text{ V}/\mu\text{s}, \qquad t_{\text{mid down}} = 4.6 \text{ ns}$$

EXAMPLE 4.10

A SPICE simulation is desired to verify the results of Example 4.9 for the logic inverter of Figure 4.47. In addition, determine the behavior of the circuit of Figure 4.51 with the substrate of M_2 connected directly to ground ($\gamma = 0.4 \text{ V}^{1/2}$, $2\phi_F = 0.6$ V). To ascertain the dynamic behavior of the gates, assume v_{IN} has levels of 0 and 4 V.

SOLUTION Separate MOSFET inverter circuits will be used to determine the behavior for the two substrate connections. An input voltage pulse having a high-level duration of approximately 30 ns will ensure that v_{OUT} reaches a steady-state value of V_{OL} before the input pulse returns to zero. The upper curves of Figures 4.56 and 4.57 are for a load transistor with a direct source-to-substrate connection, and the lower curves are for the load and inverter transistors sharing a common substrate. For the direct source-to-substrate connection, $V_{OH} = 4.00$ V and $V_{OL} = 0.27$ V, values that are in agreement with those of Example 4.9. The transient solution resulted in values of 53.6 ns

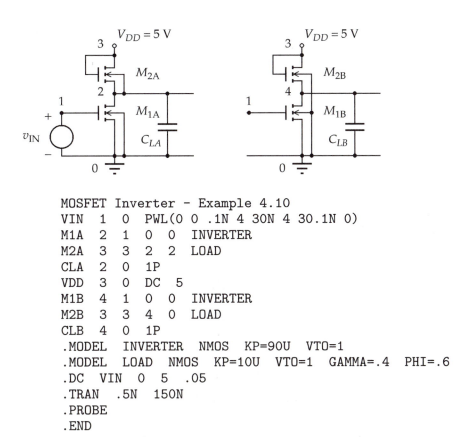

```
MOSFET Inverter - Example 4.10
VIN   1  0  PWL(0 0 .1N 4 30N 4 30.1N 0)
M1A   2  1  0  0  INVERTER
M2A   3  3  2  2  LOAD
CLA   2  0  1P
VDD   3  0  DC  5
M1B   4  1  0  0  INVERTER
M2B   3  3  4  0  LOAD
CLB   4  0  1P
.MODEL  INVERTER  NMOS  KP=90U  VTO=1
.MODEL  LOAD  NMOS  KP=10U  VTO=1  GAMMA=.4  PHI=.6
.DC  VIN  0  5  .05
.TRAN  .5N  150N
.PROBE
.END
```

Figure 4.55: SPICE circuit and file of Example 4.10.

Figure 4.56: Static SPICE solution of Example 4.10.

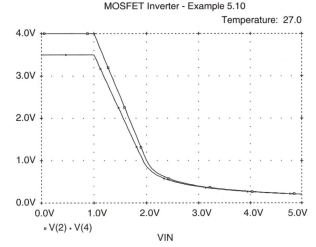

and 5.0 ns for $t_{\text{mid up}}$ and $t_{\text{mid down}}$, respectively. The fall time is very close to the value obtained using the initial derivative (4.6 ns), whereas the rise time is considerably longer than the 26.8 ns predicted using the initial derivative. For devices with a common substrate, $V_{OH} = 3.50$ V and $V_{OL} = 0.31$ V (the

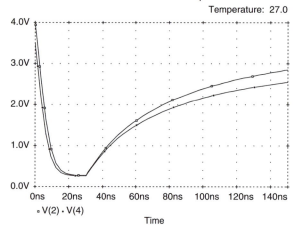

Temperature: 27.0

4.0V
3.0V
2.0V
1.0V
0.0V

0ns   20ns   40ns   60ns   80ns   100ns 120ns 140ns

□ V(2) ⬩ V(4)

Time

Figure 4.57: Dynamic SPICE solution of Example 4.10.

value of v_{OUT} for $v_{IN} = 3.50$ V). Midvoltage rise and fall times of 67.8 and 3.9 ns were obtained.

4.5 COMPLEMENTARY METAL-OXIDE SEMICONDUCTOR LOGIC GATES: AN ENERGY-EFFICIENT LOGIC FAMILY

Static and dynamic limitations constrain the performance of the MOSFET logic gates discussed in the previous sections. When the static output of one of these gates is low, there is a current of the load device that must be supplied by V_{DD}. Power is thus dissipated by the gate for this condition. The dynamic limitation may be seen on the SPICE response of Figure 4.57 (Example 4.10). A rapid downward transition of the output voltage is achieved as a consequence of the large drain current of the active device that discharges the load capacitance. The upward transition of the output voltage, however, is much slower as a result of the much smaller current of the load device. Although a larger load device (a device with a larger transconductance parameter) could be used to decrease the rise time, this would result in a larger current for a low-level output voltage of the gate. The power dissipated by the gate would be increased.

Logic circuits using complementary metal-oxide semiconductor (CMOS) field-effect transistors provide an alternative circuit configuration that overcomes these limitations of conventional MOSFET logic gates. For static conditions, the power dissipated by the circuit is essentially zero. Furthermore, an active device is used to produce the upward as well as the downward transition of the output voltage. Upward and downward transition times of the output voltage are, as a result, comparable.

THE *p*-CHANNEL MOSFET DEVICE

Up to this point, only *n*-channel MOSFET devices have been discussed. With an alternative fabricating process in which the semiconductor doping is reversed, a *p*-channel device can be formed (Figure 4.58) with an *n*-type substrate and *p*-type wells for the drain and source. Because the polarity of the drain- and

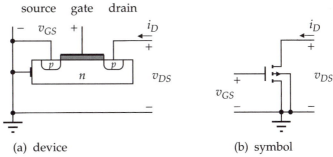

(a) device　　　　　　　　　　　　　　(b) symbol

Figure 4.58: Metal-oxide p-channel field-effect transistor. The direction of the substrate arrow is reversed to distinguish the symbol of this gate from that of an n-channel device.

source-to-substrate diodes is reversed from those of an n-channel device, the potentials of the drain and source relative to the substrate need to be negative (or zero) for normal operation. By means of a negative gate-to-substrate potential, positive changes are induced at the oxide–substrate boundary. For sufficiently negative gate voltages, mobile holes result that provide a current path between the source and drain of the device. Even though holes, rather than free electrons, are the current carriers, the operation of this device is similar to that of an n-channel device. However, because the mobile charges have a positive rather than negative charge, the polarity of all currents and voltages is reversed.

As can be seen from the drain characteristic of Figure 4.59, the threshold voltage of the device is negative (gate-to-source voltages more negative than this voltage are required for a device current). The transconductance parameter of a p-channel device (defined as a positive quantity) depends on the surface mobility of holes μ_p.

$$k = \mu_p C_{ox} W/L \ \text{A/V}^2 \tag{4.57}$$

Because the surface mobility of holes is less than that of free electrons, the transconductance of a p-channel device is less than that of an n-channel device with the same dimensions ($\mu_p/\mu_n \approx 0.4$). The following is obtained for the drain current of a p-channel device ($v_{DS} \leq 0$):

cutoff: $\quad v_{GS} > V_T, \qquad i_D = 0$

triode: $\quad v_{GS} \leq V_T \quad$ and $\quad v_{GS} - V_T < v_{DS} \leq 0$

$$i_D = -k\left[(v_{GS} - V_T)v_{DS} - \tfrac{1}{2}v_{DS}^2\right] \tag{4.58}$$

saturation: $\quad v_{GS} \leq V_T \quad$ and $\quad v_{DS} \leq v_{GS} - V_T$

$$i_D = -\frac{k}{2}(v_{GS} - V_T)^2$$

Other than for the reversal of the inequalities and the minus signs of the current expressions, these equations are essentially the same as those of an n-channel device (Table 4.2). For normal operation, all voltages of Eq. (4.58) are negative.

In place of the negative voltages of Eq. (4.58) (a situation that frequently leads to errors), a new set of device voltages and currents that are positive will

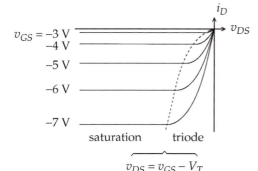

Figure 4.59: Drain characteristic of a p-channel MOS-FET device.

be introduced. The source current of a p-channel device i_S is positive, and the source-to-gate and source-to-drain voltages are also positive for normal operation.

$$i_S = -i_D, \qquad v_{SG} = -v_{GS}, \qquad v_{SD} = -v_{DS} \tag{4.59}$$

For a negative threshold voltage (that occurring for an enhancement-type p-channel device), the following may be used:

$$V_T = -|V_T| \tag{4.60}$$

Substituting the transformations of Eqs. (4.59) and (4.60) into Eq. (4.58) yields the relations of Table 4.2. It will be noted that the relations of Table 4.2 for a p-channel device may be derived from those of Table 4.1 for an n-channel device by interchanging drain and source subscripts (D and S) and introducing the magnitude of the p-channel threshold voltage.

A CMOS INVERTER GATE

A logic inverter gate using complementary MOSFET devices was first proposed by Sah and Wanlass in 1962 (Davies 1983; Sah 1988). As a result of numerous improvements, there are several families of CMOS gates and CMOS integrated circuits (Shoji 1988; Uyemura 1988). CMOS logic gates are extensively used for battery-powered electronic systems such as electronic watches and laptop computers. Consider the basic inverter circuit of Figure 4.60 with devices that have complementary symmetry, that is, devices with equal transconductance parameters and threshold voltages with equal magnitudes.

The regions of operation of the n-channel device are the same as in circuits in which an inverter is used with a load resistor or device (Figure 4.61(a)). The

Table 4.2 Source Current of an enhancement-type p-channel MOSFET

Region	Source-Drain Voltage	Source Current
Cutoff $\quad v_{SG} < |V_T|$	$v_{SD} \geq 0$	$i_S = 0$
Conduction $\left\{ \begin{array}{l} \text{Triode} \\[1em] v_{SG} \geq |V_T| \quad \text{Saturation} \end{array} \right.$	$0 < v_{SD} < v_{SG} - |V_T|$ $v_{SD} \geq v_{SG} - |V_T|$	$i_S = k\left[(v_{SG} - |V_T|)v_{SD} - \frac{1}{2}v_{SD}^2\right]$ $i_S = \dfrac{k}{2}(v_{SG} - |V_T|)^2$

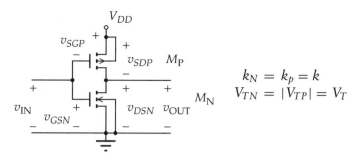

$$k_N = k_p = k$$
$$V_{TN} = |V_{TP}| = V_T$$

Figure 4.60: Complementary metal-oxide semiconductor field-effect logic inverter gate.

source-to-gate and source-to-drain voltages of the p-channel device, v_{SGP} and v_{SDP}, may readily be related to the input and output voltages of the inverter circuit by

$$v_{SGP} = V_{DD} - v_{IN}, \qquad v_{SDP} = V_{DD} - v_{OUT} \tag{4.61}$$

A cutoff condition occurs for $v_{SG} < V_T$:

$$V_{DD} - v_{IN} < V_T, \qquad v_{IN} > V_{DD} - V_T \tag{4.62}$$

The dividing line between the triode and saturation regions of the p-channel device corresponds to $v_{SDP} = v_{SGP} - V_T$ as follows:

$$V_{DD} - v_{OUT} = V_{DD} - v_{IN} - V_T, \qquad v_{OUT} = v_{IN} + V_T \tag{4.63}$$

This condition for the p-channel device is indicated in Figure 4.61(b).

The static voltage transfer characteristic of the gate will be determined for an open-circuit output condition. This requires that the currents of the devices be equal as given by

$$i_{SP} = i_{DN} \tag{4.64}$$

For $v_{IN} < V_T$, the n-channel device is cut off ($i_{DN} = 0$). Because the source-to-gate of the p-channel device is large, this device could be conducting, but because its source current is zero, its source-to-drain voltage must be zero. Hence,

Figure 4.61: Regions of operation for the devices of a CMOS inverter circuit.

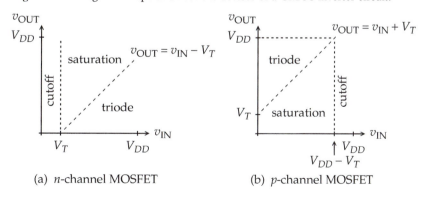

(a) n-channel MOSFET

(b) p-channel MOSFET

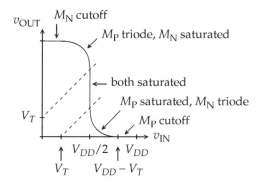

Figure 4.62: Voltage transfer characteristic of a CMOS inverter gate.

$v_{OUT} = V_{DD}$ for $v_{IN} < V_T$. A similar condition occurs for $v_{IN} > V_{DD} - V_T$ with the p-channel device being cut off. The gate-to-source voltage of the n-channel device is large, and because $i_{DN} = 0$, the output voltage is zero ($v_{OUT} = 0$ for $v_{IN} > V_{DD} - V_T$). These results are indicated by the voltage transfer characteristic of Figure 4.62.

Consider the situation for which both devices are in their saturated region of operation (the region between the dashed lines of Figure 4.62). Equating the device currents results in the following:

$$\frac{k}{2}(v_{GSN} - V_T)^2 = \frac{k}{2}(v_{SGP} - V_T)^2$$

$$v_{GSN} - V_T = v_{SGP} - V_T, \qquad v_{IN} - V_T = V_{DD} - v_{IN} - V_T \qquad (4.65)$$

$$v_{IN} = V_{DD}/2$$

This implies a vertical line for v_{OUT} over the range for which the devices are saturated. Finally, there are two transition regions in which one device is saturated and the other is in its triode region of operation. Consider the upper transition region in which the p-channel device is in its triode region and the n-channel device is saturated ($V_T < v_{IN} < V_{DD}/2$).

$$k[(v_{SGP} - V_T)v_{SDP} - \tfrac{1}{2}v_{SDP}^2] = \frac{k}{2}(v_{GSN} - V_T)^2$$

$$(V_{DD} - v_{IN} - V_T)(V_{DD} - v_{OUT}) - \tfrac{1}{2}(V_{DD} - v_{OUT})^2 = \tfrac{1}{2}(v_{IN} - V_T)^2 \qquad (4.66)$$

When this quadratic relationship is solved, the upper transition portion of the characteristic results. There is a similar set of expressions for the lower transition region in which the p-channel device is saturated and the n-channel device is in its triode region of operation:

$$\frac{k}{2}(v_{SGP} - V_T)^2 = k[(v_{GSN} - V_T)v_{DSN} - \tfrac{1}{2}v_{DSN}^2]$$

$$\tfrac{1}{2}(V_{DD} - v_{IN} - V_T)^2 = [(v_{IN} - V_T)v_{OUT} - \tfrac{1}{2}v_{OUT}^2] \qquad (4.67)$$

Although not immediately obvious from the preceding relations, the voltage transfer characteristic is symmetric about its midpoint ($v_{IN} = v_{OUT} = V_{DD}/2$).

The current supplied by V_{DD} is the source current of the p-channel device i_{SP} ($= i_{DN}$). Because for $v_{IN} < V_T$ and $v_{IN} > V_{DD} - V_T$ one of the devices is cut off, i_{SP} is zero. The supply current and power supplied by V_{DD} are therefore zero for these input voltages that correspond to low- and high-input logic levels. Therefore, for static conditions, the power supplied and hence dissipated by the inverter gate is essentially zero. A current occurs for the transition region $V_T < v_{IN} < V_{DD} - V_T$

when conduction occurs. When both devices are saturated $v_{IN} = V_{DD}/2$, the supply current is a maximum:

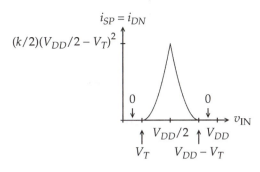

$$i_{SP\,max} = \frac{k}{2}(V_{DD}/2 - V_T)^2 \qquad (4.68)$$

For rapid transition of the input voltage, the energy dissipated by the MOSFET devices as a result of their current is generally small. The dependence of the device currents on v_{IN} is indicated in Figure 4.63.

Figure 4.63: Device currents as a function of input voltage.

In addition to a low power consumption, a CMOS logic gate has a more rapid dynamic response than a logic gate consisting of an inverter device and pull-up load. Consider the circuit of Figure 4.64 with a load capacitor C_L, which accounts for the drain-to-substrate capacitances of the devices as well as the capacitance of the circuit to which it is connected. Assume the input voltage has been zero for a long time (static conditions prevail) and that it is switched to V_{DD} at $t = 0$. Because the p-channel device is cut off for $t > 0$, the circuit of Figure 4.65 with only the n-channel device applies. The drain current of the n-channel device i_{DN} rapidly discharges the capacitor. Initially the MOSFET device will be saturated, resulting in the following:

$$i_{DN} = \frac{k}{2}(V_{DD} - V_T)^2 = -C_L\frac{dv_{OUT}}{dt}, \qquad \frac{dv_{OUT}}{dt} - \frac{k}{2C_L}(V_{DD} - V_T)^2 \quad (4.69)$$

This initial value of the derivation of v_{OUT} is valid until the device enters its triode region of operation, that is, until v_{OUT} falls below $V_{DD} - V_T$. The initial slope, projected to the time axis corresponding to $v_{OUT} = 0$, yields a useful time parameter t_0:

$$-\frac{V_{DD}}{t_0} = -\frac{k}{2C_L}(V_{DD} - V_T)^2, \qquad t_0 = \frac{2C_L V_{DD}}{k(V_{DD} - V_T)^2} \qquad (4.70)$$

A time of approximately $t_0/2$ is required for v_{OUT} to fall to $V_{DD}/2$, whereas the 90-to-10-percent fall time is somewhat larger than t_0.

A similar situation occurs when an input voltage of V_{DD} is switched to zero at $t = 0$ (Figure 4.66). For this case, the source current of the p-channel device charges the capacitor. The initial time derivation of v_{OUT} has the same magnitude for complementary devices except that it is positive. The time parameter t_0 corresponding to a projection of the initial derivative to $v_{OUT} = V_{DD}$ is the same. As a result of the p-channel device, the upward transition of v_{OUT} is considerably less than that when it is charged with a MOSFET load device. For complementary symmetry, the rise and fall times of the output are equal.

Figure 4.64: CMOS logic inverter gate with a load capacitance.

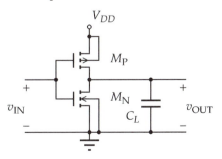

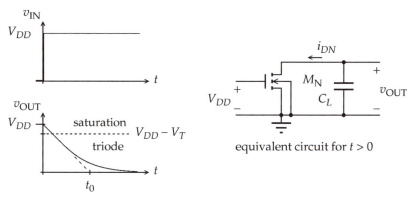

equivalent circuit for $t > 0$

Figure 4.65: An upward transition of v_{IN}.

CMOS LOGIC GATES

Logical **NOR** and **NAND** operations may readily be performed with CMOS gates; each logic input is used to switch both an n-channel and a p-channel device. The circuit for a **NOR** gate is shown in Figure 4.67(a). If both inputs are low ($v_A = v_B = 0$), the n-channel devices are cut off while the p-channel devices have a large source-to-gate voltage – they would provide a source current if a load were connected to the output of the circuit. The behavior of the circuit for this condition approximates that of a single p-channel device with a transconductance one-half that of the individual devices (a channel twice as long). Hence, for a zero load current, the output voltage is V_{DD}. If either input is high (V_{DD}), one of the p-channel devices will be cut off. For this condition, $v_{OUT} = 0$ because one of the parallel-connected n-channel devices will be conducting.

For the **NAND** gate of Figure 4.67(b), the parallel and series connections of the devices are reversed. If either input is low, one of the series-connected n-channel devices is cut off while one of the parallel-connected p-channel devices has a large source-to-gate voltage. Hence, $v_{OUT} = V_{DD}$ for no load current. Only if both inputs are high ($v_A = v_B = V_{DD}$) will the series n-channel devices conduct, resulting in a low value for v_{OUT}.

Figure 4.66: A downward transition of v_{IN}.

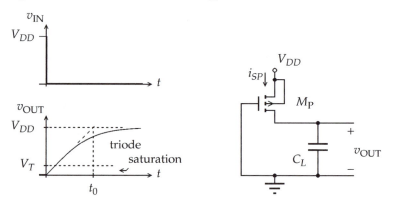

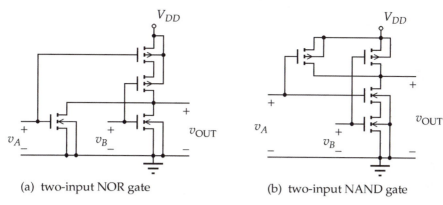

Figure 4.67: CMOS logic gates.

(a) two-input NOR gate (b) two-input NAND gate

EXAMPLE 4.11

A CMOS logic gate is connected to a load that can be simulated by two resistors (Figure 4.68). The two load resistors and V_{DD} may be replaced by a Thévenin equivalent circuit.

a. Determine v_{OUT} for $v_{IN} = V_{DD}$.

b. What is v_{OUT} for $v_{IN} = 0$?

c. What is v_{OUT} for $v_{IN} = 5$ V?

d. As a result of the load resistors, the output transition that occurs when both devices are saturated will no longer be vertical. Estimate the slope of the transition by determining v_{OUT} for $v_{IN} = 4.8$ and 5.2 V. Verify that the MOSFET devices are indeed saturated for these input voltages.

SOLUTION The load resistors and V_{DD} result in a Thévenin equivalent circuit with $V_{Th} = 5.0$ V and $R_{Th} = 10$ kΩ (Figure 4.69).

a. Only the *n*-channel device conducts for $v_{IN} = V_{DD}$ (Figure 4.69(a)). For v_{OUT} small, the drain-to-source circuit may be replaced by an equivalent resistance r_n as follows:

$$r_n = \frac{1}{k(V_{DD} - V_T)} = 1.25 \text{ k}\Omega, \qquad v_{OUT} = \frac{r_n V_{Th}}{r_n + R_{Th}} = 0.56 \text{ V}$$

Figure 4.68: CMOS logic inverter of Example 4.11.

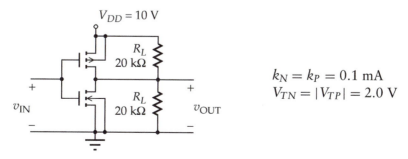

$$k_N = k_P = 0.1 \text{ mA}$$
$$V_{TN} = |V_{TP}| = 2.0 \text{ V}$$

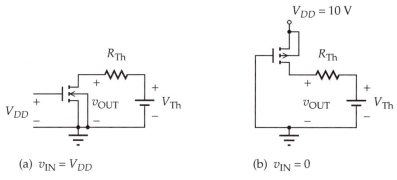

(a) $v_{IN} = V_{DD}$ (b) $v_{IN} = 0$

Figure 4.69: Equivalent circuits for $v_{IN} = V_{DD}$ and $v_{IN} = 0$.

b. For $v_{IN} = 0$, only the p-channel device is conducting (Figure 4.69(b)). For v_{SDP} small, the MOSFET device may be replaced by an equivalent resistance r_p as follows:

$$i_{SP} \approx k(V_{DD} - V_T)v_{SDP}, \qquad r_p = \frac{1}{k(V_{DD} - V_T)} = 1.25 \text{ k}\Omega$$

The voltage across r_p is 0.56 V, resulting in a value of 9.44 V for v_{OUT}.

c. Using the Thévenin equivalent output circuit, the following is obtained for the output voltage:

$$i_{SP} = i_{DN} + (v_{OUT} - V_{Th})/R_{Th}$$
$$v_{OUT} = V_{Th} + R_{Th}(i_{SP} - i_{DN})$$

For v_{IN} = 5.0 V, the devices have equal currents as follows:

$$i_{SP} = \frac{k}{2}(V_{DD} - v_{IN} - V_T)^2 = 0.80 \text{ mA}$$
$$i_{DN} = \frac{k}{2}(v_{IN} - V_T)^2 = 0.80 \text{ mA}$$

Therefore, $v_{OUT} = V_{Th} = 5.0$ V.

d. For $v_{IN} = 4.8$ V, $i_{SP} = 0.882$ mA and $i_{DN} = 0.722$ mA. From the preceding expression for v_{OUT}, a value of 6.6 V is obtained. For $v_{IN} = 5.2$ V, $i_{SP} = 0.722$ mA and $i_{DN} = 0.882$ mA, resulting in a value of 3.4 V for v_{OUT}.

$$\text{slope} = \frac{\Delta v_{OUT}}{\Delta v_{IN}} = \frac{3.4 - 6.6}{5.2 - 4.8} = -8.0$$

When $v_{IN} = 4.8$ V, both devices will be saturated if v_{OUT} is between 2.8 and 6.8 V ($v_{IN} \pm V_T$). When $v_{IN} = 5.2$ V, saturation occurs for v_{OUT} between 3.2 and 7.2 V. These conditions were fulfilled – the devices were saturated.

EXAMPLE 4.12

A CMOS logic inverter gate has an output transition that occurs for a moderately large change in v_{IN} (Figure 4.62). A much more abrupt transition can be achieved using an additional set of inverter gates – a total of three logic inverter gates. Assume the MOSFET devices have the following parameters:

$$k_N = 0.1 \text{ mA/V}^2, \qquad V_{TN} = 1.0 \text{ V}, \qquad \lambda = 0.02$$
$$k_P = 0.1 \text{ mA/V}^2, \qquad V_{TP} = -1.0 \text{ V}, \qquad \lambda = 0.02$$

A supply voltage of $V_{DD} = 5$ V is used for the gates. Use a SPICE simulation to determine the static transfer characteristic of one, two, and three gates. Consider the transition width of v_{IN} to be defined by the points at which the magnitude of the slope of the characteristic is 1. What are the transition widths for the circuits?

SOLUTION The circuit and file of Figure 4.70 will be used for a solution. It should be noted that for the p-channel devices, the drain is the lower terminal and the source is the upper terminal. The voltage transfer characteristics of Figure 4.71 are obtained. A single gate has the response having "rounded corners," V(2). Two gates form a buffer (output logic equal input logic) and have a voltage transfer characteristic V(3) that has nearly "square corners." The transfer characteristic for three gates is essentially a "perfect" response having square corners. From plots of the derivative of v_{OUT}, DV(2), DV(3), and DV(4), the

Figure 4.70: SPICE diagram and circuit for Example 4.12.

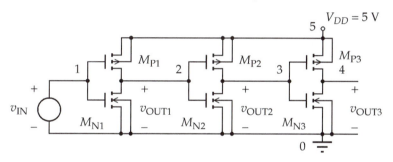

```
CMOS Inverters
VIN    1  0
MP1    2  1  5  5  MP
MN1    2  1  0  0  MN
MP2    3  2  5  5  MP
MN2    3  2  0  0  MN
MP3    4  3  5  5  MP
MN3    4  3  0  0  MN
VDD    5  0  DC  5
.MODEL  MP  PMOS  KP=.1M  VTO=-1  LAMBDA=.02
.MODEL  MN  NMOS  KP=.1M  VTO=1   LAMBDA=.02
.DC  VIN  0  5  .01
.PROBE
.END
```

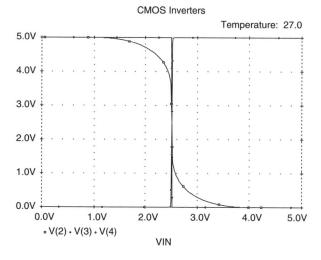

Temperature: 27.0

5.0V

4.0V

3.0V

2.0V

1.0V

0.0V

0.0V   1.0V   2.0V   3.0V   4.0V   5.0V

▫ V(2) ▫ V(3) ▫ V(4)

VIN

Figure 4.71: SPICE solution for Example 4.12.

following is obtained for the transition widths:

	V_{IL}	V_{IH}	Δv_{IN}
v_{OUT1}	2.092	2.908	0.816 V
v_{OUT2}	2.468	2.532	0.064 V
v_{OUT3}	2.480	2.520	0.040 V

4.6 LOGIC MEMORIES: THE BASIS OF MEGABYTES OF STORAGE

The static behavior of the bipolar and MOSFET logic circuits that have been discussed was specified by means of a transfer characteristic. An output of these circuits at a particular time was uniquely determined by the value of its input signals at that time. This is not the case for a memory circuit because its output depends not only on its present input signals but also on its earlier inputs. In Chapter 1 MOSFET circuits (Figure 1.27) were briefly discussed. Because the operation of these circuits depends on positive feedback, they are classified as regenerative circuits. The concept of regeneration, that is, regenerating an input signal by feeding back a portion of the output signal to the input, was utilized to enhance the sensitivity of the early radio receivers (Armstrong 1915, 1922). Although regenerative circuits are now seldom used for radio receivers or other analog systems, they are extensively used for logic circuits. The bistable flip-flop, which will be discussed in this section, is the modern descendant of this early concept, which dates back to nearly the beginning of electronics.

A MOSFET BISTABLE CIRCUIT

Consider the MOSFET circuit of Figure 4.72, which consists of two basic inverter circuits with pull-up resistors. The static transfer characteristic of a single inverter gate, v_{OUT1} versus v_{IN}, is shown in Figure 4.73(a). From the output of the first gate for the input of the second gate, the overall transfer characteristic of the logic buffer, v_{OUT2} versus v_{IN}, is obtained (Figure 4.73(b)). The characteristic is readily obtained with a point-by-point procedure in which changes in the regions of operation of the devices are taken into account. The buffer characteristic, it will be noted, has a much more abrupt output voltage transition than that of

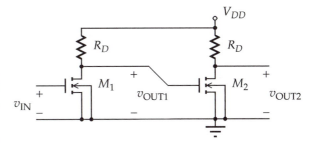

Figure 4.72: A MOSFET logic buffer (two inverters).

a logic inverter gate using a single MOSFET device. Its transfer characteristic is much closer to that which might be considered an "ideal" response.

The abrupt transition is the result of the amplification of the individual MOSFET logic inverter circuits. Consider the case for $v_{OUT2} = v_{OUT1} = v_{IN}$. A single logic inverter gate for this input has a transfer characteristic with a slope of m (a negative quantity with a magnitude that is greater than unity). For the two logic inverter gates that form a logic buffer gate, the slope of the overall characteristic is m^2, a positive quantity that is larger than the magnitude of m.

A line, $v_{OUT2} = v_{IN}$, is also indicated in Figure 4.73(b). At its intersections with the response of the gate, v_{OUT2}, the output voltage of the buffer, is equal to its input voltage. Hence, if the output of the logic buffer gate is used to provide its input, there are three equilibrium conditions: A, B, and C. A logic buffer circuit with its output connected to its input is shown in Figure 4.74. The circuit has been redrawn to emphasize its physical symmetry. Although there are three equilibrium voltages and currents for this circuit, only two of the equilibrium conditions, A and C, are stable. For either of these conditions, the circuit will return to its previous equilibrium condition if a small disturbance (for example, thermal or shot noise) should move it from its initial equilibrium condition. This is not the case for condition B, which corresponds to the abrupt transition of v_{OUT2}. A slight disturbance will propagate through the circuit, causing v_{OUT2} to move to the condition corresponding to either A or C. Although A and C are

Figure 4.73: Transfer characteristic of a MOSFET logic inverter and buffer.

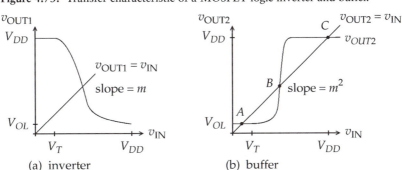

(a) inverter

(b) buffer

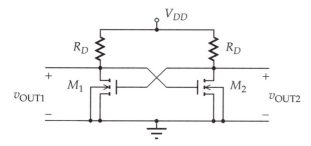

Figure 4.74: MOSFET bistable circuit.

stable equilibrium conditions, B is an *unstable* equilibrium condition; a physical circuit will never be found in an unstable equilibrium condition.

The circuit of Figure 4.74 has two stable equilibrium states: $v_{OUT1} = V_{DD}$ and $v_{OUT2} = V_{OL}$ or $v_{OUT1} = V_{OL}$ and $v_{OUT2} = V_{DD}$. These are indicated in Table 4.3. Because the circuit will remain indefinitely in a particular state as long as the supply voltage is maintained, this is a bistable circuit. It is a basic flip-flop memory circuit that is occasionally referred to as an Eccles–Jordan flip-flop after its inventers (Eccles and Jordan 1919).

A FLIP-FLOP MEMORY ELEMENT

To serve a useful memory function, it is necessary to be able to change the state of a flip-flop. This can be achieved using a second set of MOSFET devices that have set and reset input voltages v_S and v_R (Figure 4.75). The outputs of the set and reset devices are in parallel with the outputs of the MOSFET devices of the flip-flop. When v_S and v_R are less than the threshold voltage of the devices V_T, devices M_3 and M_4 are cut off, and they will have no effect on the operation of the flip-flop circuit. In accordance with the initial state of the flip-flop, high-level logic input voltages for v_S or v_R may be used to change the state. Consider the case for which M_1 is conducting, that is, $v_{OUT1} = V_{OL}$. A high-level logic input for v_S, $v_S = V_{DD}$ will have only a small effect on v_{OUT1}, reducing it slightly. The other output voltage, v_{OUT2}, will be unaffected. On the other hand, a high-level logic input voltage for v_R, $v_R = V_{DD}$ will cause M_4 to conduct. This will reduce v_{OUT2}, which, in turn, will result in a cut-off condition for M_1 if v_{OUT2} falls below V_T. As a consequence, M_2 will conduct ($v_{OUT2} \approx V_{OL}$). The result is that the flip-flop memory changes state – it will remain in the new state even after v_R is reduced to a low-level logic voltage ($v_R < V_T$).

For this memory, a high-level logic voltage for v_R produces a low-level logic voltage for v_{OUT2}. Owing to the physical symmetry of the circuit, a high-level logic voltage for v_S results in a low-level logic voltage for v_{OUT1}, which implies a high-level logic voltage for v_{OUT2}. If both inputs are high ($v_S = v_R = V_{DD}$), both outputs will be low. When the inputs are removed, the resultant state of the flip-flop memory will depend on which input was removed last. For the case in which both inputs are removed simultaneously, the resultant state of the memory will depend on both the electrical noise of circuit and possible asymmetries of the circuit. As a consequence, the resultant state tends to be

TABLE 4.3 MOSFET BISTABLE CIRCUIT

	v_{OUT1}	M_1	v_{OUT2}	M_2
State 1	V_{DD}	cutoff	V_{OL}	conducting
State 2	V_{OL}	conducting	V_{DD}	cutoff

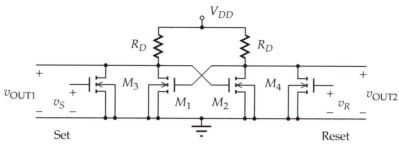

Figure 4.75: MOSFET flip-flop circuit with set and reset inputs.

unpredictable. To avoid this uncertain outcome, the condition of having both inputs simultaneously high is excluded, that is, the logic circuits producing v_S and v_R must be designed so that simultaneous high-level logic voltages of the input signals do not occur.

The circuit of Figure 4.75 is generally known as an **RS** flip-flop. If the logic variable Q is associated with v_{OUT2}, then, except when the memory is changing state, $\overline{Q}$ can be associated with v_{OUT1}. Concurrently, the logic variables S and R can be associated with v_S and v_R, respectively. This results in the logic truth table, Table 4.4.

The MOSFET device circuit of Figure 4.75 can be recognized as consisting of two logic **NOR** gates, M_1 and M_3, which, along with their common drain resistor, form one logic **NOR** gate; M_2 and M_4 form the other gate. The output of each **NOR** gate serves as an input of the other **NOR** gate (Figure 4.76(a)). As would be expected, the truth table (Table 4.4) is consistent with that of the **NOR** gate logic circuit. The same logic truth table is obtained regardless of the devices and circuit configurations used to implement the logic **NOR** gates. Therefore, MOSFET devices could have been used in place of the pull-up resistors (R_D) or, alternatively, CMOS, TTL, or any other type of logic gates could have been used. An alternative implementation of the **RS** flip-flop memory using **NAND** gates is indicated in Figure 4.76(b); complemented input logic levels are required for the set and reset inputs.

A MEMORY ARRAY

Memories consisting of flip-flops are extensively used in logic systems. A microprocessor, for example, requires numerous flip-flop–type storage registers. In addition, flip-flops are used for addressable memory arrays such as that of Figure 4.77, for which four memory cells of a much larger array are shown (simplified device symbols). Generally, a square array is utilized; a 16-bit memory consists of 4 rows and columns, a 16-kbit array (16,384 bits) consists of 128 rows and columns, and so forth. Although a device circuit is shown for only $Cell_{11}$, all memory cells have the same device circuit. Both outputs of the flip-flops are connected to

TABLE 4.4 RS FLIP-FLOP MEMORY

R	S	Q
0	0	$Q_{previous}$
1	0	0
0	1	1
1	1	Excluded

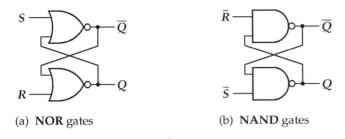

(a) **NOR** gates   (b) **NAND** gates

Figure 4.76: Logic gate implementation of an *RS* flip-flop memory.

column data lines ($\overline{D}_1$ and D_1 for *Cell*$_{11}$) through enhancement-mode MOSFET devices. The gate voltages of these devices are determined by the row address line to which they are connected.

When a memory is inactive, that is, it is neither being read or written to, all row address lines ($A_1, A_2, \ldots$) are at a potential less than the threshold voltage of the enhancement-mode devices. Hence, the devices connecting each memory cell to column data lines will be cut off. Therefore, each cell will remain in its present state, thus "storing" a single bit of data.

To read or write to a particular cell, the cell must first be addressed. A high-level row address line, for example A_1, will result in high gate voltages for the MOSFET devices connected to the address line. As a result, each memory cell of the first row will be connected to the data lines ($\overline{D}_1$ and D_1 for *Cell*$_{11}$, etc.). The column address determines which column will be read or written to – only a single row and column is simultaneously addressed.

The circuit of Figure 4.78 in which the address voltage is V_{DD} corresponds to a read operation. The resistors r represent extremely large MOSFET circuit

Figure 4.77: A memory array consisting of flip-flops using MOS-FET devices. A simplified symbol is used for the devices. Although while not shown, all substrates are connected to ground. The circuit of each cell is identical to that indicated for *Cell*$_{11}$.

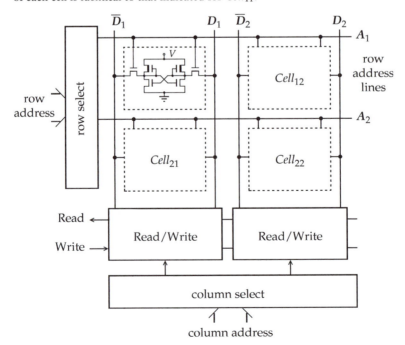

resistances. Consider the case for which the memory state is such that M_2 is conducting and M_1 is cut off: $v_{OUT2} \approx V_{OL}$ and $v_{OUT1} \approx V_{DD}$. The circuit for the $\overline{D}$ data bus is a conducting enhancement-mode device M_5 in series with the depletion mode device M_3. These devices, in turn, are in series with r and the supply voltage V_{DD}. For a very small series current (due to the large r), the drain-to-source voltage of M_3 will be very small, and that of M_5 will be approximately its threshold voltage V_{T5}. Hence, $v_{Cell1} \approx V_{DD} - V_{T5}$, a high-level logic voltage. For the circuit of

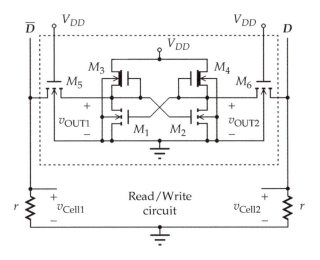

Figure 4.78: Reading a logic memory cell.

the other data line D, MOSFET device M_6 is also conducting. However, v_{OUT2} is very small ($\approx V_{OL}$), resulting in a very small voltage for v_{Cell2} ($\approx V_{OL}$). For the other memory state, the read voltages across the resistors are reversed. With a suitable MOSFET logic circuit controlled by the column address, the state of the selected memory cell can be transferred to the read output of the memory circuit. It should be noted that the voltages are such that the terminals of M_5 and M_6 that are connected to the flip-flop behave as drains, and the terminals connected to the data lines behave as sources.

A MOSFET circuit is used for writing to a selected memory cell (Figure 4.79). The set and reset devices of the read–write circuit perform a function similar to those of an **RS** flip-flop memory in which MOSFET devices M_5 and M_6 provide a series connection to the memory. If v_{OUT2} is high ($\approx V_{DD}$), a high-level logic voltage for v_R will result in the conduction of both the reset MOSFET device and M_6. This will result in a low-level logic voltage for v_{OUT2}, thus changing the state of the memory. If v_{OUT2} is initially a low-level logic voltage ($\approx V_{OL}$), the reset operation would not affect the state of the memory. A set input, however, will change the state of the memory.

The memory cell that has been discussed used enhancement-mode active MOSFET devices and depletion-mode loads. Alternatively, enhancement-mode load devices could have been used. Although the high-level logic voltages would be reduced, fewer fabrication steps would be required (no channel doping for the load devices). Memory arrays using CMOS devices

Figure 4.79: Writing to a logic memory cell.

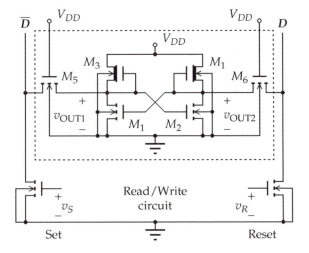

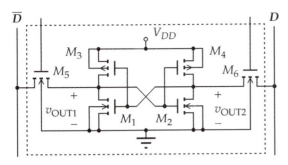

Figure 4.80: A CMOS memory cell. Transistors M_3 and M_4 are p-channel MOSFET devices.

are also common (Figure 4.80). For this memory, CMOS inverter gates are connected to form a flip-flop. Either n-channel or p-channel enhancement-mode devices may be used for connecting the flip-flop to the column data lines (n-channel devices are shown). An advantage of a CMOS memory is its very low power consumption; for static conditions the currents of the memory cells as well as those of the row and column circuits are essentially zero. Although bipolar junction transistors were used for early memory arrays, MOSFET circuits are generally used for memories now being produced.

THE DYNAMIC MEMORY ARRAY

The type of memory that has been discussed is generally known as a random-access memory (RAM). Although the term *random* is universally used, it is used in a sense different from its more conventional meaning, which is lacking a definite plan or order. A circuit external to the memory reads or writes data according to a very definite plan or order. It is from the perspective of a memory cell that the reading or writing appears to be random. The term *static* is also used, that is, static RAM or SRAM. The term static is also missapplied because the state of the memory, for most applications, is repeatedly changed (through write operations). The memory is used in dynamic systems in which its dynamic behavior is of particular importance. For these memories, static is used to describe a memory's ability to store, indefinitely, a bit of data as long as its supply voltage is maintained.

The memory cells that have been discussed require six transistors per cell. The fewer transistors required per cell, the more readily can a particular size array be fabricated. Alternatively, the fewer transistors required for each cell, the larger the array (and hence the number of bits that can be stored) that can be fabricated on a given size semiconductor chip using a particular technology. To reduce the transistor count of a cell, an alternative data storage scheme utilizing the charge storage of a capacitor has been developed (Dennard 1984; Rideout 1979; Sah 1988; Schroder 1987; Terman 1971). The voltage of a capacitor depends on its stored charge q according to the relation

$$v = q/C \tag{4.71}$$

To change the voltage, the charge must be changed, that is, a current is required:

$$\frac{dv}{dt} = \frac{1}{C}\frac{dq}{dt} = \frac{i}{C} \tag{4.72}$$

For an ideal (lossless) capacitor, the charge and hence voltage will remain unchanged for a zero current condition (an open circuit). Hence, a capacitor's charge, or lack of charge, can be associated with a bit of data. Unfortunately, the utility of this simple data storage scheme is limited by physically realizable capacitors – they are not lossless. As a result of an unavoidable dielectric leakage current, the charge of a capacitor diminishes with time – it "leaks" off. Hence, data may be stored only temporarily unless a provision is included to refresh the memory, that is, to restore the charge that leaks off.

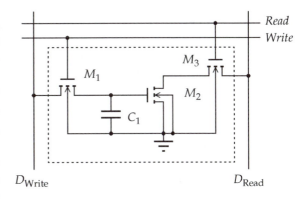

Figure 4.81: Three transistor dynamic memory cell.

Consider the three-transistor dynamic memory cell of Figure 4.81. To read the cell, its read address line is set to a high-level logic voltage (for example, V_{DD}). In addition, a load MOSFET device connected between D_{Read} and V_{DD} is activated (with a high-level gate voltage). This load device, as well as M_3, serves as a load for the MOSFET inverter M_2. The column voltage will depend on the voltage of C_1, that is, the gate-to-source voltage of M_2. For a small capacitor voltage (less than V_T), M_2 will be cut off and the voltage of the read column will be high. Conversely, a high capacitor voltage will result in the conduction of M_2, thus tending to reduce the column voltage.

To write to the cell, its address line is set to a high-level logic voltage (for example, V_{DD}). A high-level logic voltage for D_{Write} will result in a charging of C_1. The left-hand terminal of M_1 will function as a drain, and the right-hand terminal as a source (it may be treated as a source follower with a load of C_1). A low-level logic voltage (0 or V_{OL}) for D_{Write} will discharge C_1 if it is initially charged. The source and drain terminals of M_1 are reversed – the MOSFET current is in the opposite direction from that when D_{Write} is high.

As a result of charge leakage, a refresh operation is required. For this operation, the read address line is first set to a high-level logic voltage, and the state of the memory is determined by a read operation. The column write line is then set either high or low, corresponding to the memory state determined by the read operation. The write address line is then set to a high-level logic voltage and the capacitor, if it was at a high-level logic voltage, is recharged. It is necessary to "refresh" this cell while the capacitor's charge remains adequate to indicate its original state. Because there is a separate circuit for each column of cells, an entire row of cells can be refreshed simultaneously. To refresh the entire array, it is necessary to sequence through each row of the array. The cell structure of the preceding dynamic memory is simpler than that of a static memory (only three as compared with six transistors). The price "paid" is that more complex column and row circuits are required. Furthermore, the memory cannot be accessed during a refresh interval.

The circuit of a dynamic memory cell may be further simplified, that is, it can be reduced to a single transistor and capacitor. A one-transistor memory cell was mentioned in the introduction to the chapter (Figure 4.6). For this cell, a high-level logic address line voltage is used to read a cell. This operation, in effect, connects the capacitor of the cell to the column data line. Unfortunately, the data line, owing to its physical length (it connects to all cells of a particular column), has a much larger capacitance than that of the cell. Hence, the stored charge of C_1 will result in only a very small change in voltage of the data line. Not only must this small voltage change be detected, but the voltage of the cell's capacitor must be restored to its original value. Each read operation must be followed by a write operation to restore the original state of the memory. A write operation consists of applying a high-level or low-level logic voltage to the column line. Because an entire row is addressed for each read or write operation, all other cells of that row must be refreshed.

A considerable effort is being expended to produce ever larger memory arrays. This includes the development of static memories with small access times that have storage capacities in excess of 1 Mbit (Flannagan 1992). Very large dynamic memory arrays are also being developed (Itoh 1990). The complexity of modern memory arrays and other very large-scale integrated circuits is considerably beyond that which could have been imagined when early integrated circuits were fabricated. Although further improvements are expected, physical limitations will eventually set constraints on the ultimate level of complexity (Keys 1987, 1992).

EXAMPLE 4.13

Consider the logic buffer circuit of Figure 4.82, which will be used to construct a flip-flop. A SPICE simulation is to be used to verify that the equilibrium condition corresponding to point B of Figure 4.73 is unstable.

a. Determine the static transfer characteristic of the circuit v_{OUT2} versus v_{IN}. Ignore the "dashed" feedback circuit of Figure 4.82.

b. The behavior of the circuit with the "dashed" feedback circuit is to determined. A voltage source v_N is included in the feedback circuit to simulate

Figure 4.82: MOSFET circuit of Example 4.13. The node numbers of the circuit correspond to those of the SPICE program.

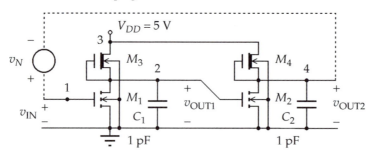

M_1, M_2		M_3, M_4	
$k = 50 \ \mu A/V^2$	$V_T = 1.0$ V	$k = 50 \ \mu A/V^2$	$V_{T0} = -1.0$ V
		$\gamma = 0.37 \ V^{1/2}$	$2\phi_F = 0.6$ V

```
MOSFET flip-flop
VIN  1  0
M1  2  1  0  0   MOSACTIVE
M3  3  2  2  0   MOSLOAD
C1  2  0  1P
M2  4  2  0  0   MOSACTIVE
M4  3  4  4  0   MOSLOAD
C2  4  0  1P
VDD  3  0  DC  5
.MODEL  MOSACTIVE  NMOS  KP=50U  VTO=1
.MODEL  MOSLOAD  NMOS  KP=50U  VTO=-1  GAMMA=.37  PHI=.6
.DC  VIN  0  5  .025
.PROBE
.END
```

Figure 4.83: Circuit file for static transfer characteristic of Example 4.13.

the effect of a noise pulse. Assume v_N is a single pulse with an amplitude of 0.1 V and a duration of 10 ns. Use an initial condition corresponding to the unstable equilibrium condition to determine the time dependence of v_{OUT2} for noise pulses with both positive and negative polarities.

SOLUTION

a. The circuit file of Figure 4.83 results in the static transfer characteristic of Figure 4.84, which yields the following equilibrium voltages ($v_{OUT2} = v_{IN}$):

$$A: 0.120 \text{ V} \qquad B: 1.722 \text{ V} \qquad C: 5.0 \text{ V} \qquad (4.73)$$

b. Although an unstable equilibrium point will not occur for a physical circuit, it can occur for a SPICE transient simulation; frequently the initial condition found for a transient simulation will be the unstable equilibrium point. Furthermore, the SPICE transient solution will tend to be stable, that

Figure 4.84: SPICE static solution for Example 4.13.

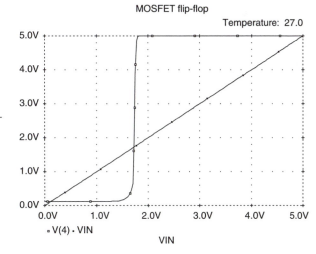

```
MOSFET flip-flop
VN   1   4   PWL(0  0  10N  .1  20N  0)
M1   2   1   0   0   MOSACTIVE
M3   3   2   2   0   MOSLOAD
C1   2   0   1P
M2   4   2   0   0   MOSACTIVE
M4   3   4   4   0   MOSLOAD
C2   4   0   1P
VNA  10  40  PWL(0  0  10N  -.1  20N  0)
M1A  20  10  0   0   MOSACTIVE
M3A  3   20  20  0   MOSLOAD
C1A  20  0   1P
M2A  40  20  0   0   MOSACTIVE
M4A  3   40  40  0   MOSLOAD
C2A  40  0   1P
VDD  3   0   DC  5
.NODESET  V(4) = 1.722
.NODESET  V(40) = 1.722
.MODEL  MOSACTIVE  NMOS  KP=50U  VTO=1
.MODEL  MOSLOAD  NMOS  KP=50U  VTO=-1  GAMMA=.37  PHI=.6
.TRAN  1N  1000N
.PROBE
.END
```

Figure 4.85: Circuit file for noise pulse simulation of Example 4.13.

is, the equilibrium will persist unless intentionaly perturbed. To ensure a desired equilibrium condition, a .NODESET command can be used. The .NODESET voltage of the circuit file of Figure 4.85 achieves this. (Without the .NODESET command, the initial condition found by the author's program corresponds to the unstable condition. There is, however, no

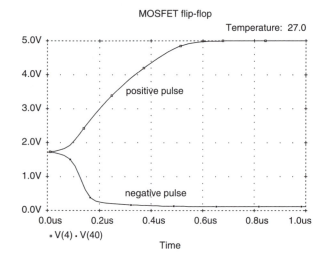

Figure 4.86: SPICE dynamic solution for Example 4.13.

certainty that all versions of SPICE will produce this initial condition.) To obtain simultaneous solutions for both pulse polarities, a second circuit file with a negative pulse voltage is included. The transient results of Figure 4.86 are obtained. A positive pulse voltage causes the output of the flip-flop circuit to go to its high-level voltage (5.0 V), whereas a negative pulse results in a low-level voltage (0.12 V). Approximately 0.6 μs is required to reach the stable equilibrium voltages for the circuit.

REFERENCES

Armstrong, E. H. (1915). Some recent developments in the audion receiver. *Proceedings of the Institute of Radio Engineers*, **3**, 3, 215–47.

Armstrong, E. H. (1922). Some recent developments of regenerative circuits. *Proceedings of the Institute of Radio Engineers*, **10**, 4, 244–60.

Davies, R. D. (1983). The case for CMOS. *IEEE Spectrum*, **20**, 10, 26–32.

Dennard, R. H. (1984). Evolution of the MOSFET dynamic RAM – A personal view. *IEEE Transactions on Electron Devices*, **ED-31**, 11, 1549–55.

Eccles, W. H. and Jordan, F. W. (1919). A trigger relay using three-electrode thermionic vacuum tubes. *Radio Review*, **1**, 143–6.

Elmasry, M. I. (1992). Digital MOS integrated circuits: A tutorial. In *Digital MOS Integrated Circuits II, with Applications to Processors and Memory Design*, M.I. Elmasry, ed. New York: IEEE Press, 3–33.

Flannagan, S. (1992). Future technology trends for static RAMs. In *Digital MOS Integrated Circuits II with Applications to Processors and Memory Design*. M. I. Elmasry, ed. New York: IEEE Press, 319–22.

Haznedar, H. (1991). *Digital Microelectronics*. Redwood City, CA: The Benjamin/Cummings Publishing Co.

Hittinger, W. C. (1973). Metal-oxide-semiconductor technology. *Scientific American*, **229**, 2, 48–57.

Hodges, D. A. and Jackson, H. G. (1988). *Analysis and Design of Digital Integrated Circuits* (2d ed.). New York: McGraw-Hill.

Hofstein, S. R. and Heiman, F. P. (1963). The silicon insulated-gate field-effect transistor. *Proceedings of the Institute of Electrical and Electronic Engineers*, **51**, 9, 1190–1202.

Itoh, K. (1990). Trends in megabit DRAM circuit design. *IEEE Journal of Solid-State Circuits*, **25**, 3, 778–89.

Kahng, D. (1976). A historical perspective of the development of MOS transistors and related devices. *IEEE Transactions on Electronic Devices*, **ED-23**, 7, 655–57.

Keyes, R. W. (1987). *The Physics of VLSI Systems*. Wokingham, England: Addison–Wesley Publishing Company.

Keyes, R. W. (1992). The future of solid-state electronics. *Physics Today*, **45**, 8, 42–8.

Mackintosh, A. R. (1988). Dr. Atanasoff's computer. *Scientific American*, **259**, 2, 90–6.

Milnes, A. G. (1980). *Semiconductor Devices and Integrated Electronics*. New York: Van Nostrand Reinhold Co.

Mukherjee, A. (1986). *Introduction to nMOS and CMOS VLSI Systems Design*. Englewood Cliffs, NJ: Prentice–Hall.

Pierret, R. F. (1990). *Field Effect Devices* (2d ed.). Reading, MA: Addison–Wesley.

Pulfrey, D. L. and Tarr, N. C. (1989). *Introduction to Microelectronic Devices*. Englewood Cliffs, NJ: Prentice–Hall.

Rideout, V. L. (1979). One-device cells for dynamic random-access memories: A tutorial. *IEEE Transactions on Electron Devices*, **ED-26**, 6, 839–52.

Sah, C.-T. (1988). Evolution of the MOS transistor – From conception to VLSI. *Proceedings of the Institute of Electrical and Electronic Engineers*, **76**, 10, 1280–1326.

Schroder, D. K. (1987). *Advanced MOS Devices*. Reading, MA: Addison–Wesley.

Shockley, W. (1952). Transistor electronics: Imperfections, unipolar and analog transistors. *Proceedings of the Institute of Radio Engineers*, **40**, 11, 1289–1313.

Shockley, W. (1976). The path to the conception of the junction transistor. *IEEE Transactions on Electron Devices*, **ED-23**, 7, 597–620.

Shockley, W. and Pearson, G. L. (1948). Modulation of conductance of thin films of semiconductors by surface charges. *Physical Review*, **74**, 2, 232–3.

Shoji, M. (1988). *CMOS Digital Circuit Technology*. Englewood Cliffs, NJ: Prentice–Hall.

Streetman, B. G. (1990). *Solid State Electronic Devices* (3d ed.). Englewood Cliffs, NJ: Prentice–Hall.

Sze, S. M. (1981). *Physics of Semiconductor Devices* (2d ed.). New York: John Wiley & Sons.

Taur, Yuan and Ning, Tak H. (1998). *Fundamentals of Modern VLSI Devices*. Cambridge: Cambridge University Press.

Terman, L. M. (1971). MOSFET memory circuits. *Proceedings of the Institute of Electrical and Electronic Engineers*, **59**, 7, 1044–58.

Tsividis, Y. (1987). *Operation and Modeling of the MOS Transistor*. New York: McGraw–Hill.

Uyemura, J. P. (1988). *Fundamentals of MOS Digital Integrated Circuits*. Reading, MA: Addison–Wesley.

PROBLEMS

4.1 The source and the substrate of an *n*-channel silicon MOSFET device are connected.

$$W = 50 \ \mu m, \qquad L = 25 \ \mu m, \qquad V_T = 2.0 \text{ V}$$
$$t_{ox} = 0.1 \ \mu m, \qquad \epsilon_{ox} = 3.9\epsilon_0, \qquad \mu_n = 800 \text{ cm}^2/\text{V} \cdot \text{s}$$

A gate-to-source voltage v_{GS} of 4 V is applied to the device.

a) What is the transconductance k of the device?
b) What is the value of v_{DS} that results in a pinch-off condition?
c) What is i_D for the pinch-off condition?
d) What are the values of i_D for v_{DS} 1 V less and 1 V greater than the potential corresponding to pinch-off?

4.2 The threshold voltage of a MOSFET device may be modified by a doping of the oxide–substrate surface with donor atoms. Repeat Problem 4.1 for $V_T = 1.0$ V.

4.3 Repeat Problem 4.1 for $V_T = 1.5$ V.

4.4 An *n*-channel MOSFET device with a transconductance of 0.1 mA/V^2 is

required for a particular application.

$$t_{ox} = 0.2 \ \mu m, \qquad \epsilon_{ox} = 3.9\epsilon_0, \qquad \mu_n = 800 \ cm^2/V \cdot s$$

a) What is the width required for a device with a length of 10 μm?
b) What is the gate-to-substrate capacitance of the device?

4.5 In a particular circuit, a drain current of 1.0 mA is required for an n-channel MOSFET device with a drain-to-source voltage of 3 V and a gate-to-source voltage of 5 V (substrate connected to source). Determine the values of k required for $V_T = 1, 2,$ and 3 V.

4.6 The gate and drain as well as the source and substrate of a particular n-channel MOSFET device are connected ($V_T = 1.0$ V, $k = 10 \ \mu A/V^2$).

a) What is i_D for $v_{DS} = 2, 3,$ and 4 V?
b) As a result of a variation in manufacturing processes (larger t_{ox}), a device has a threshold voltage of 1.5 V. What are the currents for this device with $v_{DS} = 2, 3,$ and 4 V?

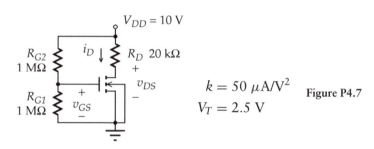

$$k = 50 \ \mu A/V^2$$
$$V_T = 2.5 \ V$$

Figure P4.7

4.7 An n-channel MOSFET device is used in the circuit of Figure P4.7.

a) What is the gate-to-source voltage v_{GS} of the MOSFET?
b) What is i_D?
c) What is v_{DS}?

4.8 Repeat Problem 4.7 for $V_T = 1.5$ V.

4.9 Repeat Problem 4.7 for $R_{G1} = R_{G2} = 100$ kΩ.

4.10 Repeat Problem 4.7 for $R_D = 100$ kΩ.

4.11 A 20 kΩ resistor R_L is connected between the drain of the MOSFET and ground in the circuit of Figure P4.7. Determine $v_{GS}, i_D,$ and v_{DS} for this circuit.

4.12 In the circuit of Figure P4.7, the upper end of R_{G2} is connected to the drain of the MOSFET device rather than to V_{DD}. Determine $v_{GS}, i_D,$ and v_{DS} for this circuit.

4.13 Repeat Problem 4.12 for $R_D = 10$ kΩ.

4.14 Repeat Problem 4.12 for $R_D = 50$ kΩ.

4.15 An n-channel MOSFET device is used in the circuit of Figure P4.15. Determine $v_{GS}, i_D,$ and v_{DS} for this circuit.

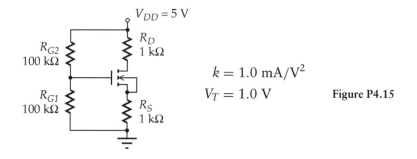

$$k = 1.0 \text{ mA/V}^2$$
$$V_T = 1.0 \text{ V}$$

Figure P4.15

4.16 Repeat Problem 4.15 for $R_D = 5 \text{ k}\Omega$.

4.17 Repeat Problem 4.15 for $R_S = 2 \text{ k}\Omega$.

4.18 Repeat Problem 4.15 for $R_D = 5 \text{ k}\Omega$ and $R_S = 2 \text{ k}\Omega$.

4.19 An *n*-channel MOSFET device is used in the circuit of Figure P4.19.

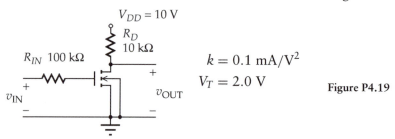

$$k = 0.1 \text{ mA/V}^2$$
$$V_T = 2.0 \text{ V}$$

Figure P4.19

a) Determine the range of v_{IN} for which the MOSFET is in its cutoff region of operation. What is v_{OUT} for this condition?

b) What is the value of v_{IN} for which $v_{\text{OUT}} = v_{\text{IN}}$?

c) What is v_{IN} for $v_{\text{OUT}} = 5 \text{ V}$?

d) What is v_{OUT} for $v_{\text{IN}} = 10 \text{ V}$?

4.20 Repeat Problem 4.19 for a load resistor $R_L = 20 \text{ k}\Omega$ connected in parallel with v_{OUT}.

4.21 Consider the MOSFET inverter circuit of Problem 4.19. Determine the sets of voltages v_{IN} and v_{OUT} for which $\frac{dv_{\text{OUT}}}{dv_{\text{IN}}} = -1$.

4.22 An *n*-channel MOSFET is used in the circuit of Figure P4.22.

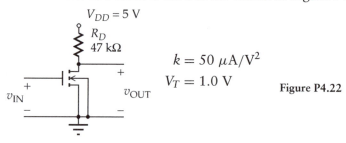

$$k = 50 \text{ }\mu\text{A/V}^2$$
$$V_T = 1.0 \text{ V}$$

Figure P4.22

a) What is the value of v_{IN} for which the device is at pinch-off (the transition from the saturation to triode region)?

b) Determine v_{IN} for which $v_{OUT} = v_{IN}$.

c) What is v_{OUT} for $v_{IN} = V_{DD}$?

d) What is v_{OUT} for a value of v_{IN} equal to v_{OUT} determined in part **(c)**?

4.23 Determine for Problem 4.22 the sets of voltages v_{IN} and v_{OUT} for which $\frac{dv_{OUT}}{dv_{IN}} = -1$.

4.24 Repeat Problem 4.22 for $R_D = 27\ \text{k}\Omega$.

4.25 Repeat Problem 4.23 for $R_D = 27\ \text{k}\Omega$.

4.26 A drain resistor of $R_D = 10\ \text{k}\Omega$ is used for the MOSFET circuit of Figure P4.22. What is v_{OUT} for $v_{IN} = V_{DD} = 5$ V? Suppose that this voltage is used as the input of a second gate. What is v_{OUT} of the second gate?

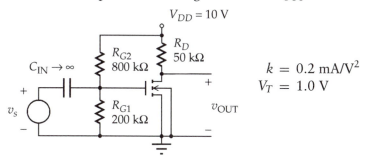

Figure P4.27

4.27 A MOSFET device is used in the amplifier circuit of Figure P4.27.

a) Determine the quiescent quantities I_D and V_{DS}.

b) What is the mutual conductance g_m of the MOSFET?

c) Draw the small-signal equivalent circuit of the amplifier and determine v_{out}/v_s.

4.28 Repeat Problem 4.27 for a load resistor $R_L = 100\ \text{k}\Omega$ connected in series with a capacitor to the drain of the MOSFET device and ground. The voltage v_{OUT} is that across R_L and the capacitance may be treated as being infinite.

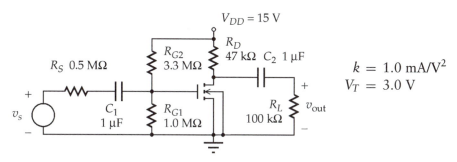

Figure P4.29

4.29 An n-channel MOSFET device is used in the amplifier circuit of Figure P4.29.

a) Determine the quiescent drain current I_D and the quiescent device voltages V_{GS} and V_{DS}.

b) What is the mutual conductance g_m of the device?

c) What is the small-signal voltage gain v_{out}/v_s of the amplifier?

d) What is v_{out}/v_s for $R_L \to \infty$?

4.30   A sinusoidal signal with a frequency of 500 Hz is used for the input signal v_s of Problem 4.29. What is the reactance of the capacitors for this frequency? Would the solution obtained for v_{out}/v_s, if infinite values of capacitance, were assumed, be expected to be valid for this sinusoidal signal?

4.31   Consider the MOSFET circuit of Figure P4.29.

a) At what frequency is the magnitude of the reactance of C_1 equal to R_S?

b) At what frequency is the magnitude of the reactance of C_2 equal to R_L?

c) On the basis of the results of the previous parts, what is the lowest frequency for which the amplifier would be expected to perform reasonably well?

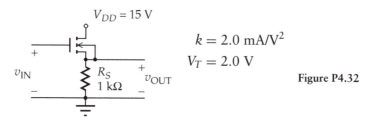

$k = 2.0 \ \text{mA/V}^2$

$V_T = 2.0 \ \text{V}$

Figure P4.32

4.32   An n-channel MOSFET device is used in the source-follower circuit of Figure P4.32.

a) What is v_{OUT} for $v_{IN} < V_T$?

b) What is v_{OUT} for $v_{IN} = 5 \ \text{V}$?

c) What is v_{OUT} for $v_{IN} = 10 \ \text{V}$?

4.33   Determine v_{IN} for $v_{OUT} = 5 \ \text{V}$ of Problem 4.32. What is $\frac{dv_{OUT}}{dv_{IN}}$ for this condition?

4.34   A biasing voltage $V_{GG} = 5 \ \text{V}$ and a signal source v_s are used for the input voltage of the MOSFET circuit of Figure P4.32.

a) Determine the quiescent drain current of the MOSFET.

b) What is the mutual conductance of the MOSFET?

c) What is the small-signal voltage gain of the circuit?

4.35   Repeat Problem 4.34 for $V_{GG} = 10 \ \text{V}$.

4.36   Repeat Problem 4.34 for $V_{GG} = 15 \ \text{V}$.

4.37   Repeat Problem 4.34 for $V_{GG} = 5 \ \text{V}$ and $R_S = 2 \ \text{k}\Omega$.

4.38   An RC circuit is excited with an input step function voltage. Assume $v_{OUT} = 0$ for $t < 0$.

a) Obtain an analytic expression for $v_{OUT}(t)$ for $t > 0$.

b) Determine $i(t)$.

c) What is the instantaneous power dissipated by R?

d) Show that the energy dissipated by R from $t = 0$ to $t \to \infty$ is equal to $\frac{1}{2}CV_0^2$.

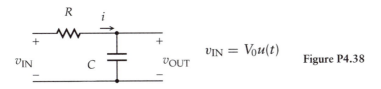

$v_{IN} = V_0 u(t)$

Figure P4.38

4.39 Repeat Problem 4.38 for $v_{IN} = V_0(1 - u(t))$. Assume $v_{OUT} = V_0$ for $t < 0$.

4.40 Assume that the circuit of Example 4.7 has the following parameters:

$$R_{on} = 100\,\Omega, \qquad R = 2\,k\Omega, \qquad V_{DD} = 10\,V, \qquad C_L = 100\,pF$$

Repeat Example 4.7 for this circuit.

4.41 Repeat Problem 4.40 for a load resistor of 5 kΩ connected in parallel with C_L.

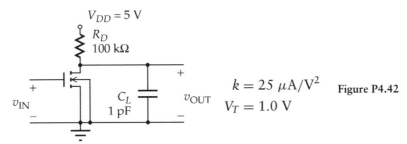

$k = 25\,\mu A/V^2$ **Figure P4.42**

$V_T = 1.0\,V$

4.42 An n-channel MOSFET device is used for the logic inverter of Figure P4.42.

a) What are the static low- and high-level output voltages V_{OL} and V_{OH}, respectively?

b) Suppose that v_{IN} has been equal to V_{OH} for a very long time and that it is suddenly switched to V_{OL} at $t = 0$. Determine the time required for v_{OUT} to increase to its midvalue of $(V_{OL} + V_{OH})/2$.

c) Suppose that v_{IN} has been equal to V_{OL} for a very long time and is suddenly switched to V_{OH} at $t = 0$. This causes the MOSFET device to be turned on rapidly. Estimate the time required for v_{OUT} to fall to its midvalue.

4.43 Repeat Problem 4.42 for $V_T = 1.5$ V.

4.44 Repeat Problem 4.42 for $k = 50\,\mu A/V^2$.

4.45 Repeat Problem 4.42 for $V_T = 1.5$ V and $k = 50\,\mu A/V^2$.

4.46 Suppose a MOSFET device with $V_T = 0$ V is used in the circuit of Figure P4.42. Determine V_{OL} and V_{OH} for this circuit. Note: A trial-and-error approach will be necessary because the device is not cut off for $v_{IN} = V_{OL}$.

4.47   A MOSFET logic inverter is used to drive an *NPN* bipolar junction transistor logic inverter (Figure P4.47).

a) What are v_{OUT1} and v_{OUT2} for $v_{IN} = 0$ V?
b) What are v_{OUT1} and v_{OUT2} for $v_{IN} = 2.5$ V?
c) What are v_{OUT1} and v_{OUT2} for $v_{IN} = 5.0$ V?

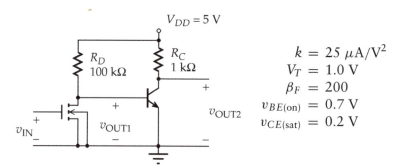

Figure P4.47

4.48   Repeat Problem 4.47 for $\beta_F = 100$.

4.49   Repeat Problem 4.47 for $k = 50\ \mu A/V^2$.

4.50   Repeat Problem 4.47 for a silicon junction diode in series with the base of the BJT device ($v_{D(on)} = 0.7$ V). The diode is oriented such that its forward-biased current is into the base of the BJT device.

4.51   Consider the circuit of Figure P4.47. Determine $\frac{dv_{OUT2}}{dv_{IN}}$ for a value of v_{IN} that results in $v_{OUT2} = 2.5$ V. Show that the derivative is constant for $0.2 < v_{OUT2} < 5.0$ V.

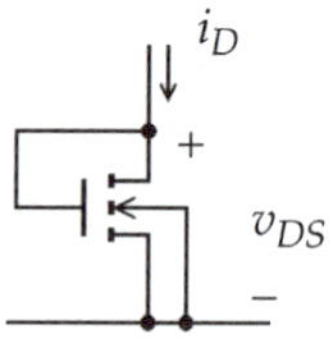

$$i_D$$
$$k = 10\ \mu A/V^2$$
$$V_T = 1.0\ V$$

Figure P4.52

4.52   The source and drain of a MOSFET device are connected together (Figure P4.52).

a) What is i_D for $v_{DS} = 1,\ 2,\ 3,\ 4,$ and 5 V?
b) What is v_{DS}/i_D, a resistance, for each of the voltages of part (a)?

4.53   Repeat Problem 4.52 for $V_T = 2.0$ V.

4.54   Suppose that a 1.0-V battery is inserted between the drain and the gate of the device of Problem 4.52. The polarity of the battery is such that $v_{GS} = v_{DS} + 1.0$ V. Repeat Problem 4.52 for this circuit.

4.55   The device of Problem 4.52 is used as the load of a MOSFET logic inverter (Figure P4.55). Use the results of Problem 4.52 and determine v_{IN} for $v_{OUT} = 1,\ 2,\ 3,$ and 4 V.

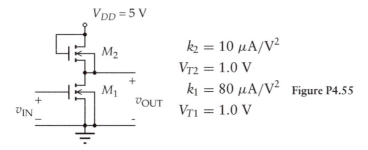

$$k_2 = 10\ \mu\text{A/V}^2$$
$$V_{T2} = 1.0\ \text{V}$$
$$k_1 = 80\ \mu\text{A/V}^2 \quad \textbf{Figure P4.55}$$
$$V_{T1} = 1.0\ \text{V}$$

4.56 Solve Problem 4.55 using analytic expressions for the dependence of v_{OUT} on v_{IN}. What is v_{IN} for $v_{\text{OUT}} = v_{\text{IN}}$? What is $\frac{dv_{\text{OUT}}}{dv_{\text{IN}}}$ for the region over which v_{OUT} has a linear dependence on v_{IN}?

4.57 Consider the MOSFET circuit of Figure P4.55.

a) What is v_{OUT} for $v_{\text{IN}} < 1.0$ V?
b) What is v_{OUT} for v_{IN} equal to the value of v_{OUT} determined in part (a)?

4.58 For the MOSFET circuit of Figure P4.55, a value of $v_{\text{OUT}} = 0.25$ V is desired for $v_{\text{IN}} = 4$ V. What is the minimum value of k_1 (M_1) that will achieve this voltage? Assume all other parameters are unchanged. What is the power supplied by V_{DD} for $v_{\text{IN}} = 4$ V?

4.59 For the MOSFET circuit of Figure P4.55, a value of $v_{\text{OUT}} = 0.25$ V is desired for $v_{\text{IN}} = 4$ V. What is the maximum value of k_2 (M_2) that will achieve this? Assume all other parameters are unchanged. What is the power supplied by V_{DD} for $v_{\text{IN}} = 4$ V?

4.60 Repeat Problem 4.57 for $V_{T2} = 0$ V. Assume all other parameters are unchanged.

4.61 Repeat Problem 4.57 for a load resistor of 1.0 MΩ connected in parallel with v_{OUT}.

4.62 In the circuit of Figure P4.55 the substrate of M_2 is connected to ground. Assume $\gamma = 0.37$ V$^{1/2}$ and $2\phi_F = 0.6$ V.

a) Determine v_{OUT} for $v_{\text{IN}} < 1.0$ V.
b) What is v_{OUT} for the input voltage determined in part (a)?

4.63 What is v_{IN} for $v_{\text{OUT}} = 2.0$ V of Problem 4.62?

4.64 Repeat Example 4.9 for $C_L = 10$ pF and all other parameters unchanged.

4.65 Repeat Example 4.9 for $k_1 = 50$ μA/V^2 and all other parameters unchanged.

4.66 A depletion-type MOSFET is to be used as a two-terminal nonlinear resistor (Figure P4.66).

a) Determine a set of expressions for the dependence of i_D on v_{DS} ($v_{DS} > 0$).
b) What is i_D for $v_{DS} = 1, 2, 3, 4$ and 5 V?

c) What is v_{DS}/i_D for each of the voltages of part (b)?

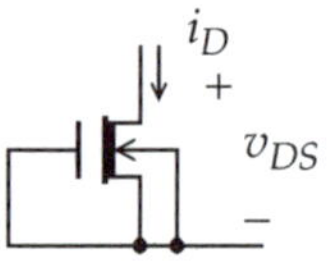

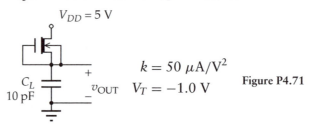

$$k = 10 \ \mu A/V^2$$
$$V_T = -1.0 \ V$$

Figure P4.66

4.67 Repeat Problem 4.66 for $V_T = -2.0$ V.

4.68 Consider the MOSFET circuit of Figure P4.66. What is v_{DS}/i_D for $v_{DS} \rightarrow 0$?

4.69 A depletion-type MOSFET ($v_{GS} = 0$) is to be used to replace the "conventional" resistor R_D of the circuit of Figure P4.42. Its gate, source, and substrate are connected. Assume the depletion-type MOSFET has a threshold voltage V_T of -1.0 V and it is desired that v_{DS}/i_D be equal to R_D (100 kΩ) for $v_{DS} = 5.0$ V.

a) Determine the required value of k for the depletion-type device.
b) What are V_{OL} and V_{OH} of the circuit?
c) What is v_{IN} for $v_{OUT} = v_{IN}$?
d) What is v_{IN} for $v_{OUT} = 4.0$ V?
e) What is v_{IN} for $v_{OUT} = 1.0$ V?

4.70 Repeat Problem 4.69 for $V_T = -2.0$ V.

$$V_{DD} = 5 \ V$$

$$C_L \quad 10 \ pF$$
$$v_{OUT}$$

$$k = 50 \ \mu A/V^2$$
$$V_T = -1.0 \ V$$

Figure P4.71

4.71 Consider the depletion-type MOSFET device of Figure P4.71 that is used to charge a capacitor. This is the effective circuit of a MOSFET inverter gate for a condition in which the active device is cut off, that is, for a low-level input voltage. Assume $v_{OUT} = 0$ at $t = 0$. What is the range of v_{OUT} for which the MOSFET device is in its saturation region? Determine the time dependence of v_{OUT} for this region. What is the time required for v_{OUT} to rise to a value that results in the MOSFET's entering its triode region of operation?

4.72 Determine, for Problem 4.71, the time required for v_{OUT} to increase from 0 to 4.5 V.

4.73 Repeat Problem 4.71 for a threshold voltage $V_T = -2.0$ V.

4.74 Consider Problem 4.71. Show that the time required for v_{OUT} to rise to a value that results in the device's entering its triode region is linearly dependent on the parameter C_L/k.

4.75 Consider the logic **NOR** gate of Figure P4.75. Determine the voltage "truth table" that gives v_{OUT} for high and low values of v_A and v_B. Assume that $V_{OH} = v_{OUT}$ for $v_A = v_B = V_{OL}$ and that $V_{OL} = v_{OUT}$ for $v_A = V_{OH}$ and $v_B = V_{OL}$.

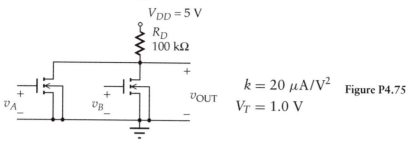

$k = 20\ \mu A/V^2$ **Figure P4.75**

$V_T = 1.0\ V$

4.76 Suppose R_D of Figure P4.75 is replaced with an enhancement-type MOSFET load device (gate connected to drain and source connected to substrate). Assume a load device with $V_T = 1.0$ V and a transconductance parameter that yields a value of v_{DS}/i_D equal to R_D for $v_{DS} = 5$ V. Repeat Problem 4.75 for this device.

4.77 Suppose R_D of Figure P4.75 is replaced with a depletion-type MOSFET load device (gate, source, and substrate connected). Assume a load device with $V_T = -1.0$ V and a transconductance parameter that yields a value of v_{DS}/i_D equal to R_D for $v_{DS} = 5$ V. Repeat Problem 4.75 for this device.

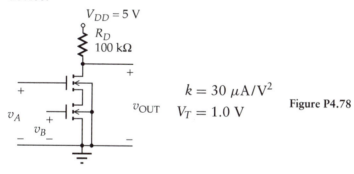

$k = 30\ \mu A/V^2$ **Figure P4.78**

$V_T = 1.0\ V$

4.78 Consider the logic **NAND** gate of Figure P4.78. Determine the voltage "truth table" that gives v_{OUT} for high and low values of v_A and v_B. Assume that $V_{OH} = v_{OUT}$ for $v_A = v_B = V_{OL}$ and that $V_{OL} = v_{OUT}$ for $v_A = v_B = V_{OH}$. The effect of a substrate bias may be ignored.

4.79 Suppose R_D of Figure P4.78 is replaced with an enhancement-type MOSFET load device (gate connected to drain and source to substrate). Assume a load device with $V_T = 1.0$ V and a transconductance parameter that yields a value of v_{DS}/i_D equal to R_D for $v_{DS} = 5$ V. Repeat Problem 4.78 for this device.

4.80 Suppose R_D of Figure P4.78 is replaced with a depletion-type MOSFET load device (gate, source, and substrate connected together). Assume a load device with $V_T = -1.0$ V and a transconductance parameter that yields a value of v_{DS}/i_D equal to R_D for $v_{DS} = 5$ V. Repeat Problem 4.78 for this device.

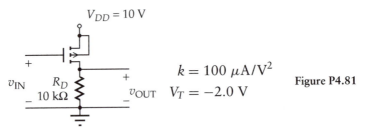

$V_{DD} = 10$ V

v_{IN} R_D $10\ k\Omega$ v_{OUT}

$k = 100\ \mu A/V^2$

$V_T = -2.0$ V

Figure P4.81

4.81 A p-channel MOSFET device is used in the circuit of Figure P4.81.

a) What is v_{OUT} for $v_{IN} = 0$?
b) What is v_{OUT} for v_{IN} equal to the value of v_{OUT} determined in part (a)?
c) What are V_{OL} and V_{OH}?

4.82 Consider the circuit of Figure P4.81.

a) What are the ranges of v_{OUT} for which the MOSFET is cut off, is in its triode region, and is in its saturation region?
b) What is v_{IN} for $v_{OUT} = v_{IN}$?
c) What is v_{IN} for $v_{OUT} = 2.0$ V?
d) What is v_{IN} for $v_{OUT} = 6.0$ V?

4.83 Suppose that a load capacitor of 50 pF is connected across v_{OUT} in the circuit of Figure P4.81.

a) Assume the input voltage has been equal to zero for a very long time and is suddenly switched to V_{DD} at $t = 0$. What is the time required for v_{OUT} to reach 5.0 V?
b) Assume the opposite for v_{IN}, that it has been equal to V_{DD} for a very long time and that is suddenly switched to zero. Determine the time required for v_{OUT} to reach 5.0 V for this condition.

4.84 Consider the CMOS logic inverter gate of Figure 4.60 ($k = 100\ \mu A/V^2$, $V_T = 1.0$ V, and $V_{DD} = 5.0$ V).

a) What is the range of v_{IN} for which one or the other of the devices is cut off? What are the corresponding values of v_{OUT}?
b) What is v_{IN} for both devices being saturated? What is the range of v_{OUT} for saturation? What is the supply current for this condition?
c) Determine v_{IN} for $v_{OUT} = 1.0$ V.

4.85 Repeat Problem 4.84 for $V_{DD} = 10$ V.

4.86 Repeat Problem 4.84 for $V_{DD} = 15$ V.

4.87 A 10-pF capacitor is connected across the output of the CMOS logic inverter gate of Figure 4.60 ($k = 100\ \mu A/V^2$, $V_T = 1.0$ V, and $V_{DD} = 5.0$ V). Assume the input voltage has been equal to zero for a very long time and it is suddenly switched to V_{DD}. Estimate the time required for v_{OUT} to change to $V_{DD}/2$.

4.88 Repeat Problem 4.87 for $V_{DD} = 10$ V.

4.89 Repeat Problem 4.87 for $V_{DD} = 15$ V.

COMPUTER SIMULATIONS

C4.1 Use a SPICE simulation to verify the results obtained for the MOSFET inverter of Example 4.3. In addition to obtaining the values needed to answer the questions of the example, determine the transfer characteristic of the circuit. What are the values of v_{IN} for which $\frac{dv_{OUT}}{dv_{IN}} = -1$?

C4.2 Use a SPICE simulation to determine the transfer characteristic of the MOSFET circuit of Example 4.4. Verify that the analytic results of the example are valid.

C4.3 Use a set of SPICE simulations to ascertain the sensitivity of the results of Example 4.4 on the parameters of the MOSFET device. Assume for the device that $0.3 \leq k \leq 1.0$ mA/V^2 and $0.75 \leq V_T \leq 1.5$ V.

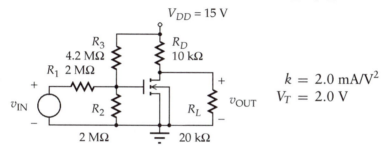

Figure C4.4

C4.4 Use a SPICE simulation to determine the behavior of the MOSFET circuit of Figure C4.4.

a) Obtain a graph (.DC) of v_{OUT} versus v_{IN} for an input voltage range of ± 5 V. What are the input and output voltages at which v_{OUT} limits?

b) What is v_{OUT} for $v_{IN} = 0$? What is the slope of the v_{OUT}-versus-v_{IN} characteristic for $v_{IN} = 0$?

c) Obtain a graph of the drain current of the MOSFET device i_D. What is i_D for $v_{IN} = 0$?

C4.5 Use SPICE simulations to determine the dynamic behavior of the circuit of Figure C4.4. Determine the behavior of the circuit for symmetrical input voltages with sinusoidal, triangular, and square waveforms ($f = 1$ kHz). Consider voltages with peak-to-peak values of 1.0 and 2.0 V.

C4.6 Use an .AC Spice simulation to determine the small-signal voltage gain of the MOSFET amplifier of Figure P4.29. Following the method of Example 4.6, determine the harmonic distortion for an input sinusoidal signal with a peak amplitude of 0.3 V. What is the harmonic distortion for an input signal that has an amplitude of only 0.1 V?

C4.7 Use a .DC Spice simulation to determine the transfer characteristic of the two-transistor circuit of Problem 4.47. For the BJT device, assume a value of I_s that results in a value of 0.7 V for v_{BEon} at the edge of saturation ($i_C \approx V_{DD}/R_C$) and $n_F = 1$. What are the values of v_{IN} for which $\frac{dv_{OUT2}}{v_{IN}} = 1$?

C4.8 A SPICE simulation is to be used to determine the static and dynamic behavior of the logic inverter gate of Problem 4.55.

a) Use a .DC simulation and determine the transfer characteristic of the gate. What are the values of v_{IN} for which $\frac{dv_{OUT}}{dv_{IN}} = -1$?

b) Assume the gate has a load capacitance C_L of 2 pF. Use an appropriate input pulse (levels of 0 and 4 V) and determine the rise and fall times of v_{OUT} (10-to-90-percent). What are the times required for v_{OUT} to reach its mid value?

C4.9 Repeat C4.8 for the substrate of M_2 connected directly to ground. Assume for the devices that $\gamma = 0.4$ V$^{1/2}$, $2\phi_F = 0.6$ V, and $V_{T0} = 1.0$ V. For the dynamic response, the high-level voltage is the static value of v_{OUT} for $v_{IN} = 0$.

C4.10 To improve the static transfer characteristic of a logic circuit, logic inverter gates are frequently connected in cascade, that is, two logic inverter gates are used to form a buffer and three logic inverter gates are used to form an improved inverter. Use a SPICE simulation to determine the static transfer characteristic of three cascaded logic inverter circuits of Figure P4.42. Obtain the transfer characteristic for one, two, and three gates. What is $\frac{dv_{OUT}}{dv_{IN}}$ for $v_{OUT} = 2.5$ V for one, two, and three gates? Determine the dynamic behavior of the gates using an input 0.5-μs pulse with a height of 5.0 V. What are the times required for v_{OUT} to reach 2.5 V for one, two, and three gates? Note: The times for upward and downward transitions will differ.

C4.11 Use a transient SPICE simulation to determine the dynamic behavior of the CMOS logic inverter gate of Example 4.12. The circuit has 5-pF capacitive loads at v_{OUT1} and v_{OUT2} and 10 pF at v_{OUT3}. In addition, each device has a gate-drain capacitance of 0.5 pF. Use an input voltage that has abrupt transitions (levels of 0 and V_{DD}). What are the rise and fall times of the output (10-to-90-percent)? What are the delay times, that is, the times necessary for the output to reach $V_{DD}/2$?

C4.12 Repeat Example 4.12 for the case in which the MOSFET devices do not have complementary symmetry. Assume for the n-channel device that $k_N = 0.2$ mA/V^2 and $V_{TN} = 0.75$ V. Also determine the response of this circuit for $V_{DD} = 10$ V.

DESIGN EXERCISES

D4.1 A common-source amplifier is indicated in Figure D4.1. A value of $v_{OUT} = 5$ V and $\frac{dv_{OUT}}{dv_{IN}} = -5$ is desired for $v_{IN} = 0$. Determine the resistance values required for the circuit for the condition that the input resistance is at least 1 MΩ. What are the values of v_{OUT} for $v_{IN} = \pm 0.5$ V?

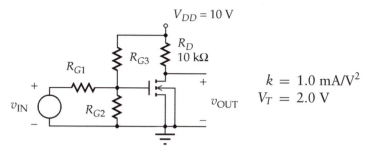

Figure D4.1

D4.2 Repeat D4.1 for a circuit with a 1-kΩ resistor in series with the common source–substrate connection of the MOSFET and ground.

D4.3 Consider the MOSFET circuit of Figure D4.3. It is desired that for $v_{IN} = 0$, $v_{OUT1} = -v_{OUT2}$ and $\frac{dv_{OUT1}}{dv_{IN}} = -0.75$. What is $\frac{dv_{OUT2}}{dv_{IN}}$ for this circuit? What are v_{OUT1} and v_{OUT2} for $v_{IN} = \pm 1.0$ V?

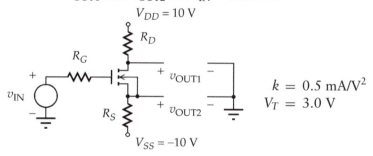

Figure D4.3

D4.4 A drain current I_D of 0.5 mA is desired for the MOSFET circuit of Figure 4.29 ($V_{DD} = 10$ V). The MOSFET device has parameters of $k = 0.5$ mA/V² and $V_T = 1.0$ V. A drain–source voltage V_{DS} of 5 V is desired. Use values of R_{G1} and R_{G2} that result in $R_{G1} \| R_{G2} = 1$ MΩ. What is the effect of a ±50% variation in k on the drain current and V_{DS} for this circuit? Suppose that a maximum variation of 20 percent (either direction) is acceptable for I_D. Redesign the circuit using a resistor in series with the common source–substrate connection of the MOSFET device and ground to achieve this. Note: It may be necessary to use a trial-and-error type solution.

D4.5 The MOSFET amplifier of Example 4.5 (Figure 4.28) is to be modified by adding a source resistor R_S in series with the source–substrate connection of the device and ground. Use a value of resistance that results in $I_D R_S = 1$ V. The new values of R_{G1} and R_{G2} should result in the same input resistance $R_{G1} \| R_{G2}$. Therefore, if the source resistor is properly bypassed, the small-signal gain of the circuit will be unchanged. However, the behavior of the amplifier will be less sensitive to parameter variations of the MOSFET device. Determine the quiescent drain current and the small-signal gain of the circuit for a MOSFET device with $V_T = 1.5$ V and for a device with $V_T = 2.5$ V.

D4.6 A MOSFET logic inverter circuit with the configuration of Figure 4.38 is to be designed. It has a load capacitance C_L of 1.0 pF, and the MOSFET device has parameters of $k = 100 \ \mu A/V^2$ and $V_T = 1.0$ V. The supply voltage V_{DD} is 5 V. The rise and fall times (10-to-90-percent) are to be no greater than 100 ns, and V_{OL} is to be no greater than 0.5 V. Determine the maximum value of R_D for the circuit (smallest power dissipation). What are the smallest values of rise and fall times that can be achieved for the condition that V_{OL} be no greater than 0.5 V? What is R_D of the circuit?

D4.7 Both devices of the MOSFET logic inverter of Figure 4.47(a) have threshold voltages of 0.8 V. Determine the ratio of the transconductance parameters k_1/k_2 required for $V_{OL} = 0.4$ V. The supply voltage V_{DD} is 5 V. What are the values of the transconductance parameters required to achieve transition times to $(V_{OL} + V_{OH})/2$ that are no greater than 20 ns. The load capacitance C_L is 2 pF.

D4.8 Repeat D4.7 for the MOSFET circuit of Figure 4.51(a) in which the devices have a common substrate-to-ground connection. Assume $\gamma = 0.4 \ V^{1/2}$, $2\phi_F = 0.6$ V, and $V_{T0} = 0.8$ V.

D4.9 Repeat D4.7 for the MOSFET circuit of Figure 4.53(a), which has a depletion-type pull-up MOSFET device. For the depletion-type device, assume $\gamma = 0.4 \ V^{1/2}$, $2\phi_F = 0.6$ V, and $V_{T0} = -1.5$ V.

D4.10 A CMOS logic inverter gate using the circuit of Figure 4.64 is to be designed that has upward and downward output voltage transitions to $V_{DD}/2$ that are no greater than 100 ns. The load capacitance C_L is 50 pF and $5 \leq V_{DD} \leq 15$ V. The devices have threshold voltages with a magnitude of 1.0 V. A design based on minimum transconductance parameters is desired.

CHAPTER FIVE

NEGATIVE FEEDBACK AND OPERATIONAL AMPLIFIERS

Negative feedback, when used with an amplifier, reduces the gain of the overall circuit because part of the output signal is used to "negate" a portion of the input signal. If properly designed, negative feedback circuits can result in improved performance characteristics – in particular, lower distortion, improved frequency and impedance characteristics, and a smaller dependence on supply voltages. To realize these benefits, an amplifier is required that has a gain considerably in excess of that which would otherwise be needed. With the advent of commercially produced integrated circuits, high-gain, low-cost amplifiers suitable for negative feedback circuits became readily available. Integrated circuit operational amplifiers (IC op amps) are now widely used "building blocks," both as individual integrated circuits (replacing discrete transistors for many applications) and within more complex integrated circuits.

The concept of positive feedback electronic circuits predates that of negative feedback (Tucker 1972). Positive feedback was initially used to increase the gain of early low-gain vacuum tube circuits. With positive feedback (regenerative circuits), an enormous increase in the sensitivity of radio receivers was achieved. Only after high-gain amplifier circuits were developed in the 1920s did the concept of using negative feedback emerge. Harold Black is credited with having first proposed this concept in 1927. According to published accounts, the idea of an electronic amplifier with negative feedback was the result of a sudden insight that Black had while crossing the Hudson river by ferry on his way to work in Manhattan (Mabon 1975; O'Neill 1985). As is generally the case, this sudden insight did not happen in an intellectual vacuum ("out of the blue"); it was the result of a succession of attempts to solve a telephone transmission problem that had engaged his attention since starting to work at Bell Telephone Laboratories in 1921. Given the importance of this invention, a brief account of the circumstances that led to its discovery may be of interest.

Long-distance telephone lines require electronic amplifiers at periodic intervals to compensate for transmission losses due to the resistance of wires and the conductive losses of insulating materials. An early application of vacuum tubes was for telephone repeater amplifiers, which were first used in 1913 and were an important component of the first transcontinental line of 1914 (Fagen 1975). To minimize transmission losses, early long-distance telephone lines used fairly large diameter copper wires (the New York–San Francisco line used 1/6-inch-diameter wire, a total of 2500 tons). With improved amplifiers, smaller diameter, less expensive but higher resistance wires could be used. Another reduction in wire requirements had been achieved through the development of multiplex systems in which a single pair of wires was used to carry several telephone conversations simultaneously. (Alexander Graham Bell was attempting to develop a multiplex system for telegraph communication, the "harmonic" telegraph, when he strayed from this task and invented the telephone.) For a telephone carrier multiplexing system, each telephone conversation is transmitted with a different carrier frequency in the same fashion that radio stations use different carrier frequencies. Active electronic devices (then vacuum tubes) are used to generate the carriers, to modulate and demodulate the carriers, and to amplify the signals that, as a result of transmission line losses, are attenuated.

Carrier multiplexing (also known as frequency multiplexing) is achieved by moving signals to frequencies higher than those associated with a single base-band telephone signal. These higher-frequency carrier signals are, in essence, "stacked" in frequency, one above another. Higher-frequency signals, however, are attenuated much more than a base-band telephone signal because transmission line losses tend to increase with frequency. Not only are more and higher-gain amplifiers needed to compensate for the attenuation, but amplifiers with very low levels of distortion are required. The instantaneous transmission line voltage is the sum of the instantaneous voltages of each signal. If this voltage is not uniformly amplified, regardless of its level, each telephone signal will tend to be distorted. Furthermore, interfering interactions between the signals occur. Amplifiers with extremely low levels of distortion, that is, highly linear amplifiers, are required for this particular application. It was with this need in mind that Black, after many other less successful attempts, conceived of using negative feedback.

Black proposed the basic symbolic circuit of Figure 5.1 in which a portion of the output voltage β is returned to the input of the amplifier (Black 1934). The amplifier is assumed to have a gain of A, that is, its output voltage is A times its input voltage v_{Error}. This results in the following:

Figure 5.1: A symbolic representation of a negative feedback system.

$$v_{\text{OUT}} = A v_{\text{Error}} = A(v_{\text{IN}} - \beta v_{\text{OUT}})$$

$$= \frac{A v_{\text{IN}}}{1 + \beta A} = \frac{v_{\text{IN}}}{\beta + 1/A} \qquad (5.1)$$

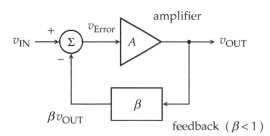

If the amplifier gain is very large, a relationship involving only β is obtained:

$$v_{\text{OUT}} = v_{\text{IN}}/\beta \quad \text{for } A \to \infty \qquad (5.2)$$

Hence, the gain of this feedback circuit tends to be determined entirely by the feedback network, a network that can be constructed from purely passive components (resistors for a simple amplifier circuit). The preceding result (Eq. (5.1)) is for an ideal linear amplifier in which $v_{OUT} = Av_{IN}$ (no distortion). This is not the case for a physically realizable amplifier in which the transfer response deviates from a linear dependence. However, as for the linear case, negative feedback reduces the dependence of the response of the overall circuit on the transfer response of the amplifier. Through this process, very low levels of distortion can be achieved with realizable amplifiers.

Despite the apparent simplicity of the circuit of Figure 5.1, much effort was required before useful negative feedback amplifier circuits emerged. Early circuits would often break into oscillation, a phenomenon then referred to as "singing." This unacceptable result was due to the delay introduced by electronic amplifiers. This delay, a phase shift for sinusoidal signals, can result in positive feedback and an oscillating behavior of the circuit. It is necessary that those amplifiers that are to be used with negative feedback circuits have especially constrained responses. An analysis of these constraints led to the development of stability theory, an understanding of which is necessary for designing negative feedback amplifier circuits (Nyquist 1932).

It is the integrated-circuit operational amplifier that has revolutionized the design of many electronic circuits. The term *operational amplifier*, however, predates integrated circuits. High-gain amplifiers were used in analog computers to perform various mathematical operations such as summing, scaling, and integration. The particular operation depended on the external feedback circuit, and analog computers were programmed by changing external components. Although digital computers that can readily simulate the behavior of analog computers have made analog computers obsolete, the descriptive term operational amplifier has remained. Integrated-circuit operational amplifiers not only have large voltage gains but are designed to be stable when used with negative feedback circuits.

In 1964, Robert Widlar, while working at Fairchild Semiconductor, developed the first commercially available IC op amp (Solomon 1991). This was the μA 709, an amplifier fabricated on a single silicon chip having an area of less than 2 mm^2. Although these early amplifiers often behaved erratically (as is the case for many early inventions), subsequent integrated circuits, the LM 101 of 1967 and the μA 741 of 1968, behaved much more predictably (more than three decades after its invention, the μA 741 was still widely used). The early IC op amps utilized bipolar junction transistors. Subsequently, op amps using junction field-effect transistors as well as MOSFET and CMOS devices have been developed.

A simplified circuit of a bipolar junction transistor op amp (the prefix integrated circuit will be assumed to apply whenever op amp is used) is shown in Figure 5.2. A three-stage amplifier circuit is common. The first stage, using a *PNP* differential amplifier circuit, has an inverting and a noninverting input (its output is the difference of its input voltages). This high-gain amplifier stage, along with the second stage, an emitter–follower driving a common–emitter amplifier, results in a very large voltage gain (10^5 to 10^6). The last stage, emitter-followers with complementary symmetry, provides a current gain and results in an output circuit

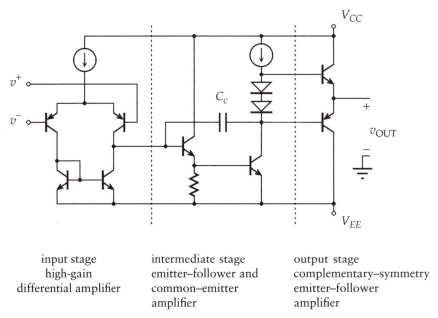

Figure 5.2: Simplified bipolar junction transistor operational amplifier circuit.

input stage	intermediate stage	output stage
high-gain	emitter–follower and	complementary–symmetry
differential amplifier	common–emitter	emitter–follower
	amplifier	amplifier

with a very low equivalent resistance. A compensating capacitance C_c is required to constrain the amplifier response so that it will be stable when used with typical feedback circuits. Twenty to thirty discrete devices are generally required for a complete low-power operational amplifier.

An advantage of a well-designed operational amplifier circuit is that both its static and dynamic transfer characteristics can be made nearly independent of the characteristics of the op amp. Hence, the design of op amp circuits is generally reduced to the design of the feedback circuit, whereas the op amp is treated as being ideal, that is, having an infinite gain. Two common amplifier circuits are shown in Figure 5.3.

An introductory treatment of negative feedback will include a discussion of using negative feedback to reduce the distortion caused by a nonlinear amplifier. The important topic of stability will next be addressed. Various linear amplifier

Figure 5.3: Typical linear operational amplifier circuits.

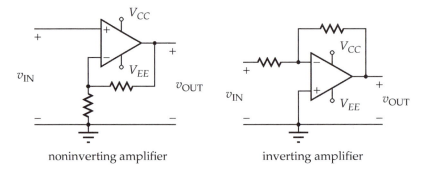

noninverting amplifier   inverting amplifier

configurations using op amps will be considered, and the concept of an ideal response will be introduced. The constraints imposed by the frequency-dependent gain of actual op amps as well as by their slew-rate limiting will be discussed. The chapter will conclude with two sections devoted to designing circuits using op amps.

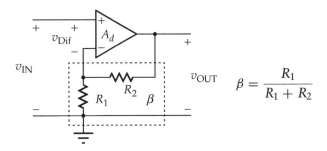

Figure 5.4: A basic feedback circuit using a difference amplifier.

5.1 NEGATIVE FEEDBACK: A KEY CONCEPT

Negative feedback is used with many different electronic amplifying circuits to achieve a desired set of overall characteristics. Although only a few basic circuits will be considered, an analysis of these circuits will establish the general principles that govern the operation of amplifier circuits with negative feedback.

The behavior of the basic circuit of Figure 5.4 with a linear amplifier will initially be determined. Although power supply connections are obviously necessary for the amplifier, these connections have not been shown; it is the role of signal voltages that is important. To simplify the analysis, the resistances of the input terminals of the amplifier will be assumed to be infinite, thereby implying that currents into the input terminals are zero. In addition, the amplifier's output resistance will be assumed to be zero. The equivalent circuit of Figure 5.5 with the amplifier replaced by an ideal voltage-dependent voltage source therefore applies.

In Figures 5.4 and 5.5, two resistors are shown for the feedback circuit (the dashed box labeled β). The portion of the output voltage fed back to the input circuit β is readily determined through the following expression:

$$\beta = \frac{R_1}{R_1 + R_2} \tag{5.3}$$

Although this basic two-resistor feedback circuit is common, other types of feedback circuits are also used. The input voltage of the amplifier v_{Dif}, is the difference between the input voltage v_{IN} and the portion of the output voltage fed back to the input βv_{OUT} as given by

Figure 5.5: Equivalent circuit for the amplifier of Figure 5.4.

$$v_{\text{Dif}} = v_{\text{IN}} - \beta v_{\text{OUT}}$$

$$v_{\text{OUT}} = A_d v_{\text{Dif}} = A_d(v_{\text{IN}} - \beta v_{\text{OUT}})$$

$$A_{fb} = \frac{v_{\text{OUT}}}{v_{\text{IN}}} = \frac{A_d}{1 + \beta A_d} \tag{5.4}$$

$$= \left(\frac{1}{\beta}\right)\frac{\beta A_d}{1 + \beta A_d}$$

The term A_{fb} has been introduced for the overall voltage gain of the amplifier with feedback. If βA_d is very large compared with

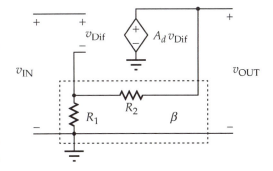

1, an "ideal response" is obtained as follows:

$$\left(\frac{v_{OUT}}{v_{IN}}\right)_{Ideal} = \frac{1}{\beta} \quad \text{for } \beta A_d \gg 1 \tag{5.5}$$

The ideal response depends only on the feedback network; it is the result that occurs for what may be characterized as an ideal amplifier, that is, an amplifier with infinite gain.

For the two-resistor feedback network, a fairly simple expression is obtained for the ideal response of the circuit as follows:

$$\left(\frac{v_{OUT}}{v_{IN}}\right)_{Ideal} = \frac{R_1 + R_2}{R_1} = 1 + \frac{R_2}{R_1} \tag{5.6}$$

In general, the actual response of the circuit may be written in terms of its ideal response as follows:

$$A_{fb} = \left(\frac{v_{OUT}}{v_{IN}}\right)_{Ideal} \left(\frac{\beta A_d}{1 + \beta A_d}\right) \tag{5.7}$$

It is the parameter βA_d (a quantity known as the open-loop gain) that yields the factor by which the actual gain deviates from the ideal gain corresponding to $\beta A_d \to \infty$.

DECIBEL NOTATION

At this point it is convenient to introduce a decibel (dB) scale for expressing voltage gains. For a voltage gain (or ratio) of A, its decibel value depends on the base ten logarithm of A:

$$A_{dB} = 20 \log |A| \tag{5.8}$$

TABLE 5.1 DECIBEL SCALE

A	0.01	0.1	0.5	1.0	2.0	10	20	100	10^3	10^4	10^5
A_{dB}	−40	−20	−6	0	6	20	26	40	60	80	100

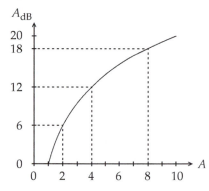

Figure 5.6: A decibel scale for $1 \le A \le 10$.

If A is a positive quantity, the magnitude operation is not needed. Table 5.1 provides a few decibel values for a wide range of A, and Figure 5.6 gives decibel values for a range of 1 to 10 for A. It will be noted that for each doubling of A, A_{db} increases by 6 dB. Concurrently, for each factor of $\frac{1}{2}$ for A, A_{db} decreases by 6 dB. This is a result of the logarithmic dependence – the logarithm of the product of two quantities being the sum of the logarithms of the individual quantities. A doubling of voltage (a 6-dB increase) implies a quadrupling of power delivered to a resistive load. To double the power, an increase of $\sqrt{2}$ is required. This implies a decibel change of

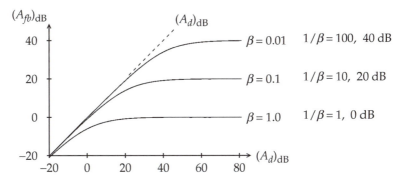

Figure 5.7: Gain with negative feedback versus amplifier gain.

3 dB ($20 \log \sqrt{2} = 10 \log 2 = 3$). Therefore, a 3-dB increase can be associated with a doubling of power, that is, a voltage change by a factor of $\sqrt{2}$. A 3-dB decrease can be associated with a halving of power, that is, a voltage change by a factor of $1/\sqrt{2}$ or 0.707.

A plot of the gain with feedback, Eq. (5.4), in which both A and A_{fb} are expressed in decibels (dB), is given in Figure 5.7. It will be noted that a value of βA_d is required that is considerably greater than that needed to achieve a gain that corrrresponds to its ideal value ($1/\beta$). Expressed in decibels, the gain A_{fb} may be written as follows:

$$(A_{\text{Ideal}})_{\text{dB}} = 20 \log\left(\frac{1}{\beta}\right) = 20 \log\left(\frac{v_{\text{OUT}}}{v_{\text{IN}}}\right)_{\text{Ideal}}$$

$$(A_{fb})_{\text{dB}} = 20 \log A_{fb} = 20 \log\left(\frac{1}{\beta}\right) + 20 \log\left(\frac{\beta A_d}{1 + \beta A_d}\right) \tag{5.9}$$

$$= (A_{\text{Ideal}})_{\text{dB}} + 20 \log\left(\frac{\beta A_d}{1 + \beta A_d}\right)$$

The last term of Eq. (5.9) is the decibel gain error (A_{Error})$_{\text{dB}}$. It is a negative quantity because, for negative feedback ($\beta A_d > 0$), the argument of the logarithm is less than 1.

$$(A_{\text{Error}})_{\text{dB}} = 20 \log\left(\frac{\beta A_d}{1 + \beta A_d}\right) \tag{5.10}$$

If, for example, $\beta A_d = 10$, the gain error is -0.83 dB, that is, the actual gain is 0.83 dB less than its ideal value. This implies the gain with feedback is 90.9 percent of the gain that would ocur with an infinite gain amplifier. If $\beta A_d = 100$, the gain error is only -0.086 dB; the gain is 99.0 percent of that for an ideal amplifier. To achieve this last condition, an amplifier with a gain 100 times that produced with the feedback is required.

REDUCING DISTORTION

The reduction of distortion is an important feature of electronic amplifier circuits with negative feedback. Although amplifiers designed for low-level output signals generally introduce very little distortion, this is not the case for amplifiers that deliver high-level output signals. Distortion is caused by the nonlinearity

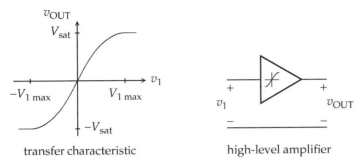

transfer characteristic   high-level amplifier

Figure 5.8: Response of a typical amplifier with a high-level output voltage.

of the electronic devices of the amplifier. The static transfer characteristic of a typical amplifier designed to produce a high-level output voltage is shown in Figure 5.8. A deviation from an ideal linear response occurs before the output voltage saturates (at $v_1 = \pm V_{1\,max}$). Although a symmetrical characteristic is shown, this is not always the case.

The feedback circuit of Figure 5.9 will be analyzed to show the effect of negative feedback on the response of the amplifier. The additional amplifier gain of the input difference amplifier (A_d) is required to achieve the same low-level response as that without feedback. Because the difference amplifier provides the input signal of the high-level amplifier, its output voltage v_1 will tend to be small. Hence, an assumption of linear behavior for the difference amplifier is not unreasonable. The feedback circuit R_1 and R_2 is placed around both amplifier stages.

An analysis of a general nonlinear amplifier system is not possible. Therefore, to illustrate the effect of negative feedback, a nonlinear amplifier with a quadratic-type response will be assumed as follows:

$$v_{OUT} = \begin{cases} -A_1 V_{1\,max} + A_2 V_{1\,max}^2 & \text{for } v_1 \leq -V_{1\,max} \\ A_1 v_1 + A_2 v_1^2 & \text{for } -V_{1\,max} < v_1 < 0 \\ A_1 v_1 - A_2 v_1^2 & \text{for } 0 \leq v_1 < V_{1\,max} \\ A_1 V_{1\,max} - A_2 V_{1\,max}^2 & \text{for } v_1 \geq V_{1\,max} \end{cases} \qquad (5.11)$$

Figure 5.9: A feedback circuit for reducing distortion of a high-level amplifier.

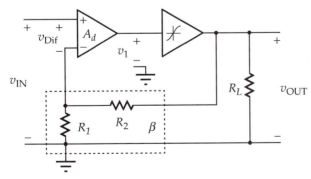

For $|v_1|$ small, a linear response $v_{OUT} = A_1 v_1$ is obtained – A_1 is the low-level voltage gain of the amplifier. The quadratic term with a coefficient of A_2 accounts for the distortion of the amplifier, and the parameter $V_{1\,max}$ corresponds to the input voltage at which the output voltage saturates, that is, the input voltage for which the derivative of v_{OUT} with respect to v_1 is zero. Consider the case for

$v_1 > 0$:

$$\frac{dv_{\text{OUT}}}{dv_1} = A_1 - 2A_2 v_1 \tag{5.12}$$

$$A_1 - 2A_2 V_{1\,\text{max}} = 0, \qquad V_{1\,\text{max}} = A_1/2A_2$$

A limiting of algebraic details is necessary; otherwise, these details will tend to obscure the development. Rather than all four cases of Eq. (5.11) for v_{OUT}, only the case for a nonsaturated amplifier with a positive input voltage, $0 \le v_1 < V_{1\,\text{max}}$, will be considered. Therefore, until stated otherwise, it will be assumed that v_1 falls in this range even though it will not be explicitly stated for each equation of the development.

If a zero output resistance and infinite input resistance is assumed for the amplifiers, the following is obtained:

$$v_{\text{OUT}} = A_1 v_1 - A_2 v_1^2 = A_d A_1 v_{\text{Dif}} - A_d^2 A_2 v_{\text{Dif}}^2 \tag{5.13}$$

The input difference voltage v_{Dif} depends on the input signal and that of the feedback network as follows:

$$v_{\text{Dif}} = v_{\text{IN}} - \beta v_{\text{OUT}} \tag{5.14}$$

If one substitutes this dependence into Eq. (5.13), the following expression that relates v_{OUT} and v_{IN} is obtained:

$$v_{\text{OUT}} = A_d A_1 (v_{\text{IN}} - \beta v_{\text{OUT}}) - A_d^2 A_2 (v_{\text{IN}} - \beta v_{\text{OUT}})^2 \tag{5.15}$$

If $v_{\text{IN}} - \beta v_{\text{OUT}}$ is treated as the dependent variable instead of v_{OUT}, an algebraic simplification is possible as follows:

$$v_{\text{IN}} - \beta v_{\text{OUT}} = v_{\text{IN}} - \beta A_d A_1 (v_{\text{IN}} - \beta v_{\text{OUT}}) + \beta A_d^2 A_2 (v_{\text{IN}} - \beta v_{\text{OUT}})^2 \tag{5.16}$$

This equation is quadratic in the term $(v_{\text{IN}} - \beta v_{\text{OUT}})$:

$$\beta A_d^2 A_2 (v_{\text{IN}} - \beta v_{\text{OUT}})^2 - (1 + \beta A_d A_1)(v_{\text{IN}} - \beta v_{\text{OUT}}) + v_{\text{IN}} = 0$$

$$(v_{\text{IN}} - \beta v_{\text{OUT}})^2 - \frac{(1 + \beta A_d A_1)}{\beta A_d^2 A_2}(v_{\text{IN}} - \beta v_{\text{OUT}}) + \frac{v_{\text{IN}}}{\beta A_d^2 A_2} = 0 \tag{5.17}$$

The quadratic formula may be used to find $v_{\text{IN}} - \beta v_{\text{OUT}}$. After simplification, the following is obtained:

$$v_{\text{IN}} - \beta v_{\text{OUT}} = \frac{(1 + \beta A_d A_1)}{2\beta A_d^2 A_2}\left(1 - \sqrt{1 - \frac{4\beta A_d^2 A_2 v_{\text{IN}}}{(1 + \beta A_d A_1)^2}}\right) \tag{5.18}$$

Only one additional step is needed to obtain an expression for v_{OUT} in terms of v_{IN}.

$$v_{\text{OUT}} = \frac{v_{\text{IN}}}{\beta} - \frac{(1 + \beta A_d A_1)}{2\beta^2 A_d^2 A_2}\left(1 - \sqrt{1 - \frac{4\beta A_d^2 A_2 v_{\text{IN}}}{(1 + \beta A_d A_1)^2}}\right) \tag{5.19}$$

This is the desired result that will now be used to illustrate the effect of negative feedback on distortion.

For a check of the validity of Eq. (5.19), the result for $A_2 \to 0$ may be obtained by expanding the square root and then taking the limit as A_2 goes to zero:

$$v_{OUT} = \frac{v_{IN}}{\beta} - \frac{v_{IN}}{\beta(1 + \beta A_d A_1)} = \left(\frac{1}{\beta}\right) \frac{\beta A_d A_1 v_{IN}}{(1 + \beta A_d A_1)} \quad A_2 \to 0 \qquad (5.20)$$

This, not surprisingly, is the result expected for a linear amplifier with a gain of $A_d A_1$ ($A_2 = 0$ implies linear behavior). To show the distortion-reducing effect of the negative feedback circuit, a numerical example is necessary. Suppose the output amplifier has a low-level gain of 10 ($A_1 = 10$) and that the amplifier saturates for an input voltage of $v_1 = \pm 2.0$ V ($V_{1\,max} = 2.0$ V). From Eq. (5.12), a value for A_2 is obtained as follows:

$$A_2 = A_1/2V_{1\,max} = 2.5 \text{ V}^{-1} \qquad (5.21)$$

The difference amplifier will also be assumed to have a gain of 10 ($A_d = 10$), and a feedback network will be used that results in an overall low-level gain of 10, the same low-level gain of the output amplifier without feedback. The feedback factor may be determined from Eq. (5.20) as follows:

$$\frac{A_d A_1}{(1 + \beta A_d A_1)} = A_1, \quad \beta = (A_d - 1)/A_d A_1 = 0.09 \qquad (5.22)$$

These numerical values may now be introduced into the equation for v_{OUT}, Eq. (5.19):

$$v_{OUT} = 11.11\, v_{IN} - 2.469\,(1 - \sqrt{1 - 0.9\, v_{IN}}) \qquad (5.23)$$

This expression is valid for a value of v_{IN} up to $v_{OUT} = 10$ V. This implies that $v_1 = 2$ V and $v_{Dif} = 0.2$ V.

$$v_{Dif} = v_{IN} - \beta v_{OUT}, \quad v_{IN} = v_{Dif} + \beta v_{OUT} = 1.10 \text{ V} \qquad (5.24)$$

Hence, the result of Eq. (5.23) is valid for $0 \le v_{IN} \le 1.10$ V. If all values of v_{IN} are taken into account, the following (complete) expression for v_{OUT} is obtained:

$$v_{OUT} = \begin{cases} -10 \text{ V} & \text{for } v_{IN} \le -1.10 \text{ V} \\ 11.11\, v_{IN} + 2.469\,(1 - \sqrt{1 + 0.9\, v_{IN}}) & \text{for } -1.10 \text{ V} < v_{IN} < 0 \\ 11.11\, v_{IN} - 2.469\,(1 - \sqrt{1 - 0.9\, v_{IN}}) & \text{for } 0 \le v_{IN} < 1.10 \text{ V} \\ 10 \text{ V} & \text{for } v_{IN} \ge 1.10 \text{ V} \end{cases}$$
$$(5.25)$$

A plot of this response, along with that for the high-level amplifier without feedback, is given in Figure 5.10(a), and the amount by which the output voltage differs from that for an ideal linear response, that is, $10v_{IN}$, is given in Figure 5.10(b). The distortion is significantly reduced even for the very modest difference amplifier gain of 10. Using a much larger difference amplifier gain, as would normally be the case, will further reduce the distortion of the amplifier (Example 5.3).

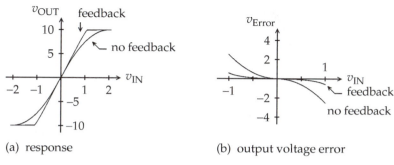

(a) response (b) output voltage error

Figure 5.10: Amplifer response with negative feedback and output voltage error.

ADDITIONAL BENEFITS OF NEGATIVE FEEDBACK

Although the reduction in amplitude distortion is important for many applications, amplifier circuits utilizing negative feedback offer many additional benefits.

Gain Sensitivity: Closely related to amplitude distortion is the sensitivity of the overall small-signal gain to the gain of the amplifier. With negative feedback, the dependence of the overall gain on that of the amplifier is reduced. For a large open-loop gain, the overall gain tends to depend primarily on the feedback fraction β and only slightly on the gain of the amplifier.

Reduced Output Resistance: The output resistance of an amplifier with negative feedback is smaller than that of an amplifier without feedback. For circuits with large open-loop gains, it is frequently possible to ignore the effect of the output resistance of an amplifer ($R_{OUT} \approx 0$).

Increased Input Resistance: In the feedback amplifier circuits of Figures 5.4 and 5.5, an infinite input resistance is implied for the amplifier (an input open circuit for v_{Dif}). Although a realizeable amplifier has a finite input resistance, the equivalent resistance for the input signal v_{IN} is larger than that of the amplifier without negative feedback. Furthermore, with large open-loop gains, the input resistance is often sufficiently large so that it can be treated as being infinite.

Reduced Dependence on Power Supply Fluctuations: For the transistor amplifier circuits considered in the previous chapters, the output signal voltage, current, or both, is dependent on the supply voltage (V_{CC} or V_{DD}). With negative feedback, this dependence is reduced. With a large open-loop gain, the effect of power supply variations, such as hum, is often reduced to a negligible level.

Reduced Dependence on Ambient Conditions: The values of components used in amplifiers often have a dependence on ambient conditions – most notably temperature. But, to the extent that negative feedback reduces the dependence of the overall gain on that of the amplifier, the dependence on ambient conditions will also be reduced.

Noise Reduction: The sensitivity of all amplifers is limited by internal noise, the result of thermal effects, the discrete nature of electronic charges, and other less understood effects. Although negative feedback reduces the output noise of an amplifier, it also reduces the output signal produced by a low-level input signal. Hence, the signal-to-noise ratio at the amplifier's output may, or may not, be improved. However, with a well-designed feedback network and a low-noise input amplifier, the noise performance of the amplifier circuit can be improved.

Simplified Design: With a large open-loop gain, the overall response of an amplifier with negative feedback tends to depend only on the feedback network. Hence, for the amplifier of Figures 5.4 and 5.5, the gain tends to depend only on the two resistors of the circuit. By using precision resistance values, the gain can be accurately controlled. Furthermore, reactive elements may be used in the feedback circuit to obtain a desired frequency response (a filter). Alternatively, nonlinear elements (for example, diodes) can be used in the feedback circuit to obtain a desired output-versus-input voltage functional dependence (for example, a limiter).

As a result of negative feedback, the gain of an amplifier circuit is reduced. Hence, for a required overall gain, a much higher amplifer gain is necessary. A more complex, higher-gain amplifer is the price paid for obtaining the benefits of negative feedback. When amplifiers utilized discrete components, negative feedback was used only for critical applications. However, with the advent of readily available high-gain integrated circuit operational amplifiers, amplifier complexity is no longer a limitation. Amplifier circuits using integrated circuit operational amplifiers are generally much simpler and less expensive than conventional lower-gain amplifier circuits.

EXAMPLE 5.1

A determination of the effect on the behavior of the circuit of Figure 5.4 of a difference amplifier with a finite input resistance R_i is desired.

a. Determine an expression for the equivalent input resistance R_{IN} of the amplifier.

b. Assume $R_i = 1$ MΩ. Evaluate R_{IN} for difference amplifiers with gains of 50, 100, and 1000. The feedback circuit for each case is to be such that v_{OUT}/v_{IN} is equal to 10 ($A_{fb} = 10$).

SOLUTION

a. The equivalent circuit of Figure 5.5 will be modified to account for the amplifier input resistance R_i (Figure 5.11). The feedback factor β is the fraction of v_{OUT} fed back to the input of the difference amplifier.

$$v_{Dif} = v_{IN} - \beta v_{OUT} = -\beta v_{OUT} \quad \text{for } v_{IN} = 0$$

For $v_{IN} = 0$, the circuit of Figure 5.12 applies. Because $\beta = -v_{Dif}/v_{OUT}$ for $v_{IN} = 0$, the following is obtained for β:

$$\beta = \frac{R_i \| R_1}{R_i \| R_1 + R_2}$$

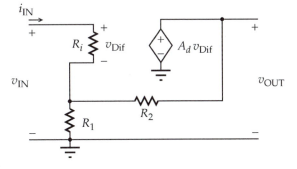

Figure 5.11: Equivalent circuit for a difference amplifier with a finite input resistance.

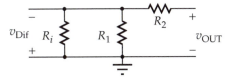

Figure 5.12: Equivalent feedback circuit for a difference amplifer with a finite input resistance.

If R_i is very large compared with R_1, then $\beta = R_1/(R_1 + R_2)$, which is the feedback fraction obtained for an infinite input resistance. Other than for taking into account the effect of R_i on β, the voltage-gain expression remains unchanged.

$$\frac{v_{OUT}}{v_{IN}} = A_{fb} = \frac{A_d}{1 + \beta A_d}$$

The input current i_{IN} depends on the voltage across R_i, that is, v_{Dif} as follows:

$$i_{IN} = \frac{v_{Dif}}{R_i} = \frac{v_{IN} - \beta v_{OUT}}{R_i}$$

$$= \frac{v_{IN}}{R_i}\left(1 - \frac{\beta A_d}{1 + \beta A_d}\right) = \frac{v_{IN}}{R_i}\left(\frac{1}{1 + \beta A_d}\right)$$

$$R_{IN} = v_{IN}/i_{IN} = R_i(1 + \beta A_d)$$

From the recognition that $1 + \beta A_d = A_d/A_{fb}$ (from the expression for v_{OUT}/v_{IN}), the following is obtained:

$$R_{IN} = R_i A_d/A_{fb}$$

Because $A_d > A_{fb}$, the equivalent input resistance is larger than R_i.

b. Because $A_{fb} = 10$ for all cases, $R_{IN} = (0.1 \text{ M}\Omega)/A_d$. Hence, for $A_d = 50$, $R_{IN} = 5$ MΩ; for $A_d = 100$, $R_{IN} = 10$ MΩ; and for $A_d = 1000$, $R_{IN} = 100$ MΩ.

EXAMPLE 5.2

A determination of the effect on the behavior of the feedback circuit of Figure 5.4 of a difference amplifier with a nonzero output resistance R_O is desired.

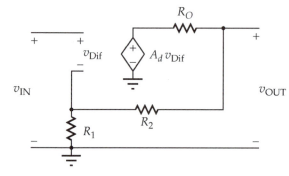

Figure 5.13: Equivalent circuit for a difference amplifer with a nonzero output resistance.

a. Determine an expression for the equivalent output resistance of the amplifier R_{OUT} for feedback resistances R_1 and R_2 that are large compared with R_O.

b. Assume $R_O = 100\ \Omega$. Evaluate R_{OUT} for $A_d = 50$, 100, and 1000. The feedback circuit for each case is to be such that v_{OUT}/v_{IN} is equal to 10 $(A_{fb} = 10)$.

SOLUTION

a. The equivalent circuit of Figure 5.5 will again be modified; this time an amplifier output resistance of R_O will be added to the circuit (Figure 5.13). The equivalent output resistance is the Thévenin equivalent resistance seen looking into the output terminals when the independent voltage source v_{IN} is properly removed, that is, replaced by a short circuit. As a result of the external voltage v_X of Figure 5.14, a current i_X results; the ratio v_X/i_X is the equivalent output resistance R_{OUT}. This use of an external voltage is essentially the same method by which an ohmmeter determines an unknown resistance.

$$v_{Dif} = -\beta v_{OUT} = -\beta v_X$$

$$i_X = \frac{v_X}{R_1 + R_2} + \frac{v_X - A_d v_{Dif}}{R_O}$$

$$= \frac{v_X}{R_1 + R_2} + \frac{v_X(1 + \beta A_d)}{R_O}$$

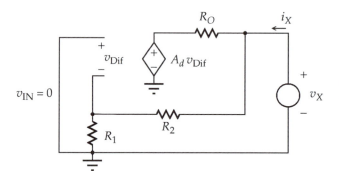

Figure 5.14: A circuit for determining R_{OUT}.

If $R_1 + R_2 \gg R_O$, the first term of the above expression may be ignored.

$$R_{OUT} = v_X/i_X = R_O/(1 + \beta A_d) = R_O A_{fb}/A_d$$

Because $A_{fb} < A_d$, the equivalent output resistance is less than R_O.

b. Because $R_O = 100 \ \Omega$ and $A_{fb} = 10$ for all cases, $R_{OUT} = 1000 \ \Omega/A_d$. Hence, for $A_d = 50$, $R_{OUT} = 20 \ \Omega$; for $A_d = 100$, $R_{OUT} = 10 \ \Omega$; and for $A_d = 1000$, $R_{OUT} = 1.0 \ \Omega$.

EXAMPLE 5.3

A SPICE simulation of the nonlinear amplifier with feedback, discussed in this section, is desired (quadratic dependence on input voltage, $A_1 = 10$, $A_2 = 2.5$). Evaluate the behavior of these circuits for difference amplifiers with gains of 10, 100, and 1000. For each case, the feedback circuit R_1 and R_2 is to be such that $v_{OUT}/v_{IN} = 10$ for a low-level output signal (linear behavior). Determine the static transfer characteristic v_{OUT} versus v_{IN}, for each circuit $(-1.0 \leq v_{IN} \leq 1.0 \text{ V})$. Use these characteristics to determine the error in v_{OUT} from that expected for an ideal amplifier $(10 \ v_{IN})$ for input voltages of 0.8 and 0.9 V. Using a transient-type solution, determine the harmonic distortion for a sinusoidal input voltage with an amplitude of 0.9 V and a frequency of 1 kHz.

SOLUTION The behavior of the nonlinear amplifier may be simulated with a polynomial controlled voltage source. If the magnitude of the amplifier's input voltage is used, a single equation is obtained for v_{OUT} that is valid for positive as well as negative values of v_1 as follows:

$$v_{OUT} = A_1 v_1 - A_2 v_1 |v_1| \quad \text{for } |v_1| < V_{1\,max}$$

A full-wave rectifier circuit is convenient for obtaining the magnitude of a voltage (subcircuit of Figure 5.15). Ideal diodes may be simulated by using a very small value for the ideality factor n of the diodes. Because SPICE is a purely numerical analyzer, parameters are not restricted to those of physically realizable devices (n of 1 to 2). If $n = 0.001$ and $I_s = 10^{-10}$ A, a forward-biased voltage of only 0.42 mV results in a diode current of 1.0 mA, and the diode's reverse saturation current is only 0.1 nA. A polynomial dependence is used for the voltage source; the two sets of node numbers for the two input voltages follow the POLY(2) specification. The coefficients b_n correspond to the following for two inputs x_1 and x_2:

$$y = b_0 + b_1 x_1 + b_2 x_2 + b_3 x_1^2 + b_4 x_1 x_2 + b_5 x_2^2 + \cdots$$

Figure 5.15: Subcircuit for nonlinear amplifier.

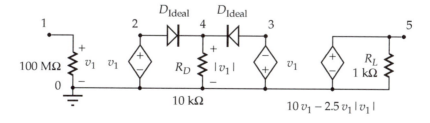

Different feedback circuits are required for each value of difference amplifier gain:

$$\beta = (A_d - 1)/A_d A_1$$

$A_d = 10$	$\beta = 0.09$	$R_1 = 0.9\ \text{k}\Omega$	$R_2 = 9.1\ \text{k}\Omega$
$A_d = 100$	$\beta = 0.99$	$R_1 = 0.99\ \text{k}\Omega$	$R_2 = 9.01\ \text{k}\Omega$
$A_d = 1000$	$\beta = 0.0999$	$R_1 = 0.999\ \text{k}\Omega$	$R_2 = 9.001\ \text{k}\Omega$

The resistance values of R_1 and R_2 are arbitrary; it is only their ratio that determines β. However, the values specified are values that would very likely be appropriate for an actual amplifier. The SPICE circuit and corresponding circuit file are given in Figure 5.16. The .DC simulation results in the static

Figure 5.16: SPICE circuit and file for Example 5.3.

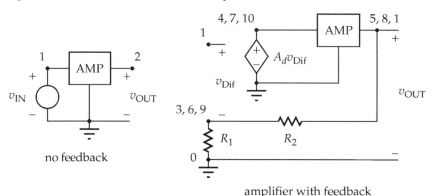

no feedback

amplifier with feedback

```
Feedback Amplifier


VIN   1   0   SIN(0 .9 1000)
X1    1   2   AMP


R11   3   0   .9K
R21   5   3   9.1K
E11   4   0   1   3   10
X2    4   5   AMP


R12   6   0   .99K
R22   8   6   9.01K
E12   7   0   1   6   100
X3    7   8   AMP


R13   9   0   .999K
R23   11  9   9.001K
E13   10  0   1   9   1000
X4    10  11  AMP
```

```
.DC   VIN   -1   1   .02
.TRAN   2U   1M   0   2U
.FOUR 1000 V(2) V(5) V(8) V(11)
.PROBE


.SUBCKT   AMP   1   5
RI    1   0   100MEG
E2    2   0   1   0   1
E3    0   3   1   0   1
D1    2   4   DIDEAL
D2    3   4   DIDEAL
RD    4   0   10K
.MODEL DIDEAL D N=.001 IS=1E-10
E4 5 0 POLY(2) 1 0 4 0
+0 10 0 0 -2.5
RL    5   0   1K
.ENDS
.END
```

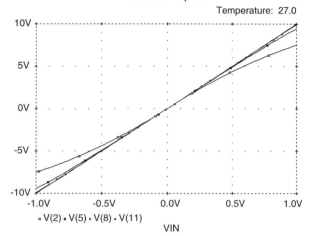

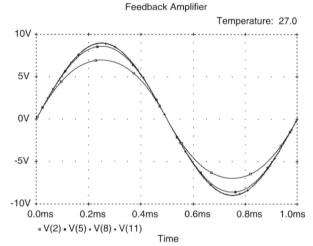

Figure 5.17: SPICE simulations for Example 5.3.

characteristic, and the .TRAN simulation results in the dynamic response of Figure 5.17. The .FOUR statement yields a Fourier analysis of the output voltages (an output listing similar to that of Figure 5.33 is obtained). The voltage errors for $v_{IN} = 0.8$ and 0.9 V as well as the harmonic distortion are summarized in Table 5.2.

TABLE 5.2 VOLTAGE ERRORS AND HARMONIC DISTORTION

A_d	$v_{IN} = 0.8$ V v_{Error}	$v_{IN} = 0.9$ V v_{Error}	$V_m = 0.9$ V Distortion
0	−1.60 V	−2.024 V	4.8%
10	−0.274 V	−0.393 V	1.02%
100	−30.2 mV	−45.7 mV	0.12%
1000	−3.05 mV	−4.66 mV	0.012%

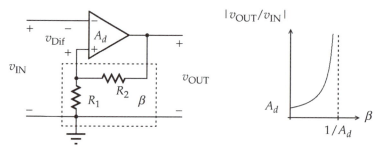

Figure 5.18: An amplifier with positive feedback.

5.2 STABILITY: NOT ALL AMPLIFIERS ARE EQUAL

In the analysis of the previous section, it was assumed that the difference amplifier's output voltage was dependent on only the instantaneous value of its input voltage. As a result of unavoidable circuit and device capacitances, the output of a physically realizable amplifier depends on derivatives and integrals, with respect to time, of the input voltage. Variations of an input voltage tend to be attenuated, and the resultant variations in output voltage are delayed. These capacitive effects can result in an unstable circuit when negative feedback is used. An unstable circuit may oscillate – a condition that is unacceptable for an amplifying circuit. As a consequence, the design of an amplifier for use with negative feedback is critical; only with a well-designed amplifier can the benefits of negative feedback be realized.

Oscillations can occur as a result of positive feedback (regeneration). Consider the circuit of Figure 5.18 in which the feedback signal βv_{OUT} is returned to the positive terminal of a difference amplifier. If it is assumed that $\beta = R_1/(R_1 + R_2)$ and the amplifier is linear, the following is obtained:

$$v_{Dif} = \beta v_{OUT} - v_{IN}$$
$$v_{OUT} = A_d v_{Dif} = A_d(\beta v_{OUT} - v_{IN})$$

$$\frac{v_{OUT}}{v_{IN}} = \frac{-A_d}{1 - \beta A_d} \quad \text{positive feedback}$$

(5.26)

As indicated in Figure 5.18, positive feedback increases the magnitude of the gain of the circuit ($\beta < 1/A_d$). For $\beta = 1/A_d$, the magnitude of the gain is infinite, that is, an output voltage can exist for a zero input voltage. As a result of transients associated with turning the circuit on, as well as electrical noise, many circuits will tend to oscillate (some circuits may go to a "latched up" condition in which no amplification occurs). In general, the output voltage will tend to increase for $\beta = 1/A_d$ until nonlinear effects limit the response.

Although a reversal of the difference amplifier's input terminals results in positive feedback, this also occurs if the gain of the amplifier should become negative. In the conventional feedback amplifier circuit of Figure 5.19, positive feedback occurs

Figure 5.19: A conventional feedback amplifier circuit.

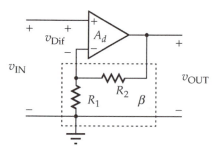

for βA_d negative, the result of either a negative gain or a negative feedback fraction, which becomes possible if other than a simple resistor network is used for the feedback circuit. An infinite response (possible oscillations) occurs for $\beta A_d = -1$.

AMPLIFIER PHASE SHIFT

With sinusoidal signals, a change in the sign of the amplifier gain occurs when the phase of the output signal differs by 180° from its input signal. A phase shift occurs for circuits with a capacitor such as the basic RC circuit of Figure 5.20. This circuit could be a part of the equivalent circuit of an amplifier. A phasor analysis based on a time dependence of $e^{j\omega t}$ is convenient for analyzing this circuit. Introducing an impedance of $1/j\omega C$ for the capacitor, the following is obtained:

$$V_2 = \frac{(1/j\omega C)V_1}{R + 1/j\omega C} = \frac{V_1}{1 + j\omega RC} \tag{5.27}$$

The ratio V_2/V_1 has both magnitude and phase as follows:

$$\frac{V_2}{V_1} = Ae^{j\theta}$$

$$A = \left| \frac{V_2}{V_1} \right| = \frac{1}{\sqrt{1 + (\omega RC)^2}} \tag{5.28}$$

$$\theta = -\tan^{-1}(\omega RC)$$

For a radian frequency of $1/RC$, the magnitude of V_2/V_1 is $1/\sqrt{2}$, and its phase is $-45°$. For higher frequencies, the magnitude decreases, and the phase tends toward $-90°$.

A circuit such as that of Figure 5.20 may exist in the signal path of a difference amplifier. If a linear amplifier is assumed, V_1 is proportional to the phasor representing the amplifier's input voltage, and the phasor output voltage of the amplifier is proportional to V_2. Hence, the amplifier's output voltage not only decreases with increasing frequency, but it is also shifted in phase as a result of the RC circuit. The first case that will be considered is that of a difference amplifier that has only these two elements affecting its frequency-dependent behavior. With the introduction of ω_b for $1/RC$ (the subscript b denoting break – a "break frequency"),

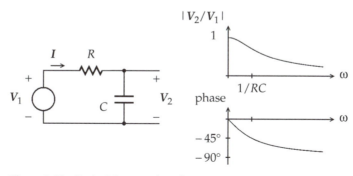

Figure 5.20: Basic RC network and response.

the following is obtained for the phasor response of the difference amplifier:

$$\frac{V_{out}}{V_{dif}} = A_d = \frac{A_{d0}}{1 + j\omega/\omega_b} \tag{5.29}$$

In this expression, A_{d0} is the gain for zero frequency, a quantity that will be assumed to be real and positive. The parameter ω_b is the radian break frequency, the frequency at which the magnitude of the gain is $A_{d0}/\sqrt{2}$ and the phase of the gain is $-45°$.

The phasor response of the feedback circuit of Figure 5.19 for an amplifier with a response given by Eq. (5.29) may now be determined as follows:

$$\frac{V_{out}}{V_{in}} = A_{fb} = \frac{A_d}{1 + \beta A_d} = \left(\frac{1}{\beta}\right)\frac{1}{1 + 1/\beta A_d} \tag{5.30}$$

$$\frac{1}{\beta A_d} = \frac{1 + j\omega/\omega_b}{\beta A_{d0}}$$

When the preceding equations are combined, the following is obtained:

$$\frac{V_{out}}{V_{in}} = \left(\frac{1}{\beta}\right)\frac{\beta A_{d0}}{\beta A_{d0} + 1 + j\omega/\omega_b}$$

$$= \left(\frac{1}{\beta}\right)\left(\frac{\beta A_{d0}}{1 + \beta A_{d0}}\right)\frac{1}{1 + j\omega/\omega_b(1 + \beta A_{d0})} \tag{5.31}$$

This expression may be simplified by introducing a new break frequency ω_h (h denotes high).

$$\omega_h = \omega_b(1 + \beta A_{d0})$$

$$\frac{V_{out}}{V_{in}} = \left(\frac{1}{\beta}\right)\left(\frac{\beta A_{d0}}{1 + \beta A_{d0}}\right)\frac{1}{1 + j\omega/\omega_h} \tag{5.32}$$

The overall feedback circuit has the same type of response as that of the amplifier, albeit its break frequency ω_h is higher than that of the difference amplifier.

The significance of the preceding theoretical result can best be illustrated with a numerical example. Consider the situation for which the difference amplifier has a low-frequency gain of 1000 ($A_{d0} = 1000$) and a Hertzian break frequency of 1 kHz ($f_b = \omega_b/2\pi = 1$ kHz):

$$A_d = \frac{A_{d0}}{1 + j\omega/\omega_b} = \frac{A_{d0}}{1 + jf/f_b} = \frac{1000}{1 + jf/1000} \tag{5.33}$$

A feedback network with $\beta = 0.1$ will be assumed. For a difference amplifier with an infinite gain, the gain with feedback A_{fb} would be $1/\beta$, that is, 10. For a difference amplifier with a finite zero frequency gain of 1000, the resultant gain is slightly less than 10.

$$\left(\frac{V_{out}}{V_{in}}\right)_{\omega=0} = \left(\frac{1}{\beta}\right)\left(\frac{\beta A_{d0}}{1 + \beta A_{d0}}\right) = 9.90 \tag{5.34}$$

The Hertzian break frequency for the response of the amplifier with feedback f_h may also be obtained as follows:

$$f_h = f_b(1 + \beta A_{d0}) = 101 \text{ kHz} \tag{5.35}$$

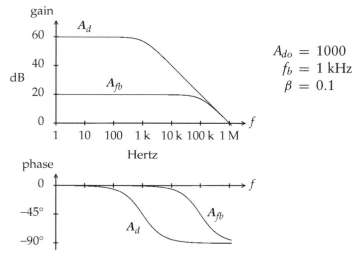

Figure 5.21: Response of a feedback circuit for a difference amplifier with a single break frequency.

The amplitude (expressed in decibels) and phase of A_d and A_{fb} are given in Figure 5.21 (for convenience, a logarithmic scale has been used for the frequency). The gain with feedback, not surprisingly, is less than that without feedback; the response curve for feedback is constrained by the response curve of the difference amplifier.

The response of Figure 5.21 for the difference amplifier with feedback can be described as "well behaved" – with increasing frequency the magnitude of the gain A_{fb} has a smooth falloff, and its phase approaches $-90°$ in a gradual fashion. If the upper break frequency ω_b is adequate, this difference amplifier and feedback circuit has a response characteristic that is adequate for most applications. Unless special care is exercised in the design of a difference amplifier, a response with more than a single break frequency will occur. In essence, most amplifiers tend to have more than a single equivalent RC circuit that determines their frequency-dependent behavior. Consider the case for an amplifier with two break frequencies ω_{b1} and ω_{b2}:

$$A_d = |A_d|e^{j\theta} = \frac{A_{d0}}{(1 + j\omega/\omega_{b1})(1 + j\omega/\omega_{b2})}$$

$$|A_d| = \frac{A_{d0}}{\sqrt{1 + (\omega/\omega_{b1})^2}\sqrt{1 + (\omega/\omega_{b2})^2}} \qquad (5.36)$$

$$\theta = -\tan^{-1}(\omega/\omega_{b1}) - \tan^{-1}(\omega/\omega_{b2})$$

The response of an amplifier of this type is given in Figure 5.22(a); $f_{b1} = 1$ kHz and $f_{b2} = 1$, 10, and 100 kHz. As a result of the two break frequencies, the magnitude of the response falls off more rapidly with increasing frequency than the magnitude of an amplifier with a single break frequency. Furthermore, with increasing frequency, the phase shift for the two break frequencies tends toward $-180°$. Because a phase shift of $-180°$ corresponds to a reversal of the sign of

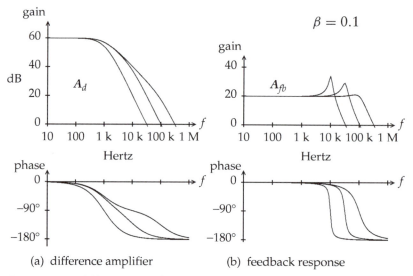

(a) difference amplifier   (b) feedback response

Figure 5.22: A difference amplifier with two break frequencies. For all responses, the left-hand curve is for $f_b = 1$ kHz, the middle curve is for $f_b = 10$ kHz, and the right-hand curve is for $f_b = 100$ kHz.

the gain of the difference amplifier (positive feedback), this difference amplifier would be expected to have a feedback response that reflects this effect.

The response of the amplifier of Fig. 5.22(a) for a $\beta = 0.1$ feedback circuit is given in Figure 5.22(b) (a numerical calculation or SPICE simulation is necessary to obtain these results; see Example 5.4). With increasing frequency, the gain of the feedback amplifier circuit tends to increase to a peak before it falls off. This effect, which is very pronounced for $f_{b2} = 1$ and 10 kHz, is predicted by the following feedback expression:

$$A_{fb} = \frac{A_d}{1 + \beta A_d} \qquad (5.37)$$

The peak in response occurs as a result of the denominator of this expression becoming small, that is, when βA_d approaches -1. Because the phase angles for the difference amplifiers with two break frequencies approach $-180°$, the peak in the response tends to occur at the frequency for which $|\beta A_d| = 1$. Because $\beta = 0.1$, this occurs for $|A_d| = 10$ (20 dB). The frequencies at which this occurs, as well as the corresponding phase angles, are summarized in Table 5.3. The magnitude of the gain, $|A_{fb}|$, for $f_{b2} = 1$ and 10 kHz, is considerably larger than its low-frequency value of approximately 10. For $f_{b2} = 100$ kHz, the gain with

TABLE 5.3   THE FEEDBACK RESPONSE FOR $|\beta A_D| = 1$

f_{b2}	f	θ	$|A_{fb}|_{dB}$	$|A_{fb}|$
1 kHz	9.96 kHz	$-168.5°$	33.9	49.6
10 kHz	30.8 kHz	$-160.1°$	29.2	28.8
100 kHz	89.5 kHz	$-127.4°$	21.1	11.4

feedback, $|A_{fb}|$, is only slightly larger than 10, which is the result of having a phase angle of $-127.4°$ – an angle considerably removed from $-180°$.

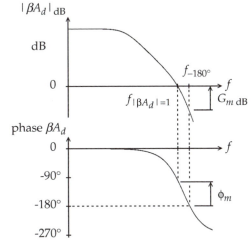

Figure 5.23: The response of βA_d.

STABILITY

A determination of the precise conditions that result in a stable or unstable feedback amplifier configuration is beyond the scope of the present treatment (see Bode 1975; Gray and Meyer 1992; Millman and Grabel 1987; Nyquist 1932; Sedra and Smith 1991; Wait, Huelsman, and Korn 1992). For most applications, not only is a stable feedback amplifier required, but a well-behaved response is also desired. It is the behavior of βA_d for frequencies that cause it to have a value close to -1 that is important.

A set of criteria based on the gain and phase margins of a difference amplifier and its feedback circuit is generally used for predicting whether the response with feedback A_{fb} will be acceptable. Both the magnitude and phase for a typical difference amplifier βA_d are indicated in Figure 5.23. This response might be obtained through a detailed theoretical analysis of the difference amplifier or from a set of experimental measurements. Two frequencies are indicated on the figure: $f_{|\beta A_d|=1}$, the frequency for which the magnitude of βA_d is unity, and $f_{-180°}$, the frequency for which the phase of βA_d is $-180°$. These frequencies tend to be close to the frequency at which $|1 + \beta A_d|$ is a minimum. Hence, the value of βA_d at these frequencies tends to be useful for estimating the behavior of the feedback circuit.

The gain margin $G_{m\,dB}$ is defined as the amount by which $|\beta A_d|$, expressed in decibels, is less than zero for $f = f_{-180°}$.

$$G_{m\,dB} = -|\beta A_d|_{dB} \quad \text{for the frequency corresponding}$$
$$\text{to the phase of } \beta A_d = -180° \qquad (5.38)$$

For this frequency, $\beta A_d = -|\beta A_d|$, that is, it is a negative real quantity. A stable feedback amplifier circuit generally requires that $|\beta A_d|$ be less than 1 for this frequency ($G_{m\,dB} > 0$). The phase margin ϕ_m is the phase angle by which the phase of βA_d deviates from $-180°$ for $f = f_{|\beta A_d|=1}$ defined by

$$\phi_m = \text{phase } \beta A_d + 180° \quad \text{for the frequency corresponding}$$
$$\text{to } |\beta A_d| = 1 \qquad (5.39)$$

The $+180°$ is a result of an extra minus sign introduced when obtaining the difference quantity. For a stable amplifier, it is generally required that the phase margin be greater than zero, that is, the phase shift should not have reached $-180°$ for the frequency at which $|\beta A_d| = 1$. In general, both the gain and phase margin conditions need to be satisfied for the feedback circuit to be stable.

The gain and phase margins are also useful for predicting the response of a stable amplifier feedback circuit. For most difference amplifiers, the response with feedback will be relatively well behaved if the gain margin is at least 10 dB and the phase margin is at least 45°, that is,

well-behaved response: $G_{m\,dB} \geq 10$ dB, $|\beta A_d| \leq 3.16$

$$\phi_m \geq 45°, \quad \text{phase } \beta A_d \geq -135° \tag{5.40}$$

It is these criteria that are generally used when designing an amplifier that is to be used with negative feedback circuits. Although an ideal response (for example, no peak in the response) is not assured, a response acceptable for many applications (a very small peak) is obtained.

The gain and phase margins depend on the feedback fraction β of the circuit. If β is a real quantity (such as that for a resistor network), a change in β affects only the $|\beta A_d|$ response, not the phase response. Increasing β moves the $|\beta A_d|$ curve upward; decreasing β moves the $|\beta A_d|$ curve downward. A change of β alters both the gain and phase margins. A decrease in β will generally result in an increase in the gain and phase margins. A smaller value of β implies a larger low-frequency gain for the feedback circuit.

EXAMPLE 5.4

A SPICE simulation of the difference amplifier with two break frequencies that produced the feedback response of Figure 5.22(b) is desired.
a. Verify that its frequency response is indeed that of Figure 5.22b.
b. Determine the time-dependent response of the output voltage for an input voltage pulse with an amplitude of 0.1 V and a duration of 0.5 μs.

SOLUTION Two basic RC circuits are required to simulate the behavior of an amplifer with two break frequencies:

$$f_b = \frac{\omega_b}{2\pi} = \frac{1}{2\pi RC}, \quad RC = \frac{1}{2\pi f_b}$$

For a break frequency of 1 kHz, a time constant RC of 0.1592 ms is required. Any combination of resistance and capacitance could be used – a 1-kΩ resistor and a 0.1592-μF capacitor are indicated in the circuit file of Figure 5.24. Break frequencies of 10 kHz and 100 kHz are achieved with capacitances of 0.01592 and 0.001592 μF, respectively ($R = 1$ kΩ). A dependent voltage source is required to isolate the two RC networks – the output of one cannot be directly connected to the input of the other. A dependent voltage source is also required for the output of the amplifier. Three independent circuits are included in the circuit file so that the response for the three values of f_{b2} can be obtained simultaneously.
a. The frequency-dependent response of the feedback amplifier circuit is indeed that given in Figure 5.22(b). The peak in the frequency response for $f_{b2} = 1$ kHz occurs at a frequency of about 10 kHz and that for $f_{b2} = 10$ kHz at approximately 31 kHz.

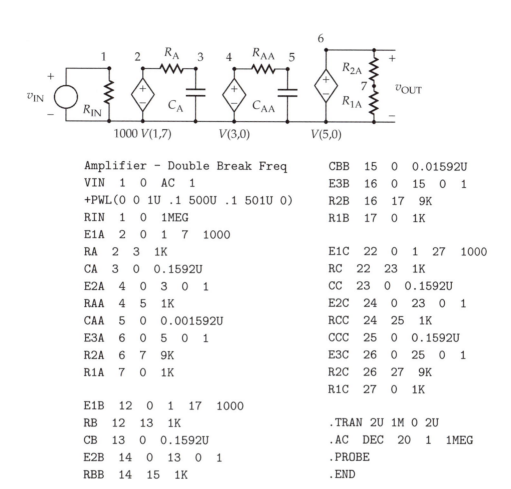

1000 $V(1,7)$ $V(3,0)$ $V(5,0)$

```
Amplifier - Double Break Freq        CBB   15   0   0.01592U
VIN   1   0   AC   1                  E3B   16   0   15   0   1
+PWL(0 0 1U .1 500U .1 501U 0)        R2B   16   17   9K
RIN   1   0   1MEG                    R1B   17   0   1K
E1A   2   0   1   7   1000
RA    2   3   1K                      E1C   22   0   1   27   1000
CA    3   0   0.1592U                 RC    22   23   1K
E2A   4   0   3   0   1               CC    23   0   0.1592U
RAA   4   5   1K                      E2C   24   0   23   0   1
CAA   5   0   0.001592U               RCC   24   25   1K
E3A   6   0   5   0   1               CCC   25   0   0.1592U
R2A   6   7   9K                      E3C   26   0   25   0   1
R1A   7   0   1K                      R2C   26   27   9K
                                      R1C   27   0   1K
E1B   12   0   1   17   1000
RB    12   13   1K                    .TRAN  2U  1M  0  2U
CB    13   0   0.1592U                .AC    DEC  20  1   1MEG
E2B   14   0   13   0   1             .PROBE
RBB   14   15   1K                    .END
```

Figure 5.24: SPICE circuit and file for Example 5.4. Three circuits are included so that responses can be obtained for $f_{b2} = 1$, 10, and 100 kHz simultaneously.

Figure 5.25: Transient solution for Example 5.4. The largest amplitude oscillations occur for $f_{b2} = 1$ kHz, whereas the response with only a small overshoot is for $f_{b2} = 100$ kHz.

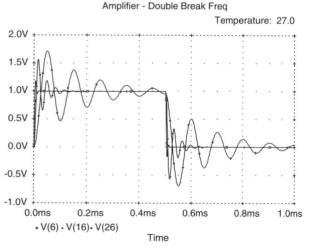

Amplifier - Double Break Freq

Temperature: 27.0

□ V(6) · V(16)· V(26)

Time

b. The time-dependent response of the feedback amplifier circuit is indicated in Figure 5.25. The damped oscillations triggered by the rising and falling edges of the input pulse have a frequency very close to the frequency for which the peak in the frequency response occurred. It should be noted that the response for the second break frequencies of 1 and 10 kHz would not be acceptable for most applications. On the other hand, the very small overshoot of the response occurring for $f_{b2} = 100$ kHz would be acceptable for most applications.

EXAMPLE 5.5

Consider the case for which a difference amplifier has three identical break frequencies as follows:

$$A = \frac{A_{d0}}{(1 + jf/f_b)^3}, \quad A_{d0} = 1000, \quad f_b = 1 \text{ kHz}$$

The difference amplifer is used with a feedback circuit having $\beta = 0.1$.

a. Show that the gain and phase margins of this circuit are negative, that is, the overall circuit is unstable.

b. Use SPICE to determine the frequency response of the circuit. This is possible because the .AC command does not check if a circuit is stable.

c. Determine the response of the amplifier circuit for an input step function voltage of 0.1 V. Because this circuit is unstable, the .TRAN command should result in an output voltage that grows indefinitely with time.

SOLUTION

a. Both the magnitude and phase of βA_d may be readily obtained.

$$|\beta A_d| = \frac{\beta A_{d0}}{(1 + f^2/f_b^2)^{3/2}}$$

$$\text{phase } \beta A_d = -3 \tan^{-1}(f/f_b)$$

The phase margin corresponds to the frequency for which $|\beta A_d| = 1$.

$$1 + f^2/f_b^2 = (\beta A_{d0})^{2/3}$$

$$f = f_b\sqrt{(\beta A_{d0})^{2/3} - 1} = 4.533 \text{ kHz} = f_{|\beta A_d| = 1}$$

$$\phi_m = -3 \tan(f/f_b) + 180° = -52.7°$$

The gain margin corresponds to the frequency for which $\theta = -180°$.

$$-3 \tan(f/f_b) = -180°, \quad f/f_b = \sqrt{3}$$

$$|\beta A_d| = \frac{\beta A_{d0}}{(1 + f^2/f_b^2)^{3/2}} = 12.5$$

$$G_{m\text{dB}} = -20 \log |\beta A_d| = -21.9 \text{ dB}$$

Because the phase and gain margins are negative, the circuit may be judged to be unstable.

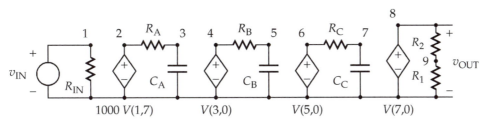

```
Amplifier - Tripple Break Freq
VIN  1  0  AC  1  PWL(0 0 1U .1)
RIN  1  0  1MEG
EA   2  0  1  9  1000
RA   2  3  1K
CA   3  0    0.1592U
EB   4  0  3 0 1
RB   4  5  1K
CB   5  0  0.1592U
EC   6  0  5  0  1
RC   6  7  1K
CC   7  0  0.1592U
ED   8  0  7  0  1
R2   8  9  9K
R1   9  0  1K
.TRAN  1U  500U  0  1U
.AC  DEC  40  1  1MEG
.PROBE
.END
```

Figure 5.26: SPICE circuit and file for Example 5.5. Three RC circuits are needed to simulate the amplifer with three break frequencies.

Figure 5.27: Frequency-dependent response of Example 5.5.

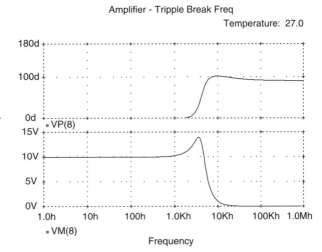

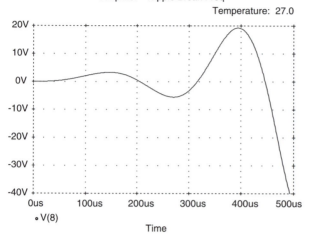

Amplifier - Tripple Break Freq
Temperature: 27.0

∘ V(8)

Time

Figure 5.28: Time-dependent response of Example 5.5.

b. The SPICE circuit and file of Figure 5.26 will be used to obtain the phasor frequency response and the time-dependent response for a step function input voltage. Even though the feedback amplifier is unstable – an experimentally determined frequency response would not be possible – a simulation response is readily obtained. Although the response of Figure 5.27 appears "well behaved," the phase increase with frequency implies a negative time delay, a physically unrealizable result.

c. The transient response is indicated in Figure 5.28. The input step function results in an exponentially growing oscillating output voltage. Although the output voltage would be limited by the saturation of a physically realizable amplifier, the simulation result continues to grow with time (it would eventually be limited by the numerical limit of the simulation program).

EXAMPLE 5.6

A stable feedback amplifier can be obtained for the difference amplifier of Example 5.5 if β is reduced sufficiently. Determine the maximum value of β that ensures a gain margin of at least 10 dB and a phase margin of at least 45°. What is the low-frequency gain of the feedback amplifier with this value of β?

SOLUTION The gain margin occurs for a frequency at which the phase of βA_d is equal to $-180°$.

$$-3 \tan^{-1} f/f_b = -180°, \qquad f/f_b = \tan 60° = \sqrt{3}$$

$$|\beta A_d| = \frac{\beta A_{d0}}{\left(1 + f^2/f_b^2\right)^{3/2}} = \frac{\beta A_{d0}}{8}$$

For a gain margin of 10 dB, $|\beta A_d| = 0.316$.

$$\beta = 8 \, |\beta A_d|/A_{d0} = 2.528 \times 10^{-3}$$

A larger value of β results in a gain margin that is less than 10 dB. A phase margin of 45° occurs for the frequency at which the phase is $-135°$.

$$-3 \tan^{-1} f/f_b = -135°, \qquad f/f_b = \tan 45° = 1$$

For this frequency, $|\beta A_d| = 1$.

$$|\beta A_d| = \frac{\beta A_{d0}}{(1 + f^2/f_b^2)^{3/2}} = \frac{\beta A_{d0}}{2^{3/2}} = \frac{\beta A_{d0}}{2.828} = 1$$
$$\beta = 2.828 \times 10^{-3}$$

Hence, the largest acceptable value of β is 2.528×10^{-3}. This results in a gain margin of 10 dB and a phase margin that is slightly larger than $45°$. The low-frequency gain is readily obtained as follows:

$$(A_{fb})_{\omega \to 0} = \frac{A_{d0}}{1 + \beta A_{d0}} = 261$$

Only a moderate reduction in gain occurs (-11.7 dB) for this value of β. Because β is small, only a modest improvement in the performance of the circuit over the performance of the amplifer by itself would be expected.

5.3 ANALYSIS OF OPERATIONAL AMPLIFIER CIRCUITS: BASIC CONSIDERATIONS

The concept of an ideal response, the response of a circuit with an op amp that has an infinite gain (an "ideal op amp"), has been discussed in Section 5.1. For the circuit considered in Figure 5.4, it was shown that the overall response of a noninverting op amp configuration could be expressed as the product of the ideal response and an error term. Often the ideal response is sufficiently accurate in predicting the behavior of an op amp circuit; thus, the effect of the finite gain of the an op amp is negligible. However, for other circumstances the behavior of the op amp is important. In either case, an analysis to determine the ideal response of the circuit is needed.

When an op amp is treated as having an infinite gain, its input difference voltage v_{Dif} is zero for a finite output voltage. In essence, the signal fed back to the input of the op amp from its output is such as to result in a zero difference input voltage. This suggests an analysis of the circuit based on the concept of a "virtual short" for the input terminals of the op amp (Figure 5.29). Viewed from the external network, the input voltage of the op amp is zero, as would be expected if its input were a short circuit. However, the input current of the op amp is also zero. Hence, the virtual-short description applies – a zero voltage as well as a zero current.

IDEAL OP AMP – INPUT VIRTUAL SHORT

Consider the amplifier circuits of Figure 5.30; virtual shorts ($\updownarrow$) are indicated for the inputs of the op amps. For the noninverting amplifier of Figure 5.30(a), the input voltage v_{IN} is equal to the voltage across R_1 as follows:

Figure 5.29: The concept of a virtual short.

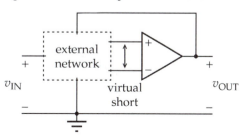

$$v_{\text{IN}} = \frac{R_1 v_{\text{OUT}}}{R_1 + R_2}, \qquad \frac{v_{\text{OUT}}}{v_{\text{IN}}} = 1 + \frac{R_2}{R_1}$$

$$(5.41)$$

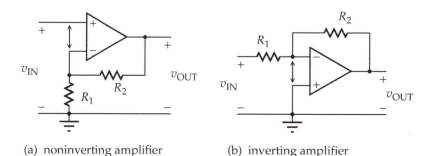

(a) noninverting amplifier   (b) inverting amplifier

Figure 5.30: Basic op amp circuits illustrating the use of a virtual short for obtaining their ideal response.

A zero input current has been assumed for the op amp. This is the ideal response that was previously obtained by solving the circuit for an op amp with a finite gain and then, after an expression for v_{OUT}/v_{IN} was obtained, letting the gain become infinite. The ideal response for the inverting amplifier (Figure 5.30(b)) is obtained by summing the currents into the inverting input of the amplifier; they must sum to zero because the input current of the op amp is zero. In addition, the voltage at the inverting terminal of the op amp is zero because the noninverting terminal is connected to the common point of the circuit (ground).

$$\frac{v_{IN}}{R_1} + \frac{v_{OUT}}{R_2} = 0, \qquad \frac{v_{OUT}}{v_{IN}} = -\frac{R_2}{R_1} \tag{5.42}$$

This is the ideal response of the inverting amplifier circuit.

The virtual-short concept is convenient for obtaining the ideal response of the amplifier circuit of Figure 5.31, which has two input voltages, v_1 and v_2. For this circuit, the input voltages of the op amp, v^+ and v^-, must be equal. If it is assumed that the input currents of the op amp are zero, these voltages are readily obtained.

$$v^- = \frac{R_4 v_2}{R_3 + R_4} + \frac{R_3 v_{OUT}}{R_3 + R_4}$$

$$v^+ = \frac{R_2 v_1}{R_1 + R_2} \tag{5.43}$$

Figure 5.31: An op amp with an inverting and a noninverting input.

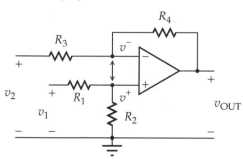

By equating these voltages, an expression for v_{OUT} in terms of v_1 and v_2 is obtained:

$$\frac{R_4 v_2}{R_3 + R_4} + \frac{R_3 v_{OUT}}{R_3 + R_4} = \frac{R_2 v_1}{R_1 + R_2}$$

$$v_{OUT} = \frac{R_2(R_3 + R_4)}{R_3(R_1 + R_2)} v_1 - \frac{R_4}{R_3} v_2 \tag{5.44}$$

$$= \frac{R_4(1 + R_3/R_4)}{R_3(1 + R_1/R_2)} v_1 - \frac{R_4}{R_3} v_2$$

If $R_1/R_2 = R_3/R_4$, the output voltage depends only on the difference of the input voltages, and thus

$$v_{OUT} = \frac{R_4}{R_3}(v_1 - v_2) \tag{5.45}$$
$$\text{for } R_1/R_2 = R_3/R_4$$

This circuit may therefore be used for obtaining a response that depends on the difference of two voltages.

An op amp circuit used for summing several input voltages is shown in Figure 5.32. Summing the currents into the inverting node of the op amp yields the following:

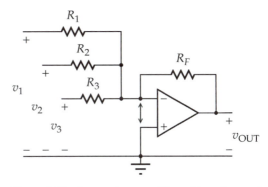

Figure 5.32: A summing amplifier with an inverted output.

$$\frac{v_1}{R_1} + \frac{v_2}{R_2} + \frac{v_3}{R_3} + \frac{v_{OUT}}{R_F} = 0$$
$$v_{OUT} = -\frac{R_F}{R_1}v_1 - \frac{R_F}{R_2}v_2 - \frac{R_F}{R_3}v_3 \tag{5.46}$$

The preceding result may be extended to any number of input voltages.

The virtual-short concept may also be used to determine the ideal response of an op amp circuit with a capacitive or inductive element such as the circuit of Figure 5.33. Again, a solution is obtained by summing the currents into the inverting node of the op amp as follows:

$$\frac{v_{IN}}{R_1} + \frac{v_{OUT}}{R_2} + C\frac{dv_{OUT}}{dt} = 0$$
$$\frac{dv_{OUT}}{dt} + \frac{v_{OUT}}{R_2C} = -\frac{v_{IN}}{R_1C} \tag{5.47}$$

A solution for v_{OUT} depends on the explicit time dependence of v_{IN}. Consider the case for an input that is a step function voltage occurring at $t = 0$ and has an amplitude of V_p:

$$v_{IN}(t) = V_p u(t)$$
$$\frac{dv_{OUT}}{dt} + \frac{v_{OUT}}{R_2C} = -\frac{V_p}{R_1C} \quad \text{for } t > 0 \tag{5.48}$$
$$v_{OUT} = Ae^{-t/R_2C} - R_2V_p/R_1$$

Figure 5.33: An op amp with a capacitive feedback circuit.

The quantity A is the constant of integration. If $v_{OUT} = 0$ at $t = 0$ (capacitor initially uncharged), the following is obtained:

$$v_{OUT} = -R_2V_p/R_1(1 - e^{-t/R_2C}) \tag{5.49}$$

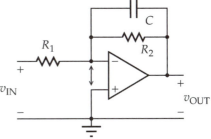

This solution is indicated in Figure 5.34. The time constant of the response, R_2C, is the time constant

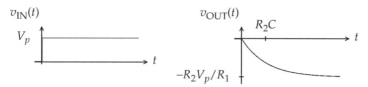

Figure 5.34: The response of the op amp circuit of Figure 5.33 for an input step function voltage.

of the circuit connected between the output of the op amp and its inverting input.

The op amp circuit of Figure 5.33 is often used to approximate the response of an integrating circuit. If R_2 is very large, the following is obtained from Eq. (5.47):

$$\frac{dv_{OUT}}{dt} \approx -\frac{v_{IN}}{R_1 C}, \quad v_{OUT} \approx -\frac{1}{R_1 C} \int v_{IN} \, dt \tag{5.50}$$

For $v_{IN} = V_p u(t)$, this expression yields the initial response of Figure 5.34, that is, the response for $t \ll R_2 C$.

The response of the op amp circuit of Figure 5.33 for a steady-state sinusoidal input voltage is also of interest. This response can be obtained using a phasor-type analysis in which the input voltage is represented by V_{in} and the output voltage by V_{out} as follows:

$$v_{IN}(t) = \mathrm{Re}\,(V_{in}e^{j\omega t}), \quad v_{OUT}(t) = \mathrm{Re}\,(V_{out}e^{j\omega t}) \tag{5.51}$$

By introducing an imaginary impedance of $1/j\omega C$ for the capacitor, the following is obtained:

$$\frac{V_{in}}{R_1} + \frac{V_{out}}{R_2} + j\omega C V_{out} = 0$$

$$\frac{V_{out}}{V_{in}} = -\left(\frac{R_2}{R_1}\right)\frac{1}{1 + j\omega R_2 C} = -\left(\frac{R_2}{R_1}\right)\frac{1}{1 + j\omega/\omega_h} \tag{5.52}$$

$$= -\left(\frac{R_2}{R_1}\right)\frac{1}{1 + jf/f_h}$$

$$\omega_h = 1/R_2 C, \quad f_h = 1/2\pi R_2 C$$

The frequency dependence of this response is similar to that of a simple RC circuit (Figure 5.20) already discussed. The magnitude expressed in decibels and the phase are indicated in Figure 5.35. The response for an input signal

Figure 5.35: Response of the circuit of Figure 5.33 for a sinusoidal input signal.

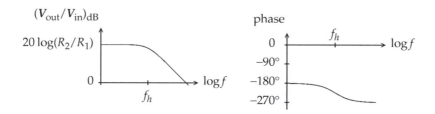

with a frequency of f_b is $1/\sqrt{2}$ (-3 dB) of that of the response for a very-low frequency signal. As a result of the minus sign of Eq. (5.52), a phase shift of $-180°$ (or $+180°$) occurs for low-frequency signals. Because the capacitor has the effect of reducing the gain of high-frequency signals, the overall circuit is generally described as a low-pass filter. If the frequency components of an input signal are less than some upper frequency, a circuit of this type may be used to reduce extraneous effects caused by noise.

OP AMP LIMITATIONS

The behavior of an actual integrated circuit op amp deviates from that of an ideal op amp – particularly in that its voltage gain, although large for low-frequency signals, decreases as the frequency is increased. This is the result of the frequency compensation used to achieve stable operation when feedback is used (Section 5.2). The frequency response of the op amp must be constrained, that is, the op amp must not have an excessive phase shift that could result in a distorted response or oscillations. It is primarily the compensating capacitor C_c of the op amp circuit of Figure 5.2 that is used to achieve an acceptable frequency response for the op amp.

A sufficiently large compensating capacitor is generally used, and thus the response of the op amp tends to be dominated by a single break frequency for frequencies at which the magnitude of its gain is greater than 1.

$$A_d = \frac{A_{d0}}{1 + j\omega/\omega_b} = \frac{A_{d0}}{1 + jf/f_b} \tag{5.53}$$

The amplitude expressed in decibels and the phase of this response are indicated in Figure 5.36. For a typical op amp, the low-frequency gain A_{d0} is 10^5 to 10^6 (100 to 120 dB). The break frequency f_b is quite small – less than 100 Hz ($f_b \approx 5$ Hz for a 741 and 25 Hz for a 356). This implies that for most signals of interest, the magnitude of the gain of the op amp is considerably less than A_{d0}, and its phase shift is close to $-90°$. Therefore, an ideal response, that corresponding to an op amp with an infinite gain, does not occur for most signals – the gain of the op amp affects the behavior of the overall circuit.

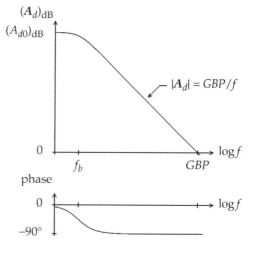

Figure 5.36: Frequency-dependent response of an integrated circuit op amp.

If $f \gg f_b$, then the op amp response of Eq. (5.53) may be approximated by the following expression:

$$A_d \approx -j A_{d0} f_b/f = -jGBP/f$$
$$GBP = A_{d0} f_b \quad \text{(the gain–bandwidth product)} \tag{5.54}$$

Manufacturers generally specify the gain–bandwidth product of op amps; seldom are its factors A_{d0} or f_b specified. Typical integrated circuit op amps have gain–bandwidth products in the range of 1–10 MHz. When $f = GBP$, then the magnitude of A_d is 1 or, expressed in decibels, 0 dB. The approximation for the response of the op amp, $-jGBP/f$, tends to be valid for most frequencies of interest.

If the frequency-dependent behavior of an op amp is known, the response of a circuit using the op amp may be obtained. Consider the noninverting amplifier of Figure 5.30(a). Its response can be expressed in terms of its ideal response and an error term. From a phasor notation, the following is obtained:

$$\frac{V_{out}}{V_{in}} = \left(\frac{V_{out}}{V_{in}}\right)_{Ideal} \frac{\beta A_d}{1 + \beta A_d}, \qquad \left(\frac{V_{out}}{V_{in}}\right)_{Ideal} = 1 + \frac{R_2}{R_1} \tag{5.55}$$

The feedback fraction, it will be recalled, is the fraction of the output voltage returned to the inverting input of the op amp. The approximate expression for the op amp gain yields the following:

$$\beta = R_1/(R_1 + R_2), \qquad \beta A_d = -j\beta GBP/f$$
$$\frac{V_{out}}{V_{in}} = \left(1 + \frac{R_2}{R_1}\right) \frac{1}{1 + 1/\beta A_d} = \left(1 + \frac{R_2}{R_1}\right) \frac{1}{1 + jf/\beta GBP} \tag{5.56}$$
$$= \left(1 + \frac{R_2}{R_1}\right) \frac{1}{1 + jf/f_h}, \quad f_h = \beta GBP$$

The upper half-power frequency for this circuit is f_h, a quantity considerably larger than f_b but less than the gain–bandwidth product of the op amp. The response of this amplifier, given in Figure 5.37, is constrained by the response of the op amp – the gain of the overall circuit falls below (or is approximately equal to) that of the op amp.

The result of Eq. (5.56), $f_h = \beta GBP$, and the corresponding response of Figure 5.37, illustrate an important concept of amplifying systems. The expression for f_h of the noninverting amplifier yields the following:

$$f_h/\beta = f_h(1 + R_2/R_1) = f_h (V_{out}/V_{in})_{Ideal} = GBP \tag{5.57}$$

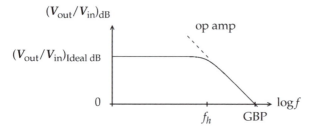

Figure 5.37: Frequency response of a noninverting amplifier.

The ideal gain, $(V_{out}/V_{in})_{Ideal}$, is the low-frequency voltage gain of the amplifier. The product of f_b and the low-frequency gain, the gain–bandwidth product of the amplifier circuit, is equal to the gain–bandwidth product of the op amp. If, for example, $GBP = 1$ MHz, an amplifier circuit with $(V_{out}/V_{in})_{Ideal} = 10$ will have an upper half-power frequency f_b of 100 kHz, whereas if the circuit is designed for a larger low-frequency gain of 100, $f_b = 10$ kHz. Although the gain of the amplifier may be arbitrarily chosen (by selecting appropriate values of R_1 and R_2), the resultant bandwidth of the circuit, f_b, is constrained by the gain–bandwidth product of the op amp. It is the active device, the op amp, that establishes the gain–bandwidth product of the amplifier circuit.

An expression that includes the effect of a finite op amp gain has been used for the preceding analysis.

$$\frac{V_{out}}{V_{in}} = \left(\frac{V_{out}}{V_{in}}\right)_{Ideal} \frac{\beta A_d}{1 + \beta A_d} \tag{5.58}$$

Although this expression was derived for a noninverting amplifier configuration, it may be shown that this expression holds for all linear amplifier circuits (Example 5.8). Consider the inverting amplifier circuit of Figure 5.30(b) – its ideal response has been determined:

$$\left(\frac{V_{out}}{V_{in}}\right)_{Ideal} = -\frac{R_2}{R_1} \tag{5.59}$$

The feedback fraction β is the fraction of the output voltage fed back to the inverting terminal (Figure 5.38). It may be determined by setting the input voltage v_{IN} to zero and removing the op amp (indicated by dashed lines) from the circuit as follows:

$$\beta = v^-/v_{OUT} = R_1/(R_1 + R_2) \tag{5.60}$$

This is the same expression as that for the noninverting amplifier. Therefore, the error term involving βA_d is the same:

$$\frac{V_{out}}{V_{in}} = \left(-\frac{R_2}{R_1}\right) \frac{1}{1 + jf/f_b}, \quad f_b = \beta GBP \tag{5.61}$$

For an ideal response of -10 (the low-frequency response), $R_2 = 10R_1$. This implies that $\beta = 1/11$. For $GBP = 1$ MHz, the upper half-power frequency of this amplifier is 91 kHz, as compared with 100 kHz for a noninverting amplifier with a gain of 10.

Figure 5.38: Determining β for an inverting amplifier configuration.

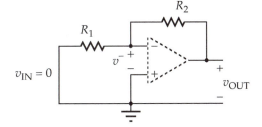

Closely related to the behavior associated with the frequency response of an op amp is its slew-rate limitation. An op amp is limited as to how rapidly its output voltage can change in an upward and downward direction.

$$SR = \left| \frac{dv_{OUT}}{dt} \right|_{max} \qquad \text{slew rate} \tag{5.62}$$

The slew rate, as with the frequency response, tends to depend on the compensating capacitor of the op amp. The slew rate arises as a result of a current limitation of the amplifying circuits that charge (or discharge) the compensating capacitor.

$$i = C_c \frac{dv_C}{dt}, \qquad \frac{dv_C}{dt} = \frac{i}{C_c} \tag{5.63}$$

The circuit sets a limit on how fast the voltage across the capacitor can change, that is, its time derivative. This derivative, in turn, establishes a limit for the derivative of the output voltage. General-purpose op amps have slew rates of 1 to 20 V/μs, whereas special-purpose, high-speed op amps have slew rates as high as 100 V/μs.

The combined effect of an op amp's frequency response (a linear effect) and a slew-rate limitation (a nonlinear effect) can be illustrated with an example. Consider a noninverting amplifier circuit with a low-frequency gain of 10 (ideal response). The op amp has a gain–bandwidth product of 1 MHz and a slew rate of 1 V/μs. Because $\beta = 1/10$, this results in an upper half-power frequency f_h of 100 kHz ($f_h = \beta GBP$). Suppose the amplifier has an input sinusoidal signal with a frequency of 100 kHz and a peak amplitude of 0.1 V. On the basis of the ideal response of this circuit, the output would be a sinusoidal signal with an amplitude of 1.0 V and would be in phase with the input signal. Because $f = f_h$, the actual response, expressed in terms of phasors, is the following:

$$\frac{V_{out}}{V_{in}} = \frac{10}{1 + j} = \frac{10 \, e^{-j45°}}{\sqrt{2}} = 7.07 \, e^{-j45°} \tag{5.64}$$

Hence, for $V_{in} = 0.1$ V, the amplitude of V_{out} is only 0.707 V, and it lags the input signal by 45°.

$$\begin{aligned} v_{IN}(t) &= (0.1 \text{ V}) \cos(2\pi f t) \\ v_{OUT}(t) &= (0.707 \text{ V}) \cos(2\pi f t - 45°) \end{aligned} \tag{5.65}$$

If it is assumed that this solution, which is based on the assumption of linear behavior, is valid, the derivative of $v_{OUT}(t)$ may be obtained as follows:

$$\frac{dv_{OUT}}{dt} = -2\pi f (0.707) \sin(2\pi f t - 45°) \tag{5.66}$$

For $f = f_h = 100$ kHz, the derivative varies between ±0.44 V/μs. The maximum magnitude is less than the slew rate of the op amp, namely 1 V/μs, and therefore the response of Eq. (5.84), based on an assumption of linear behavior, is valid (Figure 5.39(a)).

If, however, the amplitude of $v_{IN}(t)$ is increased, a slew-rate limitation occurs. Consider the case for a sinusoidal input voltage with a peak amplitude of 0.5 V. On the basis of an assumption of linear behavior, the output voltage would have

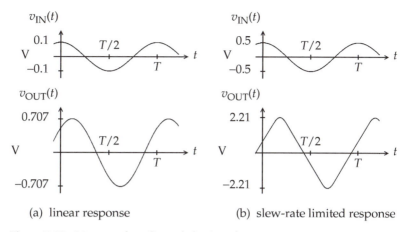

(a) linear response   (b) slew-rate limited response

Figure 5.39: Linear and nonlinear behavior of an op amp circuit.

a peak value of 3.54 V, that is, both voltages of Eq. (5.65) and Figure 5.39(a) would be magnified by a factor of 5. This implies a derivative of v_{OUT} that varies between ± 2.22 V/μs. Because the op amp has a slew rate of only 1 V/μs, a linear response does not occur. A nonlinear analysis (using SPICE) results in the output voltage of Figure 5.39(b). The magnitude of the derivative is constrained to 1 V/μs, thus resulting in a distorted output voltage waveform.

EXAMPLE 5.7

An inverting amplifier with a voltage gain of -100 and an input resistance of 10 kΩ is desired.

a. Design a conventional amplifier using an op amp circuit with two resistors. Assume an ideal response for the circuit (an op amp with an infinite gain).

b. Design an amplifier using the modified feedback network of Figure 5.40. No resistances are to be larger than 10 kΩ.

SOLUTION

a. The circuit of Figure 5.30(b) and the solution of Eq. (5.42) apply. The input resistance v_{IN}/i_{IN}, is equal to R_1. Therefore, $R_1 = 10$ kΩ and for a gain of -100, $R_2 = 100\,R_1 = 1$ MΩ.

Figure 5.40: Inverting op amp circuit of Example 5.7.

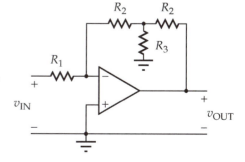

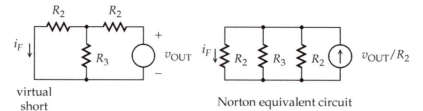

Norton equivalent circuit

Figure 5.41: Equivalent circuit for the feedback network of Example 5.7.

b. A virtual short may be assumed for the input terminals of the op amp, that
is, $v^- = 0$. The current of the feedback network i_F may be determined
using the Norton equivalent circuit of Figure 5.41 as follows:

$$i_F = \frac{G_2}{G_2 + G_3 + G_2}\left(\frac{v_{OUT}}{R_2}\right) = \frac{1/R_2}{1/R_2 + 1/R_3 + 1/R_2}\left(\frac{v_{OUT}}{R_2}\right)$$

$$= \frac{R_3 v_{OUT}}{2\,R_2\,R_3 + R_2^2}$$

As a result of the virtual short, the input resistance remains equal to R_1.
Therefore, $R_1 = 10\ \text{k}\Omega$.

$$\frac{v_{IN}}{R_1} + \frac{R_3 v_{OUT}}{2\,R_2\,R_3 + R_2^2} = 0$$

$$\frac{v_{OUT}}{v_{IN}} = -\frac{2\,R_2\,R_3 + R_2^2}{R_1\,R_3} = -\left(\frac{2\,R_2}{R_1} + \frac{R_2^2}{R_1\,R_3}\right)$$

If $R_2 = 10\ \text{k}\Omega$ (a maximum value), then $R_2 = R_1$.

$$\frac{v_{OUT}}{v_{IN}} = -(2 + R_2/R_3)$$

This implies that $R_2/R_3 = 98$ or $R_2 = 10\ \text{k}\Omega/98 = 102\ \Omega$. Although this
circuit has the same response as that of part (a), a very large resistance
(1 MΩ) is not required. This circuit would be preferred if an integrated
circuit were to be fabricated.

EXAMPLE 5.8

Use the circuit of Figure 5.29 to verify that the expression of Eq. (5.58) is valid
for an external network with linear elements.

SOLUTION The input difference voltage of the op amp is linearly dependent on
v_{IN} and v_{OUT}:

$$v_{Dif} = K v_{IN} - \beta v_{OUT}$$

The quantity K is equal to v_{Dif}/v_{IN} for $v_{OUT} = 0$, and β is equal to $-v_{Dif}/v_{OUT}$
for $v_{IN} = 0$. The quantity β is the negative feedback fraction of the circuit.
The ideal response corresponds to $v_{Dif} = 0$.

$$\left(\frac{v_{OUT}}{v_{IN}}\right)_{Ideal} = \frac{K}{\beta}$$

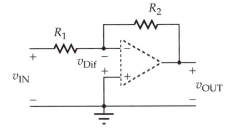

Figure 5.42: Noninverting amplifier of Example 5.8.

For a finite op amp gain of A_d, the following is obtained:

$$v_{\text{Dif}} = v_{\text{OUT}}/A_d = K v_{\text{IN}} - \beta v_{\text{OUT}}$$

$$\frac{v_{\text{OUT}}}{v_{\text{IN}}} = \frac{K A_d}{1 + \beta A_d} = \left(\frac{K}{\beta}\right) \frac{\beta A_d}{1 + \beta A_d} = \left(\frac{v_{\text{OUT}}}{v_{\text{IN}}}\right)_{\text{Ideal}} \frac{\beta A_d}{1 + \beta A_d}$$

A similar expression is obtained for phasor quantities. To illustrate the use of this expression, consider the inverting amplifier configuration of Figure 5.42. With the polarity of v_{Dif} taken into account, the following is obtained when the op amp is removed from the circuit:

$$K = v_{\text{Dif}}/v_{\text{IN}} = -R_2/(R_1 + R_2) \quad \text{for } v_{\text{OUT}} = 0$$
$$\beta = -v_{\text{Dif}}/v_{\text{OUT}} = R_1/(R_1 + R_2) \quad \text{for } v_{\text{IN}} = 0$$
$$\left(\frac{v_{\text{OUT}}}{v_{\text{IN}}}\right)_{\text{Ideal}} = \frac{K}{\beta} = -\frac{R_2}{R_1}$$

This is the same result as that obtained when an analysis based on a virtual short is used.

EXAMPLE 5.9

Obtain a SPICE simulation of a noninverting amplifier circuit having an op amp with a single break frequency; $GBP = 1$ MHz and $SR = 1$ V/μs.
a. Obtain a simulation model for the op amp that includes the slew-rate limitation.
b. The op amp is used in a circuit with a voltage gain of 10 (ideal response). Determine $v_{\text{OUT}}(t)$ for an input signal with a symmetrical square waveform $v_{\text{IN}}(t)$ that has a peak voltage of 0.1 V, a minimum voltage of 0 V, and a frequency of 50 kHz.

SOLUTION
a. The gain–bandwidth product is equal to $A_{d0} f_b$ (Eq. (5.54)). If $A_{d0} = 10^5$, then $f_b = 10$ Hz. (The behavior of the circuit may be shown to be independent of these values. For example, $A_{d0} = 10^6$ and $f_b = 1$ Hz will produce essentially the same result.) An RC circuit may be used to achieve the break frequency of 10 Hz:

$$f_b = 1/(2\pi RC), \qquad C = 1/(2\pi f_b R)$$

If $R = 1$ MΩ, then $C = 15.92$ nF. The circuit of Figure 5.43, if the diodes

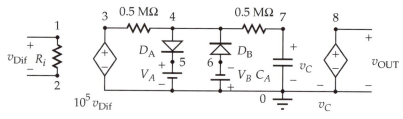

Figure 5.43: Op amp simulation circuit for Example 5.9.

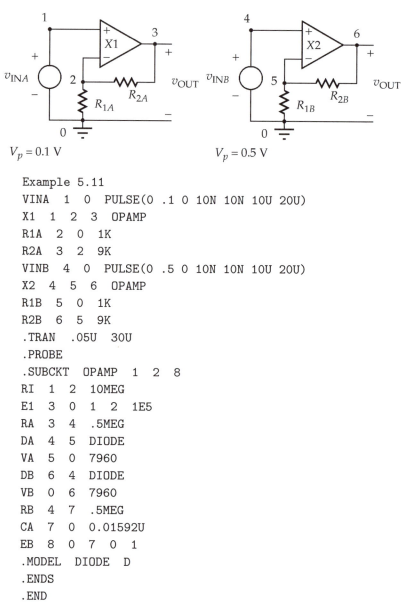

$V_p = 0.1$ V

$V_p = 0.5$ V

```
Example 5.11
VINA   1  0  PULSE(0 .1 0 10N 10N 10U 20U)
X1  1  2  3  OPAMP
R1A  2  0  1K
R2A  3  2  9K
VINB   4  0  PULSE(0 .5 0 10N 10N 10U 20U)
X2  4  5  6  OPAMP
R1B  5  0  1K
R2B  6  5  9K
.TRAN   .05U  30U
.PROBE
.SUBCKT  OPAMP  1  2  8
RI  1  2  10MEG
E1  3  0  1  2  1E5
RA  3  4  .5MEG
DA  4  5  DIODE
VA  5  0  7960
DB  6  4  DIODE
VB  0  6  7960
RB  4  7  .5MEG
CA  7  0  0.01592U
EB  8  0  7  0  1
.MODEL  DIODE  D
.ENDS
.END
```

Figure 5.44: SPICE circuit and file for Example 5.9.

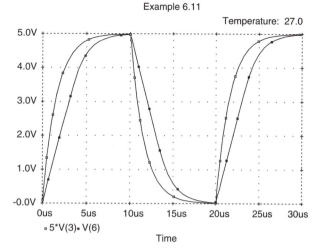

Figure 5.45: SPICE solution of Example 5.9.

of the circuit are not conducting, will produce the appropriate linear response of the op amp. For this circuit, the capacitor's voltage v_C is equal to v_{OUT}.

$$\frac{dv_{OUT}}{dt} = \frac{dv_C}{dt} = \frac{i}{C}$$

$$SR = |i|_{max}/C, \qquad |i|_{max} = C \cdot SR = 15.92 \text{ mA}$$

The diode circuits need to be such as to limit the magnitude of the current of the capacitor to 15.92 mA. This implies a voltage of 7960 V across the 0.5-MΩ resistor connected to the capacitor. Because $|v_C|$ is small, a value of $V_A = V_B = 7960$ V will achieve the limiting (for so large a voltage, the effect of the diode voltage $v_{D(on)}$, will be negligible).

b. A subcircuit will be used for the op amp to obtain $v_{OUT}(t)$ for the two input voltages (Figure 5.44). The output voltages are given in Figure 5.45. For the 0.1-V input voltage, the output voltage has been magnified by a factor of 5. If linear operation had occurred, the magnified voltage would have been identical to that for an input voltage with a peak value of 0.5 V.

5.4 PREEMPHASIS AND DEEMPHASIS CIRCUITS: DESIGN EXAMPLES

Audio signals, both voice and music, may be treated as consisting of a spectrum of signal components with different frequencies. To transmit a reasonable replica of a voice-produced signal, only a minimal bandwidth is needed. For example, telephone systems use a spectrum of approximately 300 to 3400 Hz, and specialized systems (fire, police, aircraft, etc.) use even a smaller bandwidth. A standard AM radio broadcast transmitter uses an audio spectrum of 20 to 5000 Hz, a bandwidth that, although adequate for voice, is not sufficient for high-quality music transmission. For a high-quality system (high-fidelity), a larger bandwidth, 20 Hz to 20 kHz, is needed. This is the spectrum used for commercial FM broadcasting (both monaural and stereo).

Although a wide bandwidth is needed for a high-fidelity system, the spectral distribution of the signal components is not uniform. Typical audio signals tend to have large-amplitude, low-frequency components (less than 1 kHz), whereas the amplitudes of high-frequency components (above 1 kHz) are much smaller. As a result, the wide spectrum required for a high-fidelity signal is poorly utilized – most of the signal power is concentrated in the lower part of the frequency spectrum. An overall improvement can be achieved by preemphasizing (over-amplifying) the signal's high-frequency components. At the other end of the system (for example, an FM receiver), a deemphasis circuit is used to restore the spectral amplitudes to their original values. As a result of the preemphasis–deemphasis circuits, a reduction in noise due to unavoidable electronic effects is achieved. Although these types of circuits are used for phonograph and tape recording (analog systems), they are not used for compact discs (a digital system).

PREEMPHASIS CIRCUIT

Circuits using operational amplifiers are ideally suited for modifying the frequency spectrum of signals. Consider the inverting op amp circuit of Figure 5.46 in which phasors are indicated for the input and output sinusoidal voltages. For very low-frequency sinusoids, the capacitor tends to behave as an open circuit, that is, its current is negligible compared with that of R_2.

$$\left(\frac{V_{\text{out}}}{V_{\text{in}}}\right)_{\text{low}} = -\frac{R_3}{R_1 + R_2} \quad \text{low-frequency sinusoids} \tag{5.67}$$

At higher frequencies, the magnitude of the capacitor's impedance becomes small and, in effect, tends to short out R_2. A larger magnitude of gain occurs for this condition.

$$\left(\frac{V_{\text{out}}}{V_{\text{in}}}\right)_{\text{high}} = -\frac{R_3}{R_1} \quad \text{high-frequency sinusoids} \tag{5.68}$$

The preceding equation gives the ideal response (infinite op amp gain). For much higher frequency signals, the finite gain of the op amp affects the response.

The frequency-dependent response of the circuit of Figure 5.46 may be obtained by introducing a complex impedance Z_I for the input network.

$$
\begin{aligned}
Z_I &= R_1 + \frac{R_2(1/j\omega C)}{R_2 + 1/j\omega C} = R_1 + \frac{R_2}{1 + j\omega R_2 C} \\
&= \frac{R_1 + R_2 + j\omega R_1 R_2 C}{1 + j\omega R_2 C} = \frac{(R_1 + R_2)\left[1 + j\omega R_1 R_2 C/(R_1 + R_2)\right]}{1 + j\omega R_2 C}
\end{aligned} \tag{5.69}
$$

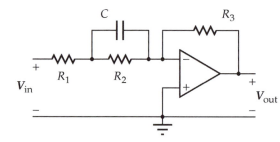

Figure 5.46: An op amp preemphasis circuit.

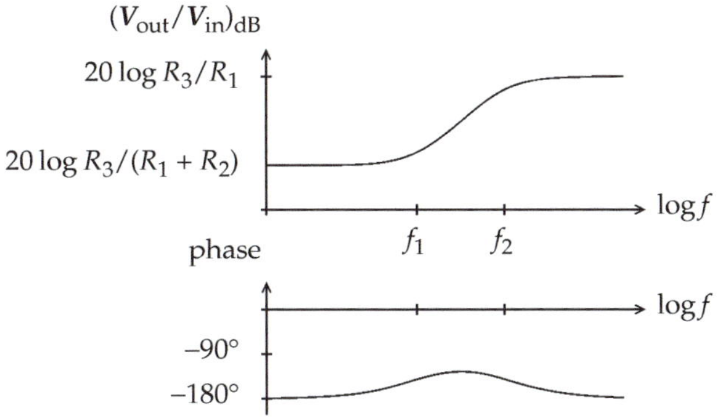

Figure 5.47: The frequency-dependent response of the preemphasis circuit of Figure 5.46.

If a virtual short is assumed for the input of the op amp, the following ideal response is obtained for the amplifier:

$$\frac{V_{in}}{Z_I} + \frac{V_{out}}{R_3} = 0$$

$$\left(\frac{V_{in}}{V_{out}}\right) = -\frac{R_3}{Z_I} = -\left(\frac{R_3}{R_1 + R_2}\right)\frac{1 + j\omega R_2 C}{1 + j\omega R_1 R_2 C/(R_1 + R_2)} \tag{5.70}$$

This expression yields the low-frequency gain (Eq. (5.86)) for $\omega \to 0$ and the high-frequency gain (Eq. (5.68)) for $\omega \to \infty$.

It is convenient to introduce a set of break frequencies as follows:

$$\omega_1 = 1/R_2 C, \qquad f_1 = \omega_1/2\pi$$

$$\omega_2 = \frac{R_1 + R_2}{R_1 R_2 C} = \frac{1}{R_1 \| R_2 C}, \qquad f_2 = \omega_2/2\pi \tag{5.71}$$

The response may then be written in terms of these break frequencies as follows:

$$\left(\frac{V_{out}}{V_{in}}\right) = -\left(\frac{R_3}{R_1 + R_2}\right)\frac{1 + jf/f_1}{1 + jf/f_2} \tag{5.72}$$

Because $f_1 < f_2$ (Eq. (5.71)), the magnitude of the numerator starts to increase with increasing frequency before the magnitude of the denominator increases (Figure 5.47). A frequency-dependent phase shift is also introduced by the preemphasis circuit.

DEEMPHASIS CIRCUIT

A deemphasis circuit is required to restore the frequency components to their original amplitude and phase. The inverting op amp configuration is ideally suited. Consider the two op amp circuits of Figure 5.48 in which both the input and feedback networks consist of complex impedances. These circuits differ only in that the impedances Z_A and Z_B are interchanged.

$$\frac{V_{out\,1}}{V_{in\,1}} = -\frac{Z_B}{Z_A}, \qquad \frac{V_{out\,2}}{V_{in\,2}} = -\frac{Z_A}{Z_B}$$

$$\left(\frac{V_{out\,1}}{V_{in\,1}}\right)\left(\frac{V_{out\,2}}{V_{in\,2}}\right) = 1 \tag{5.73}$$

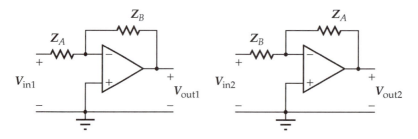

Figure 5.48: Two inverting op amp circuits with input and feedback impedances interchanged.

These circuits may be thought of as having inverse frequency responses, that is, the frequency distortion of one circuit will be compensated ("undone") by the second circuit. Hence, a deemphasis circuit can be constructed from the same elements as those of the preemphasis circuit – it is only necessary to interchange the input and feedback networks.

DESIGN

To illustrate the design of a preemphasis and a deemphasis circuit, consider the case for which the ratio of the high-frequency to the low-frequency gain is to be 5.0 (a high-frequency preemphasis of 5.0, that is, 14 dB). A lower break frequency f_1 of 500 Hz is desired, and a low-frequency gain of -1.0 will arbitrarily be assumed (it may be easily changed).

$$R_3/(R_1 + R_2) = 1 \quad \text{low-frequency gain}$$
$$R_3/R_1 = 5 \quad \text{high-frequency gain} \tag{5.74}$$
$$R_3 = 5\,R_1, \qquad R_2 = 4\,R_1$$

The value of the capacitor C may also be expressed in terms of R_1 as follows:

$$C = 1/(8\pi f_1 R_1) = 7.96 \times 10^{-5}/R_1 \tag{5.75}$$

In the preceding relationships, the values of all elements are expressed in terms of R_1. On the basis of the concept for an ideal op amp, any value of R_1 could be used (from 1 mΩ to 1 GΩ). However, on the basis of practical considerations, the choice of a value for R_1, although still arbitrary, is considerably more constrained. The output of the op amp needs to supply the current of R_3 as well as that of the input elements. General-purpose integrated-circuit op amps are usually designed to supply or sink a maximum of 10 mA. This output current must provide the input current of the next stage as well as the current of the feedback circuit R_3 (R_1 has the same current). A conservative design based on a maximum current of 1 mA is reasonable. If the circuit is to be designed for R_1 to have a maximum voltage of 10 V, a resistance of 10 kΩ is required. The following is therefore obtained for the element values:

$$R_1 = 10 \text{ k}\Omega, \qquad R_2 = 40 \text{ k}\Omega, \qquad R_3 = 50 \text{ k}\Omega$$
$$C = 7.96 \times 10^{-9} \text{ F}, \quad 7.96 \text{ nF} \tag{5.76}$$

Although larger resistance values (and concurrently a smaller value of capacitance) could be used, practical considerations also dictate an upper resistance limit. Resistance values should be small enough so that the effects of currents of unavoidable stray capacitances are negligible. For discrete circuits designed for audio signals, resistances as large as 1 MΩ are generally acceptable, whereas for higher-frequency signals, smaller resistance values are required.

Although the 10-kΩ resistance for R_1 yields a reasonable set of resistances for R_2 and R_3, the capacitance value C is not readily available. A value of 6.8 nF would be more convenient. This implies a value of 11.7 kΩ for R_1. Standard resistor values (5 percent tolerance) yield the following component values based on a resistance of 12 kΩ for R_1:

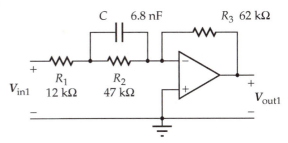

(a) preemphasis circuit

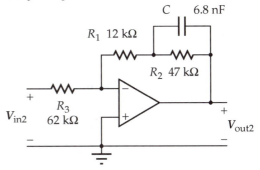

(b) deemphasis circuit

Figure 5.49: A preemphasis and deemphasis circuit using standard resistance values.

$$R_1 = 12 \text{ k}\Omega, \quad R_2 = 47 \text{ k}\Omega, \quad R_3 = 62 \text{ k}\Omega, \quad C = 6.8 \text{ nF} \qquad (5.77)$$

Circuits using these component values are indicated in Figure 5.49. It will be noted that for either circuit, a change in R_3 results in a linear scaling of its response. For example, a doubling of R_3 of the preemphasis circuit will double the gain of the circuit but will leave the relative frequency response unchanged. A doubling of R_3 of the deemphasis circuit will halve the gain of this circuit. If R_3 of the deemphasis circuit is changed to 12 kΩ, the high-frequency gain would be -1.0 (as opposed to its original value of -0.2), and the low-frequency gain would be approximately -5.0.

For an audio system, signals will have frequency components up to 20 kHz. The op amp circuits, however, will respond to signals, in particular noise, with much higher frequency components. This is generally not desirable; for a good design, the response of the circuit should be constrained to minimize the effects of high-frequency noise components. The finite gain–bandwidth product GBP of the op amp will necessarily limit the high-frequency response. Consider the preemphasis circuit of Figure 5.49(a). For a high-frequency signal, the capacitor, in effect, shorts out the resistor R_2, thus reducing the effective input circuit to a single resistor R_1. Therefore, the feedback fraction β is approximately $R_1/(R_1 + R_3) = 0.162$. The upper half-power frequency of the amplifier f_h is βGBP. Hence, a GBP of 1 MHz results in a frequency of 162 kHz for f_h. The deemphasis circuit (Figure 5.49(b)) with a value of 12 kΩ for R_3 results in a high-frequency value of $1/2$ for β and $f_h = 0.5$ MHz. Both of these upper half-power frequencies are excessively large.

A capacitive circuit external to the op amp will be required to limit the high-frequency response of the preemphasis and deemphasis circuits. Consider, initially, the preemphasis circuit of Figure 5.49(a). For high-frequency signals, the capacitance C may be treated as a short circuit. Suppose that a second capacitor C_H is connected in parallel with R_3, resulting in a complex impedance for the feedback connection $\mathbf{Z}_F$:

$$\mathbf{Z}_F = \frac{R_3(1/j\omega C_H)}{R_3 + 1/j\omega C_H} = \frac{R_3}{1 + j\omega R_3 C_H}$$

$$\frac{\mathbf{V}_{\text{in }1}}{R_1} + \frac{\mathbf{V}_{\text{out }1}}{\mathbf{Z}_F} = 0 \tag{5.78}$$

$$\frac{\mathbf{V}_{\text{out }1}}{\mathbf{V}_{\text{in }1}} = -\frac{\mathbf{Z}_F}{R_1} = -\frac{R_3}{R_1} \frac{1}{1 + j\omega R_3 C_H}$$

This relation yields an upper half-power frequency f_h as follows:

$$\frac{\mathbf{V}_{\text{out }1}}{\mathbf{V}_{\text{in }1}} = -\frac{R_3}{R_1} \frac{1}{1 + jf/f_h}, \quad f_h = 1/(2\pi R_3 C_H) \tag{5.79}$$

Because a fairly uniform response for signals with frequencies less than 20 kHz is desired, a design value of 40 kHz for f_h is reasonable.

$$C_H = 1/(2\pi f_h R_3) = 6.42 \times 10^{-11} \text{ F}, \quad 64.2 \text{ pF} \tag{5.80}$$

A standard capacitor value of 56 pF will be used, resulting in a slightly higher half-power frequency.

The high-frequency response of the deemphasis circuit of Figure 5.49(b) may be constrained in a similar fashion. For high-frequency signals, the capacitor C may again be treated as a short circuit. Therefore, a second capacitor C_H connected across the feedback elements for this circuit results in the following high-frequency behavior:

$$\frac{\mathbf{V}_{\text{out }2}}{\mathbf{V}_{\text{in }2}} = -\frac{R_1}{R_3} \frac{1}{1 + jf/f_h}, \quad f_h = 1/(2\pi R_1 C_H) \tag{5.81}$$

Again a value of capacitance will be determined that yields 40 kHz for f_h as follows:

$$C_H = 1/(2\pi f_h R_1) = 3.32 \times 10^{-10} \text{ F}, \quad 332 \text{ pF} \tag{5.82}$$

A standard capacitance value of 330 pF will be used.

SPICE VERIFICATION

To verify the behavior of the preemphasis and deemphasis circuits, a SPICE simulation will be employed (Figure (5.50)). A value of R_3 that results in a low-frequency gain of -5 for the deemphasis circuit will be used ($R_3 = 12$ kΩ). To demonstrate that the deemphasis circuit does indeed restore the input signal, the circuits will be connected in cascade, and the op amp subcircuit model of Figure 5.44 will be used for the simulation. The frequency response (expressed in decibels) of the individual circuits is given in Figure 5.51. It will be noted that a "flat" overall response corresponding to a gain of about 5, that is 14 dB, is

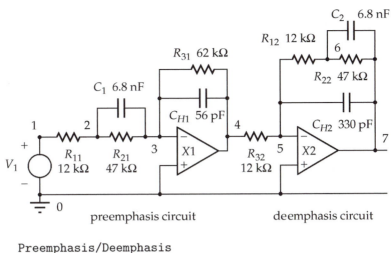

```
Preemphasis/Deemphasis
V1   1   0   AC   1   PWL(0 0 .1U 1 200U 1 200.1U 0)
R11  1   2   12K
R21  2   3   47K
C1   2   3   6.8N
X1   0   3   4   OPAMP
CH1  4   3   56P
R31  4   3   62K
R32  4   5   12K
X2   0   5   7   OPAMP
CH2  7   5   330P
R12  6   5   12K
R22  7   6   47K
C2   7   6   6.8N
.AC   DEC   20 10   100K
.TRAN  1U   400U  0  1U
.PROBE
{                        }   OPAMP subcircuit of Fig. 5.44

.END
```

Figure 5.50: SPICE circuit and file for a preemphasis and a deemphasis circuit.

achieved for frequencies less than 20 kHz. At 20 kHz the response is down by 2.4 dB from its low-frequency value.

The time-dependent behavior of the circuits for a 1-V input pulse with a duration of 200 μs is obtained from the transient analysis (Figure 5.52). It will be noted that the preemphasis circuit results in a significant distortion of the input pulse (it is also inverted). However, the deemphasis circuit restores the original time dependence of the pulse. The finite rise and fall times are a result of the limitation of the frequency-dependent response to signal components with frequencies less than about 20 kHz.

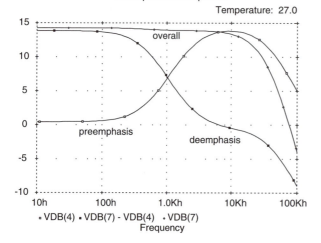

Figure 5.51: Fequency-dependent response of the circuits of Figure 5.50.

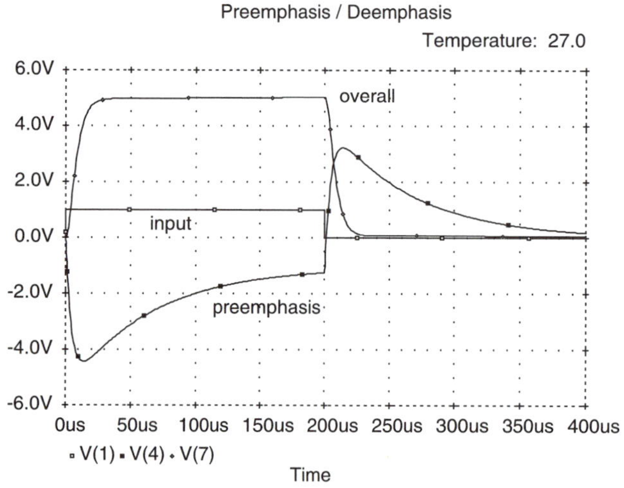

Figure 5.52: Time-dependent response of the circuits of Figure 5.50.

5.5 A WIDE-BANDWIDTH AMPLIFIER: A DESIGN EXAMPLE

Integrated-circuit op amps are extremely convenient for constructing linear amplifiers. These circuits tend to require considerably fewer components than circuits using discrete transistors. Furthermore, the design process is much easier in that only a limited knowledge of the behavior of op amps is required. Unfortunately, op amp circuits tend to have a more limited bandwidth than well-designed circuits using discrete transistors.

To illuminate the design procedure, an amplifier requirement of a low-level voltage gain of 100 will be considered. The amplifier is to be ac coupled, that is, a nonvarying (dc) input voltage should have no effect on the amplifier's output. The frequency response of the amplifier is to be uniform (within 3 dB) down to 50 Hz, whereas the upper half-power frequency is to be as high as possible. An additional condition is that the circuit is to require only a single supply voltage V_{CC} of 15 V.

SINGLE-STAGE AMPLIFIER

To achieve a wide bandwidth, several cascaded amplifier stages will be needed. However, to justify this assertion, an amplifier using only a single op amp will initially be considered. A conventional noninverting amplifier configuration that has a gain of 100 is indicated in Figure 5.53.

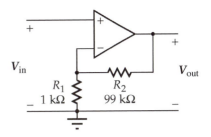

Figure 5.53: An amplifier using a single op amp.

$$\left(\frac{V_{\text{out}}}{V_{\text{in}}}\right)_{\text{Ideal}} = 1 + R_2/R_1, \tag{5.83}$$
$$R_2/R_1 = 99$$

The resistances R_1 and R_2 are reasonable design values (in constructing this circuit, a standard value of 100 kΩ would be used for R_2). The upper half-power frequency of the response depends on the gain bandwidth product GBP of the op amp as follows:

$$f_h = \beta GBP = \left(\frac{R_1}{R_1 + R_2}\right) GBP \tag{5.84}$$
$$= 0.01\, GBP$$

If, for example, $GBP = 1$ MHz, $f_h = 10$ kHz. This response is not even adequate for a high-fidelity audio amplifier. If, however, $GBP = 20$ MHz, an upper half-power frequency f_h of 200 kHz results.

Operational amplifiers that may be described as undercompensated are available with gain–bandwidth products of 20 MHz and higher. These op amps are generally designed for stable operation in circuits with feedback fractions (β) that are 0.2 or less. This implies a gain of 5 or greater for a noninverting amplifier configuration. (General-purpose op amps are usually designed to be stable in circuits with values of β up to 1.) Alternatively, an externally compensated op amp could be used. When a smaller-than-normal value of compensating capacitance C_c, is used the gain–bandwidth of the op amp is increased. Through selection of an appropriate capacitance, the response of the op amp can be optimized for the circuit in which it is used.

For the discussion that follows, it will be assumed that op amps with gain–bandwidth products of 20 MHz will be used. The high-frequency behavior of the amplifier circuits will depend on this quantity; if op amps with a different GBP are used, the high-frequency response will be different. For example, if op amps with $GBP = 10$ MHz are used, the resultant upper half-power frequency will be only one-half that of the circuit using 20-MHz op amps. For the circuit of Figure 5.53, the upper half-power frequency f_h will be 200 kHz ($GBP = 20$ MHz):

$$\frac{V_{\text{out}}}{V_{\text{in}}} = \frac{100}{1 + jf/f_h}, \quad f_h = 200 \text{ kHz} \tag{5.85}$$

This is the response of the RC equivalent circuit of Figure 5.54. The same circuit will also yield the response for an input voltage $v_{\text{IN}}(t)$ that is a step function

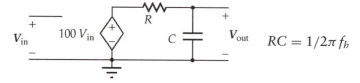

Figure 5.54: An RC circuit that simulates the response of the op amp circuit of Figure 5.53.

$V_p\mu(t)$ as follows:

$$v_{OUT}(t) = 100\,V_p(1 - e^{-t/\tau}) \quad \text{for } t > 0$$
$$\tau = RC = 1/2\pi f_h \tag{5.86}$$

The time constant of the amplifier circuit is related to its upper half-power frequency. The 10-to-90-percent rise time t_r is readily obtained as follows:

$$t_r = 2.2\,\tau = 2.2/2\pi f_h = 0.35/f_h \tag{5.87}$$

For an amplifier with $f_h = 200$ kHz, the rise time is 1.75 μs. To reduce the rise time, the bandwidth of the circuit must be increased.

TWO-STAGE AMPLIFIER

Consider, as a next step in the design process, the situation in which two amplifier circuits are used, each with a gain of 10 (Figure 5.55). Although this is a more complex circuit than that of the amplifier with a single op amp, this circuit has a much higher overall upper half-power frequency. Because the voltage gain of each stage is 10, the feedback fraction is 0.1. Hence, for a gain–bandwidth product of 20 MHz, the upper half-power frequency of each stage f_h is 2 MHz. The overall response is the product of the individual responses:

$$\frac{V_{out}}{V_{in}} = \frac{100}{(1 + jf/f_h)^2}, \qquad \left|\frac{V_{out}}{V_{in}}\right| = \frac{100}{1 + f^2/f_h^2} \tag{5.88}$$

The overall upper half-power frequency $f_{h\,\text{overall}}$, occurs for a frequency at which the denominator of the magnitude of the gain expression is equal to $\sqrt{2}$.

$$1 + f_{h\,\text{overall}}^2/f_h^2 = \sqrt{2}, \qquad f_{h\,\text{overall}} = f_h\sqrt{\sqrt{2} - 1} = 0.644\,f_h \tag{5.89}$$

Hence, the overall upper half-power frequency of the two-stage amplifier is 1.29 MHz, which is a considerable increase over that of the single-stage amplifier.

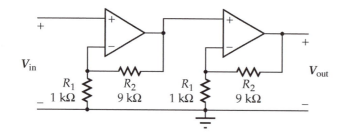

Figure 5.55: An amplifier using two op amps.

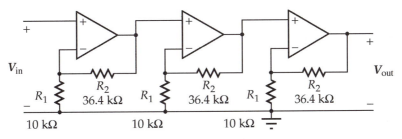

Figure 5.56: A three-stage amplifier circuit.

THREE-STAGE AMPLIFIER

For the next step of the design process, a three-stage amplifier (Figure 5.56) will be considered. To achieve an overall gain of 100, a gain of 4.64 ($\sqrt[3]{10}$) is required for each stage. Both R_1 and R_2 have been increased to reduce the currents of the circuit. Because $\beta = 1/4.64 = 0.216$, the upper half-power frequency of a single stage is 4.32 MHz. The response of three stages is the third power of that of a single stage as follows:

$$\frac{V_{out}}{V_{in}} = \frac{100}{(1 + jf/f_h)^3}, \qquad \left|\frac{V_{out}}{V_{in}}\right| = \frac{100}{(1 + f^2/f_h^2)^{3/2}} \tag{5.90}$$

Again, the overall half-power frequency occurs for a frequency at which the magnitude of the gain is equal to $\sqrt{2}$.

$$(1 + f_{h\,overall}^2/f_h^2)^{3/2} = \sqrt{2}, \qquad f_{h\,overall} = f_h\sqrt{\sqrt[3]{2} - 1} = 0.510\,f_h \tag{5.91}$$

The overall half-power frequency is 2.20 MHz, which is an improvement over the two-stage circuit.

Although additional stages might also be considered, the large op amp gain–bandwidth product of 20 MHz may not be realized for amplifier stages with smaller voltage gains. For three stages, the gain of each stage, 4.64, results in a value of 0.216 for β. This is slightly larger than the maximum value of 0.2 for which stable operation is specified for a particular op amp (a 357). If four amplifier stages were to be used, the gain of each stage would be 3.16, and β would be 0.316, an excessively large value for the preceding op amp. Although an externally compensated op amp could be used, its gain–bandwidth product would probably be less than 20 MHz. Hence, any improvement in the overall performance of the amplifier that might be achieved would be small.

A single RC simulation circuit was used to determine the rise time of an amplifier in terms of its upper half-power frequency (Eq. (5.87)). Two RC circuits would be needed to determine the step-function response of a two-stage amplifier – the exponential response of the first stage serving as the input voltage of the second stage. Three RC simulation circuits would be needed for the three-stage amplifier. The overall step-function response of these simulation circuits is not a simple exponential function. However, the rise times tend to be related to the overall half-power frequency of the amplifier. The following expression based on the response of a single amplifier stage is a reasonable approximation for relating the overall response time to the overall half-power frequency of a multistage amplifier:

$$t_{r \text{ overall}} \approx 0.35/f_{h \text{ overall}} \tag{5.92}$$

This implies rise times of approximately 0.78 and 0.45 μs for the two- and three-stage amplifiers. A SPICE simulation is recommended if a precise value of rise time is required.

FINAL DESIGN

At this point, the three-stage amplifier with an overall upper half-power frequency of 2.20 MHz will be chosen for the final design. An amplifier with a 2.20-MHz bandwidth is suitable, for example, for amplifying a moderate-quality video signal (without a color subcarrier). If a significantly wider bandwidth is needed, an amplifier using either discrete devices or alternative integrated circuits (not op amps) will be required. It is now necessary to complete the design, that is, to take into account the single voltage supply requirement and realization of the desired low-frequency response.

For a single supply voltage of V_{CC}, the negative supply connection of the op amp will be connected to the ground point of the circuit (Figure 5.57). This requires that the input voltages of the op amp as well as its output voltage be constrained to a range bounded by 0 and 15 V ($V_{CC} = 15$ V). Ideally, the signal and output voltages should have quiescent values of about 7.5 V ($V_{CC}/2$). The input voltage of Figure 5.57 may be considered as the sum of a quiescent voltage V_{IN} and a sinusoidal quantity $\text{Re}(V_{in}e^{j\omega t})$. A similar condition prevails for the output voltage.

$$v_{IN}(t) = V_{IN} + \text{Re}(V_{in}e^{j\omega t})$$
$$v_{OUT}(t) = V_{OUT} + \text{Re}(V_{out}e^{j\omega t}) \tag{5.93}$$

For the quiescent condition (no signal), the capacitor may be treated as an open circuit. Because the current of R_2 will be zero for this condition, the quiescent input and output voltages will be equal:

$$V_{OUT} = V_{IN} \quad \text{quiescent values} \tag{5.94}$$

Therefore, if $V_{IN} = V_{CC}/2$, $V_{OUT} = V_{CC}/2$, which is a condition appropriate for the single supply voltage.

For very high-frequency signals, the capacitor will behave as a short circuit, resulting in the same high-frequency response as that of a circuit without a capacitor. A phasor analysis may be used to obtain the frequency-dependent behavior of the amplifier circuit. Equating the phasor voltages at the

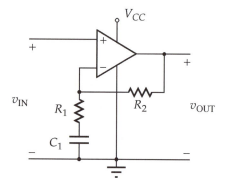

Figure 5.57: An amplifier circuit using a single supply voltage.

inputs of the op amp yields the following:

$$V_{in} = \frac{(R_1 + 1/j\omega C_1)V_{out}}{R_1 + 1/j\omega C_1 + R_2} = \frac{(1 + j\omega R_1 C_1)V_{out}}{1 + j\omega(R_1 + R_2)C_1}$$

$$\frac{V_{out}}{V_{in}} = \frac{1 + j\omega(R_1 + R_2)C_1}{1 + j\omega R_1 C_1} \tag{5.95}$$

For $\omega \to 0$ the gain is 1, whereas for $\omega \to \infty$ the gain of the preceding expression is $1 + R_2/R_1$, which is that expected for a circuit with the capacitor replaced by a short circuit. The response for Eq. (5.95) may be expressed in terms of two break frequencies f_1 and f_2 as follows:

$$\frac{V_{out}}{V_{in}} = \frac{1 + jf/f_1}{1 + jf/f_2}, \quad f_1 = 1/2\pi(R_1 + R_2)C_1, \quad f_2 = 1/2\pi R_1 C_1 \tag{5.96}$$

If this circuit is used for the three-stage amplifier, $R_2/R_1 = 3.64$, and the ratio of the break frequencies is 4.64 (Figure 5.58).

The frequency-dependent response of three amplifier stages, each using the circuit of Figure 5.57, is the third power of the expression for a single stage:

$$\left|\frac{V_{out}}{V_{in}}\right| = \frac{(1 + f^2/f_1^2)^{3/2}}{(1 + f^2/f_2^2)^{3/2}}, \quad f_2 = 4.64 f_1 \tag{5.97}$$

Because the design value of the lower half-power frequency f_ℓ is 50 Hz, a value of f_1 (or f_2) for which the magnitude of the gain is $100/\sqrt{2}$ needs to be determined. Once the break frequencies are known, the value of capacitance C_1 can be obtained. An approximation will be employed because it will be asumed that, for $f = f_\ell$, the 1 of the numerator term of Eq. (5.97) can be ignored.

$$\left|\frac{V_{out}}{V_{in}}\right| \approx \frac{(f^2/f_1^2)^{3/2}}{(1 + f^2/f_2^2)^{3/2}} = \left(\frac{f_2}{f_1}\right)^3 \frac{1}{(1 + f_2^2/f^2)^{3/2}} \tag{5.98}$$

For this expression, $(f_2/f_1)^3 = 100$. Therefore, the lower half-power frequency f_ℓ is the frequency at which the denominator of the preceding expression is equal to $\sqrt{2}$.

$$(1 + f_2^2/f_\ell^2)^{3/2} = \sqrt{2}, \quad f_2 = f_\ell\sqrt{\sqrt[3]{2} - 1} = 0.510 f_\ell \tag{5.99}$$

For a lower half-power frequency of 50 Hz, $f_2 = 25.5$ Hz. When the expression of Eq. (5.96) is used for f_2, the following is obtained for C_1 ($R_1 = 10$ kΩ):

$$C_1 = 1/2\pi f_2 R_1 = 5.24 \times 10^{-7} \text{ F}, \quad 0.624 \ \mu\text{F} \tag{5.100}$$

Figure 5.58: Frequency response for the amplifier of Figure 5.57.

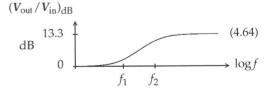

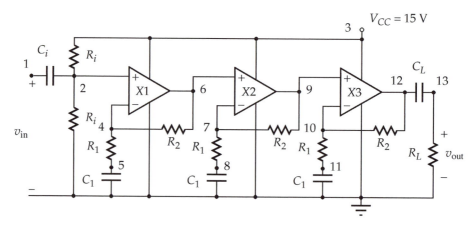

$$R_1 = 10 \text{ k}\Omega, \qquad R_2 = 36 \text{ k}\Omega, \qquad C_1 = 0.68 \ \mu F$$
$$R_i = 1.0 \text{ M}\Omega, \qquad C_i = 0.1 \ \mu F, \qquad C_L = 47 \ \mu F, \qquad R_L = 1.0 \text{ k}\Omega$$

Figure 5.59: The final design of a three-stage amplifier. Node numbers are those used for a SPICE simulation.

A standard capacitance value of 0.68 μF will be used. For this circuit, f_1 is equal to 5.50 Hz, and $f_\ell/f_1 = 9.1$. Because $f_\ell^2/f_1^2 = 82.6$, the approximation that this quantity is large compared with 1 is valid.

A capacitor C_i and a set of resistors R_i will be used for the input circuit of the amplifier (Figure 5.59). The input capacitor C_i is sufficiently large so that the magnitude of the sinusoidal voltage across C_i is negligible compared with the input voltage at a frequency of 50 Hz. An arbitrary load resistor (it was not specified), along with a coupling capacitor, is shown for the output circuit. As for the input circuit, for a frequency of 50 Hz, the sinusoidal voltage across C_L is negligible compared with the load voltage.

SPICE VERIFICATION

For a SPICE simulation of the amplifier, a modification of the previously used subcircuit is necessary (Figure 5.60). The 20-MHz gain–bandwidth product corresponds to a break frequency of 200 Hz for $A_{d0} = 10^5$. This requires a capacitance of 795.8 pF for an equivalent circuit with a series resistance of 1 MΩ. The limiting voltages of ± 19895 V result in a slew rate of 50 V/μs (the specified slew rate of a 357 op amp). Two external connections (nodes 9 and 10) have been added to the subcircuit to account for the power supply connections. Through the use of a polynomial voltage source specification (EB), the zero-signal value of the output voltage is shifted to the midvalue of the supply voltages ($V_{CC}/2$ for $V_{EE} = 0$). The supply resistor R_{Supply} has been included to account for the supply currents. For an actual device, the current supplied by the output of the op amp is derived from V_{CC} and V_{EE}. To model this behavior, a more complex circuit is necessary.

The circuit file of Figure 5.61 will be used to simulate the behavior of the three-stage amplifier. Phasor (.AC) and transient (.TRAN) analyses are included. The frequency response of the first stage of the amplifier and for the overall response of

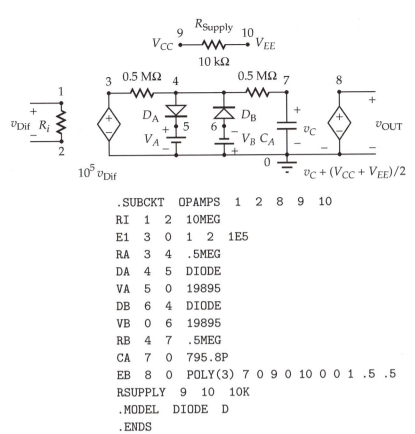

```
.SUBCKT   OPAMPS   1   2   8   9   10
RI    1   2   10MEG
E1    3   0   1   2   1E5
RA    3   4   .5MEG
DA    4   5   DIODE
VA    5   0   19895
DB    6   4   DIODE
VB    0   6   19895
RB    4   7   .5MEG
CA    7   0   795.8P
EB    8   0   POLY(3)  7 0 9 0 10 0 0 1 .5 .5
RSUPPLY   9   10   10K
.MODEL   DIODE   D
.ENDS
```

Figure 5.60: Subcircuit for op amp with supply connections.

the amplifier is given in Figure 5.62. The lower and upper half-power frequencies of the simulation are 47.7 Hz and 2.2 MHz, respectively, which are values very close to the design quantities. The transient response for a 1-μs input pulse is given in Figure 5.63. The rise time of 165 ns is very close to that predicted by the approximate relationship ($0.35/f_{b\,\text{overall}}$) of 159 ns. For the frequency-dependent and the transient response, the gain is slightly less than the design value of 100 (40 dB). This is a result of the standard resistance value of 36 kΩ (instead of 36.4 kΩ) being used for R_2. If a precise value of 100 is required for the overall voltage gain, R_2 of one stage could be increased slightly.

As a result of the overall lower half-power frequency of 50 Hz, the amplitude of the response for a 50-Hz sinusoidal input signal is down by 3 dB, that is, the output voltage is only $1/\sqrt{2}$ of that which it would be for a much higher frequency (for example, 1 kHz). This low-frequency limitation has a significant effect on the amplification of signals with other waveforms. A simulation was run for an input voltage with a square waveform and a frequency of 250 Hz (5 times the lower half-power frequency of the amplifier). A considerable distortion of the output voltage resulted (Figure 5.64). The coupling capacitors C_i and C_L along with the capacitive feedback networks result in a "sag" of the output voltage. Instead of the output voltage remaining at a constant high or

```
Three-Stage Amplifier
VIN 1 0 AC 1 PWL(0 0 1N .02 1U .02
+1.001U -.02 2U -.02 2.001U 0)
CI   1   2   0.1U
RI1  3   2   1MEG
RI2  2   0   1MEG
X1   2   4   6   3   0   OPAMPS
R11  4   5   10K
C11  5   0   0.68U
R21  6   4   36K
X2   6   7   9   3   0   OPAMPS
R12  7   8   10K
C12  8   0   0.68U
R22  9   7   36K
X3   9   10  12  3   0   OPAMPS
R13  10  11  10K
C13  11  0   0.68U
R23  12  10  36K
CL   12  13  47U
RL   13  0   1K
.AC  DEC  20  1   10MEG
.TRAN  .01U  3U  0  .01U
.PROBE
```

{ } OPAMPS subcircuit of Fig. 5.60

```
.END
```

Figure 5.61: SPICE file for the three-stage amplifier of Figure 5.59.

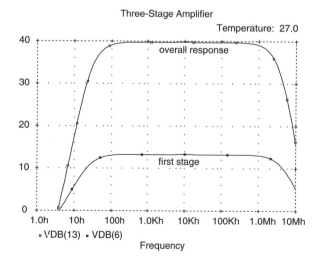

Figure 5.62: Frequency response of the three-stage amplifier.

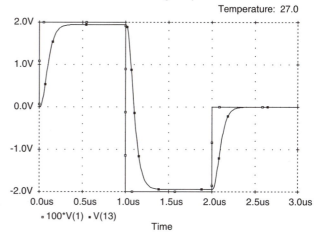

Figure 5.63: Transient response for a 1-μs input pulse. The rise time is 165 ns.

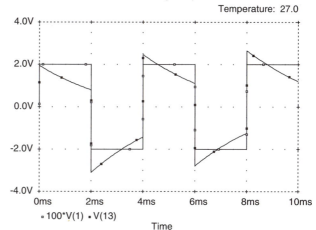

Figure 5.64: Transient response for a voltage with a 250-Hz square waveform.

low level, the magnitude of the output voltage decreases with time. For a lower-frequency square-wave voltage (a voltage with a longer period), the sag is even more pronounced.

The coverage of op amp applications in this chapter has necessarily been limited, and only a few applications have been discussed. To gain a fuller perspective on the numerous applications of op amps, other texts are recommended (Franco 1988; Stout and Kaufman 1976; Van Valkenburg 1982; Wait et al. 1992).

REFERENCES

Black, H. S. (1934). Stablized feedback amplifiers. *The Bell System Technical Journal*, **13**, 1, 1–18.

Bode, H. W. (1975). *Network Analysis and Feedback Amplifier Design*. Huntington, NY: Robert E. Krieger Publishing Company.

Fagen, M. D. (1975). *A History of Engineering and Science in the Bell System – The Early Years (1875–1925)*. Murray Hill, NJ: Bell Telephone System Laboratories, Inc.

Franco, S. (1988). *Design With Operational Amplifiers and Analog Integrated Circuits*. New York: McGraw–Hill Book Company.

Gray, Paul R. and Meyer, Robert G. (1992). *Analysis and Design of Analog Integrated Circuits* (3d ed.). New York: John Wiley & Sons.

Mabon, P. C. (1975). *Mission Communications: The Story of Bell Laboratories*. Murray Hill, NJ: Bell Telephone Laboratories, Inc.

Millman, J. and Grabel, A. (1987). *Microelectronics* (2nd ed.). New York: McGraw–Hill Book Company.

Nyquist, H. (1932). Regeneration theory. *The Bell System Technical Journal*, **11**, 1, 126–47.

O'Neill, E. F. (1985). *A History of Engineering and Science in the Bell System – Transmission Technology (1925–1975)*. Murray Hill, NJ: AT&T Bell Laboratories.

Sedra, A. S. and Smith, K. C. (1991). *Microelectronic Circuits* (3rd ed.). Philadelphia: Saunders College Publishing.

Solomon, J. E. (1991). A tribute to Bob Widlar. *IEEE Journal of Solid State Circuits*, **26**, 8, 1087–9.

Stout, D. F. and Kaufman, M. (1976). *Handbook of Operational Amplifier Circuit Design*. New York: McGraw–Hill Book Company.

Tucker, D. G. (1972). The history of positive feedback: The oscillating audion, the regenerative receiver, and other applications up to around 1923. *The Radio and Electronic Engineer*, **42**, 2, 69–80.

Van Valkenburg, M. E. (1982). *Analog Filter Design*. New York: Holt, Rinehart, and Winston.

Wait, J. V., Huelsman, L. P., and Korn, G. A. (1992). *Introduction to Operational Amplifier Theory and Applications* (2d ed.). New York: McGraw–Hill, Inc.

PROBLEMS

5.1 Consider the feedback amplifier of Figure 5.4 with $R_1 = 10$ kΩ, $R_2 = 100$ kΩ, and $A_d = 100$. What are β and A_{fb} of this circuit?

5.2 Repeat Problem 5.1 for $A_d = 10^3$ and for $A_d = 10^4$.

5.3 What is the ideal value of A_{fb} for the circuit of Problem 5.1? What is the value of A_d required for A_{fb} to be within 10 percent of its ideal value? What is the value required for A_{fb} to be within 1 percent of its ideal value?

5.4 In the feedback amplifier of Figure 5.4, $R_1 = 5.6$ kΩ, $R_2 = 150$ kΩ, and $A_d = 100$. What is A_{fb} of this circuit?

5.5 What is the ideal value of A_{fb} for the circuit of Problem 5.4? What is the value of A_d required for A_{fb} to be within 5 percent of its ideal value? What is the value required for A_{fb} to be within 1 percent of its ideal value?

5.6 With Eq. (5.4) as a starting point, obtain an expression for $\frac{dA_{fb}}{dA_d}$. What is the sensitivity of A_{fb}, that is, $\left(\frac{A_d}{A_{fb}}\right)\left(\frac{dA_{fb}}{dA_d}\right)$?

5.7 For the feedback amplifier circuit of Figure 5.4, $R_1 = 4.7$ kΩ, $R_2 = 100$ kΩ, and $A_d = 500$. What is A_{fb} of the amplifier circuit? What is the percentage change of A_{fb} for a 10 percent change of A_d? (Hint: The result of Problem 5.6 is applicable.)

5.8 What is the value of A_d required for a 10 percent change in A_d to result in only a 1 percent change in A_{fb} of Problem 5.7?

5.9 A particular amplifier has a voltage gain of A. What is its decibel gain for $A = 5, 15, 25$, and 50?

5.10 An amplifier has a decibel gain A_{dB} of 21 dB. What is the magnitude of the voltage gain? What is the magnitude of the voltage gain for $A_{dB} = 12$, 18, and 36 dB?

5.11 An amplifier has a decibel gain of 19 dB. What is the magnitude of the voltage gain? Suppose that the voltage gain increases by 5 percent. What is the corresponding change in the decibel gain? What is the change in decibel gain for increases in voltage gain of 10, 20, and 50 percent? What is the change in decibel gain for decreases in voltage gain of 10, 20, and 50 percent?

5.12 It is desired that the feedback amplifier of Figure 5.4 have an ideal voltage gain of 33 dB. Determine the required resistance ratio R_2/R_1. Suppose that the maximum magnitude of the output voltage is 10 V and the magnitude of the current of the feedback circuit is not to exceed 1 mA. Determine values of R_1 and R_2 that satisfy this condition.

5.13 Determine the minimum value of A_d, expressed in decibels, required for the amplifier of Problem 5.12 if the actual voltage gain is not to deviate by more than 1 dB from the ideal value. What is the minimum value of A_d, expressed in decibels, for the gain to be within 0.2 dB of its ideal value?

5.14 Consider the feeback amplifier of Figure 5.4 in which the phasor gain of the difference amplifier is $-j100$. Determine the complex value of A_{fb} for $R_1 = 10$ kΩ and $R_2 = 100$ kΩ. What is A_{fb} expressed in decibels?

5.15 Repeat Problem 5.14 for $R_1 = 4.7$ kΩ and $R_2 = 240$ kΩ.

5.16 The difference amplifier of Figure 5.4 has an input resistance R_i of 1 MΩ and a gain A_d of 1000. The feedback network has resistances of $R_1 = 100$ kΩ and $R_2 = 200$ kΩ. What are the values of R_{IN} and A_{fb} of the circuit?

5.17 Repeat Problem 5.16 for $R_1 = 10$ kΩ.

5.18 Repeat Problem 5.16 for $R_i = 10$ kΩ.

5.19 Consider the equivalent circuit for a difference amplifier with a nonzero output resistance Figure 5.13. Obtain an expression for R_{OUT} for the case in which $R_1 + R_2$ is not large compared with R_{OUT}.

5.20 A difference amplifier with an equivalent circuit of Figure 5.13 has the following parameters: $R_1 = 1$ kΩ, $R_2 = 22$ kΩ, and $R_O = 500$ Ω. What is R_{OUT} for $A_d = 100, 1000$, and 10^4? What are the corresponding values of A_{fb}?

5.21 Repeat Problem 5.20 for $R_O = 10$ kΩ.

5.22 Consider the result of Eq. (5.25), the dependence of v_{OUT} on v_{IN} for the nonlinear amplifier with feedback. Show that v_{OUT} is very nearly equal to $10\, v_{IN}$ for a small magnitude of v_{IN}.

5.23 The nonlinear feedback amplifier of Figure 5.9 has the following parameters: $R_1 = 1\ k\Omega$, $R_2 = 10\ k\Omega$, $A_d = 10$, $A_1 = 10$, and $A_2 = 2.5\ V^{-1}$.

a) Determine $V_{1\,max}$.

b) What is v_{OUT}/v_{IN} for $|v_{IN}|$ small?

c) What is v_{Error} for $v_{IN} = 0.9\ V$?

5.24 Repeat Problem 5.23 for $A_d = 100$.

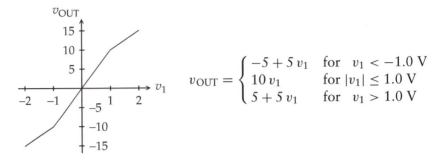

$$v_{OUT} = \begin{cases} -5 + 5\,v_1 & \text{for}\quad v_1 < -1.0\ V \\ 10\,v_1 & \text{for}\quad |v_1| \leq 1.0\ V \\ 5 + 5\,v_1 & \text{for}\quad v_1 > 1.0\ V \end{cases}$$

Figure P5.25

5.25 The nonlinear response of a particular amplifier may be approximated by the piecewise linear response of Figure P5.25 ($|v_1| \leq 2\ V$). The amplifier is used in the circuit of Figure 5.9 with $A_d = 10$ and with R_1 and R_2 chosen so that $v_{OUT} = 10\,v_{IN}$ for $|v_1|$ small. Determine expressions for the piecewise linear dependence of v_{OUT} on v_{IN}.

5.26 Repeat Problem 5.25 for a difference amplifier with a voltage gain of 100 ($A_d = 100$).

5.27 A difference amplifier with $A_{d0} = 10^5$ and $f_b = 10\ Hz$ is used in a feedback circuit with $\beta = 0.02$.

a) What is the low-frequency gain of the feedback amplifier circuit?

b) What is the frequency for which the response A_{fb} is 1.0 dB less than its low-frequency value?

c) What is the frequency for which A_{fb} is 10 dB less than its low-frequency value?

5.28 Repeat Problem 5.27 for $A_{d0} = 10^4$ and $f_b = 100\ Hz$.

5.29 Repeat Problem 5.27 for a feedback circuit with $\beta = 0.01$.

5.30 Repeat Problem 5.27 for a feedback circuit with $\beta = 0.1$.

5.31 A difference amplifier with $A_{d0} = 10^4$ and $f_b = 100\ Hz$ is used in a feedback amplifier circuit with $\beta = 0.1$.

a) What is the low-frequency gain of the feedback circuit?

b) What is the frequency for which the gain is 3 dB less than its low-frequency value?

c) What is A_d for this frequency?

d) What is the magnitude and phase of A_{fb} for $f = 100\ kHz$?

5.32 Consider the case for which the difference amplifier of Problem 5.31 has a second break frequency of $f_{b2} = 100$ kHz.

  a) What is the low-frequency gain of the feedback circuit?
  b) What is A_d for $f = 100$ kHz?
  c) What is A_{fb} for $f = 100$ kHz?

5.33 What is the phase margin of the op amp circuit of Problem 5.32 ($\beta = 0.1$)?

5.34 What is the phase margin of the op amp circuit of Problem 5.32 for a feedback circuit with $\beta = 0.2$?

5.35 What is the phase margin of the op amp circuit of Problem 5.32 for a feedback circuit with $\beta = 0.05$?

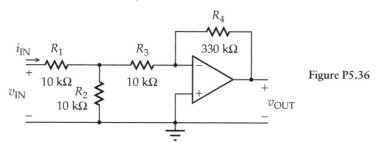

Figure P5.36

5.36 Consider the inverting amplifier circuit of Figure P5.36.

  a) Determine the voltage gain v_{OUT}/v_{IN} with ideal behavior of the op amp assumed.
  b) What is the input resistance of the amplifier v_{IN}/i_{IN}?
  c) A variable-gain amplifier can be realized by using a variable resistor for R_2. Determine the range of R_2 necessary to vary the magnitude of the gain from 1 to 10.

5.37 Suppose $R_3 = 0$ in the circuit of Figure P5.36. Determine v_{OUT}/v_{IN} and v_{IN}/i_{IN} for this condition. Assume ideal behavior of the op amp.

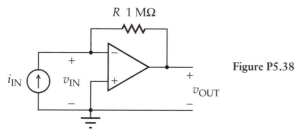

Figure P5.38

5.38 The input signal of the op amp circuit of Figure P5.38 is an ideal current source.

  a) Assume ideal behavior of the op amp and determine an equivalent transfer resistance of the circuit, that is, v_{OUT}/i_{IN}.
  b) What is v_{IN}/i_{IN} for a finite difference amplifier gain A_d of 10^5?
  c) What is v_{IN}/i_{IN} for $A_d = 10^3$?

5.39 Repeat Problem 5.38 for a current source with an equivalent resistance of 1 kΩ.

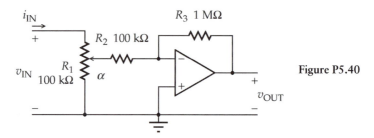

Figure P5.40

5.40 A linear potentiometer is used for an input level control (volume control if an audio amplifier) of an inverting op amp circuit (Figure P5.40). The parameter α corresponds to the setting of the variable resistance tap $(0 \leq \alpha \leq 1)$. This results in a resistance of αR_1 for the lower portion of the potentiometer and a resistance of $(1 - \alpha) R_1$ for the upper portion. Assume ideal behavior of the op amp. Determine v_{OUT}/v_{IN} as a function of the potentiometer setting α. What is the dependence of the input resistance of the circuit v_{IN}/i_{IN}, on α?

5.41 Modify the circuit of Figure P5.40 so that the minimum value of $|v_{OUT}/v_{IN}|$ is 0.1. Assume ideal behavior of the op amp. (Hint: Use a resistance in series with the lower end of R_1.)

5.42 A noninverting op amp circuit is used in conjunction with an input level control (Figure P5.42). Assume ideal behavior of the op amp. Determine the voltage gain v_{OUT}/v_{IN} and the input resistance v_{IN}/i_{IN} as a function of α.

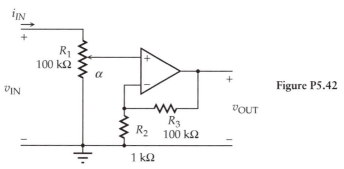

Figure P5.42

5.43 Design a circuit similar to that of Figure P5.42 that has a minimum voltage gain of 0.1, a maximum gain of 20, and an input resistance of 1 MΩ.

5.44 The op amp circuit of Figure P5.44 has two inputs. Determine v_{OUT} as

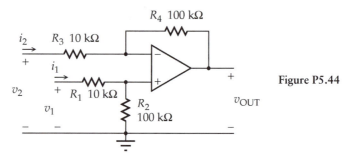

Figure P5.44

a function of v_1 and v_2. Suppose 5 percent tolerance resistors are used for the circuit. Determine the maximum value of $|2 v_{OUT}/(v_1 + v_2)|$, the common-mode voltage gain, that could occur as a result of resistance variations.

5.45 For the op amp circuit of Figure P5.44, determine i_1 and i_2 as a function of v_1 and v_2. Why is v_2/i_2 not a simple resistance?

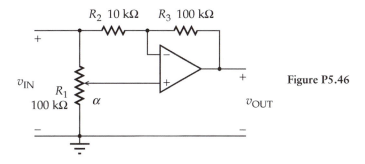

Figure P5.46

5.46 Ideal behavior may be assumed for the op amp of Figure P5.46. Determine v_{OUT}/v_{IN} as a function of α. What is v_{OUT}/v_{IN} for $R_2 = R_3$?

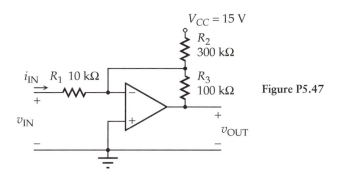

Figure P5.47

5.47 Determine v_{OUT} as a function of v_{IN} for the op amp circuit of Figure P5.47. Assume ideal behavior of the op amp. What is v_{IN}/i_{IN}?

5.48 Repeat Problem 5.47 with R_2 connected to a supply of -10 V.

5.49 Consider the circuit of Figure 5.33 with $R_1 = 10$ kΩ and $R_2 = 1$ MΩ.

a) With ideal behavior of the op amp assumed, determine a value for C that results in an upper half-power break frequency of 20 kHz.
b) Suppose the op amp has a gain–bandwidth product of 1 MHz. For the capacitor determined in part(a), determine V_{out}/V_{in} for a frequency of 20 kHz.
c) What is V_{out}/V_{in} for $f = 20$ kHz, $C = 0$, and a gain–bandwidth product of 1 MHz?

5.50 Repeat Problem 5.49 for $R_2 = 100$ kΩ.

5.51 Repeat Problem 5.49 for $R_2 = 100$ kΩ and an op amp with $GBP = 5$ MHz.

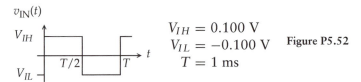

$V_{IH} = 0.100$ V
$V_{IL} = -0.100$ V   **Figure P5.52**
$T = 1$ ms

5.52 For the op amp circuit of Figure 5.33, $R_1 = 10$ kΩ, $R_2 = 1$ MΩ, and $C = 1.0$ nF. Assume ideal behavior of the op amp and that the current of R_2 can be ignored. Determine and sketch $v_{OUT}(t)$ for the input voltage of Figure P5.52. As a result of R_2, the steady-state value of $v_{OUT}(t)$ will have a zero average value for an input voltage with a zero average value.

5.53 Repeat Problem P5.52 with the current of R_2 taken into account.

5.54 Suppose that the input square-wave voltage of Problem P5.52 does not have symmetry, that is, $V_{IH} = 0.110$ V, and $V_{IL} = -0.090$ V. Assume an ideal op amp and that steady-state conditions prevail. Determine $v_{OUT}(t)$. (Hint: Use superposition.)

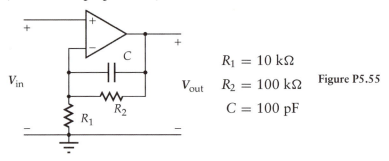

$R_1 = 10$ kΩ
$R_2 = 100$ kΩ   **Figure P5.55**
$C = 100$ pF

5.55 Consider the noninverting op amp circuit of Figure P5.55. Ideal behavior of the op amp may be assumed.

a) Determine V_{out}/V_{in} as a function of frequency.
b) Determine the approximate frequency for which V_{out}/V_{in} is 3 dB down from its zero frequency value.
c) Suppose the op amp has a gain–bandwith product of 1 MHz. What is the actual response for the frequency determined in part(b)?

5.56 Repeat Problem 5.55 for $C = 20$ pF.

5.57 Repeat Problem 5.55 for $C = 20$ pF and $GBP = 5$ MHz.

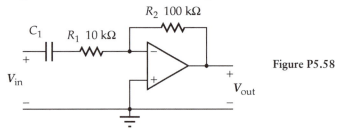

Figure P5.58

5.58 Consider the op amp circuit of Figure P5.58 in which ideal behavior of the op amp may be assumed. What is V_{out}/V_{in} for very high-frequency signals? What is it for very low-frequency signals? Obtain a frequency-dependent expression for V_{out}/V_{in}. What is the value of C_1 that results in a response that is 3-dB down from its high-frequency value for a frequency of 50 Hz?

5.59 For the circuit of Figure P5.58, assume $C_1 = 1\ \mu F$ and ideal behavior of the op amp. Determine an expression for V_{out}/V_{in} and the frequency for which V_{out}/V_{in} is down by 3 dB from its high-frequency value. What is the frequency at which the gain is down by 5 dB from its high-frequency value?

5.60 For the circuit of Figure P5.58 assume $C_1 = 5\ \mu F$ and ideal behavior of the op amp. Determine a capacitance C_2 connected in parallel with R_2 that results in an upper half-power frequency of 20 kHz. What is the lower half-power frequency of the amplifier?

5.61 Consider the op amp circuit of Figure 5.55 in which $R_1 = 4.7\ k\Omega$ and $R_2 = 100\ k\Omega$. What is the upper half-power frequency of the overall response of the circuit for op amps with gain–bandwidth products of 3 MHz. What is the frequency at which the overall response is 1 dB less than its low-frequency value?

5.62 The individual stages of a two-stage op amp circuit (Figure 5.55) have different gains. The gain of the first stage is 10, whereas the gain of the second stage is 5. What is the overall half-power frequency of the two-stage amplifier? Assume identical op amps with gain–bandwidth products of 5 MHz.

COMPUTER SIMULATIONS

C5.1 An ideal difference amplifier was used for the feedback amplifier circuit of Example 5.3. Consider the case for which the difference amplifier has the same transfer characteristic as the power amplifier, namely, $v_1 = A_1 v_{Dif} - A_2 v_{Dif}^2$ for $v_{Dif} > 0$. Repeat Example 5.3 for this difference amplifier. Compare the results obtained ($v_{IN} = 0.8$ V, 0.9 V, and for a sinusoidal input, $V_m = 0.9$ V) with those for the linear difference amplifier.

C5.2 Consider the case for which an amplifier has a saturation that depends on the cube of its input voltage, that is, $v_{OUT} = A_1 v_1 - A_3 v_1^3$. This function is valid for both positive and negative values of v_1. Repeat Example 5.3 for this amplifier in which $A_1 = 10$ and A_3 is chosen such as to result in $\frac{dv_{OUT}}{dv_1} = 0$ for $v_{OUT} = 10$ V.

C5.3 A source-follower amplifier using MOSFET devices with complementary symmetry is used to drive an 8 Ω loudspeaker (Figure C5.3). With negative feedback and an input difference amplifier, the inherent distortion of the source followers can be significantly reduced.

    a) Obtain a plot of the static transfer characteristic of the MOSFET source-follower amplifier, that is, v_{OUT} versus v_1 for a ± 15 V range of v_1.

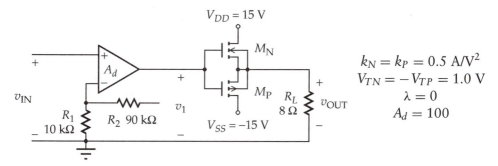

Figure PC5.3

b) Consider the case for which R_2 of the difference amplifier feedback network is connected to v_1 (no feedback for the MOSFET devices). Obtain a plot of v_{OUT} versus v_{IN} for a ± 2 V range of v_{IN}. Except for the gain of the circuit, this characteristic should have the same distortion as the MOSFET devices of the previous part.

c) Repeat part (b) with R_2 connected to v_{OUT}, thus including the MOSFET devices in the feedback loop. A considerable reduction in distortion should occur.

d) Repeat part (c) using a difference amplifier with a larger gain, $A_d = 1000$.

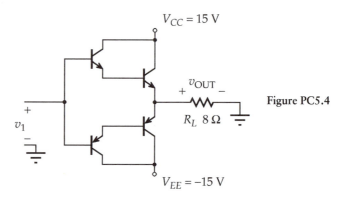

Figure PC5.4

C5.4 Repeat Simulation C5.3 for a BJT emitter–follower amplifier with complementary symmetry (Figure C5.4). The cascaded emitter followers are used to obtain sufficient output current. Assume $\beta_F = 100$ for all transistors and that their current scale factors I_s are such as to result in base–emitter voltages with a magnitude of 0.75 V when the devices are conducting. For conduction, assume the magnitude of v_{OUT} is 12 V.

C5.5 A simulation of the feedback amplifier of Problem 5.25 is desired. The circuit of Figure C5.5 with ideal diodes is suggested for modeling the response of the nonlinear amplifier. Determine the component values for the model and then obtain its static transfer characteristic to verify the design. Use the difference amplifier and the feedback circuit of the problem to

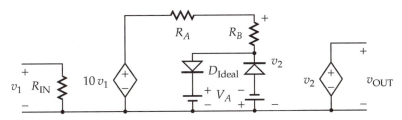

Figure PC5.5

determine the static transfer characteristic of the circuit. Also determine, through a .TRAN simulation, the harmonic distortion for an input signal that results in a peak output voltage of 15 V.

C5.6 An operational amplifier has a gain–bandwidth product GBP of 5 MHz and a low-frequency gain A_{d0} of 2×10^5. The op amp is used in a non-inverting feedback amplifier circuit with $R_1 = 10 \text{ k}\Omega$ and $R_2 = 90 \text{ k}\Omega$ – a circuit that results in an ideal gain of 10 for low-frequency signals.

a) With a SPICE simulation, show that the frequency-dependent response of this circuit is well behaved for signals with frequencies up to 10 MHz. Verify that the upper half-power frequency f_h is equal to βGBP.

b) Suppose that as a result of a manufacturing defect the op amp has a second break frequency of 200 kHz. Determine the frequency response of a feedback circuit using this op amp.

c) Consider the case in which the op amp of part (**b**) is used in a feedback circuit with $R_1 = 1 \text{ k}\Omega$ and $R_2 = 99 \text{ k}\Omega$, which results in an ideal low-frequency gain of 100. Show that the response of this circuit is well behaved.

d) Determine the phase margins of the circuits of parts (**b**) and (**c**). Because the phase of A_d does not cross $-180°$, it is not possible to determine a gain margin.

C5.7 Consider the case for a conventional amplifier (no feedback) with a double break frequency of 100 kHz and a low-frequency gain of 100. Use an .AC SPICE simulation to determine the upper half-power frequency of the amplifier. Use a .TRAN simulation to determine the rise time (10-to-90 percent) of an input step function. What is the time required for the output to reach 99 percent of its final value?

C5.8 The behavior of a particular amplifier can be characterized by three break frequencies as follows:

$$A_d = \frac{A_{d0}}{(1 + jf/f_1)(1 + jf/f_2)^2}$$

$$A_{d0} = 1000, \quad f_1 = 1 \text{ kHz}, \quad f_2 = 100 \text{ kHz}$$

Suppose that the amplifier is used in a feedback circuit with $\beta = 0.1$.

a) Use an .AC spice simulation to determine the frequency response of the overall feedback circuit. What is the amount (expressed in decibels) by which the peak in response exceeds the low-frequency gain? What are the gain and phase margins of the feedback amplifier?

b) From the plots used to determine the gain and phase margins, determine a reduced value of β that results in acceptable values of gain and phase margins (Eq. (5.40)). Determine the frequency response of a feedback circuit using this value of β. What is the low-frequency gain of the amplifier? What is the amount by which the peak in the response exceeds the low-frequency gain?

C5.9 A SPICE simulation to determine the small- and large-signal behavior of the difference amplifier circuit of Example 5.7 is desired. Assume identical transistors with $\beta_F = 100$, $n_F = 1.0$, and $I_s = 5 \times 10^{-16}$ A. With a .DC SPICE simulation, determine the dependence of v_{01} and v_{02} on v_1. From these plots, determine the small-signal voltage gains of the amplifier, v_{01}/v_1 and v_{02}/v_1. What is the range of v_1 over which the output voltages are within 10 percent of the values predicted by a linear dependence on v_1?

C5.10 Consider the high-pass inverting op amp circuit of Figure P5.58 of the problem set. For a particular audio application, the capacitor C_1 has a value of 0.2 μF. The op amp has a gain–bandwidth product GBP of 1 MHz.

a) On the basis of analytic considerations, what is the lower half-power frequency f_ℓ of the circuit? Use a SPICE .AC solution to verify this result.

b) A .TRAN solution is desired for a periodic square-wave input voltage. The input voltage varies between -0.5 and $+0.5$ V and has a frequency equal to the lower half-power frequency of the circuit f_ℓ. Obtain a solution for $v_{OUT}(t)$ for at least four periods of the input voltage.

c) Repeat part (b) for square-wave input voltages with frequencies of $3 f_\ell$ and $10 f_\ell$.

d) Comment on the preceding results. On the basis of these results, estimate the lower half-power frequency required for an amplifier that is to provide a reasonable 100-Hz square-wave output voltage.

DESIGN EXERCISES

D5.1 The frequency response of a noninverting operational amplifier circuit is to be within 1 dB of its low-frequency value at a frequency of 100 kHz. A low-frequency gain of 20 (26 dB) is required. Determine the minimum value of GBP required for the op amp and design a circuit to achieve the desired response. Verify, using a SPICE simulation, that the response specification is indeed satisfied for an op amp with the minimum GBP. Use SPICE to determine the 10-to-90 percent rise time of the amplifier circuit.

D5.2 Design a deemphasis circuit similar to that of the one shown in Figure 5.49(b) in which the magnitude of the high-frequency gain is 0.1

that of the low-frequency response. An input resistance of 100 kΩ, is desired and the magnitude of the low-frequency gain is 5. The lower break-frequency f_1 is 500 Hz. Use SPICE to verify that the response does indeed fulfill the requested design parameters. Assume an op amp with a gain–bandwidth product of 1 MHz. What is the upper half-power frequency of the circuit?

D5.3 An amplifier with an upper half-power frequency f_{h1} of 10 kHz has been used to amplify a pulse. This results in a response with a time constant of $1/2\pi f_{h1}$. Design an amplifier circuit similar to the preemphasis circuit of Figure 5.46 to "restore" partially the initial waveform of the pulse. A resultant time constant that is only 10 percent of that of the first amplifier is desired. The gain of the circuit used to restore the pulse should be unity for low-frequency signals. Use a SPICE simulation to verify that your circuit does indeed improve the waveform of the pulse.

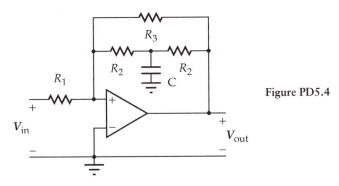

Figure PD5.4

D5.4 A preemphasis circuit has the configuration of Figure D5.4. It will be noted that the low-frequency gain $-(2\,R_2)\|R_3/R_1$ occurs while the high-frequency gain is $-R_3/R_1$. Design a circuit that has an input resistance of 50 kΩ and break frequencies (Figure 5.47) of $f_1 = 1\,\text{kHz}$ and $f_2 = 3\,\text{kHz}$. A low-frequency gain with a magnitude of 1 is desired. Suppose an op amp with $GBP = 5$ MHz is used. What is the upper half-power frequency of the circuit? Determine the value for a capacitance C_H placed in parallel with R_3 that reduces the upper half-power frequency to 50 kHz.

D5.5 A deemphasis circuit is to be designed that can be used in conjunction with the preemphasis circuit of Design Exercise D5.4. Design a circuit using the same configuration of elements. The magnitude of the input impedance is to be no less than 10 kΩ at any frequency, and a high-frequency gain with a magnitude of 1 is desired.

D5.6 Design a deemphasis circuit to be used in conjunction with the preemphasis circuit of Design Exercise D5.4 that has the configuration of Figure 5.63(b). An input resistance of 50 kΩ and a high-frequency gain with a magnitude of 1 are desired.

D5.7 If Design Exercises D5.4 and D5.5 (or D5.6) were done, use SPICE to verify that the designs are valid. Also determine the response of the preemphais circuit and the overall circuit for a 200-μs pulse.

D5.8   A low-level ac-coupled amplifier with overall lower and upper half-power frequencies of 40 Hz and 100 kHz, respectively, is desired. To operate the amplifer from a single supply of only 5 V, an LM 358 with $GBP = 1$ MHz has been chosen for the design. The input resistance of the amplifier (1 kHz) is to be 100 kΩ. Design a two-stage amplifier that meets the design requirements and results in the maximum overall possible decibel gain. Design a three-stage amplifier that provides the maximum possible decibel gain. Use SPICE to verify the designs.

D5.9   A two-stage broad-band amplifer using noninverting op amps is to be designed. Its lower and upper half-power frequencies are to be 20 Hz and 2 MHz, respectively. The midfrequency gain (1 kHz) is to be within 1 dB of 20 dB. It is to have an input resistance of 10 kΩ (1 kHz), and it is to operate from a single supply voltage. Determine a circuit that achieves this and specify the minimum GBP required for the op amps. Determine the tolerance required for the feedback resistors (R_1 and R_2 of Figure 5.55) so that the gain is maintained within ± 1 dB of 20 dB for the worst-case situations. Use SPICE to verify the design for the nominal component values (minimum GBP) as well as for the worst-case resistance values that result in gains of 19 and 21 dB.

CHAPTER SIX

ELECTRONIC POWER SUPPLIES

Essentially all electronic systems require a nonvarying supply voltage (or current), that is, a dc voltage (or dc current). On the other hand, the electric power supplied by utilities is characterized by an alternating voltage and current having a sinusoidal time dependence. In North America, a frequency of 60 Hz is common, whereas 50 Hz is used in most other areas of the world. Utility potentials depend on the usage: residential service is 120 V (rms) in North America, whereas 220–240 V is common for residential service elsewhere.

A semiconductor junction diode allows a current in only one direction; its reverse-biased current is negligibly small and can be ignored for nearly all applications. Hence, a diode may be used to convert an alternating source of current to a current with a single direction – a process generally referred to as rectification. For many electronic applications it is also necessary to transform the utility voltage to a desired voltage using an iron-core transformer.

The resistor R_L of the power supply of Figure 6.1 represents the load to which electrical power is to be supplied. The secondary voltage of the transformer $v_{\text{Trans}}(t)$ is rectified by the diode, resulting in a load voltage $v_{\text{Load}}(t)$ that has a single polarity. The load current $v_{\text{Load}}(t)/R_L$ also has a single polarity.

The usefulness of the supply shown in Figure 6.1 is very limited because the load voltage is zero for a significant portion of each period of the input voltage. Although the supply could be used for charging a battery, a supply voltage with extended off intervals is unsuitable for most electronic applications (imagine a microprocessor trying to process data that are "lost" during each off interval). A storage of electrical energy is required to sustain a load voltage (or current) during these off periods. Both capacitors and inductors store electrical energy as follows:

$$E_{\text{Capacitor}} = \tfrac{1}{2}Cv_C^2, \qquad E_{\text{Inductor}} = \tfrac{1}{2}Li_L^2 \tag{6.1}$$

Energy can be supplied to a capacitor, for example, when the transformer voltage is positive, and when the transformer voltage is negative, the energy of the capacitor may be used to supply power to the load resistor.

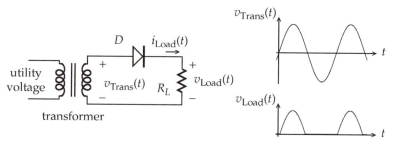

Figure 6.1: An elementary power supply.

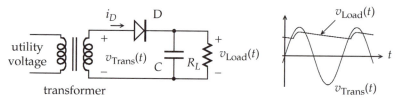

Figure 6.2: An elementary power supply with a capacitor filter.

Consider the circuit of Figure 6.2 in which a large capacitor has been connected in parallel with the load resistor of the previous circuit. Whenever the transformer voltage is greater than the load voltage, the diode conducts, resulting in a diode current i_D. This current (a flow of charge) will charge the capacitor. However, when $v_{Trans}(t)$ falls below $v_{Load}(t)$, the diode will be reverse biased and its current will be zero. During this interval, the capacitor, if sufficiently large, will be slowly discharged by the load resistor. Although the load voltage shown in Figure 6.2 has a small variation with time, it is suitable for many electronic applications. Large electrolytic capacitors (capacitances of hundreds if not thousands of microfarads are common) are generally used to minimize the amplitude of the voltage fluctuations. This function performed by the capacitor is known as filtering, that is, reducing the unacceptable fluctuations of a supply without an energy storage element.

A modern power supply will generally utilize an electronic regulator, a circuit using transistors (often fabricated as a single integrated circuit), that will not only further reduce the load voltage fluctuations of the filter but will tend to compensate for changes in load current. The basic elements of an electronic power supply are indicated in Figure 6.3. Often the rectifier will be more complex than a single diode – a bridge rectifier consisting of four diodes (usually on a single integrated circuit) is common. In addition to filters using a single capacitor, a resistor- or inductor-capacitor combination may be used.

Figure 6.3: Basic elements of an electronic power supply.

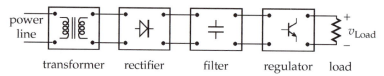

Chemical batteries will be discussed in the last section of the chapter. With the development of extremely low-power integrated circuits, numerous electronic systems, including remote control units, tape and CD players, cellular phones, and note-book computers are now powered by self-contained batteries. As power requirements are further reduced for electronic circuits, more battery-powered electronic systems can be expected.

6.1 RECTIFIERS: FROM ALTERNATING TO DIRECT CURRENT

The importance of being able to convert alternating current supplied by electric utilities to direct current required by electronic circuits was stressed in the introduction to the chapter. Although in the past vacuum tube diodes were used for rectification, now it is the silicon semiconductor junction diode that is used for essentially all electronic applications. Power levels of rectifier circuits vary from a few milliwatts to thousands of watts. Semiconductor junction diodes range from a microscopic size for microampere currents of integrated circuits to large waste-basket size discrete diodes for currents of thousands of amperes.

THE HALF-WAVE RECTIFIER

The circuit of Figure 6.4, with a transformer that has a sinusoidal secondary voltage of $v_{\text{Trans}}(t)$, will be used to develop a quantitative perspective of the rectification process.

$$v_{\text{Trans}}(t) = V_m \sin \omega t \tag{6.2}$$

It will initially be assumed that the diode of the circuit can be replaced with an ideal diode model, that is, the forward-biased voltage of the diode will be assumed to be zero. Hence, the load voltage will be equal to the transformer voltage whenever the transformer voltage is positive and zero otherwise.

$$v_{\text{Load}}(t) = \begin{cases} V_m \sin \omega t & \text{for } V_m \sin \omega t > 0 \\ 0 & \text{otherwise} \end{cases} \tag{6.3}$$

Because only the positive "half" of the transformer voltage is, in effect, used, this circuit is known as a half-wave rectifier circuit, and the corresponding output voltage of Figure 6.4 is described as a half-wave rectified voltage.

The average value of the load voltage V_{av} and load current I_{av}, which are the voltage and current that a conventional digital multimeter would indicate (dc range), are of interest. The average current, if multiplied by the time for which the circuit operates, yields the quantity of charge that flows through the load.

Figure 6.4: A half-wave rectifier circuit and load voltage for an ideal diode.

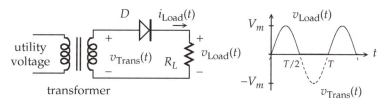

This quantity is of interest when designing a rectifer to charge a battery. The average value of a periodic quantity such as $v_{\text{Load}}(t)$, is defined in terms of the area under the time-dependent curve as follows:

$$V_{\text{av}} T = \int_0^T v_{\text{Load}}(t)\,dt \tag{6.4}$$

The left-hand side of Eq. (6.4), $V_{\text{av}} T$, is the area of a rectangle of height V_{av} and width of one period. The right-hand side is the area under the load voltage curve (both have the dimension of volt-seconds). Although the limits of 0 and T have been used for the integral, any interval of T seconds (from t_0 to $t_0 + T$) may be used because the function is periodic. For the half-wave rectified voltage obtained using an ideal diode model, V_{av} may readily be determined as follows:

$$\begin{aligned} V_{\text{av}} &= \frac{1}{T}\int_0^T v_{\text{Load}}(t)\,dt = \frac{V_m}{T}\int_0^{T/2} \sin \omega t\,dt \\ &= \frac{V_m}{\omega T}\int_0^{T/2} \sin \omega t\,d\omega t = \frac{V_m}{\omega T}\int_0^{\pi} \sin \theta\,d\theta \\ &= 2V_m/\omega T = V_m/\pi \end{aligned} \tag{6.5}$$

In the preceding expression, it was recognized that the period T is defined such that when $t = T$, the argument of the trigonometric function specifying the transformer voltage is 2π. Hence ωT is equal to 2π. The average value of the load voltage is slightly less than one-third the peak amplitude of the load voltage.

For the ideal diode model to yield a reasonably accurate result, it is necessary that the peak transformer voltage V_m be large compared with the diode's forward-biased voltage. If this is not the case, the constant forward-biased voltage diode model should be used for a more accurate result. This reduces the load voltage (Figure 6.5).

$$v_{\text{Load}}(t) = \begin{cases} V_m \sin \omega t - v_{D(\text{on})} & \text{for } V_m \sin \omega t - v_{D(\text{on})} > 0 \\ 0 & \text{otherwise} \end{cases} \tag{6.6}$$

The average load voltage is also reduced.

Figure 6.5: The effects of a forward-biased diode voltage of $v_{D(\text{on})}$ on the load voltage of a half-wave rectifier circuit.

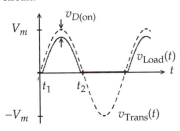

$$\begin{aligned} V_{\text{av}} &= \frac{1}{T}\int_0^T v_{\text{Load}}(t)\,dt \\ &= \frac{1}{T}\int_{t_1}^{t_2} \left(V_m \sin \omega t - v_{D(\text{on})} \right) dt \\ &= \frac{2}{T}\int_{t_1}^{T/4} \left(V_m \sin \omega t - v_{D(\text{on})} \right) dt \end{aligned} \tag{6.7}$$

The quantities t_1 and t_2 are the times at which the argument of the integrals is zero.

$$V_m \sin \omega t_1 - v_{D(\text{on})} = 0,$$

$$\theta_1 = \omega t_1 = \sin^{-1} \left(v_{D(\text{on})}/V_m \right) \tag{6.8}$$

A substitution of $\theta = \omega t$ and $\theta_1 = \omega t_1$ results in the following:

$$V_{av} = \frac{1}{\pi} \int_{\theta_1}^{\pi/2} \left(V_m \sin\theta - v_{D(on)} \right) d\theta = \frac{1}{\pi} \left[-V_m \cos\theta - v_{D(on)}\theta \right]_{\theta_1}^{\pi/2}$$

$$= \frac{1}{\pi} \left[V_m \cos\theta_1 - v_{D(on)}(\pi/2 - \theta_1) \right] \tag{6.9}$$

It should be noted that for $v_{D(on)} = 0$, $\theta_1 = 0$, and $V_{av} = V_m/\pi$, the same result as for an ideal diode. For small values of transformer voltage, the forward-biased diode voltage is significant. For $V_m = 7$ V and $v_{D(on)} = 0.7$ V, $v_{D(on)}/V_m = 0.1$ and $\theta_1 = 0.1$ (5.74°). This results in a value of 1.94 V for V_{av} as compared with the 2.22 V (V_m/π) obtained if the ideal diode model is used.

FULL-WAVE RECTIFIER – A CENTER-TAPPED TRANSFORMER

A significant improvement in the load voltage may be achieved by utilizing the negative half of the transformer voltage. Consider the circuit of Figure 6.6 in which a transformer with a center-tapped secondary winding is used in conjunction with two diodes. If this circuit did not include D_2 (an open circuit), the same behavior as that of the half-wave rectifier of Figure 6.4 would be expected. The transformer voltage $v_B(t)$ is positive during the interval when $v_A(t)$ is negative, and $v_{Load}(t)$ would be zero for D_2 removed from the circuit. However, with the diode D_2 in the circuit, D_2 will be forward biased when $v_B(t)$ is positive, resulting in a load voltage $v_{Load}(t) = v_B(t)$ during this interval (ideal diode assumption). The net result, as shown in Figure 6.7, is a load voltage that is described as a full-wave rectified voltage. The half period during which $v_{Load}(t) = 0$ for the half-wave rectifier circuit is eliminated.

Figure 6.6: A full-wave rectifier using a transformer with a center-tapped secondary winding.

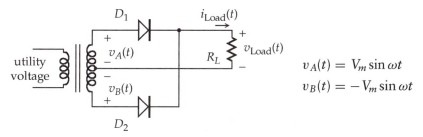

$$v_A(t) = V_m \sin\omega t$$
$$v_B(t) = -V_m \sin\omega t$$

Figure 6.7: The transformer voltages and load voltage of a full-wave rectifier circuit.

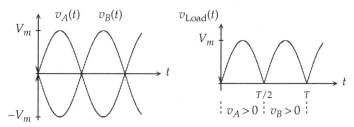

The average load voltage, if ideal behavior of the diodes is assumed, may readily be determined.

$$V_{av}T = \int_0^T v_{Load}(t)\, dt = V_m \int_0^{T/2} \sin \omega t \, dt - V_m \int_{T/2}^T \sin \omega t \, dt$$

$$(6.10)$$

$$V_{av} = \frac{2V_m}{T} \int_0^{T/2} \sin \omega t \, dt = 2V_m/\pi$$

The average load voltage for a full-wave rectified sinusoidal voltage is, not surprisingly, twice that of a half-wave rectified voltage (the area under the load voltage curve for one period of the transformer voltage is twice as great). It should also be noted that the period of the load voltage is $T/2$; hence, its periodic frequency is twice that of the transformer voltage. For a 60-Hz utility voltage, the frequency of the load voltage of a full-wave rectifier circuit is 120 Hz.

Although the forward-biased voltage of the diodes reduces the average load voltage, the average full-wave rectified voltage for a diode voltage of $v_{D(on)}$, is twice that of the half-wave rectified voltage of Eq. (6.9).

$$V_{av} = \frac{2}{\pi} \left[V_m \cos \theta_1 - v_{D(on)}(\pi/2 - \theta_1) \right]$$

$$(6.11)$$

$$\theta_1 = \sin^{-1}\left(v_{D(on)}/V_m \right)$$

As for the case of a half-wave rectifier circuit with small values of V_m, the forward-biased voltage of the diodes has a significant effect.

FULL-WAVE RECTIFICATION – A BRIDGE RECTIFIER

Another rectifier circuit that is used to obtain full-wave rectification is the bridge rectifier. The term *bridge* is derived from the similar configuration of a 19th-century measuring circuit invented by Samuel Christie and improved upon by Charles Wheatstone. Consider the circuit of Figure 6.8, which uses a bridge rectifier consisting of four diodes. (It is common for the diodes of a bridge rectifier to be drawn at a 45° angle as opposed to being vertical or horizontal as is the generally accepted custom for circuit diagrams.) The operation of the circuit may best be understood by determining the current paths that correspond to $v_{Trans}(t)$ being positive or negative. For a positive value of $v_{Trans}(t)$ and hence, $i_{Trans}(t)$, the current path is through D_1 (D_4 will not allow a current in this direction), through the load resistor (a downward direction on the circuit diagram), and through diode D_3, thus completing the current's path back to the transformer.

Figure 6.8: A full-wave bridge rectifier circuit.

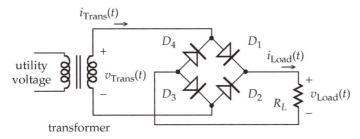

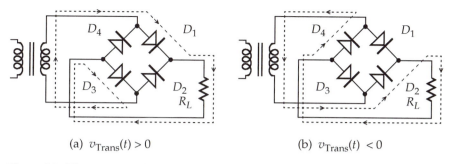

(a) $v_{Trans}(t) > 0$ (b) $v_{Trans}(t) < 0$

Figure 6.9: The current paths of a bridge rectifier.

This path, shown in Figure 6.9(a), results in a positive load voltage. It may readily be demonstrated that diodes D_2 and D_4 are reverse biased for this condition, which justifies the assumption that their currents are zero.

Consider now the case for a negative transformer voltage (Figure 6.9(b)). For a transformer current in the opposite direction, that is for $i_{Trans}(t)$ negative, the current path from the transformer is through D_2 to the load resistor, through the load resistor (again, in a downward direction on the circuit diagram), and through D_4 back to the transformer. For this case, diodes D_1 and D_3 are reverse biased. The load voltage, owing to the altered circuit path, is again positive. Hence, the load voltage is the same as that of Figure 6.7 (ideal diode behavior). The bridge rectifier diode circuit, in effect, switches the current path according to the polarity of the transformer voltage. As a result, the direction of the load current is always the same ($i_{Load}(t) \geq 0$).

Figure 6.10 is an alternative representation of a bridge rectifier circuit. Through a comparison of the circuits of Figures 6.8 and 6.10, one should be able to verify that they are indeed electrically identical. From Figure 6.10, it may be seen that when $v_{Trans}(t)$ is positive, the current path is through the outer two diodes, D_1 and D_3, and the other diodes are reverse biased. However, when $v_{Trans}(t)$ is negative, the path is through the diagonally drawn diodes D_2 and D_4, thus altering the current path to the load. Diodes D_1 and D_2 are reverse biased for this condition.

The advantage of a bridge rectifier circuit is that a center-tapped transformer secondary winding, a winding that requires twice the number of turns of a noncenter-tapped winding, is not required. On the other hand, regardless of

Figure 6.10: An alternative representation of a bridge rectifier circuit.

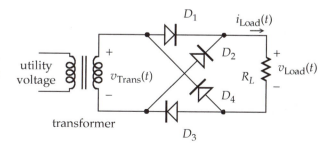

the current path, a bridge rectifier has two forward-biased diodes in series with the load resistor. Using the constant voltage model results in a load voltage reduction of $2v_{D(\text{on})}$ as opposed to only $v_{D(\text{on})}$ for the center-tapped transformer circuit. This

$$v_{\text{Load}}(t) = \begin{cases} |V_m \sin \omega t| - 2v_{D(\text{on})} & \text{for } |V_m \sin \omega t| - 2v_{D(\text{on})} > 0 \\ 0 & \text{otherwise} \end{cases} \qquad (6.12)$$

The magnitude operation yields the correct load voltage for both polarities of transformer voltage.

The average load voltage of a bridge rectifier may be obtained by replacing $v_{D(\text{on})}$ of Eq. (6.21) with $2v_{D(\text{on})}$.

$$V_{\text{av}} = \frac{2}{\pi} \left[V_m \cos \theta_1 - 2v_{D(\text{on})}(\pi/2 - \theta_1) \right] \qquad (6.13)$$
$$\theta_1 = \sin^{-1}(2v_{D(\text{on})}/V_m)$$

For $V_m = 7.0$ V and $v_{D(\text{on})} = 0.7$ V (the same values as used with the half-wave rectifier circuit), $\theta_1 = 0.201$ rad ($11.54°$) and $V_{\text{av}} = 3.15$ V. From the ideal diode model, an average load voltage of 4.46 V is obtained.

EXAMPLE 6.1

For diode circuits, a knowledge of the diode voltage when it is reverse biased is important because diodes are limited by the voltage that they can withstand without being permanently damaged. It is important that a diode with a sufficiently large inverse voltage capability be utilized for rectifier circuits. Assume ideal diode behavior for the following circuits:

a. Determine and sketch the diode voltage $v_D(t)$ of the half-wave rectifier circuit of Figure 6.4. What is the peak inverse voltage of the diode?

b. Repeat part (a) for the diodes of the full-wave rectifier circuit of Figure 6.6.

c. Repeat part (a) for the diodes of the bridge rectifier circuit of Figure 6.8.

SOLUTION

a. The following is obtained for the instantaneous diode voltage $v_D(t)$:

$$v_{\text{Trans}}(t) = V_m \sin \omega t, \qquad v_D(t) = v_{\text{Trans}}(t) - v_{\text{Load}}(t)$$
$$\text{for } 0 < t < T/2 : v_{\text{Load}}(t) = v_{\text{Trans}}(t), \ v_D(t) = 0$$
$$\text{for } T/2 < t < T : v_{\text{Load}}(t) = 0, \ v_D(t) = V_m \sin \omega t$$

The peak value of the diode inverse voltage is V_m (Figure 6.11).

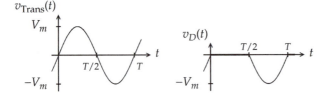

Figure 6.11: Diode voltage of half-wave rectifier of Example 6.1(a).

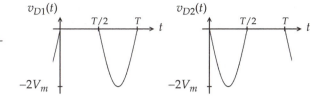

Figure 6.12: Diode voltage of full-wave rectifier of Example 6.1(b).

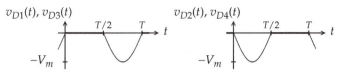

Figure 6.13: Diode voltage of full-wave bridge rectifier of Example 6.1(c).

b. The following is obtained for $v_{D1}(t)$, the instantaneous voltage across D_1 of Figure 6.6:

$$v_A(t) = V_m \sin \omega t, \quad v_B(t) = -V_m \sin \omega t, \quad v_{D1}(t) = v_A(t) - v_{\text{Load}}(t)$$

for $0 < t < T/2 : v_{\text{Load}}(t) = v_A(t), \; v_{D1}(t) = 0$

for $T/2 < t < T : v_{\text{Load}}(t) = v_B(t), \; v_{D1}(t) = 2V_m \sin \omega t$

The voltage $V_{D2}(t)$ is obtained in a similar fashion.

$$v_{D2}(t) = v_B(t) - v_{\text{Load}}(t)$$

for $0 < t < T/2 : v_{\text{Load}}(t) = v_A(t), \; v_{D2}(t) = -2V_m \sin \omega t$

for $T/2 < t < T : v_{\text{Load}}(t) = v_B(t), \; v_{D2}(t) = 0$

The peak value of the diode inverse voltage for both diodes is $2V_m$ (Figure 6.12).

c. For the interval $0 < t < T/2$, diodes D_1 and D_3 of the bridge rectifier are conducting. Hence, $v_{D1}(t) = v_{D2}(t) = 0$. As a result, D_2 and D_4 are connected in parallel with the transformer and load resistor

$$v_{D2}(t) = v_{D4}(t) = -v_{\text{Trans}}(t) = -V_m \sin \omega t$$

For the interval $T/2 < t < T$, diodes D_2 and D_4 are conducting. Hence, $v_{D2}(t) = v_{D4}(t) = 0$, and diodes D_1 and D_3 are now in parallel with the transformer and load resistor.

$$v_{D1}(t) = v_{D3}(t) = v_{\text{Trans}}(t) = V_m \sin \omega t$$

The peak value of the diode inverse voltage of all diodes is V_m (Figure 6.13).

EXAMPLE 6.2

Consider the half-wave rectifier of Figure 6.4 with $V_m = 10$ V, $R_L = 100 \; \Omega$, and $v_{D(\text{on})} = 0.7$ V. Because the peak transformer voltage is not very large, use the constant forward-biased voltage diode model.

a. Determine the average load voltage V_{av} and load current I_{av}.

b. Determine the average power dissipated by the load resistor.

c. Determine the average power dissipated by the diode.

SOLUTION

a. Equations 6.8 and 6.9 will be used to determine the average load voltage and current.

$$\theta_1 = \sin^{-1}\left(v_{D(\text{on})}/V_m\right) = 4.01° \quad (0.070\text{ rad})$$

$$V_{\text{av}} = \frac{1}{\pi}\left[V_m \cos\theta_1 - v_{D(\text{on})}(\pi/2 - \theta_1)\right] = 2.84\text{ V}$$

$$I_{\text{av}} = V_{\text{av}}/R_L = 28.4\text{ mA}$$

b. The average power P_{av} dissipated by the load resistor is the average of the instantaneous load power $i_{\text{Load}}(t)v_{\text{Load}}(t)$:

$$P_{\text{av}} = \frac{1}{T}\int_0^T i_{\text{Load}}(t)v_{\text{Load}}(t)\,dt = \frac{1}{R_L T}\int_0^T v_{\text{Load}}^2(t)\,dt$$

It should be noted that the average power is not equal to the product of the average voltage and current. From the expression for $v_{\text{Load}}(t)$ of Eq. (6.6), the following is obtained:

$$
\begin{aligned}
P_{\text{av}} &= \frac{1}{R_L T}\int_{t_1}^{t_2}\left(V_m\sin\omega t - v_{D(\text{on})}\right)^2 dt \\
&= \frac{1}{R_L \pi}\int_{\theta_1}^{\pi/2}\left(V_m\sin\theta - v_{D(\text{on})}\right)^2 d\theta \\
&= \frac{1}{R_L\pi}\left[\frac{V_m^2}{2}\left(\pi/2 - \theta_1 + \frac{1}{2}\sin 2\theta_1\right) - 2V_m v_{D(\text{on})}\cos\theta_1\right.\\
&\quad \left.+ v_{D(\text{on})}^2(\pi/2 - \theta_1)\right] = 0.208\text{ W}
\end{aligned}
$$

c. The average power dissipated by the diode $P_{D\text{av}}$ depends upon its instantaneous current and voltage as follows:

$$
\begin{aligned}
P_{D\text{av}} &= \frac{1}{T}\int_0^T i_D(t)v_D(t)\,dt = \frac{1}{T}\int_{t_1}^{t_2} i_D(t)v_D(t)\,dt \\
&= v_{D(\text{on})}\left[\frac{1}{T}\int_{t_1}^{t_2} i_D(t)\,dt\right] = v_{D(\text{on})}I_{\text{av}} = 19.9\text{ mW}
\end{aligned}
$$

To obtain the final result, it was noted that $i_D(t) = i_{\text{Load}}(t)$.

EXAMPLE 6.3

A SPICE simulation of the full-wave bridge-rectifier circuit of Figure 6.8 is desired.

$$v_{\text{Trans}}(t) = V_m\sin 2\pi ft, \quad V_m = 16\text{ V}, \quad f = 60\text{ Hz}$$
$$R_L = 50\ \Omega, \quad I_s = 5\times 10^{-11}\text{ A}, \quad n = 1.4$$

Identical diodes with the preceding parameters are used in the circuit. Determine the average load voltage V_{av} and the average power dissipated by the load resistor and by each of the diodes.

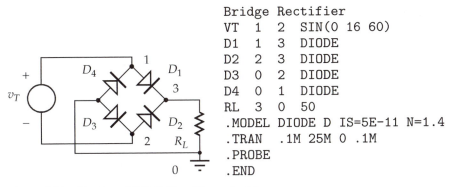

```
Bridge Rectifier
VT   1   2   SIN(0 16 60)
D1   1   3   DIODE
D2   2   3   DIODE
D3   0   2   DIODE
D4   0   1   DIODE
RL   3   0   50
.MODEL DIODE D IS=5E-11 N=1.4
.TRAN  .1M 25M 0 .1M
.PROBE
.END
```

Figure 6.14: Circuit and SPICE file of Example 6.3.

SOLUTION   The circuit and corresponding SPICE file of Figure 6.14 will be used. A sinusoidal voltage specification is required for the transient analysis (0 offset, 16-V peak value, and 60-Hz frequency). The .TRAN statement yields data points every 0.1 ms for a duration of 25 ms ($3T/2 = 25$ ms for $f = 60$ Hz). A zero no-point value has been specified, and the step ceiling was set to 0.1 ms. It has been the author's experience that for a reasonably accurate simulation of a sinusoidal signal, a step ceiling of one-hundredth of a period or less is required. Figure 6.15 is obtained for the load voltage and its average. The peak load voltage of 14.4 V implies a peak forward-biased diode voltage of 0.8 V. The "running" average, AVG, of a function $x(t)$ is given by the following intergal:

$$\mathrm{AVG}(x) = \frac{1}{t} \int_0^t x(t')\, dt'$$

Although the period of the transformer voltage is $1/f$ (16.67 ms), that of the load voltage is one-half this value, namely, 8.33 ms. An evaluation of the AVG term at $t = 8.33$ ms or any integer multiple of 8.33 ms results in the actual

Figure 6.15: SPICE solution of Example 6.3.

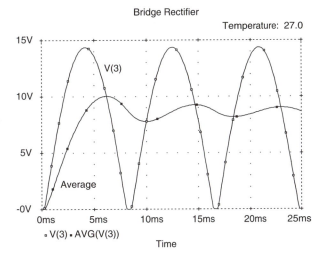

average value of the term, that is, a value of 8.66 V for V_{av}. The average load power is obtained from the running average of $v_{Load}(t)i_{Load}(t)$, that is, V(3)*I(RL). Its value is 1.96 W. The average power dissipated by D_1 is obtained from the running average of V(1,3)*I(D1). A time interval of T, 16.67 ms, is required for this condition because this is the period of the diode voltage and current. A value of 69.4 mW is obtained for the average power dissipated by D_1 as well as for each of the other diodes.

6.2 FILTERS: REDUCING LOAD-VOLTAGE FLUCTUATIONS

In the introduction to the chapter it was pointed out that an energy storage element is required for an electronic power supply. A capacitor, for example, can supply energy to a load during intervals when the voltage of a rectifier circuit would otherwise be inadequate. This process of alternately storing and then supplying energy to minimize load-voltage fluctuations is known as filtering.

CAPACITOR FILTERS – HALF-WAVE RECTIFIERS

The half-wave rectifier circuit of Figure 6.2 with a capacitor filter will initially be considered to obtain a quantitative perspective of its behavior. In the previous section, the transformer secondary voltage of a power supply was assumed to have a sinusoidal time dependence. This tends to be a reasonable approximation for rectifier circuits without a filter.

Capacitor filter circuits, however, tend to result in large peak values of diode currents that distort the secondary transformer voltage (an ideal transformer model is not a reasonable approximation for most power supply circuits). The distortion is the result of an equivalent inductance and resistance of the secondary winding of the transformer. Furthermore, as a consequence of the nonlinear magnetization characteristic of an iron core, these equivalent elements have a nonlinear characteristic. Even though a knowledge of these effects is important for analyzing and designing power supply circuits, adequate data are generally not available. As is not infrequently the case for electronic circuits, an experimental procedure must ultimately be used.

Although it is recognized that the input voltage provided by a transformer of a rectifier–filter circuit is not sinusoidal, a sinusoidal input voltage will be used for an initial analysis. The result of this analysis will provide a perspective from which a more refined treatment will be possible. Furthermore, a sinusoidal input voltage approximation yields a reasonable quantitative measure of the fluctuation in the output voltage. Consider the circuit of Figure 6.16. A constant forward-biased voltage diode model (a voltage of $v_{D(on)}$) will be used, and it will be, for convenience, assumed that the capacitor voltage $v_{Load}(t)$, is zero at $t = 0$. As a result of the step function $u(t)$, the ideal voltage source is

Figure 6.16: A half-wave rectifier circuit with a capacitor filter.

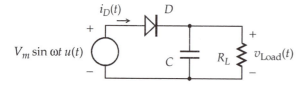

zero for $t < 0$ and "turns on" at $t = 0$. When the voltage reaches $v_{D(on)}$, the diode will conduct, and $v_{Load}(t)$ will initially be equal to $V_m \sin \omega t - v_{D(on)}$. During this interval the capacitor will be charged (its current, which is proportional to the derivative of its voltage, will be positive). Its stored energy, $\frac{1}{2} C v_{Load}^2(t)$, will increase. This occurs for the time interval up to $T/4$ (Figure 6.17).

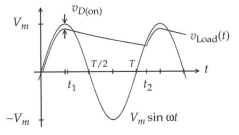

Figure 6.17: Input and load voltage of a half-wave rectifier and filter.

If the diode continues to conduct after $t = T/4$, the load voltage will tend to follow the input voltage back downward. Except for extremely small values of capacitance, a negative diode current would be required for this to occur. At a time just slightly greater than $T/4$ (indicated at t_1), the diode current becomes zero, and the diode ceases to conduct – its voltage becomes less than $v_{D(on)}$. When this occurs, the diode and input voltage source are, in effect, removed from the circuit, resulting in the equivalent circuit of Figure 6.18.

The capacitor is discharged by the load resistor during the interval over which its voltage $v_{Load}(t)$ remains greater than $V_m \sin \omega t - v_{D(on)}$, that is, up to the time t_2.

$$v_{Load}(t) = V_p e^{-(t-t_1)/R_L C}, \quad t_1 < t < t_2 \tag{6.14}$$

The time t_1 and peak value V_p can be obtained by determining the time at which the diode current becomes zero. For the large capacitance values generally required for a power supply, the voltage V_p will be extremely close to $V_m - v_{D(on)}$, and t_1 will be extremely close to $T/4$. Although the maximum value of $v_{Load}(t)$, V_p, occurs at $t = t_1$, its minimum value V_{min} occurs at $t = t_2$.

$$V_{min} = V_p e^{-(t_2-t_1)/R_L C}$$
$$v_{L\,p-p} = V_p - V_{min} = V_p \left(1 - e^{-(t_2-t_1)/R_L C}\right) \tag{6.15}$$

The quantity $v_{L\,p-p}$ is the peak-to-peak variation in the load voltage – a quantity generally known as the peak-to-peak ripple voltage. After the first period of the input voltage, the load voltage will also be periodic.

The peak-to-peak value of the ripple voltage depends on the filter capacitor of the circuit (the larger its capacitance, the smaller the ripple). Concurrently, as the capacitance is increased, the discharge interval $t_2 - t_1$ also increases, approaching, but not quite reaching, a full period of the input voltage T ($t_2 \approx 5T/4$). A generally accepted approximation used for analyzing and designing power supplies is

Figure 6.18: The exponential discharge of a filter capacitor.

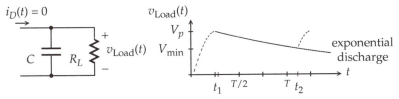

to let $t_2 - t_1$ equal T as follows:

$$v_{L\,p-p} \approx V_p(1 - e^{-T/R_LC}) \tag{6.16}$$

Because T is slightly greater than $t_2 - t_1$, this approximation predicts a peak-to-peak voltage slightly larger than its actual value. This is appropriate when designing a power supply because one usually needs to ascertain the value of capacitance that results in a peak-to-peak ripple voltage that is not greater than a prescribed value.

$$1 - e^{-T/R_LC} \approx v_{L\,p-p}/V_p, \qquad e^{T/R_LC} \approx \frac{1}{1 - v_{L\,p-p}/V_p}$$

$$C \approx \frac{T}{R_L \ln\left(\frac{1}{1 - v_{L\,p-p}/V_p}\right)} \tag{6.17}$$

Although the value of capacitance given by Eq. (6.17) will be slightly in error, the error will always be on the high side. Hence, if a capacitor of this value is used, the resultant value of $v_{L\,p-p}$ will fulfill the design requirement.

A simplified expression, requiring another approximation, is normally used for relating the peak-to-peak ripple voltage to the filter capacitance. For most electronic applications a value of $v_{L\,p-p}$ that is small compared with V_p is desired. This requires that the exponential term of Eq. (6.16), e^{-T/R_LC}, be close to unity, which, in turn, requires the exponent T/R_LC to be small compared with unity. Hence, the following approximations may be used:

$$e^{-T/R_LC} \approx 1 - T/R_LC \quad \text{for } T/R_LC \ll 1$$

$$v_{L\,p-p} \approx \frac{V_pT}{R_LC}, \qquad C \approx \left(\frac{V_p}{v_{L\,p-p}}\right)\frac{T}{R_L} \tag{6.18}$$

The same result is obtained from Eq. (6.17) if $v_{L\,p-p}/V_p$ is assumed small compared with unity:

$$\ln\left(\frac{1}{1 - v_{L\,p-p}/V_p}\right) \approx \ln(1 + v_{L\,p-p}/V_p) \approx v_{L\,p-p}/V_p$$

$$C \approx \left(\frac{V_p}{v_{L\,p-p}}\right)\frac{T}{R_L} \tag{6.19}$$

Because relatively large capacitances are usually required, electrolytic-type capacitors are generally used for power supplies. Commercially available electrolytic capacitors have relatively large capacitance values even though they are physically fairly small (electrolytic capacitors with values of 1 μF up to 10,000 μF and greater are common). These capacitors, however, are characterized by very wide tolerance specifications. Although a tolerance of -10 percent, $+50$ percent is not uncommon, many catalogues do not even give a tolerance specification. Furthermore, only a very limited set of capacitance values are generally available (1, 2.2, 4.7, 10 μF, etc., may characterize a particular manufacturer's offering). Therefore, it may be necessary to use a capacitance with a nominal value that is as much as 100 percent greater than actually needed. Hence, the use of approximations is justified.

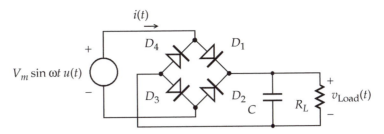

Figure 6.19: A full-wave rectifier with a capacitor filter.

CAPACITOR FILTERS – FULL-WAVE RECTIFIERS

The peak-to-peak ripple voltage of a power supply can be reduced by increasing the size of the filter capacitor. Concurrently, a decrease in the period T will also reduce the peak-to-peak value of the ripple voltage. Although the frequency ($f = 1/T$) of a supply voltage may be a given (for example, a particular utility system), a full-wave rectifier has a similar effect in that the time interval over which the capacitor needs to supply energy is reduced.

Consider the full-wave rectifier and filter circuit of Figure 6.19 that results in the load voltage of Figure 6.20. Up to $t_1 (\approx T/4)$, diodes D_1 and D_3 conduct, resulting in a load voltage that reaches a peak value of $V_m - 2v_{D(\text{on})}$. During the conduction interval, the capacitor is charged (its current is positive). From t_1 to t_2, the load voltage is greater than $|V_m \sin \omega t| - 2v_{D(\text{on})}$ and, as a result, none of the diodes will be conducting. The same type capacitor discharge as for the half-wave rectifier occurs except that the time t_2 is now less than $3T/4$, and the interval $t_2 - t_1$ is less than $T/2$. From an approximation of $t_2 - t_1 \approx T/2$, the following is obtained for the peak-to-peak ripple voltage:

$$v_{L\,p-p} = V_p - V_{\text{min}} = V_p\left(1 - e^{-(t_2-t_1)/R_L C}\right) \approx V_p\left(1 - e^{-T/2R_L C}\right)$$
$$\approx \frac{V_p T}{2 R_L C} \quad \text{for } R_L C \gg T/2 \tag{6.20}$$

On the basis of these approximations, the peak-to-peak ripple voltage for a given capacitance and other circuit components, is only one-half that of the half-wave rectifier. As a result, full-wave rectifiers are nearly always used for electronic power supplies.

The next step in developing an understanding of the operation of a rectifier–filter circuit is a determination of the currents of the circuit. Consider the full-wave rectifier circuit of Figure 6.19. The current of C and R_L is equal to the currents of diodes D_1 and D_2 as follows:

$$i_{D1}(t) + i_{D2}(t) = C\frac{dv_{\text{Load}}(t)}{dt} + \frac{v_{\text{OUT}}}{R_L} \tag{6.21}$$

When diode D_1 is conducting, $v_{\text{Load}}(t) = V_m \sin \omega t - 2v_{D(\text{on})}$, and the current of D_2 is

Figure 6.20: Load voltage of a full-wave rectifier and filter.

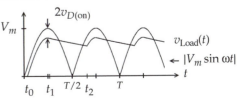

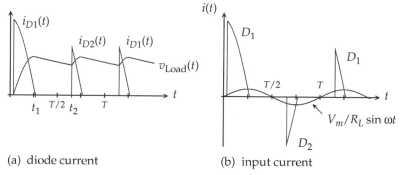

(a) diode current   (b) input current

Figure 6.21: Diode and input current of a bridge rectifier and filter circuit.

zero:

$$i_{D1}(t) = \omega C V_m \cos \omega t + \left(V_m \sin \omega t - 2v_{D(\text{on})}\right)/R_L \quad \text{for } D_1 \text{ conducting} \quad (6.22)$$

This occurs for the initial increase of the input voltage ($t_0 < t < t_1$) as well as for succeeding brief intervals when D_1 is conducting (Figure 6.21(a)). When D_2 is conducting, $v_{\text{Load}}(t) = -V_m \sin \omega t - 2v_{D(\text{on})}$, and the current of D_1 is zero:

$$i_{D2}(t) = -\omega C V_m \cos \omega t + \left(-V_m \sin \omega t - 2v_{D(\text{on})}\right)/R_L \quad \text{for } D_2 \text{ conducting}$$
$$(6.23)$$

This current is also indicated in Figure 6.21(a).

After the initial transient charging current ($t < t_1$), the diode currents are periodic. The currents alternate: $i_{D1}(t) > 0$ when the input voltage $V_m \sin \omega t$ is near its maximum value and $i_{D2}(t) > 0$ when the input voltage is near its minimum value. The current of the voltage source $i(t)$ depends on the diode currents as follows:

$$i(t) = i_{D1}(t) - i_{D4}(t) \tag{6.24}$$

Because D_2 and D_4 conduct simultaneously,

$$i_{D2}(t) = i_{D4}(t), \qquad i(t) = i_{D1}(t) - i_{D2}(t) \tag{6.25}$$

This results in the input current of Figure 6.21(b). After the initial transient, the current is periodic and consists of narrow "spikes" as opposed to the sinusoidal current that would result if R_L were connected directly to the voltage source.

A NONIDEAL TRANSFORMER

An elementary equivalent circuit model of an iron-core transformer includes an equivalent secondary resistance and inductance (Figure 6.22). The series elements R_T and L_T affect the transformer's terminal voltage $v_{\text{Trans}}(t)$.

$$v_{\text{Trans}}(t) = V_m \sin \omega t - R_T i_{\text{Trans}}(t) - L_T \frac{di_{\text{Trans}}(t)}{dt} \tag{6.26}$$

The transformer current, however, is not that of Figure 6.21(b) because the

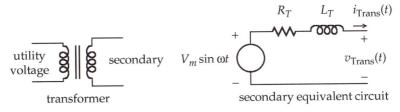

Figure 6.22: Elementary model of a power transformer.

input voltage of the rectifier is $v_{\text{Trans}}(t)$, not $V_m \sin \omega t$. A solution for the currents and voltages of the circuit requires that the equivalent-circuit elements of the transformer be taken into account when solving the circuit. This is most readily achieved with a numerical simulation.

Consider the circuit of Figure 6.23 in which an equivalent circuit is utilized for the transformer. A transient voltage specification and analysis statement is required with the .TRAN statement providing a solution for a time interval of 100 ms, that is, five periods of the 50-Hz supply voltage. Along with the

Figure 6.23: A rectifier–filter circuit with an equivalent transformer circuit.

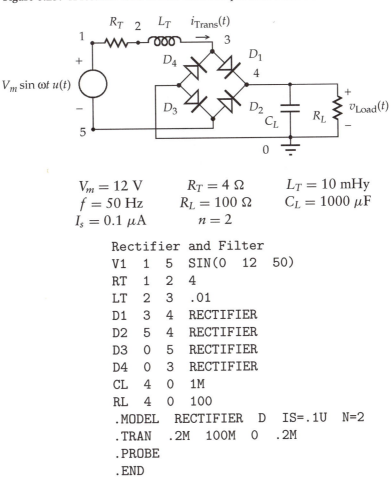

$$V_m = 12 \text{ V} \qquad R_T = 4 \ \Omega \qquad L_T = 10 \text{ mHy}$$
$$f = 50 \text{ Hz} \qquad R_L = 100 \ \Omega \qquad C_L = 1000 \ \mu\text{F}$$
$$I_s = 0.1 \ \mu\text{A} \qquad n = 2$$

```
Rectifier and Filter
V1   1  5  SIN(0  12  50)
RT   1  2  4
LT   2  3  .01
D1   3  4  RECTIFIER
D2   5  4  RECTIFIER
D3   0  5  RECTIFIER
D4   0  3  RECTIFIER
CL   4  0  1M
RL   4  0  100
.MODEL   RECTIFIER   D   IS=.1U   N=2
.TRAN   .2M  100M  0   .2M
.PROBE
.END
```

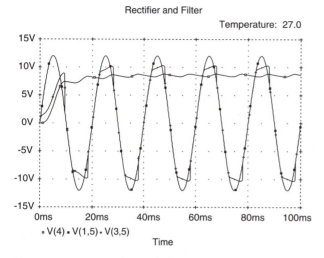

Figure 6.24: SPICE solution for load and input voltage of rectifier–filter circuit.

transformer voltage V(3,5), the sinusoidal input voltage V(1,5) is included on the Probe graph of Figure 6.24 for a comparison. The distortion, the result of the series elements R_T and L_T is readily apparent. The load voltage is also indicated – it is about 1.5 V less than the peak transformer voltage. It is not until after approximately 40 ms, that is, two periods, that a steady-state periodic solution is obtained.

Figure 6.25 provides a detailed view of the periodic response ($t > 50$ ms), which yields the following data:

$$v_{L\,\max} = 8.83 \text{ V}, \qquad v_{L\,\min} = 8.34 \text{ V}, \qquad v_{L\,p-p} = 0.49 \text{ V}$$
$$i_{\text{Trans}\,\max} = 0.294 \text{ A}, \qquad i_{\text{Trans}\,\min} = -0.294 \text{ A} \qquad (6.27)$$
$$t_{\text{pulse width}} \approx 5 \text{ ms}$$

Although the "width" of the spikes is greater than that for a sinusoidal voltage source (Figure 6.21(b)), the peak current remains considerably greater than that which would occur for a simple resistive load (0.294 A compared with a current of about 12 V/104 Ω ≈ 0.12 A for R_L connected directly to the transformer).

The SPICE simulation results of Figures 6.24 and 6.25 are typical of those that are experimentally obtained for a transformer-powered rectifier-filter circuit. As a result of the distortion of the transformer voltage, the peak load voltage is generally much less than that which would be expected based on the open-circuit voltage of the transformer ($V_m - 2v_{D(\text{on})}$).

For an actual transformer circuit, another problem arises that further complicates an analysis. The magnetic flux of the transformer for large currents is not linearly dependent on the current. Although this effect can be approximated with a nonlinear series resistance and inductance, this is seldom done when designing a power supply because the required transformer parameters are generally not available. Instead a set of laboratory tests is required. Past experience, however, does provide a reasonable starting point for the

Figure 6.25: SPICE solution for the input current of the rectifier–filter circuit.

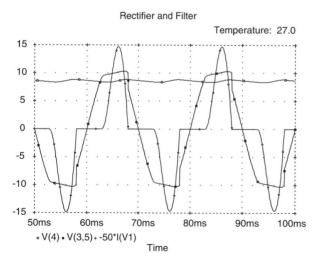

design of a power supply. For a power transformer with a power rating that is approximately equal to the power dissipated by R_L (the equivalent load resistance), the peak load voltage is found to be approximately equal to the specified rms value of the secondary voltage ($V_m/\sqrt{2}$). This cannot be "proven" from circuit considerations; it is simply an engineering rule of thumb that is reasonably valid.

EXAMPLE 6.4

The half-wave rectifier circuit of Figure 6.26 is essentially that of the diode radio detector of Figure 1.3 in which the input voltage $v_C(t)$ is an amplitude modulated signal. Assume $v_C(t) = V_c \sin 2\pi f t$, where V_c is the amplitude of the carrier that, as a result of the modulation, changes slowly with time.

a. Determine the value of C_1 for which the load voltage $v_M(t)$ has a peak-to-peak ripple voltage that is approximately 10 percent of its peak value (V_c constant).

b. What are the values of C_1 required to reduce the ripple voltage to 5 percent and 1 percent of the peak value of the load voltage?

c. To estimate the effect of C_1 on a modulating signal, consider the case for which V_c is suddenly reduced to zero (for convenience, this may be assumed to occur at $t = 0$). Determine the time required for $v_M(t)$ to fall to 10 percent of its peak value for each capacitance value of parts (a) and (b).

SOLUTION The period T of the radio-frequency carrier with a frequency of 1 MHz is 1 μs.

a. Equation (6.18) will be used to determine the capacitance C_1 ($C \to C_1$, $R_L \to R_1$, and $v_L \to v_M$). For $v_{M\,p-p}/V_p = 0.1$, the following is obtained:

$$C_1 = \left(\frac{V_p}{v_{M\,p-p}}\right)\frac{T}{R_1} = 10^{-9} \text{ F or 1 nF}$$

b. For $v_{M\,p-p}/V_p = 0.05$, $C_1 = 2$ nF and for $v_{M\,p-p}/V_p = 0.01$, $C_1 = 10$ nF.

c. It will be assumed that $v_M(t) = V_p$ at $t = 0$ when the carrier vanishes. As a result of the diode's being reverse biased, a simple RC discharge occurs:

$$v_M(t) = V_p e^{-t/R_1 C_1} \quad \text{for } t > 0$$

Let $t = t_1$ for $v_M(t_1) = 0.1\,V_p$. The following is obtained for t_1:

$$0.1\,V_p = V_p e^{-t_1/R_1 C_1}$$
$$t_1 = R_1 C_1 \ln 10 = 2.30\,R_1 C_1$$
$$= 2.3 \times 10^{-5} \text{ s} \quad \text{or} \quad 23\ \mu\text{s} \quad \text{for } C_1 = 1 \text{ nF}$$

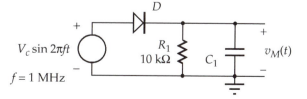

Figure 6.26: Rectifier circuit of Example 6.4.

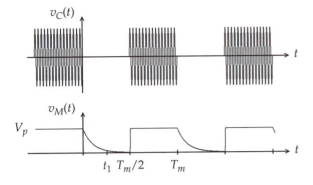

Figure 6.27: Load voltage for a periodic *on–off* carrier.

For $C_1 = 2$ nF, $t_1 = 46$ μs, and for $C_1 = 10$ nF, $t_1 = 230$ μs. The relevance of t_1 may be seen by assuming that $v_C(t)$ consists of a series of periodic carrier bursts such as would occur if V_c were an *on–off* periodic square wave voltage (Figure 6.27). For a reasonable replica of the periodic square-wave modulating voltage, it is necessary that $T_m/2$ be greater than t_1. For a "good" replica, a value of $T_m/2 = 2t_1$ is a reasonable criteria. This implies the following for the frequency of the modulating signal f_m:

$$T_m = 4\,t_1, \qquad f_m = 1/T_m = 1/4\,t_1$$

For $C_1 = 1$ nF, $f_m = 10.9$ kHz. This implies that the upper acceptable square-wave modulating frequency is about 10.9 kHz for a filter with a 1-nF capacitor. Larger capacitors reduce the acceptable modulating frequency. For $C_1 = 2$ nF, $f_m = 5.4$ kHz, and for $C_1 = 10$ nF, $f_m = 1.09$ kHz.

EXAMPLE 6.5

Consider the full-wave rectifier and filter of Figure 6.19.

$$V_m = 20 \text{ V} \qquad f = \omega/2\pi = 60 \text{ Hz}$$
$$R_L = 100 \text{ } \Omega \qquad C = 2000 \text{ } \mu\text{F} \qquad v_{D(on)} = 0.8 \text{ V}$$

a. Determine the peak load voltage V_p and the peak-to-peak ripple voltage $v_{L\,p-p}$.

b. What is the peak inverse voltage that the diodes must sustain?

c. Estimate the peak current of the diodes.

SOLUTION For a frequency of 60 Hz, $T = 16.7$ ms.

a. $V_p = V_m - 2v_{D(on)} = 18.4$ V, $\qquad v_{L\,p-p} = \dfrac{V_p T}{2\,R_L C} = 0.77$ V

b. Consider the case when D_1 is conducting for a peak input voltage of V_m. The reverse bias voltage of D_2 for this condition is $V_m - v_{D(on)} = 19.2$ V. The peak inverse voltage of the diodes is thus 19.2 V.

c. The current of D_2 has a peak value at $t = t_2$ (Figure 6.21). From Eq. (6.23), the following is obtained for the current:

$$i_{D2}(t_2) = -\omega C V_m \cos \omega t_2 + (-V_m \sin \omega t_2 - 2v_{D(on)})/R_L$$

The time t_2 is the time at which D_2 turns on. It corresponds to the intersection of the $R_L C$ discharge (Eq. (6.24)) and the magnitude of the input voltage less $2v_{D(\text{on})}$:

$$v_L(t_2) = V_p e^{-(t_2-t_1)/R_L C} \approx V_p[1 - (t_2 - T/4)/R_L C]$$

The linear approximation for the exponential is justified because the ripple voltage is small. Furthermore, $t_1 \approx T/4$.

$$V_p[1 - (t_2 - T/4)/R_L C] = -V_m \sin \omega t_2 - 2v_{D(\text{on})}$$

To solve this equation, let $x = (t_2 - T/4)/T$, the fraction of a period over which the discharge occurs. This results in the following:

$$\sin \omega t_2 = \sin(2\pi x + \pi/2) = \cos 2\pi x$$
$$f(x) = V_p(1 - xT/R_L C) + V_m \cos 2\pi x + 2v_{D(\text{on})}$$

A solution is desired for which $f(x) = 0$. It will be noted from Figure 6.25 that $t_2 - T/4$ is less than one-half period, $T/2$. Hence, $x < 0.5$. Through a trial-and-error numerical evaluation of $f(x)$, a solution of $x = 0.458$ is obtained.

$$t_2 = xT + T/4 = 0.708T$$
$$\omega t_2 = 0.708\omega T = 0.708(2\pi) \text{ radians or } 254.9°$$
$$i_{D2}(t_2) = 4.1 \text{ A}$$

The peak current of each diode for a sinusoidal input voltage is thus 4.1 A.

EXAMPLE 6.6

Design a power supply that produces an average load voltage of 12 V and an average load current of 0.5 A. The peak-to-peak ripple load voltage is to be no greater than 1.0 V. The power line frequency is 60 Hz.

SOLUTION An average load voltage of 12 V and a current of 0.5 A imply an equivalent load resistance R_L of 24 Ω. As a result of the 1 V peak-to-peak ripple voltage, it will be assumed that the load voltage varies from 11.5 to 12.5 V ($V_p = 12.5$ V). From Eq. (6.18), the value of the filter capacitance may be determined as follows:

$$C = \left(\frac{V_p}{v_{L\,p-p}}\right)\frac{T}{R_L} = 8.7 \times 10^{-3} \text{ F}$$

A capacitor with a nominal value of 10,000 μF is required. To account for the transformer effects, that is, an equivalent nonlinear secondary circuit, a transformer with an rms secondary voltage of 12 V will be needed. For a primary voltage of 120 V, the turns ratio is 10:1. A secondary rms current rating of at least 0.5 A is required.

If the load should happen to be removed from the supply, the peak load voltage V_p will increase to nearly V_m. For a no-load condition, the transformer peak voltage will be $12\sqrt{2} \approx 17$ V. Because this no-load condition can not be excluded, a filter capacitor with a voltage rating in excess of 17 V is required – 25 V is a standard value.

Diodes with a peak current rating of at least five times the average current, that is 2.5 A, will be required. This is a reasonable estimate based on the SPICE simulation for the circuit of Figure 6.28. For a no-load condition, the peak inverse voltage of the diodes is approximately 17 V. Hence, diodes with a rating of at least this value are required. An integrated circuit bridge rectifier having an average current rating of 0.5 A and a peak inverse voltage rating of 25 V is recommended.

It will be noted that to achieve a smaller ripple voltage, an exceedingly large filter capacitor would be required. For example, for $V_{L\,p-p} = 0.1$ V, a capacitance of nearly 100,000 μF (0.1 F) would be required. Because this is not an acceptable design value (try to find one in an electronics catalog), an electronic regulator is needed.

6.3 ZENER DIODE REGULATOR: AN IMPROVED OUTPUT VOLTAGE

A zener diode circuit can be used to reduce the ripple voltage of a power supply and to maintain a load voltage that remains nearly constant despite changes that may occur in the load current. In the discussion of diodes of Chapter 2, it was assumed that the current of a reverse-biased semiconductor junction diode is essentially zero ($-I_s$ for $v_D \ll 0$). However, for a sufficiently large reverse-bias voltage, an abrupt breakdown process occurs that results in a negative current that changes very rapidly with diode voltage (Figure 6.28). When Shockley first observed this breakdown current, he associated it with a quantum mechanical tunneling effect previously observed in dielectrics and first described by Clarence Zener in 1934 (McAfee et al. 1951; Smits 1985; and Zener 1934). Tunneling, however, was not the primary breakdown effect for Shockley's diodes. As a result of the high electric fields produced by a large reverse-bias voltage, free electrons gain sufficient energy to generate electron–hole pairs, which is a mechanism known as impact ionization. Furthermore, an avalanche process occurs as newly produced free electrons result in additional ionizing impacts. For most semiconductor junction diodes, it is an avalanche breakdown process that tends to occur. It is only for extremely highly doped semiconductor junction diodes that tunneling is the dominant mechanism.

Especially fabricated semiconductor junction diodes that rely on an avalanche breakdown effect are frequently used for regulator circuits and other electronic applications. As a result of Shockley's original misnaming of the breakdown effect, these diodes are now universally known as zener diodes. They are available

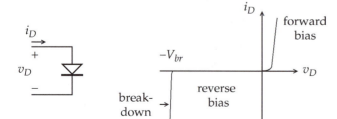

Figure 6.28: Reverse-bias breakdown behavior of a semiconductor junction diode.

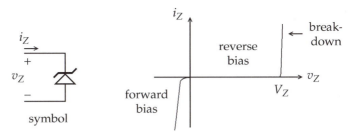

Figure 6.29: A zener diode characteristic.

in a wide range of breakdown voltages (from a few volts to tens of volts) and in a wide range of power ratings (up to several hundred watts).

Because it is the reverse-biased region of a zener diode that is generally utilized, the alternative voltage and current labeling of Figure 6.29 will be used ($v_Z = -v_D$, $i_Z = -i_D$). The zener breakdown voltage is V_Z, a positive quantity. For a zener voltage v_Z between 0 and V_Z, the current is essentially zero. However, when the diode voltage v_Z exceeds the breakdown voltage V_Z, the current of the diode rapidly increases. To model the breakdown behavior of a zener diode, an equivalent circuit model with an equivalent zener resistance is frequently used (Figure 6.30).

For the ideal diode of the reverse-biased model to conduct, the terminal voltage must exceed the zener voltage V_Z. This results in a linear dependence of the current on voltage as defined by

$$i_Z = (v_Z - V_Z)/r_z \quad \text{for } v_Z > V_Z \tag{6.28}$$

A forward-biased model is also shown in Figure 6.30. It will be noted that the current of this equivalent circuit is zero for positive values of v_Z (its ideal diode is reverse biased). Hence, the forward-biased model may be connected in parallel with the reverse-biased model if negative values of v_Z are anticipated. The equivalent circuits of Figure 6.30 are generally utilized for analytic circuit solutions. Computer simulations such as SPICE, however, tend to use an exponential current dependence on the diode voltage for breakdown (a more rapid increase in current than predicted by a linear model).

The equivalent zener resistance r_z of a zener diode is generaly quite small and is comparable with the forward-biased equivalent resistance for the same magnitude of current ($\approx V_T/i_Z$). Hence, for most applications, the voltage across the zener diode tends to remain nearly constant for $i_Z > 0$ and approximately equal to the zener breakdown voltage V_Z. Consider the zener diode regulator circuit of Figure 6.31, which has an input voltage of v_{Supply}. If the zener diode voltage v_Z is less than the zener voltage V_Z the diode current will be zero.

Figure 6.30: A zener diode equivalent circuit.

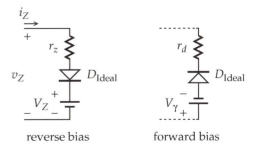

reverse bias   forward bias

$$v_{\text{Load}} = \frac{R_L v_{\text{Supply}}}{R_L + R_1} \quad \text{for } v_{\text{Load}} < V_Z \quad (6.29)$$

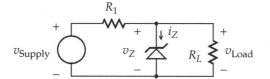

Figure 6.31: A zener diode regulator circuit.

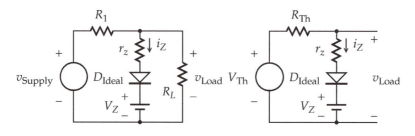

actual circuit   Thévenin equivalent circuit

Figure 6.32: Zener diode circuit and Thévenin equivalent circuit.

However, if the value of v_{Load} predicted by Eq. (6.29) exceeds V_Z, the zener diode will tend to keep $v_{\text{Load}} \approx V_Z$ regardless of v_{Supply}. This may readily be shown by using the equivalent circuit model of Figure 6.30 for the zener diode (Figure 6.32).

To simplify the analysis, v_{Supply}, R_1, and R_L have been replaced by a Thévenin equivalent circuit as follows:

$$V_{\text{Th}} = \frac{R_L v_{\text{Supply}}}{R_1 + R_L}, \qquad R_{\text{Th}} = R_1 \| R_L \tag{6.30}$$

If $V_{\text{Th}} > V_Z$, then the ideal diode of the equivalent circuit will conduct, and the following is obtained for v_{Load}:

$$i_Z = (V_{\text{Th}} - V_Z)/(R_{\text{Th}} + r_z), \qquad v_{\text{Load}} = V_Z + i_Z r_z$$
$$v_{\text{Load}} = V_Z + \frac{r_z(V_{\text{Th}} - V_Z)}{R_{\text{Th}} + r_z} \qquad \text{for} \quad V_{\text{Th}} > V_Z \tag{6.31}$$

From the expression for V_{Th} of Eq. (6.30), the following is obtained for v_{Load}:

$$v_{\text{Load}} = \frac{R_{\text{Th}} V_Z}{R_{\text{Th}} + r_z} + \left(\frac{r_z}{R_{\text{Th}} + r_z}\right)\left(\frac{R_L}{R_1 + R_L}\right) v_{\text{Supply}} \tag{6.32}$$

The load voltage, based on the results of Eqs. (6.29) and (6.32), is indicated in Figure 6.33. It will be noted that the slope of the characteristic has a linear dependence on r_z for $v_{\text{Supply}} > (1 + R_L/R_1) V_Z$. Because for many applications r_z

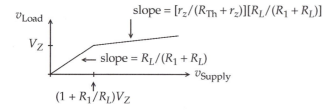

Figure 6.33: Dependence of load voltage on supply voltage for a zener diode regulator.

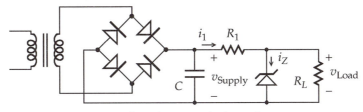

Figure 6.34: Power supply with a zener diode regulator.

is quite small compared with R_{Th}, the variation in v_{Load} with v_{Supply} is very small for this condition.

A power supply that utilizes a zener diode regulator circuit is indicated in Figure 6.34. For the zener diode to function as a regulator, its current i_Z must remain greater than zero (the diode must remain in its breakdown mode of operation). When this occurs, the load voltage remains approximately V_Z. The capacitor's voltage v_{Supply} depends on the circuit to which it is connected, the zener diode, R_1, and R_L. To determine v_{Supply}, a simplifying assumption that the load voltage remains equal to V_Z is generally justified. After v_{Supply} is determined, in particular its peak-to-peak variation, the variation in load voltage will be determined. If the variation in load voltage is indeed small, then the simplifying assumption is justified.

The instantaneous current of R_1, that is, $i_1(t)$, depends on the difference of the supply voltage and the load voltage, the latter being approximated as V_Z:

$$i_1(t) = (v_{Supply} - V_Z)/R_1 \tag{6.33}$$

During the discharge interval (rectifier diodes not conducting), $i_1(t)$ is provided by the capacitor C as follows:

$$C\frac{dv_{Supply}}{dt} = -i_1(t) = -(v_{Supply} - V_Z)/R_1$$

$$\frac{dv_{Supply}}{dt} + \frac{(v_{Supply} - V_Z)}{R_1 C} = 0 \tag{6.34}$$

A solution is readily obtained by assuming the dependent variable to be $v_{Supply} - V_Z$ as given by

$$v_{Supply} - V_Z = Ae^{-t/R_1 C} \tag{6.35}$$

A peak supply voltage of V_p will be assumed for $t = t_1$ (Figure 6.35):

$$v_{Supply} - V_Z = (V_p - V_Z)e^{-(t-t_1)/R_1 C} \tag{6.36}$$

Figure 6.35: Dependence of supply voltage and current on time.

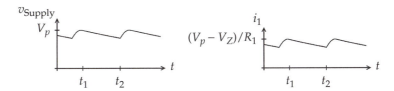

The maximum supply voltage occurs for $t = t_1$, whereas the minimum supply voltage occurs for $t = t_2$ ($t_2 - t_1 \approx T/2$):

$$v_{\text{Supply max}} = V_p$$
$$v_{\text{Supply min}} = V_Z + (V_p - V_Z)e^{-T/2R_1C} \tag{6.37}$$

The peak-to-peak ripple voltage is the difference of these two quantities.

$$v_{\text{Supply p-p}} = (V_p - V_Z)(1 - e^{-T/2R_1C}) \tag{6.38}$$

As a result of the zener diode regulator circuit, a fairly large value of $v_{\text{Supply p-p}}$ can generally be tolerated.

Equation (6.32) developed for the zener diode circuit may now be used to determine the peak-to-peak variation in the load voltage $v_{\text{Load p-p}}$ as follows:

$$
\begin{aligned}
v_{\text{Load p-p}} &= \left(\frac{r_z}{R_{\text{Th}} + r_z}\right)\left(\frac{R_L}{R_1 + R_L}\right)v_{\text{Supply p-p}} \\
&= \left(\frac{r_z}{R_{\text{Th}} + r_z}\right)\left(\frac{R_L}{R_1 + R_L}\right)(V_p - V_Z)(1 - e^{-T/2R_1C}) \tag{6.39}
\end{aligned}
$$

The load voltage ripple is much smaller than that of v_{Supply}. Therefore, it is frequently possible to use a much smaller filter capacitor than would otherwise be needed. As is often the case, a "price" must be paid for the small load voltage ripple. Both R_1 and the zener diode dissipate electrical power, producing heat (thermal power) that must be removed from the circuit. Furthermore, if the load is removed from the circuit (R_L replaced by an open circuit), the zener diode current increases ($i_Z(t) = i_1(t)$ for this condition). The zener diode must be capable of dissipating the power corresponding to this current, a power with a time average of approximately $i_1(t)V_Z$.

EXAMPLE 6.7

Consider the power supply and zener diode regulator circuit of Figure 6.34.

$$V_p = 20 \text{ V} \qquad V_Z = 12 \text{ V} \qquad f = 60 \text{ Hz}$$
$$C = 470 \ \mu\text{F} \qquad R_1 = 47 \ \Omega$$

Assume v_{Load} remains approximately equal to V_Z, that is, i_Z remains greater than zero.

a. Determine the maximum and minimum values of v_{Supply} and i_1.
b. What is the maximum load current for which the approximation that $v_{\text{Load}} = V_Z$ is valid?
c. What is the ripple load voltage for an average load current of 100 mA?
d. What is the ripple load voltage for zero load current?

SOLUTION

a. The maximum supply voltage is V_p, and the peak-to-peak ripple voltage is given by Eq. (6.38).

$$v_{\text{Supply max}} = V_p = 20 \text{ V}$$
$$v_{\text{Supply p-p}} = (V_p - V_Z)(1 - e^{-T/2R_1C}) = 2.51 \text{ V}$$
$$v_{\text{Supply min}} = v_{\text{Supply max}} - v_{\text{Supply p-p}} = 17.49 \text{ V}$$

This results in the following for the maximum and minimum values of i_1:

$$i_{1\,\text{max}} = (V_p - V_Z)/R_1 = 170 \text{ mA}$$
$$i_{1\,\text{min}} = (v_{\text{Supply min}} - V_Z)/R_1 = 117 \text{ mA}$$

b. The zener current and load current depend on i_1:

$$i_1 = i_Z + i_{\text{Load}}$$

Because the zener current must remain greater than zero, the load current must remain less than $i_{1\,\text{min}} = 117$ mA.

c. A load current of 100 mA implies an equivalent load resistance R_L of 120 Ω (12 V/.1 A). This results in the following for the zener diode current:

$$i_{Z\text{max}} = i_{1\,\text{max}} - i_{\text{Load}} = 70 \text{ mA}$$
$$i_{Z\text{min}} = i_{1\,\text{min}} - i_{\text{Load}} = 17 \text{ mA}$$

Hence, the zener diode current varies between 17 and 70 mA. By using the smallest current to estimate the equivalent zener resistance, a value of 1.47 Ω ($V_T/i_{Z\text{min}}$) is obtained. If this resistance value is used to calculate $v_{\text{Load p-p}}$, the following is obtained from Eq. (6.42):

$$R_{\text{Th}} = R_1 \| R_L = 33.8 \ \Omega$$
$$v_{\text{Load p-p}} = \left(\frac{r_z}{R_{\text{Th}} + r_z}\right)\left(\frac{R_L}{R_1 + R_L}\right) v_{\text{Supply p-p}} = 0.075 \text{ V}$$

A peak-to-peak ripple load voltage of 75 mV is obtained. It should be noted that this is an upper estimate for $v_{\text{Load p-p}}$ because the effective value of r_z is probably smaller than that corresponding to the minimum zener diode current.

d. A zero load current imples $R_L \rightarrow \infty$. For this condition, $i_Z = i_1$:

$$i_{Z\text{max}} = i_{1\,\text{max}} = 170 \text{ mA}, \qquad i_{Z\text{min}} = i_{1\,\text{min}} = 117 \text{ mA}$$

The minimum diode current will again be used for an estimate of the equivalent zener resistance ($r_z = 0.214 \ \Omega$).

$$R_{\text{Th}} = R_1, \qquad v_{\text{Load p-p}} = \left(\frac{r_z}{R_{\text{Th}} + r_z}\right) v_{\text{Supply p-p}} = 0.016 \text{ V}$$

The ripple load voltage is thus only 16 mV. Although this and the previous values of ripple voltage are only estimates based on an approximation for r_z that is open to question, the small value of ripple voltage is, nonetheless, typical of that obtained with zener diode regulator circuits.

6.4 AN ELECTRONIC REGULATOR: NEARLY IDEAL POWER SUPPLY

An electronic regulator, which relies on an internal voltage reference along with an amplifier and a feedback circuit, will result in a power supply with a nearly ideal output characteristic. The load voltage of a properly designed electronic regulator will be essentially independent of the voltage of the rectifier–filter circuit as well as the load current. Because of their low cost, electronic regulators fabricated on a single integrated circuit are extensively used for power supplies. The most common integrated circuit regulator is the three-terminal regulator designed for a fixed voltage. It has an input, an output, and a ground terminal and produces its design load voltage for currents up to its maximum rated current. Electronic regulators have an internal protection circuit that will shut down the regulator if an excessive load current or excessive power dissipation of the integrated circuit occurs. Because integrated-circuit regulators with a wide range of voltage and current specifications are available, the design of an electronic power supply is greatly simplified.

A BASIC OPERATIONAL AMPLIFIER REGULATOR

A basic electronic regulator using an operational amplifier is shown schematically in Figure 6.36. On the basis of assumed ideal behavior for the operational amplifier, the output voltage is equal to the reference voltage source V_{Ref}. Because of the op amp, the current of the reference voltage source is essentially zero. Therefore, variations in load current will not affect the reference voltage; hence, the load voltage for a current range over which ideal behavior of the op amp occurs is a reasonable assumption.

In the circuit of Figure 6.36, the output voltage is necessarily less than v_{Supply}, the voltage of the rectifier–filter circuit. Because the supply voltage for this circuit will not be the constant voltage assumed in the discussion of operational amplifier circuits (Chapter 5), it is reasonable to ask whether the fluctuations in the supply voltage will not result in comparable fluctuations in the load voltage. But, as a result of the feedback circuit ($\beta = 1$ for the circuit of Figure 6.36) and the large voltage gain of the op amp, the fluctuations in the load voltage will tend to be much smaller than those of v_{Supply}.

Figure 6.36: A basic electronic regulator using an op amp.

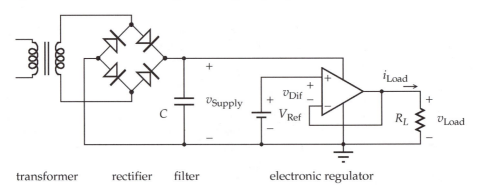

transformer   rectifier   filter   electronic regulator

In the analysis of op amps with symmetrical power supplies, it was assumed that the output voltage v_{Load} for the circuit being considered was equal to the difference voltage gain of the op amp A_d times the difference input voltage. For the situation of Figure 6.36 with a single supply, it would be reasonable to assume that the output of the op amp has a dependent source of $v_{\text{Supply}}/2$, which results in the equivalent circuit of Figure 6.37:

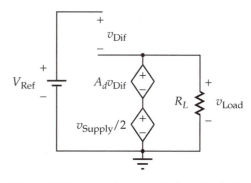

Figure 6.37: Approximate equivalent circuit for the basic electronic regulator of Figure 6.36 that accounts for its single supply voltage.

$$v_{\text{Load}} = A_d v_{\text{Dif}} + v_{\text{Supply}}/2$$
$$v_{\text{Dif}} = V_{\text{Ref}} - v_{\text{Load}} \tag{6.40}$$

The difference voltage v_{Dif} may be eliminated from the preceding equations to yield the following expression for v_{Load}:

$$v_{\text{Load}} = A_d(V_{\text{Ref}} - v_{\text{Load}}) + v_{\text{Supply}}/2$$
$$v_{\text{Load}} = \frac{A_d V_{\text{Ref}}}{1 + A_d} + \frac{v_{\text{Supply}}}{2(1 + A_d)} \tag{6.41}$$

The output voltage of Eq. (6.41) may be written in a slightly different form as follows:

$$v_{\text{Load}} = \frac{A_d}{1 + A_d}\left(V_{\text{Ref}} + \frac{v_{\text{Supply}}}{2\,A_d}\right) \tag{6.42}$$

Although v_{Supply} affects the output, because of the large difference voltage gain term of its denominator, its effect will be very small compared with V_{Ref}. For large values of A_d, the output voltage is essentially V_{Ref}, and the effect of the supply voltage is negligible (see Figure 6.38).

Supply voltage fluctuations, in addition to those that may be associated with the equivalent circuit of Figure 6.37, also affect the output voltage of the op amp. A power supply rejection ratio $PSRR$ is generally specified for an op amp to relate the magnitude of a change in its equivalent input offset voltage to the magnitude of a change in its supply voltage. Although separate quantities for the change in each supply voltage (V_{CC} or V_{EE}) as well as for a symmetrical change in these voltages may be specified, usually only a single quantity, with no specific designation, is given. A typical value for the reciprocal of $PSRR$, expressed in decibels, is 80 dB. This implies that a peak-to-peak variation of 1.0 V in v_{Supply} results in a 0.1-mV variation in the equivalent input offset voltage of the op amp. For the circuit of Figure 6.36, this equivalent voltage variation is in series with V_{Ref}. Hence, for most electronic regulator applications, a supply voltage with relatively large ripple voltage will have a negligible effect on the output voltage of an op amp regulator.

Figure 6.38: Supply and load voltages of the basic electronic regulator of Figure 6.36.

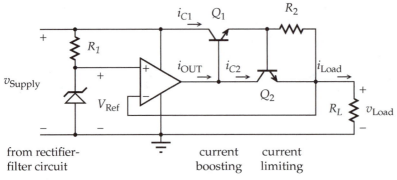

from rectifier-
filter circuit

current
boosting

current
limiting

Figure 6.39: An electronic regulator using a zener diode voltage reference. An output emitter–follower transistor amplifier Q_1 is used to boost the output current of the op amp, and a second transistor Q_2 limits the output current of the regulator.

AN ELECTRONIC REGULATOR WITH A ZENER DIODE VOLTAGE REFERENCE

At electronic regulator having a zener diode voltage reference as well as a circuit to boost and, if necessary, limit its output current, is shown in Figure 6.39. Because the input current of the op amp is essentially zero, the reference voltage V_{Ref} depends only on the supply voltage as follows:

$$V_{Ref} = V_Z + \frac{r_z(v_{Supply} - V_Z)}{R_1 + r_z} \quad \text{for } v_{Supply} > V_Z \tag{6.43}$$

Supply voltage variations, including ripple voltage, are therefore reduced by a factor of $r_z/(R_1 + r_z)$. Because the output current capability of a conventional low-power op amp is limited (10 to 20 mA being typical), a current-boosting circuit is included for the electronic regulator.

The first transistor of the output circuit Q_1 functions as an emitter–follower amplifier supplying the load current through the small current-sensing resistor R_2. The other transistor, Q_2, is active only when its base–emitter voltage, the voltage across the current-sensing resistor, is adequate to forward bias its base–emitter junction. Hence, if the voltage across R_2 is less than approximately 0.5 V, the base and collector currents of Q_2 will tend to be negligible (silicon transistor). This corresponds to $i_{Load} R_2$ being less than 0.5 V:

$$i_{C2} \approx 0 \quad \text{for } i_{Load} < 0.5 \text{ V}/R_2 \tag{6.44}$$

The current out of the emitter of Q_1 is equal to the load current i_{Load} for this condition as follows:

$$i_{Load} = (1 + \beta_F)i_{B1} = (1 + \beta_F)i_{OUT} \tag{6.45}$$

The output current of the op amp, i_{OUT}, is the base current of Q_1 when the collector current of Q_2 is negligible. Because of the current gain of Q_1, the maximum output current could be as large as $(1 + \beta_F)$ times the maximum output current of the op amp. An op amp with an output current capability of 10 mA can therefore result in a load current of 1 A for a transistor with a current transfer ratio β_F of 100.

As a result of the load current, the power dissipated by Q_1 can become rather large – a power transistor is generally required for Q_1. The following is obtained for the voltage and current of Q_1 if Q_2 is not conducting:

$$v_{CE1} = v_{\text{Supply}} - i_{\text{Load}} R_2 - v_{\text{Load}}$$
$$i_{C1} = \frac{\beta_F i_{\text{Load}}}{1 + \beta_F} \approx i_{\text{Load}} \tag{6.46}$$

The instantaneous power dissipated by the collector of Q_1, $p_D(t)$, is the product of the terms of Eq. (6.46):

$$p_D(t) = i_{C1} v_{CE1} \approx i_{\text{Load}}(v_{\text{Supply}}(t) - i_{\text{Load}} R_2 - v_{\text{Load}}) \tag{6.47}$$

Because the load current for a well-designed regulator will depend only on R_L and will have no time dependence, the average dissipated power $P_{D\text{av}}$ depends on the average value of v_{Supply}, that is, $V_{\text{Supply av}}$ as follows:

$$P_{D\text{ av}} = i_{\text{Load}}(V_{\text{Supply av}} - i_{\text{Load}} R_2 - v_{\text{Load}}) \tag{6.48}$$

A maximum dissipated power $P_{D\text{max}}$ occurs for a zero value of load voltage, that is, an output short circuit as follows:

$$P_{D\text{max}} = i_{\text{Load max}}(V_{\text{Supply av}} - i_{\text{Load max}} R_2) \approx i_{\text{Load max}} V_{\text{Supply av}} \tag{6.49}$$

Because this condition could accidentally occur, it is necessary to limit the short-circuit current of the electronic regulator to a value for which the power dissipation rating of Q_1 is not exceeded.

For an excessively large load current, the voltage developed across R_2, v_{BE2}, is such as to cause Q_2 to become active. When this occurs, the collector current of Q_2 will subtract from the base current of Q_1 as follows:

$$i_{B1} = i_{\text{OUT}} - i_{C2} \tag{6.50}$$

Hence, if R_2 is chosen so that $i_{\text{Load max}} R_2$ is approximately 0.6 to 0.7 V, Q_2 will conduct when the load current reaches its maximum value. This will limit the base current of Q_1 and therefore limit any further increase in the load current.

The base–emitter junction of a typical integrated-circuit transistor will function as a zener diode. This junction, when reverse biased, has a breakdown voltage of 6 to 8 V. Because the precise zener voltage V_Z depends on the doping of the base and emitter regions, it is difficult to control. Furthermore, the zener voltage depends on the temperature of the integrated circuit. An electronic regulator, as a result of the power it is dissipating, may operate at an elevated temperature. Hence, a large temperature sensitivity of the regulator's voltage reference, a reference fabricated on the same integrated circuit as the output power transistor, is not acceptable.

AN ELECTRONIC REGULATOR WITH A BAND-GAP VOLTAGE REFERENCE

Most integrated-circuit electronic voltage regulators rely on an internal "band-gap" voltage reference. This voltage reference designation is associated with the

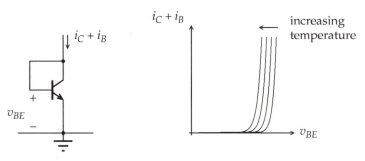

Figure 6.40: Temperature dependence of the base–emitter voltage of a bipolar junction transistor.

band-gap energy of a semiconductor, that is, the energy difference between the conduction and valence bands, which is generally expressed in electron volts. It is the effect of this energy difference on a semiconductor's intrinsic carrier concentration and hence on the diodes' voltage-versus-current characteristic that is utilized. These voltage references, if well designed, not only have a reasonably precise voltage but also a very small temperature dependence (Grebene 1984; Soclof 1991). Furthermore, these references have smaller voltages than those obtained with zener diodes.

The voltage of a band-gap reference depends on the forward-bias voltage of the base–emitter junction of a transistor as well as a second compensating voltage source. As indicated in Figure 6.40, the voltage of a diode-connected transistor has a very large temperature sensitivity. For a given current, the voltage sensitivity of the base–emitter voltage of a typical integrated-circuit transistor is -1.5 to -2 mV/C°. Therefore, a temperature increase of 50 C° will result in a voltage decrease of 75 to 100 mV, that is, a voltage decrease that is approximately 11 to 15 percent of the nominal diode voltage. If, for example, the diode voltage were to be used as a reference voltage of a 5-V supply, a temperature increase of 50 C° would result in an output voltage decrease of approximately 0.5 to 0.7 V. For most applications, this voltage change would be unacceptable.

A band-gap voltage reference uses a second voltage source with a positive temperature coefficient connected in series with a base–emitter junction. This second voltage source compensates for the negative temperature coefficient of the junction. Consider the three-transistor circuit of Figure 6.41, the basic circuit of a band-gap voltage reference. Identical transistors will be assumed for Q_1 and Q_2, and base currents will be ignored compared with the collector currents of the transistors. The following is obtained for the collector currents of Q_1 and Q_2:

Figure 6.41: A basic band-gap voltage reference.

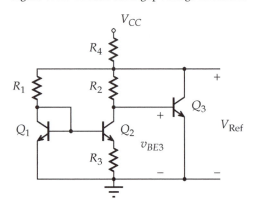

$$i_{C1} = I_s\, e^{v_{BE1}/n_F V_T},$$
$$i_{C2} = I_s\, e^{v_{BE2}/n_F V_T} \tag{6.51}$$
$$i_{C1}/i_{C2} = e^{(v_{BE1}-v_{BE2})/n_F V_T}$$

The difference of the base–emitter voltages, $v_{BE1} - v_{BE2}$, depends on the voltage across R_3 as follows:

$$v_{BE1} = v_{BE2} + i_{C2}R_3$$

$$i_{C2} = \frac{n_F V_T}{R_3}\ln(i_{C1}/i_{C2}) \tag{6.52}$$

The reference voltage of the circuit V_{Ref} is equal to the base–emitter voltage of Q_3, v_{BE3} plus the voltage across R_2:

$$V_{\text{Ref}} = \underbrace{v_{BE3}}_{\text{negative}} + \underbrace{\frac{n_F V_T R_2}{R_3}\ln(i_{C1}/i_{C2})}_{\text{positive temperature coefficient}} \tag{6.53}$$

Because V_T is equal to kT/e, the second term of Eq. (6.53) increases with temperature, whereas the first term, the base–emitter voltage of Q_3, decreases with temperature. Component values may be chosen such that the temperature sensitivity of V_{Ref} of Eq. (6.53) is zero at a particular (nominal) temperature. Furthermore, by using an appropriate amplifier and feedback regulator circuit, an output voltage equal to or greater than the reference voltage may be obtained.

The circuit of a typical low-current (100 mA) electronic voltage regulator, a 78LXX, is indicated in Figure 6.42 (the part number has a prefix that depends on its manufacturer, and the "XX" designation is the voltage of the regulator). This regulator, as indicated in Table 6.1, can be obtained for nominal output voltages of 2.6 to 15 V, the different voltages being achieved by adjusting R_1 of the voltage divider at its output when it is manufactured. Transistors Q_3, Q_4, and Q_5 form the band-gap reference voltage source ($V_{\text{Ref}} \approx 1.5$ V). The *NPN* transistors Q_9 and Q_{11}, along with the current mirror formed by the *PNP* transistors Q_7 and Q_8, form a differential amplifier. The voltage of the noninverting input, the base of Q_9, is the reference voltage V_{Ref}, whereas the voltage of the inverting input, the base of Q_{11}, is proportional to the output voltage. The output of the differential amplifier provides the base current of Q_{12}, an emitter–follower amplifier, which in turn provides the base current of Q_3, the output emitter–follower amplifier of the regulator. Transistor Q_{14} limits the output current by limiting the base current of Q_{12} when the load current becomes excessive.

The current required by the voltage reference, it will be noted, is

TABLE 6.1 THE 78LXX SERIES ELECTRONIC VOLTAGE REGULATORS

		v_{Supply}	
	v_{Load} V	min V	max V
78L02	2.6	4.75	20
78L05	5.0	7.0	20
78L06	6.2	8.5	20
78L08	8.0	10.5	23
78L09	9.0	11.5	24
78L10	10.0	12.5	25
78L12	12.0	14.5	27
78L15	15.0	17.5	30

The maximum load current is 100 mA.

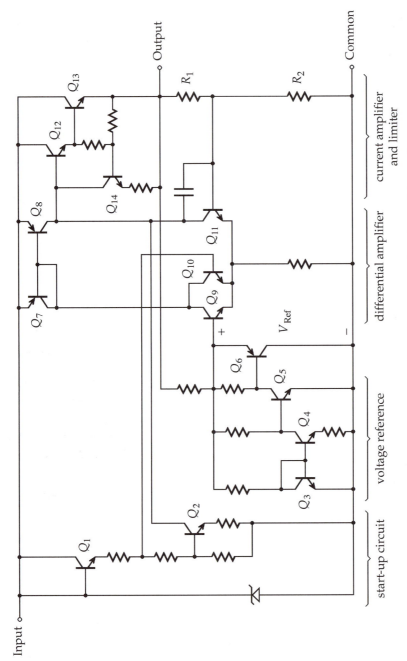

Figure 6.42: The 78LXX electronic voltage regulator. The "XX" designation is the voltage of the regulator.

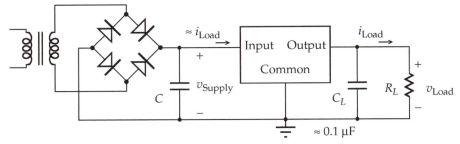

Figure 6.43: A power supply using a three-terminal electronic voltage regulator. A small output capacitor C_L is used to minimize the effect of transient output currents on the output voltage.

derived from the output terminal of the regulator. This is common for voltage regulators because the use of the output voltage minimizes the effect of variations (including ripple) of the supply voltage on the reference voltage. But, a problem arises with this scheme: a zero reference and output voltage is one possible state of the regulator. Because of this problem, a start-up circuit is required. Transistors Q_1 and Q_{10} provide an initial reference voltage for the difference amplifier, thus producing an initial output voltage when the supply is turned on. The regulator also has a thermal shutdown circuit (Q_2).

In addition to the 100-mA series regulators of Table 6.1, there is the 78MXX series with a maximum output current of 500 mA and the 78XX series with a 1.5-A current rating. A suitable heat sink is required for the high-current series. Similar negative voltage regulators, the 79LXX, 79MXX, and 79XX series, are also available. Furthermore, other electronic regulators with a wide range of specifications, including regulators with adjustable output voltages, are available.

A complete power supply using a three-terminal electronic regulator is shown in Figure 6.43. In essence, the regulator converts a fluctuating input voltage (ripple) that varies with variations in load current, v_{Supply}, to a nearly constant load voltage v_{Load}. For a typical electronic regulator, a supply voltage that is 2 V more than v_{Load} (or greater) is required (see Table 6.1 for the 78LXX series). Except for very small load currents, the input current of the regulator is approximately equal to the load current because the current of the regulator circuit tends to be small.

As a result of the regulator, the rectifier–filter circuit supplies a nearly constant current. Hence, a constant current discharges the filter capacitance C when the diodes are not conducting:

$$C \frac{dv_{\text{Supply}}}{dt} \approx -i_{\text{Load}} \qquad (6.54)$$

For a discharge time of $T/2$, approximately that of a full-wave rectifier, the following is obtained for the ripple of the supply voltage $v_{\text{Supply p–p}}$:

$$v_{\text{Supply p–p}} = \frac{i_{\text{Load}} T}{2\,C} \qquad (6.55)$$

Because of the electronic regulator, a fairly large ripple voltage can be tolerated.

The main limitation is that the instantaneous minimum value of v_{Supply} not fall below the minimum input voltage required by the regulator. Because this occurs for the maximum load current, Eq. (6.55) may be used to determine the minimum value of filter capacitor that assures a minimum input voltage for the regulator.

The regulators that have been considered function as an electronically controlled series resistance inserted between the rectifier–filter circuit and the load. As a result, the dissipated power is approximately the product of the load current and the voltage difference of the regulator $V_{\text{Supply av}} - v_{\text{Load}}$. If this voltage difference is large, the dissipated power will be large for a large load current. Hence, the regulator will not only be inefficient, but a large heat sink will be required to maintain an acceptable temperature for the integrated circuit. An alternative, more efficient electronic regulator circuit is the switching regulator.

THE ELECTRONIC SWITCHING REGULATOR

Although more complex than series electronic regulators, switching-type regulators are, for many applications, considerably more power efficient. Hence, these regulators tend to be used for supplies with moderate-to-large load currents (currents greater than 1 or 2 A). A switching regulator, as its name implies, uses a transistor (or transistors) operated in a switching mode, that is, the transistor is rapidly switched between being either cut off or saturated. When the transistor is cut off, its power dissipation is zero, and when it is saturated, its collector–emitter voltage is very small; therefore, even for a large collector current, the transistor's dissipated power is small. In addition to the transistor switch (or switches), an inductor and capacitor are used for energy storage.

A detailed analysis of switching regulators is beyond the present treatment, but several texts are available for additional information (Chetty 1986; Gottlieb 1984; Lenk 1995; Pressman 1977). Consider the basic circuit of Figure 6.44 in which the switch S is a bipolar junction transistor that alternates between being switched on (short circuit) and off (open circuit). When the switch is on, the voltage difference of the inductor, if the voltage across the switch can be ignored, is $v_{\text{Supply}} - v_{\text{Load}}$:

$$v_{\text{Supply}} - v_{\text{Load}} = L \frac{di}{dt} \tag{6.56}$$

With a positive voltage difference assumed, the current of the inductor i is an increasing function of time. When the switch is opened, the current of the inductor will not change abruptly. The current for this condition is through the diode; the voltage across the inductor, if the voltage across the diode is ignored, is $-v_{\text{Load}}$:

$$-v_{\text{Load}} = L \frac{di}{dt} \tag{6.57}$$

The inductor's current will decrease for this condition. Because a switching rate of 10 to 50 kHz is typical, a relatively

Figure 6.44: A basic step-down switching-type regulator.

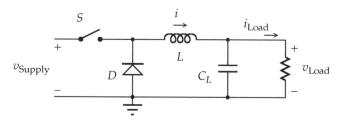

small load capacitance C_L will maintain a nearly constant load voltage over a switching period. The average load voltage v_{Load} will depend on the length of the *on* and *off* intervals of the switch. Therefore, a feedback circuit is used that senses the load voltage, compares it with a reference voltage, and then adjusts the on–off timing of the switching transistor.

Neither the load capacitance C_L nor the inductor of a switching regulator can be fabricated on an integrated circuit, and, except for low-current supplies, an external power transistor (or transistors) is required. Switching-type power supply modules complete with power transformer, rectifier, filter capacitance, and the electronic regulator are available for a wide range of voltage and current specifications. In addition to the basic circuit of Figure 6.44, alternative circuits are also used, including circuits that result in a load voltage greater than the voltage of the rectifier–filter circuit.

EXAMPLE 6.8

Design a power supply that uses an electronic voltage regulator to obtain a load voltage of 12 V and a maximum load current of 1.0 A. The input of the supply is a 120 V (rms), 60-Hz utility voltage.

SOLUTION The regulator circuit of Figure 6.43 will be utilized with a 7812 three-terminal integrated-circuit regulator to provide a 12-V load voltage and a maximum load current of 1.5 A. For proper operation of the regulator, a minimum input voltage of 14.5 V is required. On the basis of the nonideal transformer considerations of Section 6.2, a rectified voltage of approximately 18 V for a dc load current of 1 A is a reasonable expectation for a power transformer with a secondary rating of 18 V and 1 A.

Because the electronic regulator is capable of supplying a load current in excess of 1.5 A (the result of a load fault), a bridge rectifier with a current rating of 2 A or more is required. For an average input voltage of 18 V to the electronic filter, a peak-to-peak ripple voltage of 2 V is acceptable (a voltage that varies between 17 and 19 V). The regulator will also function for a smaller input voltage that may occur as a result of a lower-than-normal utility voltage. A 10-percent decrease in the utility voltage translates to an average supply voltage of 16.2 V. For this situation, a ripple voltage greater than 2 V might result in a minimum voltage uncomfortably close to the minimum input voltage needed by the regulator. Equation (6.55) is used to determine the filter capacitance C for a load current of 1 A ($T = 1/60$ s):

$$C = \frac{i_{\text{Load}} T}{2\, v_{\text{Supply p–p}}} = 4167\ \mu\text{F}$$

A standard electrolytic capacitance with a value of 4700 μF is used.

The voltage rating of the capacitor and the diodes must be adequate to withstand the voltages of the circuit for a zero load current. An 18-V (rms) secondary transformer voltage implies a peak voltage of 25.5 V. However, a transformer designed for a load current of 1 A is very likely to have a somewhat larger secondary voltage for a very small load current. Therefore, diodes with a peak inverse diode voltage of 50 V need to be used (this tends to be the smallest voltage rating of readily available bridge rectifiers). For a very small load current, the maximum voltage of the filter capacitor will be approximately 25 V. However, if the power line voltage should happen to be excessive (for example, 10-percent high), a somewhat larger voltage rating is needed. Hence, to be on the safe side, a 50-V rating is desired.

On the basis of a μA7812C regulator's specifications (Texas Instruments 1992), the load voltage of the regulator will be between 11.4 and 12.6 V (± 5 percent of the 12-V nominal value). Furthermore, the output voltage v_{Load} will vary by no more than 12 mV for a variations in load current based on its typical specification (240 mV for the worst case). A typical ripple rejection ratio of 71 dB is given. This implies that the ripple of the regulated load voltage will be 0.0028 times that of the input voltage v_{Supply}. For $v_{Supply\ p-p}$ of 2 V, the ripple of v_{Load} will be 5.6 mV. The minimum ripple rejection ratio given is 51 dB. For this worst case, the load ripple voltage will be 56 mV. On the basis of these specifications, the output of the electronic regulator will be acceptable for most applications.

6.5 BATTERIES: AN INCREASINGLY IMPORTANT ELECTRICAL ENERGY SOURCE

Throughout the 19th century, chemical batteries were used to power telegraph and telephone systems. In the first decades of the 20th century, before the widescale distribution of utility-produced electric power, batteries were used to power early radio receivers. Before the advent of the transistor, battery powered portable electronic systems, such as radios, were relatively heavy and bulky and were characterized by limited battery lives. Because transistor circuits tend to require much smaller currents and voltages than the vacuum tube circuits they replaced, small, lightweight, truly portable battery-powered systems became practicable. This spawned the widescale introduction of many new portable products, the transistor radio (originally referred to as simply a "transistor") being one of the first. We now have battery-powered Walkmans, wireless telephone handsets, and cellular phones, pagers, remote controls, hearing aids, pacemakers, and notebook computers. As a result of these applications, battery usage has been increasing.

The earliest battery appears to be that described by Alessandro Volta in 1800 – "a source of perpetual power" (Heise and Calhoon 1971). Its potential was the result of dissimilar metals that were placed in contact with an electrolyte (Volta's work led to the formulation of the electromotive series for metals). Many types of cells using a wide variety of electrodes and electrolytes were developed during the 19th century. A significant development was the wet cell invented by Georges Leclanché in 1860 that used carbon and zinc electrodes. In the 1880s, a Leclanché-type dry cell, the predecessor of the presently used carbon–zinc battery, was introduced. This cell utilizes a cathode (the positive terminal of the battery)

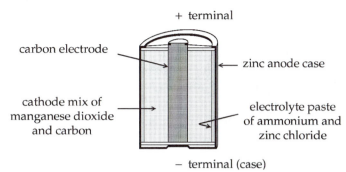

Figure 6.45: Cross-sectional view of a carbon–zinc dry cell.

consisting of a carbon rod surrounded by a mixture of powdered manganese dioxide and carbon (Figure 6.45). The cathode is surrounded by an electrolyte consisting of a paste of ammounium and zinc chloride that is in intimate contact with the zinc case that forms the anode (negative terminal) of the battery (Cahoon and Heise 1976; Crompton 1982; Mantell 1983).

The current-producing capability of a carbon–zinc dry cell depends on chemical reactions involving electronic charges at the anode and cathode of the cell:

$$Zn + 2Cl^- \longrightarrow ZnCl_2 + 2e^- \qquad \text{(anode)}$$

$$2MnO_2 + 2NH_4^+ + 2e^- \longrightarrow 2MnOOH + 2NH_3 \qquad \text{(cathode)}$$

$$2MnO_2 + 2NH_4Cl + Zn \longrightarrow ZnCl_2 \cdot 2NH_3 + H_2O + Mn_2O_3 \quad \text{(overall)}$$

$$(6.58)$$

The negative electronic charges flow from the anode to the cathode through the external circuit. However, by convention, hypothetical positive charges are envisioned as flowing in the opposite direction, resulting in an external positive current from the cathode to the anode of a battery. As the cell is discharged, the active chemical reactants (left side of the bottom reaction of Eq. (6.58)) are depleted.

Carbon–zinc dry cells are produced in a variety of sizes from that of the 1-cm-diameter AAA cell to the standard 3.4-cm diameter flashlight D cell and larger lantern-type multicell batteries. A fresh, unused cell has an open-circuit voltage that is slightly greater than 1.5 V. As the cell is discharged, its terminal voltage decreases, eventually falling to a voltage at which the circuit it is powering will no longer function properly. The internal resistance of a cell, an equivalent series resistance of a circuit model of a battery, depends on the physical size of the cell as well as its previous usage. Fresh small-size cells have an internal resistance, determined by a momentary short-circuit current measurement, of only a few to several tenths of an ohm.

Other primary (nonrechargeable) cells, in particular the zinc–chloride cell (frequently designated as "heavy duty"), and the alkaline–manganese dioxide ("alkaline") cell have also been developed; these provide larger energy outputs, albeit at a higher cost. On the basis of the application, an alkaline cell may provide several times the electrical energy of the same-size carbon–zinc cell. Specialized cells, including mercuric and silver oxide "button" cells, lithium manganese dioxide cells, and zinc air cells are used for miniaturized electronic systems.

Another type cell, the secondary cell, can be electrically recharged as a result of a reversible chemical reaction (Crompton 1982; Grant 1975). The lead–acid vented storage battery is nearly universally used for automobiles, whereas more convenient sealed versions are available for electronic applications. Rechargeable nickel–cadmium cells, however, tend to be the most commonly used rechargeable cells for electronic applications. The chemical reactions of this cell involve cadmium and cadmium hydroxide at the anode and nickelic hydroxide and nickelous hydroxide at the cathode:

$$Cd + 2OH^- \longrightarrow Cd(OH)_2 + 2e^- \qquad \text{(anode)}$$

$$2NiOOH + 2H_2O + 2e^- \longrightarrow 2Ni(OH)_2 + 2OH^- \qquad \text{(cathode)} \qquad (6.59)$$

$$2Ni(OH)_2 + Cd(OH)_2 \longrightarrow 2NiO(OH) + Cd + 2H_2O \qquad \text{(overall)}$$

The $2e^-$ term of the anode and cathode reactions results in an external electron flow from the anode to the cathode of the cell. During charging, the reactions of Eq. (6.59) are reversed. Even though a nickel–cadmium cell has an initial open-circuit terminal voltage of only about 1.2 V, compared with the 1.5 V of a conventional carbon–zinc dry cell, the voltage of the nickel–cadmium cell remains nearly constant during its discharge. It is therefore possible, for most applications, to replace individual carbon–zinc cells directly with nickel–cadmium cells that have the advantage of being able to be recharged many hundreds of times before being replaced.

The external current of a chemical cell is the result of ionic reactions at the cathode and anode of the cell (e.g., Eq. (6.58) or (6.59)). As a result of these reactions, the active compounds of the cell are depleted. One would therefore expect, at least for a first-order approximation, that the external charge transfer that occurs before a cell is depleted, Q_{Bat}, would be independent of the time dependence of the current – only its integral over time is involved.

$$Q_{Bat} \approx \int_0^{T_b} i_{Bat}(t)\,dt \qquad (6.60)$$

A cell capable of producing a charge transfer of 1 C, could therefore supply, for example, a constant current of 1 mA for 1000 s, or a constant current of 1 μA for 10^6 s. Although this tends to be approximately the case, cells generally produce a greater charge transfer if they are used intermittently.

During the discharge of a cell, its voltage tends to decline (Figure 6.46). Although a cutoff voltage is used to define the lifetime of a battery, there is not a set of standardized cutoff potentials for the various types of cells. The cutoff potential is the minimum voltage for which a particular electronic circuit application functions properly. Lowering the minimum voltage of the circuit through an appropriate design change will increase the usefull life of a cell or battery of cells.

The time integral of current (coulombs per second) is charge (coulombs). For a battery specification, a time dimension of hours rather than seconds is generally used, thus resulting in a hybrid unit for charge that has a dimension of ampere-hours (1 Ah = 3600 C). Generally, large cell currents tend to reduce the charge that a cell can supply. Cells usually function better when operated intermittently,

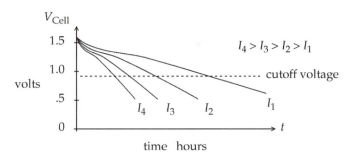

for there is a tendency for the chemicals to rejuvenate somewhat between discharge periods. However, a cell has a limited shelf life because it will tend to discharge on its own.

In addition to providing power for portable electronic equipment, batteries are frequently used in conjunction with a utility-powered rectifier and filter circuit to provide uninterrupted electrical power during utility outages. This is frequently for the backup of only a critical portion of an electronic circuit such as the clock circuit of a clock-timer or the memory circuit of a telephone dialer. If a primary (nonrechargeable) battery is used, the electronic load may be connected to the utility-powered source and the battery with a set of diodes (Figure 6.47). When the utility supply is functioning, the supply voltage V_{Supply} is such as to be larger than the battery backup voltage V_{Bat}. A current through D_1 provides the load current of R_{Load2}, the critical electronic circuit. However, for a utility outage, V_{Supply} will be reduced to zero, resulting in a current provided by the battery through D_2. Because V_{Bat} is less than V_{Supply}, the load voltage will be somewhat smaller for this condition. If it is necessary that a constant load voltage be maintained, an electronic regulator (for example, a zener diode circuit) could be incorporated as part of the load.

An alternative battery backup scheme (Figure 6.48) utilizes a rechargeable battery (for example, nickel–cadmium cells). For this application, the battery can serve as part of the filter–regulator circuit. The load resistor R_L represents the electronic circuit for which operation during a utility outage is required (for example, a fire alarm). An equivalent circuit of the battery is indicated. The battery will generally consist of several series-connected cells to achieve a desired voltage. Hence, the series resistance r_{Bat} will be the sum of the resistances of the individual cells. The equivalent circuit of Figure 6.48 may be recognized as being essentially the same as that of the zener diode regulator of Figure 6.34 with a zener voltage of V_Z and a resistance of r_z. Thus, the same solution applies. The circuit of Figure 6.48 is generally designed to maintain a small continuous battery current that serves

Figure 6.47: A battery backup for a critical electronic circuit.

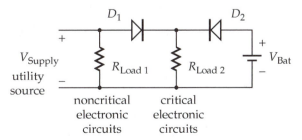

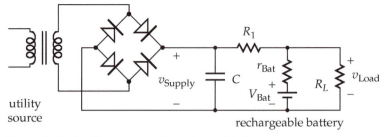

Figure 6.48: A utility supply and a rechargeable battery.

to keep the battery fully charged. Unlike the zener diode regulator, the battery circuit will supply a load current for a utility outage, for the diodes of the rectifier serve to disconnect the transformer from the circuit.

EXAMPLE 6.9

A particular Walkman AM–FM cassette tape player is powered by two series-connected AA cells. The following battery currents result for battery voltages of 2 to 3 V (independent of voltage):

	AM radio	FM radio	Tape
No or low-level output	25 mA	35 mA	120 mA
Medium-level output	35 mA	45 mA	130 mA
High-level (loud) output	50 mA	60 mA	145 mA

The variation in battery current results from power being supplied to the earphone by the audio amplifier. Normal operation of the Walkman occurs for a battery voltage of 2 V or greater. For intermittent operation, a carbon–zinc cell has a charge capacity of approximately 0.9 Ah; an alkaline cell, 2.1 Ah; and a rechargeable nickel–cadmium cell, 0.7 Ah. Estimate the number of hours that each type cell will last for each mode of operation. Estimate the energy, expressed in joules and watt-hours, supplied by each cell.

SOLUTION It will be assumed that the battery current for the medium-level output corresponds to the average battery current.

$$Q_{Ah} = I_{Bat} T_h$$

The following is obtained for carbon–zinc cells:

AM radio	$T_h = 0.9\ \text{Ah}/35\ \text{mA} = 25.7\ \text{h}$
FM radio	$T_h = 0.9\ \text{Ah}/45\ \text{mA} = 20\ \text{h}$
Tape	$T_h = 0.9\ \text{Ah}/130\ \text{mA} = 6.9\ \text{h}$

For alkaline cells (2.1 Ah), the times are increased to 60, 46.7, and 16.2 h. Rechargeable nickel–cadmium cells (0.7 Ah) reduce the times to 20, 15.6 and 15.4 h. The energy supplied by the battery E is equal to the integral over time of the instantaneous power as follows:

$$E = \int_0^{T_h} p \, dt = \int_0^{T_h} V_{\text{Bat}} I_{\text{Bat}} \, dt$$

Because the battery current was found to be independent of voltage, I_{Bat} may be removed from the integral.

$$E = I_{\text{Bat}} \int_0^{T_h} V_{\text{Bat}} \, dt = (I_{\text{Bat}} T_h) \left(\frac{1}{T_h} \int_0^{T_h} V_{\text{Bat}} \, dt \right) = Q_{Ah} V_{\text{Bat av}}$$

Hence, the energy is the charge supplied multiplied by the average battery voltage. If 2.5 V is assumed for the average battery voltage of the carbon–zinc cells (1.25 V per cell), an energy of 2.25 Wh (0.9 Ah × 2.5 V), that is 8100 J (1Ws is 1 J), is obtained. An energy value of 5.25 Wh (1.89×10^5 J) is obtained for the alkaline cells (same average voltage). The voltage of a nickel–cadmium cell during discharge remains essentially constant at 1.2 V, resulting in a battery voltage of 2.4 V. The energy supplied is 1.68 Wh (6048 J).

REFERENCES

Cahoon, N. C. and Heise, G. W. (Eds.) (1976). *The Primary Battery – Volume 2*. New York: John Wiley & Sons.

Chetty, P. R. K. (1986). *Switch-Mode Power Supply Design*. Blue Ridge Summit, PA: Tab Professional and Reference Books.

Crompton, T. R. (1982). *Small Batteries – Volume 1: Secondary Cells*. New York: John Wiley & Sons.

Crompton, T. R. (1982). *Small Batteries – Volume 2: Primary Cells*. New York: John Wiley & Sons.

Gottlieb, Irving M. (1984). *Power Supplies, Switching Regulators, Inverters, and Converters*. Blue Ridge Summit, PA: Tab Books.

Grant, J. C. (1975). *Nickel–Cadmium Battery* (2d ed.). Gainesville, FL: General Electric Co.

Grebene, Alan B. (1984). *Bipolar and MOS Analog Integrated Circuit Design*. New York: John Wiley & Sons.

Heise, G. W. and Cahoon N.C. (Eds.) (1971). *The Primary Battery – Volume 1*. New York: John Wiley & Sons.

Lenk, John D. (1995). *Simplified Design of Switching Power Supplies*. Boston: Butterworth–Heinemann.

Mantell, C. L. (1983). *Batteries and Energy Systems* (2d ed.). New York: McGraw-Hill.

McAfee, K. B., Ryder, E. J., Shockley, W., and Sparks, M. (1951). Observations of Zener current in germanium *p-n* junctions. *Physical Review*, 83, 650–1.

Pressman, Abraham I. (1977). *Switching and Linear Power Supply, Power Converter Design*. Rochelle Park, NJ: Hayden Book Co.

Smits, F. M. (Ed.) (1985). *A History of Engineering and Science in the Bell System – Electronics Technology (1925–1975)*. Indianapolis, IN: AT&T Laboratories.

Soclof, Sidney (1991). *Design and Applications of Analog Integrated Circuits*. Englewood Cliffs, NJ: Prentice–Hall.

Texas Instruments (1992). *Linear Circuits Data Book 1992* (Vol. 3). Dallas, TX: Texas Instruments, Inc.

Zener, C. (1934). A theory of the electrical brakdown of solid detectors. *Proceedings of the Royal Society of London*, **145**, 523–9.

PROBLEMS

6.1 The input voltage of the diode half-wave rectifier circuit of Figure P6.1 is $v_{Source}(t)$. Assume that the input voltage has a symmetrical square wave-form with maximum and minimum values of V_p and $-V_p$, respectively. Assume ideal behavior of the diode.

a) Determine the average value of the load voltage V_{av}.
b) Determine the rms value of the load voltage V_{rms}.
c) Determine the average power dissipated by the load resistor.

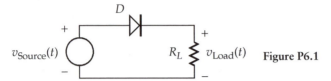

Figure P6.1

6.2 Obtain numerical values for the answers of Problem 6.1 for a peak voltage V_p of 10 V and a load resistance R_L of 1 kΩ.

6.3 Repeat Problem 6.1 for a diode with a constant forward voltage $v_{D(on)}$ of 0.75 V.

6.4 Repeat Problem 6.2 for a diode with a constant forward voltage $v_{D(on)}$, of 0.75 V.

6.5 Repeat Problem 6.1 for the polarity of the diode reversed.

6.6 Consider the case for which the input voltage of Problem 6.1 has a symmetrical triangular waveform with maximum and minimum values of V_p and $-V_p$, respectively. Repeat Problem 6.1 for this condition.

6.7 Obtain numerical values for the answers of Problem 6.6 for a peak voltage V_p of 10 V and a load resistance R_L of 1 kΩ.

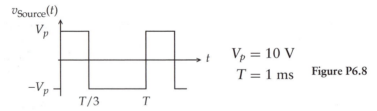

$V_p = 10$ V
$T = 1$ ms   Figure P6.8

6.8 The rectifier circuit of Figure P6.1 has the input voltage $v_{Source}(t)$ of Figure 6.8.

a) Determine the average load voltage with ideal behavior of the diode assumed.
b) Determine the rms value of the load voltage assuming ideal behavior of the diode.
c) Repeat parts (a) and (b) with $v_{D(on)} = 0.75$ V assumed.

6.9 Repeat Problem 6.8 for a positive pulse that has a time duration of $T/4$.

6.10 Consider the transformer–rectifier circuit of Figure 6.6. Assume that $V_m = 12$ V, the power-line frequency is 60 Hz, and $v_{D(on)} = 0.75$ V.

a) Determine the average value of the load voltage V_{av}.
b) Suppose it is assumed that the output voltage could be approximated as a full-wave rectified sinusoid (ideal rectifier diodes) with a peak value of $V_m - v_{D(on)}$. The average value of this voltage would be $2(V_m - v_{D(on)})/\pi$. Compare this approximation for V_{av} with that determined in part (a).
c) Repeat parts (a) and (b) for $V_m = 6$ V.

6.11 Repeat Example 6.2 for the full-wave rectifier circuit of Figure 6.6.

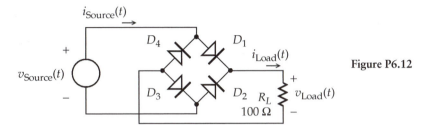

Figure P6.12

6.12 The input voltage of the bridge-rectifier circuit of Figure P6.12 is $v_{Source}(t)$. Assume $v_{Source}(t) = 10 \sin 2\pi ft$ V along with ideal behavior of the diodes.

a) What is the average value of the load voltage V_{av}?
b) Sketch $i_{Source}(t)$ and $i_{Load}(t)$. What are their average values?
c) Repeat parts (a) and (b) for diodes with a forward-biased voltage of $v_{D(on)} = 0.75$ V.

6.13 Consider the circuit of Figure P6.12 and assume ideal behavior of the rectifier diodes and an arbitrary periodic input voltage $v_{Source}(t)$. Show that the rms value of $v_{Load}(t)$ is equal to the rms value of $v_{Source}(t)$.

6.14 A full-wave rectifier such as that of Figure P6.12 is generally used for an ac voltmeter. Assume ideal behavior of the diodes (an electronic rectifier can accomplish this type of behavior) and a periodic input voltage $v_{Source}(t)$. Determine the ratio of the average of $v_{Load}(t)$ to the rms value of the input voltage $v_{Source}(t)$, for the following waveforms:

a) The input $v_{Source}(t)$ is a sinusoidal voltage with a peak value of V_m.
b) The input is a symmetrical triangular-wave voltage with a peak value of V_p (minimum of $-V_p$).
c) The input is a symmetrical square-wave voltage with a peak value of V_p.

Note: This type of ac meter is generally calibrated to read correctly for a sinusoidal voltage.

6.15 A bridge rectifier circuit is used in the circuit of Figure P6.15 to charge a

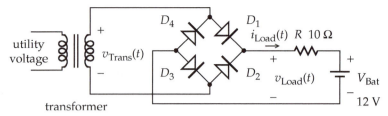

Figure P6.15

battery.

$$v_{\text{Trans}}(t) = V_m \sin \omega t, \qquad V_m = 20 \text{ V}$$

The resistor R is used to limit the current of the circuit $i_{\text{Load}}(t)$. A constant forward-biased voltage model for the diodes with $v_{D(\text{on})} = 0.8 \text{ V}$ is appropriate for this circuit. What is the average charging current I_{av} and the power supplied to the battery?

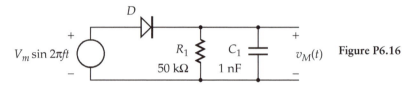

Figure P6.16

6.16 The half-wave rectifier circuit of Figure P6.16 is used for a radio detector. The frequency of the input signal f is 455 kHz (the intermediate frequency of the superheterodyne receiver of Figure 1.10) and $V_m = 1.0 \text{ V}$. Assume ideal behavior of the diode.

a) Determine the peak value of $v_M(t)$.
b) What is the peak-to-peak ripple voltage?
c) What is the peak value of the diode current?
d) What is the value of C_1 required to result in a peak-to-peak ripple voltage of 10 mV?

6.17 Repeat Problem 6.16 for a diode with a forward-biased voltage $v_{D(\text{on})}$, of 0.55 V.

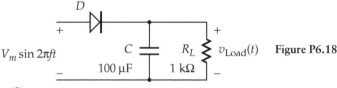

Figure P6.18

utility source

6.18 A half-wave rectifier–filter circuit is connected directly to an electric utility source with a voltage of 120 V (rms) and a frequency f of 60 Hz (Figure P6.18). For the large voltages of the circuit, an assumption of ideal behavior for the diode is readily justified.

a) Determine the peak value of $v_{\text{Load}}(t)$.
b) What is the peak-to-peak ripple of the load voltage?

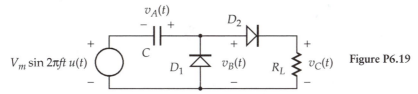

$V_m \sin 2\pi f t\, u(t)$ $v_A(t)$ C D_1 $v_B(t)$ D_2 R_L $v_C(t)$ **Figure P6.19**

c) Suppose that the load is not a resistor but instead results in a constant current of 100 mA. Determine the peak-to-peak ripple of the load voltage for this condition.

6.19 Consider the rectifier circuit of Figure P6.19, a half-wave voltage doubler. Assume ideal behavior of the diodes and that at $t = 0$ the capacitor C is uncharged.

a) Determine and sketch $v_A(t)$ and $v_B(t)$ with the assumption that the current of D_2 may be ignored (a very large R_L). A time interval of two periods is required to achieve a periodic response for $v_B(t)$.

b) Determine $v_C(t)$ based on the result of part (a). What is the peak value of $v_C(t)$?

Note: An actual power supply would use a second capacitor connected in parallel with R_L to obtain a steady voltage of approximately $2V_p$.

6.20 Consider the full-wave bridge rectifier and filter circuit of Figure 6.19. The output ripple voltage, $v_{\text{Load}\,p-p}$, depends on the time constant of the circuit, that is, $R_L C$. Assume that $t_2 - t_1 = T/2$.

a) On the basis of the linear approximation of Eq. (6.20), what is the time constant that results in a 10-percent ripple voltage ($v_{L\,p-p}/V_p = 0.1$)?

b) What is the value of $v_{L\,p-p}/V_p$ predicted by the exponential expression for the time constant of part (a)?

6.21 Consider the full-wave bridge rectifier and filter circuit of Figure 6.19. Show that after steady-state conditions are achieved, the average value of the current of D_1 is one-half the average load current.

6.22 Consider the full-wave bridge rectifier and filter circuit of Figure 6.19

$$V_m = 30 \text{ V}, \quad f = 60 \text{ Hz}, \quad R_L = 500\ \Omega, \quad C = 100\ \mu\text{F}$$

Ideal behavior of the diodes may be assumed.

a) What is the peak value of $v_{\text{Load}}(t)$?

b) What is the peak-to-peak value of the load ripple voltage?

c) What is the approximate time interval for which the diodes conduct, that is, approximately $3T/4 - t_2$ of Figure 6.21?

6.23 Repeat Problem 6.22 for $C = 1000\ \mu\text{F}$.

6.24 Repeat Problem 6.22 for $C = 1000\ \mu\text{F}$ and $R_L = 100\ \Omega$.

6.25 Repeat Problem 6.22 for a constant load current of 100 mA. What is the value of C required to reduce the peak-to-peak ripple voltage to 1 V?

6.26 A zener diode with a breakdown voltage of 12 V is used in the circuit of Figure P6.26. Assume that a minimum zener current i_Z of 5 mA is required to ensure proper operation of the zener diode.

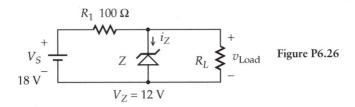

Figure P6.26

a) What is the minimum value of R_L for which proper operation of the zener diode is ensured?
b) Estimate the variation of v_{Load} that occurs for a variation in R_L between that obtained in part (a) and infinity (an open circuit). Use a value of r_z that corresponds to the minimum zener diode current.
c) What is the power dissipated by R_1 and Z for the load resistance of part (a)? What is the power dissipated by R_1 and Z for $R_L = \infty$ (an open circuit)?

6.27 Assume that the nominal load current of the circuit of Figure P6.26 is 30 mA. Determine the variation in load voltage for a ± 10 mA variation in load current.

6.28 Assume that the nominal load current of the circuit of Figure P6.26 is 30 mA. Determine the variation in load voltage for a ± 1 V variation in V_S.

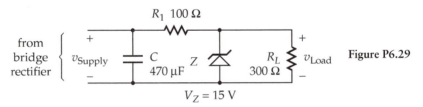

Figure P6.29

6.29 A zener diode with a breakdown voltage of 15 V is used in the circuit of Figure P6.29. The frequency of the utility supply is 60 Hz. Assume that the peak value of v_{Supply} is 25 V.

a) Determine the minimum value of v_{Supply}.
b) Determine the peak-to-peak value of the load ripple voltage.
c) Estimate the average electrical power dissipated by R_1 and Z. Estimate the power efficiency of the filter and regulator, that is, the ratio of the power dissipated by R_L to that supplied by the rectifier circuit.

6.30 Determine the minimum value of capacitance C in Problem 6.29 that assures a zener diode current of 5 mA. What is the peak-to-peak value of the load ripple voltage for this condition?

6.31 An electronic filter using a power transistor is shown schematically in Figure P6.31. As a result of the resistor and capacitor of the base circuit of the transistor, the ripple voltage at the base of the transistor will be much less than that of v_{Supply}. The supply voltage, derived from a

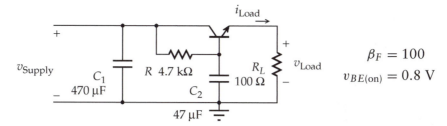

Figure P6.31

full-wave rectifier with a power-line frequency of 60 Hz, has an average value of 18.0 V.

a) Determine the average values of the load voltage and current.
b) Determine the supply ripple voltage, that is, $v_{\text{Supply p-p}}$. The discharge current of C_1 may be assumed to be constant and equal to the average load current.
c) Estimate the peak-to-peak ripple of v_{Load} by determining the ripple value of the base voltage of the transistor. The ripple component of v_{Supply} may be assumed to be a sinusoidal voltage with a peak-to-peak value equal to that determined in part (b).

6.32 Repeat Problem 6.31 for $R_L = 250\ \Omega$.

6.33 Repeat Problem 6.31 for $v_{\text{Supply av}} = 24.0$ V.

6.34 Determine the dependence of the average value of the load voltage on the average value of the load current for the circuit of Problem 6.31. Assume the average supply voltage remains a constant 18 V. What is the Thévenin voltage and resistance of the equivalent output circuit of the electronic filter?

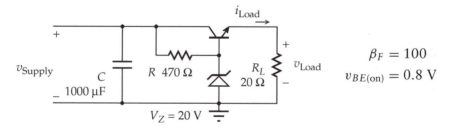

Figure P6.35

6.35 The elementary regulator of Figure P6.35 uses a power transistor and a zener diode. As a result of the zener diode, the base voltage of the transistor remains nearly constant and equal to the zener voltage. The supply voltage, derived from a full-wave rectifier with a power-line frequency of 60 Hz, has an average value of 30.0 V.

a) Determine the average values of the load voltage and current.
b) Determine the supply voltage ripple, that is, $v_{\text{Supply p-p}}$. The discharge

current of C may be assumed to be constant and equal to the average load current.

c) Estimate the peak-to-peak ripple of v_{Load} by determining the ripple value of the base voltage of the transistor.

6.36 Repeat Problem 6.35 for a value of R_L that results in an average load current of 200 mA.

6.37 Consider the op amp electronic regulator of Figure 6.36. Suppose the op amp has an equivalent output resistance R_O of 50 Ω and a gain–bandwidth product of 1 MHz. The reference voltage V_{Ref} is 5 V. Show that for frequencies above the break frequency of the op amp response, the equivalent output impedance of the regulator is inductive. Determine the value of the inductance.

6.38 The three-terminal regulator of Figure 6.43 is used to provide a load voltage of 5 V at a current of 2 A. The bridge rectifier results in a peak supply voltage of 12 V. A minimum difference of 2.5 V is required between the input and output terminals of the regulator for it to function properly. Determine the minimum-size filter capacitance that can be used (power-line frequency of 60 Hz). What is the average power dissipated by the regulator for this capacitance?

6.39 Repeat Problem 6.38 for the condition that the supply voltage ripple $v_{\text{Supply p–p}}$ not be greater than 2 V.

6.40 Repeat Problem 6.38 for a power-line frequency of 800 Hz. Not only will a much smaller filter capacitor be required, but the size of the iron-core transformer will also be reduced.

6.41 Standard D-type cells (flashlight batteries) have the following approximate charge capacity for intermittent service: carbon–zinc, 6 Ah; zinc–chloride, 9 Ah; alkaline, 15 Ah. Repeat Example 6.9 for these batteries.

6.42 Determine the life of the cells of Problem 6.41 when used to power a 0.3-A flashlight bulb. What is the life for a 0.5-A bulb?

6.43 Consider the battery backup circuit of Figure 6.47.

$$V_{\text{Supply}} = 12 \text{ V}, \qquad V_{\text{Bat}} = 9 \text{ V}, \qquad R_{\text{Load2}} = 1 \text{ k}\Omega$$

Assume $v_{D(\text{on})} = 0.7$ V for the diodes.

a) What is the voltage of R_{Load2} for the utility-supplied power and when there is a utility outage?

b) Suppose that a constant load voltage of 6 V is required. Design a zener diode regulator circuit to achieve this. Assume that a zener diode current of at least 1 mA is required for its operation.

6.44 For short-duration power outages, a large electrolytic capacitor can be used as a backup energy source (Figure P6.44). During normal operation, $V_{\text{Supply}} = 5$ V. Assume that the electronic load (for example, a clock) operates satisfactorily for a voltage of 2 V or greater. What is the time interval for which the electronic load will operate during a utility outage?

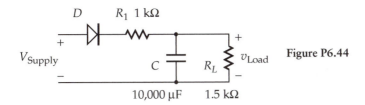

Figure P6.44

6.45 Repeat Problem 6.44 for a load that is characterized by a constant current of 2.5 mA.

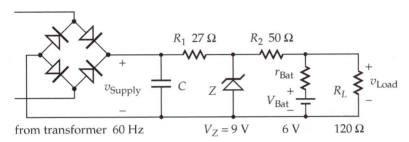

Figure P6.46

6.46 Consider the power supply of Figure P6.46 that uses a 6-V rechargeable battery. Assume the peak supply voltage is 12 V and that the battery has an equivalent resistance of 0.1 Ω. Determine the average battery current as well as the peak-to-peak ripple load voltage.

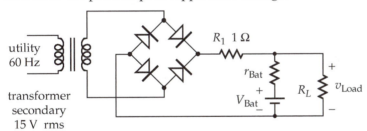

Figure P6.47

6.47 A 12-V lead–acid automobile storage battery with an equivalent resistance of 25 mΩ is used to power a load requiring a current of 3 A. The battery charger of Figure P6.47 is connected to the battery. Determine the average battery current with diode voltages v_{Don} of 0.8 V assumed. Determine and sketch the ripple load voltage. What is its peak-to-peak value?

COMPUTER SIMULATIONS

C6.1 A solution of Problem 6.15 is desired using a SPICE simulation. Assume a value of I_s for the diodes of the bridge rectifier that results in a diode voltage of approximately 0.8 V for the peak diode current ($n_F = 1.4$).

   a) Determine $v_{Load}(t)$, $i_{Load}(t)$, and I_{av} of the circuit for an ideal transformer. What is the peak current of the diodes?

b) Consider the case for a transformer with a series resistance of 0.2 Ω and a series inductance of 0.5 mH. Determine $v_{Load}(t)$, $i_{Load}(t)$, and I_{av} for this condition. What is the peak current of the diodes?

C6.2 Obtain a solution for the half-wave voltage doubler circuit of Problem 6.19 using a SPICE simulation. The input is a 120-V (rms) utility voltage source with a frequency of 60 Hz. Component values are $C = 10\ \mu F$ and $R_L = 20\ k\Omega$. Assume a value of I_s for the diodes that results in a voltage of 0.8 V for a peak current of 100 mA ($n_F = 1.2$).

a) Determine the three voltages of the circuit, $v_A(t)$, $v_B(t)$, and $v_C(t)$ for steady-state conditions. What is the average load voltage and current?

b) Consider the case for a second capacitor (10 μF) connected in parallel with R_L. Determine the three voltages of the circuit for this condition. What is the average load voltage and current? What is the peak-to-peak ripple of the load voltage? What are the voltage ratings required for the capacitors?

c) Determine the capacitances required to reduce the peak-to-peak ripple load voltage to 5 V.

C6.3 In Example 6.6, a rectifier and filter circuit were designed to produce an average load voltage of 12 V at a current of 0.5 A. The transformer used for the circuit has a series resistance of 1 Ω and a series inductance of 2 mH. The diodes may be assumed to have a value of I_s that results in a voltage of 0.75 V for a current of 0.5 A (the average current).

a) Determine the average and peak-to-peak values of the load voltage for a filter capacitance C of 10,000 μF. What is the peak diode current?

b) Repeat part (a) for $C = 20,000\ \mu F$.

C6.4 The SPICE diode simulation model includes a reverse breakdown effect that produces a diode current component $i_{Dbreakdown}$ given by

$$i_{Dbreakdown} = -I_{breakdown}e^{-(v_D + V_{breakdown})/V_T}$$
$$\text{IBV} = I_{breakdown}, \qquad \text{BV} = V_{breakdown}$$

a) Using the default value for IBV, determine v_D for reverse currents of 10 and 200 mA for a diode with a breakdown voltage of 12 V. Determine the effect of specifying a breakdown current of 10 mA.

b) Simulate the behavior of the zener diode regulator of Example 6.7 for a zero load current and a load current of 100 mA. Determine the maximum and minimum values of load voltage for the maximum and minimum supply voltages of Example 6.7.

C6.5 The band-gap voltage reference of Figure 6.41 has the following transistor and circuit parameters:

$$\beta_F = 100, \qquad I_s = 1 \times 10^{-15}\ \text{A}, \qquad n_F = 1$$
$$R_1 = 500\ \Omega, \qquad R_2 = 3\ k\Omega, \qquad R_3 = 140\ \Omega, \qquad R_4 = 1\ k\Omega$$

a) Determine the dependence of V_{Ref} on V_{CC} (0 to 10 V) for temperatures of 27, 54, 81, and 128 °C.

b) Consider the case for $V_{CC} = 5$ V. Through a trial and error process, determine the value of R_2 that minimizes the temperature dependence of V_{Ref}.

C6.6 The transformer and rectifier circuit of Figure 6.23 are used with the electronic regulator of Problem 6.31 (Figure P6.31). Use a value of 24 V for V_m and determine the steady-state, peak-to-peak ripple of v_{Supply}, v_{Load}, and the base-ground voltage of the transistor. Use a value of I_s for the transistor that results in a base–emitter voltage of 0.8 V for a collector current of 120 mA ($n_F = 1.2$).

C6.7 Determine the dependence of the average value of v_{Load} on the load current of Simulation C6.6 using load resistances of 50, 100, 500, 1000, and 2000 Ω. On the basis of these data, what are the approximate Thévenin voltage and resistance of the circuit?

DESIGN EXERCISES

D6.1 Design a transformer power supply (full-wave rectification, 60-Hz power line frequency) with a capacitor filter to provide an average load voltage of 15 V and a current of 200 mA. Although the peak-to-peak load ripple voltage is to be no greater than 0.5 V, the smallest standard-size capacitor is to be used (nominal values with multiples of 1, 2.2, 4.7). What is the approximate turns ratio of the transformer required for a primary voltage of 120 V (rms)? What is the approximate load voltage of the supply for a zero load current?

D6.2 Design a power supply regulated by a zener diode to replace a 9-V battery used to power a small radio. Assume that a maximum current of 50 mA is required and that the peak-to-peak ripple load voltage is not to exceed 10 mV. A transformer-powered (60 Hz) bridge rectifier is desired, and the power dissipated by the filter-regulator circuit should not exceed 50-percent of that supplied to the load.

D6.3 Design a rectifier, filter, and electronic regulator circuit that will produce an output voltage of 9 V for a current of 0 to 250 mA.

D6.4 Design a rectifier, filter, and electronic regulator circuit that will produce an output voltage of 15 V for a current of 0 to 100 mA.

D6.5 Modify the design of D6.4 so that a variable output voltage of 0 to 15 V is obtained. Use a standard adjustable electronic regulator and a standard-value potentiometer for the voltage adjustment.

D6.6 Design an op amp power supply that has outputs of $+12$ V and -12 V for currents of at least 50 mA. The rectifier circuit using a center-tapped transformer (Figure 6.6) will provide positive and negative rectified voltages if a second set of diodes is used. (The diodes have interconnections identical to those of a bridge rectifier.) Use two electronic regulators for the design.

D6.7 A rechargeable battery of 10 nickel–cadmium cells is used in the emergency power supply of Figure 6.48 (60-Hz utility supply). The load current is 50 mA. High capacity D cells with a 4.0 Ah charge capacity and an equivalent internal resistance of 20 mΩ are available. To maintain the battery charge, an average charging current of 80 mA is required, and the minimum instantaneous charging current is not to be less than 20 mA. The peak value of v_{Supply}, V_p, is 18 V.

a) Determine the values of R_1 and C required (maximum R_1 and minimum C).

b) What is the length of time for which a utility outage can be tolerated with the cells not being discharged by more than 50 percent of their fully charged capacity?

FABRICATION OF INTEGRATED CIRCUITS

Modern electronic systems depend on the technology developed for fabricating integrated circuits. Integrated circuit technologies have achieved complexities unimaginable using discrete components and have made economical mass production possible. Electronic circuit design and integrated circuit fabrication require highly specialized, distinct areas of technological expertise. However, to produce optimal integrated circuits, it is necessary that the practitioners of these two technologies interact. The result of this joint effort has been a cornucopia of general purpose and specialized integrated circuits.

An understanding of physical and chemical processes utilizing high-vacuum techniques and high-temperature reactions is required for designing and fabricating integrated circuits. Moreover, the physical dimensions and tolerances of integrated circuit elements are much smaller than those utilized by more conventional engineering disciplines. Thus, the design and fabrication of integrated circuits is an extremely challenging endeavor.

Figure A.1 provides a perspective on commonly used physical dimensions. For measurements associated with everyday activities, we tend to think in terms of either inches and feet or centimeters and meters, depending on our cultural background. For the elements of an integrated circuit, these are very large dimensions. On a logarithmic scale, integrated circuit dimensions, which are generally expressed in microns (μm, 10^{-6} m), tend to fall midway between the size of atoms and everyday measurements. Through conventional machining techniques, tolerances of 0.001 in., $\approx$25 μm, are not uncommon, whereas the smallest dimension of a typical machined item might be only 0.01 in., $\approx$0.25 mm. As a comparison, the wavelengths of visible light are centered around an approximate value of 0.5 μm, which is a dimension comparable to the smallest dimensions of modern integrated circuit devices. To place this in perspective, the smallest integrated circuit dimensions are about a thousand times larger than the "size" of atoms and one thousand times smaller than those of extremely small, conventionally machined parts.

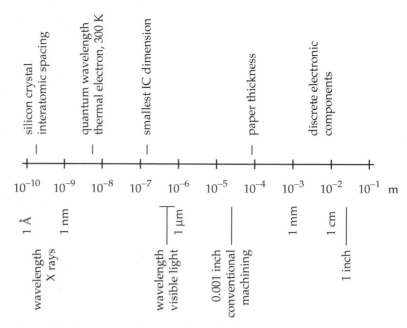

Figure A.1: Perspective of commonly used physical dimensions.

A.1 INTEGRATED CIRCUIT TRANSISTORS

Although gallium arsenide is used for specialized integrated circuits, most commercial circuits are fabricated from silicon. As indicated in Figure A.2, individual integrated circuits (generally hundreds) are fabricated on a single silicon wafer. Because the depth of the electronic components fabricated on the wafer is generally considerably less than the wafer's thickness, most of the wafer's thickness serves as the physical support of the components. Individual integrated circuits, simultaneously fabricated, are separated by either a sawing or laser cutting operation. A BJT-type integrated circuit, depending on its application, may have 10 to 1000 or more individual transistors. Typical dimensions of BJT devices are 10 to several 10s of microns. Analog and digital BJT integrated circuits will have numerous resistors and possibly a few very small-capacity capacitors.

Figure A.2: Integrated circuits fabricated on a single wafer.

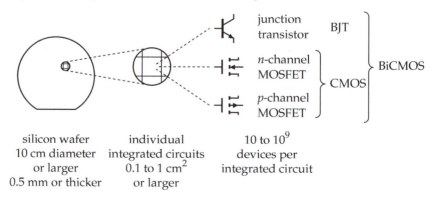

A very common logic integrated circuit design uses only *n*-channel MOSFET devices. Conventional circuits of this design may have thousands to millions of devices with device dimensions that are only a couple microns. The upper limit for the number of MOSFET devices that can be fabricated on a single integrated circuit appears to be considerably in excess of a billion. To produce these ultralarge-scale integrated circuits, device dimensions of less than a micron are required. Logic systems using *n*-channel and *p*-channel MOSFET devices, CMOS circuits, are also common. Although CMOS logic circuits are more energy efficient than those using only *n*-channel MOSFET devices, the increased complexity of fabricating both types of devices tends to result in integrated circuits with fewer devices. Another advanced integrated circuit, the BiCMOS integrated circuit, utilizes all the devices indicated in Figure A.2.

The integrated circuits are fabricated using a set of mask and optical lithography steps to define the different regions of devices. Following their fabrication, the integrated circuits are separated and tested. Leads are then attached to the good circuits, which are subsequently appropriately packaged.

A typical bipolar junction transistor is indicated in Figure A.3. This device, a planar transistor, is fabricated on a *p*-type substrate onto which an *n*-type epitaxial layer has been grown. The active transistor area is the region defined by the thin *p*-type base region below the emitter terminal; the other areas serve to connect the device terminals to the active area. For normal transistor operation, the heavily doped n^+ emitter region supplies the free electrons that cross

Figure A.3: A small-geometry *NPN* integrated-circuit transistor. The width of the device is typically less than 100 μm, and the depth of the *n*-type epitaxial layer that forms its collector is only about 15 μm.

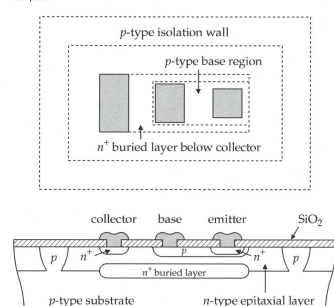

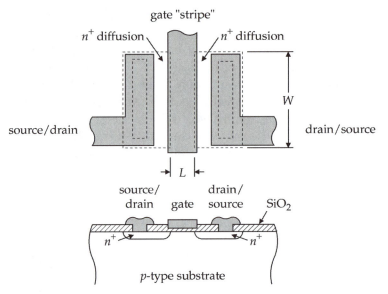

Figure A.4: Structure of a typical n-channel MOSFET integrated circuit device. For small devices, the depth of the source–drain regions is 2–3 μm or less, and the channel dimensions L and W are a few microns or less.

the forward-biased emitter–base junction in a downward direction (Figure A.3). Immediately below this junction is the collector–base junction which, for normal operation, is reverse-biased. The n^+ buried region, which extends from below the collector terminal to the active area of the transistor, ensures a low-resistance connection to the collector. The heavily doped n^+ region below the collector terminal results in an "ohmic" connection to the lightly doped n-type collector region. The transistor is surrounded by a p-type isolation wall that joins the underlying substrate. The substrate is connected to the most negative potential of the circuit. Hence, the junctions formed between the n^+ buried layer and the epitaxial layer of the collector with the substrate are reverse biased, and thus the only current between the collector and the substrate is the capacitive current of the reverse-biased junction. As can be seen from Figure A.3, relatively large isolation walls are needed to isolate the collector regions of adjacent transistors.

An integrated-circuit n-channel MOSFET device is depicted in Figure A.4. Because it is not necessary to surround this device with an isolation wall, the device area is much smaller than that of a BJT device. The MOSFET device consists of two heavily doped (n^+) wells that function as the drain and source of the device. The external voltages of the circuit determine which well is the drain or source, the more positive well being the drain. The gate is above the very thin portion of the oxide layer. When the device is operating, the channel of mobile charges is below the gate and between the two wells. An enhancement device, a device with a positive threshold voltage, is shown. Alternatively, the channel region can be lightly doped to reduce the threshold voltage of the device. Although this requires an additional processing step, depletion devices formed in this fashion are desirable for many applications. The transconductance parameter of a MOSFET device depends on processing parameters – in particular the oxide thickness and the channel dimensions W and L. Generally, different channel widths W are used to fabricate devices with different transconductance values. Thus, except for W, the MOSFET devices of an integrated circuit will be identical. Very small physical dimensions, less than a micron, are realized in modern ultralarge-scale integrated circuits.

The earliest integrated circuits, introduced in the mid-1960s, utilized bipolar junction transistors. Integrated circuits with MOSFET devices did not appear until the end of the decade. Numerous integrated circuit device configurations are now common. Bipolar analog integrated circuits may not only use low-power transistors similar to the transistor of Figure A.3 but also PNP transistors, junction field-effect transistors, and power devices. Numerous processing techniques have also been introduced to improve the basic MOSFET device of Figure A.4. Furthermore, charge storage and transfer MOSFET-type circuits are now extensively used for memory, other logic functions, and video applications.

A.2 FABRICATION PROCESSES

Although transistors have regions with different densities and types of dopants, an active semiconductor, that is, the region below the oxide layer of Figures A.3 and A.4, is a single crystal. Donor or acceptor atoms are introduced into the initial semiconductor crystal lattice to form the n-type or p-type regions. Throughout the fabrication process, many steps of which are at elevated temperatures, it is necessary to preserve the single-crystal structure of the semiconductor – a structure compromised by rapid temperature changes or by contamination. Although the discussion that follows will focus on silicon transistors, many of the same fabrication techniques are used to produce gallium–arsenide transistors. To produce an integrated circuit, a number of different sequential fabrication steps – steps that may need to be repeated multiple times – are needed.

CRYSTAL GROWTH AND WAFER FABRICATION

Quartzite, a pure form of sand (SiO_2), is the primary feedstock for integrated circuits. An initial carbon-reducing reaction in a high-temperature furnace is used to produce metallurgical-grade silicon, which is then treated with hydrogen chloride to form trichlorosilane ($SiHCl_3$). A fractional distillation process is then used to reduce the concentration of impurities, and a hydrogen reduction reaction of the purified trichlorosilane yields what is known as electronic-grade silicon, a polycrystalline material.

The Czochralski process is used to grow a single crystal from the polycrystalline silicon. A seed crystal of silicon, with the desired crystal lattice orientation, is brought into contact with a silicon melt (silicon has a melting point of 1415 °C). The seed is then slowly withdrawn from the melt so the silicon that adheres to the seed crystal (which serves as a template) forms a single crystal as it solidifies. With the slow withdrawl of the crystal, the growth progresses, forming a cylindrical crystal called a boule. A boule length of 1 to 2 m with a diameter of up to 30 cm can presently be achieved. To form p-type silicon that may be desired for the substrate of an integrated circuit, an acceptor impurity, such as boron, is added to the silicon melt before the crystal is withdrawn.

One or more flats are ground the length of the boule to indicate the crystal orientation and doping type. After etching, the boule is sliced into wafers with a steel saw blade impregnated with diamond particles. To complete the preparation of the wafers, they are mechanically lapped, etched, and then polished through an electochemical polishing process.

EPITAXIAL DEPOSITION

This process consists of growing a silicon crystal onto the surface of a silicon substrate. Because the surface atoms of the substrate serve as the template for the epitaxial growth, the growth and the substrate form a single crystal. The epitaxial layer will generally have a different doping than the substrate, thereby forming a junction at the boundary of the epitaxial layer and the substrate. This process was used to grow the n-type epitaxial layer that forms the collector of the transistor of Figure A.3, while the base and emitter of the transistor, formed by subsequent doping steps, are embedded in the epitaxial layer.

A thorough cleaning of the substrate surface onto which the growth is to occur, including the removal of the surface oxide, is important. An epitaxial deposition occurs when a vapor containing silicon atoms reacts with the substrate within a high-temperature reactor (1000 to 1200 °C). Generally, hydrogen is used as a carrier for the vapor, and either silicon tetrachloride ($SiCl_4$) or silane (SiH_4), gases that are reduced by the hydrogen at the high reactor temperature, is the source of silicon atoms. A controlled quantity of acceptor or donor atoms is used to form a p- or n-type epitaxial layer. Because this is a high-temperature process, the doping atoms of the substrate tend to migrate into the epitaxial layer, forming a graded junction.

DOPING: THERMAL DIFFUSION

A high-temperature diffusion reaction is frequently used to introduce acceptor or donor atoms into a crystal lattice. This is the process by which the base and emitter regions of the transistor of Figure A.3 were doped and the drain and source wells of the MOSFET device of Figure A.4 were formed. At elevated temperatures (above 1000 °C), a sufficient number of atoms of the crystal will have enough energy to break their lattice bonds, thereby providing vacancies for the dopant atoms. As a result, the diffusion coefficient of donor or acceptor atoms, such as phosphorus or boron, is adequate for doping the silicon crystal. Nitrogen is used to carry the dopants to the surface of the silicon wafer.

The diffusion profile of the dopant's atoms in the silicon wafer depends on the wafer temperature during the diffusion and the time interval over which it was exposed to the dopant atoms. Since diffusion occurs at elevated temperatures, careful control of the subsequent high-temperature reactions that may be required is necessary as the dopant atoms have a tendency to migrate.

DOPING: ION IMPLANTATION

Although thermal diffusion is useful for forming moderate- and large-size transistor regions, it does not provide the control necessary for producing well-defined regions required for modern small-size devices. Furthermore, it is difficult to produce lightly doped regions. Ion implantation, a process considerably more complex and expensive than diffusion, provides a means of producing accurately controlled, well-defined, doped regions. Implantation, a vacuum process, is carried out by bombarding the silicon wafer surface with high-energy dopant ions (10 to 200 keV).

The source of ions is generally a feed gas such as BF_3, AsH_3, or PH_3 that contains a gaseous compound of the desired dopant. A hot filament produces electrons which, after being accelerated, ionize the feed gas. A mass spectrometer is then used to separate the desired implant ions from impurities. The high-energy ions, at the same time as being embedded in the crystal lattice, may also result in significant damage to the crystal structure. Hence, annealing at a moderate temperature (500 to 600 °C) is necessary to mitigate the lattice defects and dislocations caused by the ion implantation. Because only a moderate annealing temperature is required, undesirable reactions that may occur at higher temperatures, such as the migration of dopant atoms, are avoided.

THERMAL OXIDATION

A thin oxide layer is used to protect the surface of an integrated circuit as well as for the gate region of MOSFET devices. In addition, an oxide layer is frequently grown following a high-temperature processing step such as thermal diffusion to protect the surface before the next step. Because the oxide is formed from the surface layer of the silicon wafer, the transition region that forms the surface of the transistor, is not exposed to atmospheric contaminants as a result of this process. Hence, surface defects that tend to degrade the performance of transistors are minimized.

Either pure oxygen (dry oxidation) or a mixture of oxygen and water vapor (wet oxidation) is passed over the silicon surface at an elevated temperature (900 to 1200 °C). Oxygen atoms move inward into the silicon, initially forming a thin-surface oxide layer. To continue the oxidation process, that is, to obtain a thicker oxide layer, oxygen atoms must diffuse through the existing oxide layer. An hour or longer may be necessary to obtain a 1-μm thick oxide layer.

OPTICAL LITHOGRAPHY

For a processing step such as introducing dopant atoms by diffusion or ion implantation, it is necessary to define the precise boundaries for the process. For the very small dimensions utilized, this is achieved through an optical lithography process. The integrated circuit's surface with the transistor and wiring details (for example, the top view of Figure A.3 or A.4) may be laid out at a convenient size, several hundred times the actual size, and then photographically reduced. Individual masks are required for the different processing steps; for example,

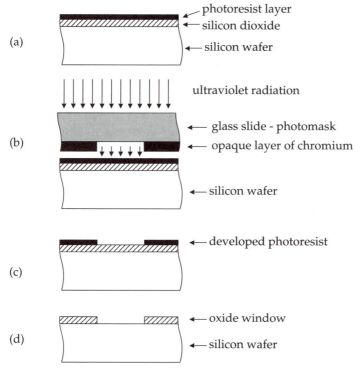

Figure A.5: The photomasking process.

one to form the emitter, another the base, and still another the buried collector layer of the transistor of Figure A.3. The image of an integrated circuit, after being photographically reduced, is replicated and transferred to a single mask that has the image of the entire wafer (generally hundreds of integrated circuits). The photomasks will be either the size of the integrated circuit's image ($1\times$) or a few times that of the image (possibly $5\times$ or $10\times$).

The masking process by which boundaries are defined relies on a photoresist, a thin photosensitive film that is applied uniformly to the entire surface of the silicon wafer and hardened by baking at a low temperature. When the photoresist is exposed to ultraviolet light, a chemical reaction occurs, thus differentiating regions that have been exposed from those that have not. In Figure A.5(a), a photoresist has been applied to an oxide layer that has previously been grown on a silicon wafer. As shown in Figure A.5(b), the photomask is placed directly above the silicon wafer below a source of ultraviolet light. On the basis of the process used, the mask may either be in contact with the wafer or there may be a slight gap between the mask and the wafer. Alternatively, an optical system may be used if the mask image is to be reduced. The opaque layer of chromium, photographically formed when the mask was fabricated, allows the ultraviolet radiation to strike selected areas of the photoresist. For the photoresist illustrated, the developing reaction removes the photoresist from the exposed region (Figure A.5(c)). This photoresist is designated as being positive, whereas an alternative-type photoresist in which developing removes the unexposed region is designated as negative.

The exposed oxide is then removed by etching, resulting in the oxide "window" of Figure A.5(d) through which, for example, doping atoms can be introduced. The unexposed photoresist is removed by what is known as a "stripper."

To illustrate a set of photolithographic processing steps, the bipolar junction transistor of Figure A.3 will be considered. Before the growth of the epitaxial layer, an oxide is grown on the wafer, a p-type substrate. A photomask, which defines the n^+ buried layer, is used to produce an oxide window corresponding to this layer. A thermal diffusion process is then used to form the heavily doped n-type region in the substrate, the oxide layer is removed by etching, and an n-type epitaxial layer is grown over the entire wafer. During this process, the donor atoms of the previously formed, heavily doped, n-type region will tend to migrate into the epitaxial layer. Again, the surface of the wafer, now the upper surface of the epitaxial layer, is oxidized. This is followed by a photomasking process that defines the p-type isolation walls. After another thermal diffusion process that forms the walls, the oxidation and photomasking steps are repeated to form the the shallower p-type base region of the transistor. Still another oxidation and photomask will be used to form the n^+ emitter region and the collector contact. Although several repetitions using different photomasks are necessary to form the various regions of an individual transistor, thousands, if not millions of transistors are formed simultaneously on the silicon wafer.

ETCHING

Etching is the process by which an entire surface layer or selected regions of a layer that have been determined by a photmasking process may be removed. A wet etching process, using a dilute hydrofluoric acid solution can be used to remove silicon dioxide. The dilute acid solution is highly selective, that is, it dissolves the silicon dioxide but has little effect on the underlying silicon wafer. The silicon wafer of Figure A.5(c) could be etched using this process to produce the oxide window of Figure A.5(d). Silicon nitride, aluminum, and polycrystalline layers can also be removed by wet etching with appropriate etchant solutions. An alternative to wet etching is a dry process known as plasma etching. Plasma etching not only provides a much greater degree of control but is less sensitive to different ambient conditions. Because much smaller features can be obtained, plasma etching tends to be preferred for modern integrated circuits.

THIN FILMS

For early integrated circuits, metallic films that formed contacts and interconnections were obtained by evaporation. This technique has been replaced by a sputtering process that results not only in a much better coverage of the film, but can also be used to deposit metal alloys. An evacuated chamber is required for the sputtering process. The silicon wafer is located near a "target" that contains the desired film material, for example, aluminum. A low-pressure inert gas is introduced, and, as a result of a potential difference between the wafer and the target, a plasma discharge is initiated. Energetic ions of the inert gas strike the target, causing atoms of the target to be ejected and deposited on the silicon wafer. Another widely used process, particularly for forming thicker films, is chemical vapor deposition. In addition to metals, aluminum oxide, dielectrics, and resistive films may be deposited by this process. If the process is carried a step further, it is possible using appropriate processing steps to deposit overlapping conducting layers insulated from each other.

A.3 SUMMARY

Although only two very basic transistor structures were considered in the previous discussion, integrated circuit fabrication techniques have been developed for producing more sophisticated types of transistors, and there are additional, more advanced processing techniques that are utilized in fabricating modern integrated circuits. Several general-coverage electronic and semiconductor texts have brief discussions of fabrication techniques (e.g., Elmasry 1983; Gray and Meyer 1992; Grebene 1984; Soclof 1991; Taur and Ning 1998; Sze 1985, 1998; Tsividis 1987). Furthermore, texts that deal explicitly with fabrication technologies are also available (e.g., Campbell 1996; Chang and Sze 1996; and Jaeger 1988).

Campbell (1996) starts his preface with the word "magic" to characterize the fabrication process. Not only might modern integrated circuits appear to be magic to those not familar with their fabrication, but they may also appear to be so to those intimately involved with their fabrication. Given the number of unique processing steps that rely on different scientific disciplines, an expert in one area may feel totally inadequate in another. To provide an historical perspective, consider the situation of those who struggled to fabricate the first somewhat erratically working transistor at the Bell Telephone Laboratories. A suggestion at that time, or even over the next decade, of simultaneously fabricating a billion interconnected transistors on a single integrated circuit, would have been viewed as pure fantasy. But, the fantasy has been realized.

REFERENCES

Campbell, Stephen A. (1996). *The Science and Engineering of Microelectronic Fabrication.* New York: Oxford University Press.

Chang, C. Y. and Sze, S. M. (Eds.) (1996). *ULSI Technology.* New York: McGraw–Hill.

Elmasry, M. I. (1983). *Digital Bipolar Integrated Circuits.* New York: John Wiley & Sons.

Gray, P. R. and Meyer, R. G. (1992). *Analysis and Design of Analog Integrated Circuits* (3d ed.). New York: John Wiley & Sons.

Grebene, A. B. (1984). *Bipolar and MOS Analog Integrated Circuit Design.* New York: John Wiley & Sons.

Jaeger, R. C. (1988). *Introduction to Microelectronic Fabrication.* Reading, MA: Addison–Wesley.

Soclof, S. (1991). *Design and Applications of Analog Integrated Circuits.* Englewood Cliffs, NJ: Prentice–Hall.

Sze, S. M. (1985). *Semiconductor Devices: Physics and Technology.* New York: John Wiley & Sons.

Sze, S. M. (Ed.) (1998). *Modern Semiconductor Device Physics.* New York: John Wiley & Sons.

Taur, Y. and Ning, T. H. (1998). *Fundamentals of Modern VLSI Devices*. Cambridge, U.K.: Cambridge University Press.

Tsividis, Y. P. (1987). *Operation and Modeling of the MOS Transistor*. New York: McGraw–Hill.

THE DESIGN PROCESS

The design process cannot be readily characterized. That is, there is not a set of step-by-step rules by which a successful design can be realized. Furthermore, even after a design that satisfies all requirements is completed, it may be difficult to judge whether it is indeed optimal. It is not uncommon at a project's completion for those involved to express the thought that if only they had done it another way, it would have been so much easier – the benefit of hindsight.

When discussing design, we often think in terms of large, complex systems. But the design of such a system might be the result of an individual's intense effort spanning several years such as the development of wide-band frequency-modulated broadcasting by Armstrong in the 1930s (Armstrong 1940). Alternatively, a design might be the result of the intense effort of a team of scientists and engineers as was, for example, the design of the compact audio disc, which was the joint effort of two normally competing corporations, Sony and Philips, on opposite sides of the world (Miyaoka 1984). A large design project, however, involves solving many small design problems – problems that at first glance may seem to be trivial but on carrying out the design prove to be otherwise. At times, an overall design may need to be modified because one or more of its components cannot be realized. The designs discussed in this appendix will be relatively basic – basic to the point that at first glance they may seem trivial. But these designs, although seemingly trivial, will illustrate principles that may be applicable to the design of other, considerably more complex systems.

It is obvious that the way one approaches the design of a system depends on the particular system. But it also depends on one's previous design experience as well as one's knowledge of published literature. Furthermore, individual temperament is important. Does one have a bent toward an analytic approach or toward a simulation approach? Very likely, both will be necessary, but the amount of effort spent on each will depend on the individual or individuals involved. Although simulations are attractive, the validity and insight that can be gained through an analytic solution can be important. Even though seldom mentioned, a knowledge of designs that did not work can be particularly valuable. Unfortunately, it is necessary to build up this body of knowledge on one's own because it is not generally deemed appropriate to publish negative results. Finally, it is worthwhile to spend some time thinking about a problem (as is commonly said, "sleep on it") before beginning. All too frequently one is either tempted to jump in immediately or may be pressured to do so to obtain a piece of hardware – but often a piece that may likely need to be redesigned.

A few simple design problems related to individual chapters of the text follow. These designs rely on the discussion of the particular chapter – not on techniques that one might use based on a more extensive knowledge of electronics gained through additional course work and experience. The design specifications may appear very limited; however, in proceeding through the design one should find that the methods employed will apply to a much wider set of design problems.

B.1 BIPOLAR JUNCTION TRANSISTOR CIRCUITS (CHAPTER 3)

A SINGLE-TRANSISTOR LOGIC INVERTER

For the transistor circuit of Figure 3.36, $V_{CC} = 5$ V, $V_{BB} = -3$ V, $\beta_F = 100$, and $v_{BE(\text{on})} = 0.7$ V. Furthermore, there is an output load capacitance C_L of 10 pF. For a no-load condition, the midpoint of the output voltage transition is to occur for $v_{IN} = 2.5$ V and is to have a slope of -5. For an abrupt change of the input voltage, the time for the output voltage to change to its midvalue of approximately $V_{CC}/2$ is to be no greater than 20 ns. Design a circuit that has a minimum power dissipation.

DESIGN

Two equations of Section 3.3, Eq. (3.21) for the dependence of v_{OUT} on v_{IN} and Eq. (3.22) for the slope of the response, are needed. The expression for the slope yields a numerical value for R_C/R_{B1} as follows:

$$\text{slope} = -\beta_F R_C/R_{B1}, \qquad \frac{R_C}{R_{B1}} = \frac{\text{slope}}{-\beta_F} = .05 \tag{B.1}$$

Equation (3.21) for v_{OUT} is used to obtain an expression involving the other base resistor, namely R_C/R_{B2}. It should be noted that, on the basis of the design requirement, $v_{OUT} = 2.5$ V for $v_{IN} = 2.5$ V.

$$v_{OUT} = V_{CC} - \left(\frac{R_{B2}v_{IN} + R_{B1}V_{BB}}{R_{B1} + R_{B2}} - v_{BE(\text{on})} \right) \frac{\beta_F R_C}{R_{Th}}$$

$$\text{where } R_{Th} = \frac{R_{B1}R_{B2}}{R_{B1} + R_{B2}} \tag{B.2}$$

Equation (B.2) can be manipulated to yield a value for R_C/R_{B2} as follows:

$$v_{OUT} = V_{CC} - \beta_F \left[v_{IN} \frac{R_C}{R_{B1}} + V_{BB} \frac{R_C}{R_{B2}} - v_{BE(\text{on})} \left(\frac{R_C}{R_{B1}} + \frac{R_C}{R_{B2}} \right) \right]$$

$$\left(V_{BB} - v_{BE(\text{on})} \right) \frac{R_C}{R_{B2}} = \frac{V_{CC} - v_{OUT}}{\beta_F} - \left(v_{IN} - v_{BE(\text{on})} \right) \frac{R_C}{R_{B1}} \tag{B.3}$$

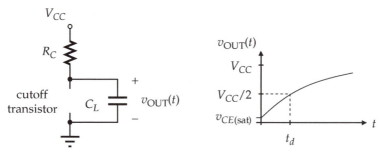

Figure B.1: Transient behavior for an input downward transition.

Upon introducing the voltage values and 0.05 for R_C/R_{B1}, a value is obtained for R_C/R_{B2} as follows:

$$R_C/R_{B2} = 0.0176 \tag{B.4}$$

If, for example, $R_C = 1\ k\Omega$, then $R_{B1} = 20\ k\Omega$ and $R_{B2} = 56.9\ k\Omega$. On the basis of static considerations alone, the resistance values can be scaled (for example, all doubled or halved) without affecting the static transfer characteristic of the gate.

As a result of the load capacitance, the dynamic response of the gate depends on the actual resistance values of the circuit. For an abrupt downward transition of v_{IN}, the transistor will be cut off, and its collector resistor will determine the time constant of the concurrent upward transition of v_{OUT} (Figure (B.1)). If $v_{CE(sat)}$ is ignored ($v_{CE(sat)} \approx 0$), the following is obtained for v_{OUT} if it is assumed that the input voltage has a downward transition at $t = 0$:

$$v_{OUT} = V_{CC}\left(1 - e^{-t/R_C C_L}\right) \qquad t \geq 0 \tag{B.5}$$

The time delay required for v_{OUT} to reach 2.5 V, that is $V_{CC}/2$, can now be obtained as follows:

$$\begin{aligned} V_{CC}/2 &= V_{CC}\left(1 - e^{-t_d/R_C C_L}\right) \\ t_d &= R_C C_L \ln 2 \end{aligned} \tag{B.6}$$

$$R_C = \frac{t_d}{0.693\ C_L} = 2.886\ k\Omega$$

The preceding value for R_C is obtained by introducing a delay time t_d of 20 ns. This results in the following resistance values for the base resistors:

$$R_{B1} = 57.7\ k\Omega, \qquad R_{B2} = 164\ k\Omega \tag{B.7}$$

If smaller resistance values are used, the power dissipated by the gate will be greater. An upward transition of v_{IN} that results in a downward transition of v_{OUT} may be shown to result in a transition time of less than 20 ns.

A SINGLE-TRANSISTOR SMALL-SIGNAL AMPLIFIER

A small-signal transistor amplifier is to be designed to supply an output signal voltage to a load with an equivalent resistance R_L of 10 kΩ. The input voltage source has an equivalent resistance of 25 kΩ, and the supply voltage V_{CC} is 10 V.

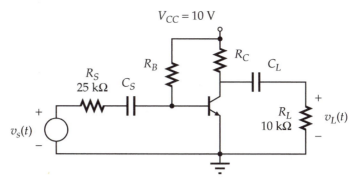

$V_{CC} = 10$ V

R_S
25 kΩ C_S

R_B

R_C

C_L

R_L
10 kΩ

$v_L(t)$

$v_s(t)$

Figure B.2: Basic transistor circuit.

The voltage gain is to have a magnitude of at least 10 and is to be reasonably uniform for sinusoidal signals with a frequency range of 50 to 20 kHz. Assume that the transistor used for the amplifier could have a current transfer ratio β_F in the range of 75 to 150 and that $n_F = 1.0$. The discussion of Section 3.5 may be considered sufficient for the design.

DESIGN

A transistor with a midvalue for β_F, that is, $\beta_F = 112$, is considered for the initial design. If necessary, the circuit will be modified to accommodate the range of transistor values for which it is to function.

The design will be based on the circuit of Figure B.2. The corresponding small-signal equivalent circuit is presented in Figure B.3. A value for the collector resistor R_C that is equal to R_L (10 kΩ) is a reasonable design choice. A quiescent collector–emitter voltage V_{CE} of $V_{CC}/2$ will tend to allow for the largest output signal variation and will tend to be least sensitive to variations in β_F. These considerations establish a design value for the quiescent value of the collector current I_C:

$$V_{CE} = V_{CC} - I_C R_C, \qquad I_C = \frac{V_{CC} - V_{CE}}{R_C} = 0.5 \text{ mA} \tag{B.8}$$

For a transistor with $\beta_F = 112$, a quiescent base current I_B of 4.46 μA is required. The base resistor R_B determines this current as follows:

$$R_B = \frac{V_{CC} - v_{BE(on)}}{I_B} = 2.09 \text{ M}\Omega \tag{B.9}$$

The quiescent base current also determines the transistor equivalent resistance

Figure B.3: Small-signal equivalent circuit of the basic transistor amplifier of Figure B.2. The equivalent device capacitance C_μ of Figure 3.77 is assumed to be small enough that its effect can be ignored.

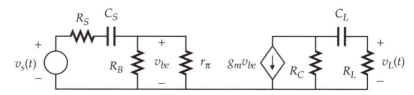

R_S C_S

$v_s(t)$

R_B v_{be} r_π $g_m v_{be}$ R_C R_L $v_L(t)$

C_L

r_π of the small-signal equivalent circuit (Eq. (3.63) because $r_\pi = r_{be}$ for the approximations of this section):

$$r_\pi = \frac{n_F V_T}{I_B} = 5.61 \text{ k}\Omega \tag{B.10}$$

A nominal value of 25 mV is used for V_T. Equation (3.69) is used to obtain a value for the mutual conductance g_m as follows:

$$g_m = \frac{I_C}{n_F V_T} = 20 \text{ mS} \tag{B.11}$$

The voltage gain of the amplifier is determined using the small-signal equivalent circuit of Figure B.3. It is assumed that the capacitances C_S, and C_L are sufficiently large that they may be treated as short circuits. Because $R_B \gg r_\pi$ (2.08 MΩ compared with 5.61 kΩ), its effect will be ignored.

$$v_{be} = \frac{r_\pi v_s(t)}{r_\pi + R_S}$$

$$v_L(t) = -g_m v_{be} R_C \| R_L = -\frac{g_m r_\pi v_s(t) R_C \| R_L}{r_\pi + R_S} \tag{B.12}$$

$$v_L(t) = -16.3 \, v_s(t)$$

Hence, the magnitude of the small-signal gain of 16.3 exceeds the minimum design requirement of 10.

Capacitance values must now be determined. Although a detailed consideration of the effect of finite capacitances is beyond the introductory treatment of the chapter, reasonable "estimates" for the capacitance can be obtained. For a sinusoidal signal, it is the reactance of the capacitance that determines the response of a circuit:

$$X_S = -\frac{1}{2\pi f C_S}, \qquad X_L = -\frac{1}{2\pi f C_L} \tag{B.13}$$

The magnitude of these quantities will be the largest for the lowest frequency of interest f_ℓ (50 Hz for the design specification). Hence, the magnitudes of the reactances should be small compared with the resistances to which they are connected. This implies the following:

$$\frac{1}{2\pi f_\ell C_S} \ll R_S, \qquad \frac{1}{2\pi f_\ell C_L} \ll R_L \tag{B.14}$$

Consider the case for which the preceding relations are equal.

$$C_S = \frac{1}{2\pi f_\ell R_S} = 0.064 \ \mu\text{F}$$

$$C_L = \frac{1}{2\pi f_\ell R_L} = 0.159 \ \mu\text{F} \tag{B.15}$$

Because the response of two circuits is involved, a doubling of capacitance values should result in an acceptable response. Doubling and using the next largest standard capacitance values yields $C_S = 0.2 \ \mu\text{F}$ and $C_L = 0.5 \ \mu\text{F}$. Furthermore, a standard resistance value of 2.0 MΩ is used for R_B.

To complete the design considerations, the effect of transistors with different values of β_F must be determined. A 2.0-MΩ resistance for R_B results in a quiescent base current of 4.65 μA and a corresponding value of 5.38 kΩ for r_π. These values do not depend on the transistor. The following is obtained for transistors with extreme values of β_F:

$$
\begin{aligned}
&\beta_F = 75 && \beta_F = 150 \\
&I_C = 0.349 \text{ mA} && I_C = 0.698 \text{ mA} \\
&V_{CE} = 6.51 \text{ V} && V_{CE} = 3.02 \text{ V} \\
&v_L(t) = -12.3\, v_s(t) && v_L(t) = -24.7\, v_s(t)
\end{aligned}
\qquad\qquad \text{(B.16)}
$$

Because transistors with extreme values of β_F result in an acceptable response, the design conditions are fulfilled.

B.2 METAL-OXIDE FIELD-EFFECT TRANSISTOR (CHAPTER 4)

BIASING A MOSFET CIRCUIT

For a small-signal amplifier, the quiescent drain current of the device should have a minimal dependence on the device's parameters. A source resistor, as shown in Figure (B.4), may be used to minimize the dependence. Determine values of the gate biasing resistors R_{G1} and R_{G2} and of the source resistor R_S that result in a drain current that does not vary by more than ± 10 percent from a nominal value of 0.25 mA. A minimum value of R_S is desired, and the parallel combination of the gate resistors is to be approximately 1.0 MΩ. Because the substrate of the device is connected to ground, its threshold voltage will depend on the quiescent source–substrate voltage.

DESIGN

Because the nominal quiescent drain current I_D is 0.25 mA, the drain–ground voltage V_{DG} is 5.0 V. For a ± 10-percent variation of I_D, the variation in V_{DG} is ∓ 0.5 V. A trial and error method of solution is necessary. Although a strictly analytic solution could be utilized, SPICE simulations used in conjunction with analytic calculations will minimize the overall effort.

Figure B.4: A MOSFET circuit with a source resistor.

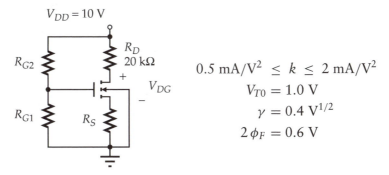

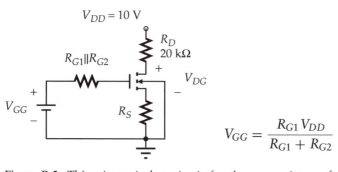

$$V_{GG} = \frac{R_{G1} V_{DD}}{R_{G1} + R_{G2}}$$

Figure B.5: Thévenin equivalent circuit for the gate resistors of Figure B.4.

The gate resistors of Figure B.4 may be replaced by a Thévenin equivalent circuit (Figure B.5). Because the quiescent gate current is zero, the design process is reduced to finding values of V_{GG} and R_S. Convenient voltage values are assumed for $I_D R_S$ corresponding to a nominal quiescent drain current of 0.25 mA. If this voltage is known, the threshold voltage of the device can be determined. The next step is to calculate the quiescent gate–source voltage V_{GS} that yields the nominal drain current of 0.25 mA. A "midvalue" of 1.0 mA/V^2 is assumed for k. The equivalent voltage V_{GG} is $V_{GS} + I_D R_S$. To determine the drain currents for devices with $k = 0.5$ mA/V^2 and 2.0 mA/V^2, a SPICE simulation can be used.

To begin, a value of 1.0 V is tried for $I_D R_S$. This corresponds to a value of 4 kΩ for R_S because $I_D = 0.25$ mA. To determine the threshold voltage V_T, Eq. (4.53) is used ($v_{SB} = I_D R_S = 1.0$ V).

$$V_T = V_{T0} + \gamma \left(\sqrt{v_{SB} + 2\phi_F} - \sqrt{2\phi_F} \right)$$
$$= 1.196 \text{ V} \tag{B.17}$$

If the device is assumed to be in its saturated region of operation, Eq. (4.7) is used to determine the quiescent gate–source voltage as follows:

$$I_D = \frac{k}{2}(V_{GS} - V_T)^2$$

$$V_{GS} - V_T = \sqrt{\frac{2 I_D}{k}} = 0.707 \text{ V} \tag{B.18}$$

$$V_{GS} = 1.903 \text{ V}$$

This requires a value of 2.903 V for V_{GG}. The MOSFET circuit and the corresponding SPICE file for a simulation are given in Figure B.6. The circuits with devices MA and MC yield the quiescent currents for $k = 0.5$ mA/V^2 and 2.0 mA/V^2, respectively. The circuit with device MB provides a check for $k = 1.0$ mA/V^2. The following is obtained from the output file of the SPICE simulation:

$k = 0.5$ mA/V^2	$k = 1.0$ mA/V^2	$k = 2.0$ mA/V^2
$I_D = 0.207$ mA	$I_D = 0.250$ mA	$I_D = 0.287$ mA
$V_{DG} = 5.87$ V	$V_{DG} = 5.00$ V	$V_{DG} = 4.26$ V

This solution is not acceptable because the variation in I_D is too large.

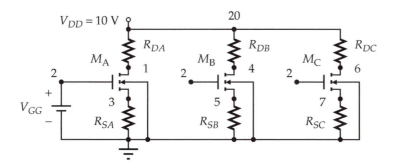

```
MOSFET Bias
VGG   2   0   2.903
VDD   20  0   10
.OP

MA    1   2   3   0   MOSA
RDA   20  1   20K
RSA   3   0   4K

MB    4   2   5   0   MOSB
RDB   20  4   20K
RSB   5   0   4K

MC    6   2   7   0   MOSC
RDC   20  6   20K
RSC   7   0   4K

.MODEL   MOSA   NMOS   KP=.5E-3 VTO=1 PHI=.6 GAMMA=.4

.MODEL   MOSB   NMOS   KP=1E-3 VTO=1 PHI=.6 GAMMA=.4

.MODEL   MOSC   NMOS   KP=2E-3 VTO=1 PHI=.6 GAMMA=.4

.END
```

Figure B.6: SPICE circuit and simulation file for the MOSFET circuit with a source resistor.

Because a larger value of R_S is needed, a circuit will be examined that has a value of 1.5 V for $I_D R_S$ ($R_S = 6$ kΩ because $I_D = 0.25$ mA). A value of 1.270 V is obtained for V_T and 3.477 V for V_{GG}. A SPICE simulation for these values yields the following:

$k = 0.5$ mA/V^2	$k = 1.0$ mA/V^2	$k = 2.0$ mA/V^2
$I_D = 0.217$ mA	$I_D = 0.250$ mA	$I_D = 0.277$ mA
$V_{DG} = 5.66$ V	$V_{DG} = 5.00$ V	$V_{DG} = 4.47$ V

A still larger value is needed for R_S.

For $I_D R_S = 2.0$ V ($R_S = 8$ kΩ because $I_D = 0.25$ mA), a value of 1.335 V is obtained for V_T and 4.042 V for V_{GG}. This yields the following:

$$k = 0.5 \text{ mA/V}^2 \quad k = 1.0 \text{ mA/V}^2 \quad k = 2.0 \text{ mA/V}^2$$
$$I_D = 0.224 \text{ mA} \quad I_D = 0.250 \text{ mA} \quad I_D = 0.271 \text{ mA}$$
$$V_{DG} = 5.53 \text{ V} \quad V_{DG} = 5.00 \text{ V} \quad V_{DG} = 4.58 \text{ V}$$

These results are extremely close to those required, for the overall variation in I_D is 0.047 mA (a $\pm 10\%$ variation is 0.050 mA). A slight increase in V_{GG} would make the variation in I_D symmetrical about 0.25 mA.

Values for R_{G1} and R_{G2} can now be obtained by using expressions for V_{GG} and $R_{G1} \| R_{G2}$ as follows:

$$V_{GG} = \frac{R_{G1} V_{DD}}{R_{G1} + R_{G2}}$$

$$R_{G1} \| R_{G2} = \frac{R_{G1} R_{G2}}{R_{G1} + R_{G2}} \tag{B.19}$$

$$\frac{V_{GG}}{R_{G1} \| R_{G2}} = \frac{V_{DD}}{R_{G2}}$$

$$R_{G2} = \frac{V_{DD}}{V_{GG}} (R_{G1} \| R_{G2}) = 2.47 \text{ M}\Omega$$

A design value of 1.0 MΩ is used for $R_{G1} \| R_{G2}$. An expression and value can be obtained for R_{G1} by considering the voltage across R_{G2}, namely $V_{DD} - V_{GG}$ as follows:

$$V_{DD} - V_{GG} = \frac{R_{G2} V_{DD}}{R_{G1} + R_{G2}}$$

$$\frac{V_{DD} - V_{GG}}{R_{G1} \| R_{G2}} = \frac{V_{DD}}{R_{G1}} \tag{B.20}$$

$$R_{G1} = \frac{V_{DD}}{V_{DD} - V_{GG}} (R_{G1} \| R_{G2}) = 1.68 \text{ M}\Omega$$

If the nearest standard 5-percent resistance values are used for R_{G1} and R_{G2}, 2.4 MΩ and 1.6 MΩ, respectively, the nominal value of V_{GG} is only slightly changed ($V_{GG} = 4.0$ V). Although the quiescent drain currents also shift slightly, their variation remains within a range of 0.05 mA.

B.3 NEGATIVE FEEDBACK AND OPERATIONAL AMPLIFIERS (CHAPTER 5)

A SMALL-SIGNAL AMPLIFIER

An amplifier is to be designed using an operational amplifier to supply an output signal voltage to a load with an equivalent resistance R_L of 10 kΩ. The input voltage source has an equivalent resistance of 25 kΩ, and the supply voltage

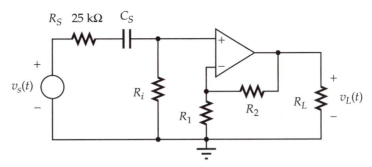

Figure B.7: Operational amplifier circuit. Both positive and negative supply voltages are necessary for the operational amplifier.

V_{CC} is 10 V. The voltage gain is to have a magnitude of at least 10 and is to be reasonably uniform for sinusoidal signals with a frequency range of 50 Hz to 20 kHz. These design requirements for an op amp circuit are the same as those for the single-transistor small-signal amplifier previously considered.

DESIGN

Although the supply voltage is only 10 V (equivalent to ±5 V supplies), it is adequate for an LF356 op amp. Alternatively, an op amp especially designed for a low supply voltage could be used. To initiate the design process, the basic noninverting amplifier circuit of Figure B.7 is considered. A capacitor C_S is in series with the input voltage source. This circuit will ensure proper operation of the amplifier if the input signal source should happen to have an offset voltage. The input resistor R_i provides a dc connection to the noninverting input of the op amp. If an op amp with field-effect input devices is used, the input resistor can be very large (1 MΩ is acceptable). Therefore, the input signal for the op amp, if the capacitance is treated as a short circuit, will be essentially $v_s(t)$. If ideal behavior is assumed for the op amp, the following is obtained for the gain of the circuit (Eq. (5.41)):

$$v_L(t) = \left(1 + \frac{R_2}{R_1}\right) v_s(t) \tag{B.21}$$

Hence, if the ratio R_2/R_1 is equal to 9, a gain of 10 will be realized. A ratio of 10 for R_2/R_1 yields a gain of 11, which is a value that should ensure an overall gain of 10 if components deviate from their nominal design values. The following resistance values are reasonable:

$$R_1 = 10 \text{ k}\Omega, \qquad R_2 = 100 \text{ k}\Omega \tag{B.22}$$

These values of resistance minimize the current that needs to be supplied by the output of the amplifier and are sufficiently small to minimize the effect of capacitive currents due to stray capacitances for the frequency range over which the amplifier needs to function.

For a single supply voltage, a circuit similar to that of Figure 5.57 is needed (Figure B.8). For a zero input signal voltage, the noninverting input voltage is $V_{CC}/2$. For the dc condition being considered, the op amp circuit is a unity gain

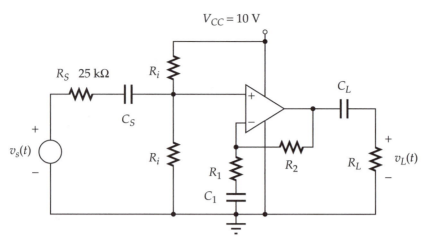

Figure B.8: An operational amplifier using a single supply.

buffer (C_1 behaves as an open circuit). Hence, the output voltage of the op amp as well as its inverting input voltage will be $V_{CC}/2$.

The amplifier circuit of Figure B.8 has three capacitances. On the basis of the discussion of Section 5.5, they result in the following break frequencies that determine the low-frequency response of the circuit:

$$f_S \approx \frac{1}{2\pi (R_i/2)C_S}, \qquad f_1 = \frac{1}{2\pi R_1 C_1}, \qquad f_L = \frac{1}{2\pi R_L C_L} \qquad \text{(B.23)}$$

The quantity $R_i/2$ for the input break frequency arises because, for small-signal behavior, the two input resistors are effectively in parallel. If these resistors have values of 1 MΩ, their parallel combination of 0.5 MΩ is large compared with R_S (25 kΩ). Because $R_i/2$ is much larger than the resistances of the expressions for the other break frequencies, an input capacitance C_S is chosen so as to have a negligible effect at the lower design frequency of 50 Hz. A value of 5 Hz will be used for f_S:

$$C_S = \frac{1}{2\pi f_S(R_i/2)} = 0.0637 \ \mu F \qquad \text{(B.24)}$$

Because R_1 and R_L are equal (10 kΩ), equal values for C_1 and C_L are used. If f_1 and f_L are equal to one-half the overall lower design frequency, an acceptable overall response for 50 Hz is obtained ($f_1 = f_L = 25$ Hz):

$$C_1 = C_L = \frac{1}{2\pi f_L R_L} = 0.637 \ \mu F \qquad \text{(B.25)}$$

Standard values are used for all capacitors.

$$C_S = 0.1 \ \mu F, \qquad C_1 = C_L = 1.0 \ \mu F \qquad \text{(B.26)}$$

To achieve an adequate response for the upper design frequency of 20 kHz, an op amp with a gain–bandwidth product of at least 220 kHz is needed (Eq. (5.84)). The gain–bandwidth product of most op amps exceeds this value.

It should be noted that the design of Figure B.8 is as complex as that of the single-transistor amplifier of Figure B.2. However, the manufacturer's variations in op amp parameters will have a much smaller effect on the performance of the circuit than will manufacturer's variations in transistor parameters. Furthermore, the op amp circuit will work equally well with much smaller values of load resistance.

B.4 ELECTRONIC POWER SUPPLIES (CHAPTER 6)

POWER SUPPLY WITH A SELECTABLE OUTPUT VOLTAGE

A power supply is to be designed that can be used instead of a battery to power a small electronic system such as a radio or CD player. Output voltages of 3, 4.5, 6, and 9 V, selected by a switch, are desired. The supply is to operate off a 120 V, 60 Hz power line and is to supply a load current of 0–500 mA.

DESIGN

A power supply with an electronic voltage regulator is likely to result in the simplest circuit. Furthermore, electronic regulation will probably be necessary to produce an output voltage with a sufficiently small voltage ripple required by systems designed to be powered by batteries. Power supply "hum," the result of the alternating current source, is often a problem for battery-powered systems.

An electronic regulator of the type discussed in Section 6.5 is shown schematically in Figure B.9. For most electronic regulators, the current of the common terminal I_0 is relatively small. This may be seen from the basic regulator circuit of Figure 6.39 in which the current of the common terminal is that of the zener diode and that of the negative supply connection of the op amp. Although more complex circuits, such as that of Figure 6.42, are used for integrated circuit regulators, the current of the common terminal remains small.

A resistor divider network R_1 and R_2 is used to produce an output load voltage that is greater than the nominal output voltage of the regulator v_{Reg}, the voltage across R_1. The following is obtained for the load voltage, v_{Load}:

$$v_{\text{Load}} = v_{\text{Reg}} + (v_{\text{Reg}}/R_1 + I_0)R_2$$
$$= (1 + R_2/R_1)v_{\text{Reg}} + I_0 R_2 \tag{B.27}$$

Figure B.9: Electronic voltage regulator with a voltage divider.

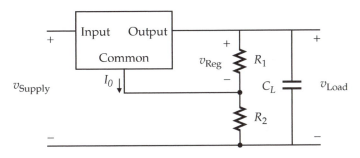

For the condition that I_0 is small compared with the current of R_1, namely v_{Reg}/R_1, the last term of Eq. (B.27) can be ignored.

$$v_{Load} \approx (1 + R_2/R_1)v_{Reg} \tag{B.28}$$

Hence, an arbitrary output load voltage can readily be obtained by means of a simple resistor voltage divider ($v_{Load} \geq v_{Reg}$).

Although one of the standard fixed voltage regulators discussed in Section 6.5 could be used, an "adjustable" voltage regulator designed for the circuit of Figure B.9 is selected. If mounted on a suitable heat sink, an LM317, a three-terminal adjustable regulator, is capable of supplying a load current of 1.5 A. Because this regulator has a regulated voltage v_{Reg}, of only 1.25 V, it can produce the design-specified voltages. Furthermore, its common-terminal current I_0 has a nominal value of only 50 μA. Hence, if the current of R_1 is sufficiently large, for example $100 I_0$ (5 mA), Eq. (B.28) applies.

A transformer, rectifier, electronic voltage regulator, and switched set of resistors composing a voltage divider, are indicated in Figure B.10. Consider, initially, the regulator and voltage divider circuit. With the switch in its uppermost position (marked 3 V), the load voltage will be the smallest. Values of R_1 and R_2 that result in $v_{Load} = 3$ V are therefore desired. Equation (B.28) is used, and an equality is assumed:

$$1 + R_2/R_1 = \frac{v_{Load}}{v_{Reg}}$$

$$R_2 = \left(\frac{v_{Load}}{v_{Reg}} - 1 \right) R_1 = \left(\frac{3.0 \text{ V}}{1.25 \text{ V}} - 1 \right) R_1 = 1.4 \, R_1 \tag{B.29}$$

For the next position of the switch (at the junction of R_3 and R_4), the effective

Figure B.10: A power supply with an adjustable regulator.

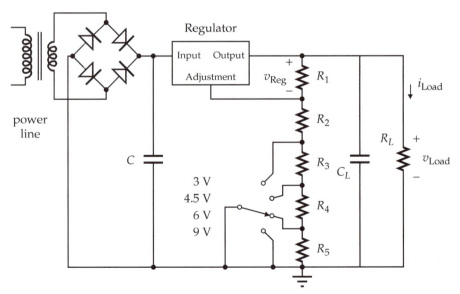

value of R_2 in Eq. (B.28) is $R_2 + R_3$ ($R_2 \rightarrow R_2 + R_3$). A value of 4.5 V is desired for v_{Load} as follows:

$$1 + \frac{R_2 + R_3}{R_1} = \frac{v_{\text{Load}}}{v_{\text{Reg}}}$$

$$R_2 + R_3 = \left(\frac{v_{\text{Load}}}{v_{\text{Reg}}} - 1\right) R_1 = \left(\frac{4.5 \text{ V}}{1.25 \text{ V}} - 1\right) R_1 = 2.6 \, R_1 \tag{B.30}$$

$$R_3 = 2.6 \, R_1 - R_2 = 1.2 \, R_1$$

In a similar fashion, for the next position of the switch, $v_{\text{Load}} = 6$ V, and $R_2 \rightarrow R_2 + R_3 + R_4$.

$$1 + \frac{R_2 + R_3 + R_4}{R_1} = \frac{v_{\text{Load}}}{v_{\text{Reg}}}$$

$$R_2 + R_3 + R_4 = \left(\frac{v_{\text{Load}}}{v_{\text{Reg}}} - 1\right) R_1 = \left(\frac{6.0 \text{ V}}{1.25 \text{ V}} - 1\right) R_1 = 3.8 \, R_1 \tag{B.31}$$

$$R_4 = 3.8 \, R_1 - R_2 - R_3 = 1.2 \, R_1$$

For $v_{\text{Load}} = 9$ V, $R_2 \rightarrow R_2 + R_3 + R_4 + R_5$.

$$1 + \frac{R_2 + R_3 + R_4 + R_5}{R_1} = \frac{v_{\text{Load}}}{v_{\text{Reg}}}$$

$$R_2 + R_3 + R_4 + R_5 = \left(\frac{v_{\text{Load}}}{v_{\text{Reg}}} - 1\right) R_1 = \left(\frac{9.0 \text{ V}}{1.25 \text{ V}} - 1\right) R_1 = 6.2 \, R_1$$

$$R_5 = 6.2 \, R_1 - R_2 - R_3 - R_4 = 2.4 \, R_1 \tag{B.32}$$

All resistance values have been specified in terms of R_1. If the current of R_1 is chosen to be $100 \, I_0$, that is 5 mA, the following obtained:

$$v_{\text{Reg}}/R_1 = 5 \text{ mA}$$

$$R_1 = \frac{1.25 \text{ V}}{5 \text{ mA}} = 250 \, \Omega \tag{B.33}$$

Resistance values of the other resistors may now be determined as follows:

$$R_2 = 350 \, \Omega$$
$$R_3 = R_4 = 300 \, \Omega \tag{B.34}$$
$$R_5 = 600 \, \Omega$$

If standard ± 5-percent tolerance resistors are used, the series combinations of two resistors can be utilized for those that are not standard values. Alternatively, if $R_1 = 150 \, \Omega$, then $R_2 = 210 \, \Omega$ (a series connection of a 100 Ω and a 110 Ω resistor), $R_3 = R_4 = 180 \, \Omega$, and $R_5 = 360 \, \Omega$. Only for R_2 are two series-connected resistors required. The current of R_1 is 8.3 mA, an acceptable value. Although 0.25-W resistors would be adequate for the circuit, 0.5-W resistors provide a greater margin of safety (a more conservative design).

The analysis of Section 6.2 covering full-wave rectifiers and nonideal transformers is used to complete the design. Because the maximum load voltage is 9 V, a minimum supply voltage (the voltage across the filter capacitor C) of about 11 V is needed for the electronic regulator to function properly. Therefore, a design value of 12 V is appropriate. A transformer with a secondary rms rating of 12 V is a reasonable choice (see concluding remarks of Section 6.2). Although a transformer with an rms current rating of 0.5 A for its secondary winding might be acceptable, a standard transformer with a 1-A rating should be used. The higher current rating, in effect a higher power rating for the transformer, is to ensure that the tranformer's power capacity will not be exceeded. In addition to the output load power, electrical power is dissipated by the diodes of the bridge rectifier as well as by the regulator.

For a bridge rectifier, the peak inverse voltage of the diodes is approximately the peak transformer voltage V_m. For a 12-V rms secondary voltage, the peak voltage is 17 V. However, the peak inverse voltage for a no-load condition is larger. Although diodes with a 25-V peak inverse rating might be adequate, a bridge rectifier with a 50-V rating is preferable. Concurrently, a 1-A bridge rectifier should be used.

A 0.1-μF load capacitor C_L is recommended by the integrated circuit's manufacturer to minimize output transient effects. Finally, a value needs to be determined for the filter capacitor C. When the diodes are not conducting, the supply voltage depends on the charge q of the filter capacitor:

$$v_{Supply} = \frac{q}{C}$$

(B.35)

$$\Delta v_{Supply} = \frac{\Delta q}{C} = \frac{i_{Supply} \Delta t}{C}$$

The current i_{Supply} is the input current of the regulator and, except for very small load currents, is essentially equal to the load current. Its maximum value is 500 mA. For a full-wave rectifier, the time interval over which the capacitor supplies current, Δt, is approximately $T/2$, where T is the period of the power line voltage ($T = 1/60$ s for a 60-Hz power line frequency). This yields the following design relationship:

$$\Delta v_{Supply} = \frac{i_{Supply} T}{2\,C}$$

(B.36)

Suppose a standard capacitance of 4700 μF is used. For a load current of 500 mA, the following is obtained:

$$v_{Supply} = \frac{(500 \text{ mA})(1/60 \text{ s})}{(2)(4700 \; \mu\text{F})} = 1.77 \text{ V}$$

(B.37)

If the supply voltage does not fall below 12 V, this peak-to-peak value of ripple voltage will be acceptable. However, on the basis of the transformer used, this may not be the case. A 9-V load voltage setting and a current of 500 mA could result in a minimum supply voltage that falls below 12 V, causing the regulator not to function properly. To ensure proper operation, a transformer with a larger secondary voltage, for example a tranformer with a secondary rms rating of 16 V, could be used. Also, a larger filter capacitor could be used that would reduce the ripple of the supply voltage – for example, a capacitance that is the parallel combination of two 4700-μF capacitors. A laboratory testing of the circuit is necessary to ensure that the supply meets the design requirements.

One final consideration is the tolerance specification of the resistors of the voltage divider that establish the output voltage. Because resistance values may vary, so too may the output load voltage. Hence, if a precise output voltage is required (a tolerance was not included as part of the design specification), a tighter resistance tolerance, for example ±1 percent, is necessary. The acceptable tolerance of the output voltage will depend on the electronic ciruits with which the supply is used.

REFERENCES

Armstrong, E. H. (1940). Evolution of frequency modulation. *Electrical Engineering*, **59**, 12, 485–93.

Miyaoka, Senri (1984). Digital audio is compact and rugged. *IEEE Spectrum*, **21**, 3, 35–9.

INDEX